LIMNOLOGICA

Ecology of Inland Waters Vol. 20, No. 1
1989

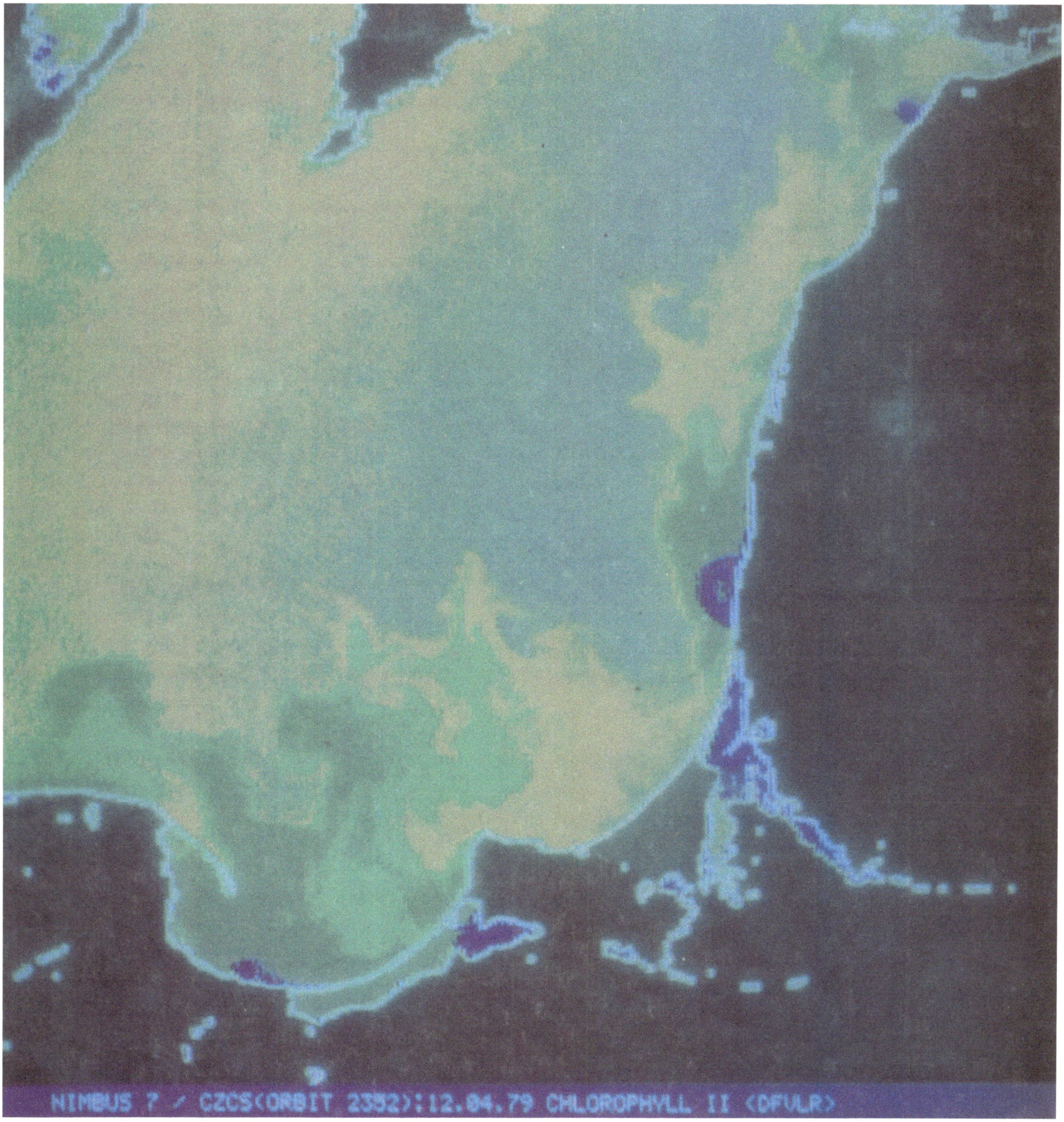

Akademie-Verlag Berlin

LIMNOLOGICA

Ecology of Inland Waters

Akademie der Wissenschaften der DDR
Forschungsbereich Biowissenschaften und Medizin
Zentralinstitut für Mikrobiologie und Experimentelle Therapie Jena
Redaktion LIMNOLOGICA
Beutenbergstraße 11
Jena DDR-6900

GENERAL INFORMATION

Scope. LIMNOLOGICA were founded by THEODOR SCHRÄDER in 1962. They will publish papers which fall in the scope of limnology, i.e. papers dealing with any aspects of ecology and hydrobiology of inland waters and adjacent biotopes, as well as reviews, discussions, and essays that develop limnological-ecological theory (including all qualitative and quantitative aspects of planktonic, periphytic, benthonic, neustonic, and nektonic organisms, nutrient and energy flow, bioindication, biomonitoring, water intern control mechanisms, ecological modelling and simulation, biological-ecotechnological water quality management). Proceedings of specialized symposia may also be accepted for publication, if the scientific standard set by the journal is met. Not acceptable is purely hydrological, hydrophysical, biochemical, physiological, genetical, or marine biological work, that, while using inland water test organisms, is unrelated to the scope of LIMNOLOGICA.

Submission of articles. Manuscripts should be submitted to the editor-in-chief, the members of the editorial board, and the secretary.

Publication schedule. LIMNOLOGICA is published in one volume of nearly 480 pages per year. Each volume comprises two issues.

Subscriptions. All issues are available upon your usual supplier all over the world. If you can not get LIMNOLOGICA on this way, please directly send subscription orders to:
Akademie-Verlag Berlin
Leipziger Straße 3—4
DDR-1086 Berlin

Readers in the FRG and Berlin(West), respectively, please send orders to:
Kunst und Wissen
Erich Bieber OHG
Wilhelmstraße 4—6
D-7000 Stuttgart 1

Readers in the other Western-European countries please send orders to:
Kunst und Wissen
Erich Bieber GmbH
General Wille-Straße 4
CH-8002 Zürich

Readers in other non-socialist countries please send orders to:
Buchexport
Volkseigener Außenhandelsbetrieb der DDR
Leninstraße 16
DDR-7010 Leipzig

Cover Photo
Spring bloom of phytoplankton. Gulf of Gdansk and Baltic Proper. Nimbus 7/12. 04. 1973. (Photo: Dr. U. HORSTMANN, Kiel, FRG).

Proceedings of the Estuary Symposium, Kühlungsborn, German Democratic Republic, November 16—21, 1987

ÄSTUAR

Limnologica (Berlin) **20** (1), 3, April 1990

Estuary Symposium, Kühlungsborn, GDR, Nov. 16—21, 1987

Preface

Estuaries and coastal waters are economically important zones in all countries. The purposes for which they are used by different interests such as shipping, fisheries, recreation, agriculture, industry and communities vary and are often contradictory. Overutilization by any of these interests restricts their use by others and in the extreme cases leads to the collapse of the ecosystem and the loss of most of the benefits it provides "free of charge."

The Baltic Sea, and in particularly its coast, is highly loaded. United in the HELCOM Convention, for the past few years the Baltic countries have been cooperating successfully to reduce the pollution load to acceptable levels. This effort calls for a great deal of research to elucidate the key processes taking place in the Baltic ecosystem and its regulating mechanisms as a first step towards defining ecosystem status.

The south and south-east coast of the Baltic are characterized by boddens and haffs which are fairly uniform in character. Therefore it would be useful to collate the results of national research projects undertaken in the region as a basis for drawing more general conclusions. This was the aim of the Estuary Symposium organized jointly by the Department of Biology of the Wilhelm-Pieck-University Rostock and the Biological and Chemistry Societies of the GDR, held at Kühlungsborn (GDR) from 16th to 21st November, 1987. The comprehensive ecological research programme conducted by the Coastal Research Unit of the Department of Biology in the GDR's inland coastal waters for more than 20 years formed the scientific background. The reasons for holding this, the first international symposium devoted to the problems of estuarine and coastal waters, arose out the necessity of preserving and managing them for the benefit of society and the concern felt by the scientists studying these waters.

In line with the aims of the CMEA programme "Ecosystem Analysis in the Baltic Sea", a project uniting scientists from the Soviet Union, Poland and the GDR and forming parts of the CMEA's "World Oceans" programme, and of the Baltic Marine Biologists (BMB), the symposium was meant simultaneously to stimulate and improve cooperation in the Baltic region.

The symposium was attended by 80 scientists, representatives of economic interests and students from the Soviet Union, Poland, the Federal German Republic, Sweden, the Netherlands and the GDR. The 41 lectures and 6 posters that were presented all dealt with four principal topics of the symposium:

— physical, chemical and biological characteristics,
— biochemical cycles and energy flow,
— ecophysiological problems, including field and laboratory experiments, and
— anthropogenic influences and management.

They considered not only the scientific background to eutrophication, but also problems relating to the assessment, prevention and reversal of eutrophication. For the first time ecologists, physiologists, microbiologists, water chemists and geologists submitted their results for joint discussion. The exchange of opinions was correspondingly broad and led to new approaches, broader cooperation and improved research programmes. A round·table discussion on "Perspectives in Estuary Research and Management" rounded off the symposium.

In view of the interest shown in the symposium it was decided to hold another. The second Estuary Symposium will be organized by the Institute of Oceanography of Gdansk University (Poland) and will be held in 1990.

The present volume of LIMNOLOGICA contains those papers that were received and accepted for presentation before the symposium deadline. Two additional papers are included which were submitted by Soviet scientists who had to cancel their participation at short notice to take part in expeditions.

At this point the Organizating Committee wishes to thank all those who contributed to the success of the symposium. We thank in particular the Rector and Prorector, the staff of the Directorate for International Affairs and the Director of the Department of Biology of the Wilhelm-Pieck-University Rostock. Our thanks are also due to the staff of the Experimental Ecology and Marine Biology Units of the Department for their help in preparing and organizing the symposium. And, finally, we are indebted to Dr. habil. S. Jost Casper, Editor in Chief of LIMNOLOGICA and Mrs. Anita Lange (Secretary) for their help and cooperation in arranging the publication of the proceedings.

U. Schiewer
E. A. Arndt
G. Schlungbaum

Limnologica (Berlin) **20** (1), 5—8, April 1990

Zoological Institute, Academy of Sciences of USSR, Leningrad, USSR

Some Physico-chemical and Biological Phenomena in the Salinity Gradient

By VLADISLAV V. KHLEBOVICH

With 5 Figures

Key words: Critical salinity; Horohalinicum; Processes in the salinity gradient of natural waters

Abstract

In waterbodies with smooth salinity gradient the change of fauna composition from sea water to freshwater or vice versa is not a gradual process but occurs in a rather narrow zone of "critical salinity" (horohalinicum by the later authors). This zone is also the place of mass death of sea- and freshwater forms both in estuaries and in the laboratory, the limit of life activity of isolated cells and tissues as well as of forms devoid of osmotic regulation. This limit in the millions years of organic evolution was not overcome by many taxons which are represented at present exclusively by the marine forms (Radiolaria, Anthozoa, Echinodermata, Ascidia ect.).

Absorption of dissolved organic matter by the organisms depends on salinity and usually stops below $5^{0}/_{00}$; thus the problem of absorption of dissolved organic substances by integuments for nutrition is rather a problem of marine hydrobiology and not freshwater hydrobiology.

At the „critical salinity" a more or less sharp change of many physico-chemical characteristics is observed — a change in the ions ratio (rapic factor), a change of charge on the particles of suspension and a change of sedimentation parameters.

The nonlinear change of biological processes at the salinity gradient seems to be historically predeterminated by an adequately nonlinear change in physico-chemical properties of natural waters.

In the world there is only a small number of water basins in which the salinity gradient may be observed to a comparatively large distance and with a more or less distinct time stability. In this respect the Baltic sea with its large bays characterized by smooth variation of salinity and by geographic "reference" of isohalines is considered to be a classical example. It is exactly here that the qualitative variation of fauna of the water basins in the river-sea section was thoroughly studied for the first time. The result of these studies expressed as a well-known REMANE's curve (REMANE 1934, 1940) is shown in Fig. 1 (left part). From the Figure it is clearly seen that the variation in the salinity does not lead to a gradual displacement of the freshwater species by the marine ones (or vice versa) but results in a clear zone of minimum of species (Artenminimum) corresponding to salinity about $5-8^{0}/_{00}$. The marine fauna corresponds to the salinity above this level while the predominantly freshwater fauna is met below this level. Later ZENKEVITSCH (1959) and JÄRVEKÜLG (1979) paid attention to the fact that actually the general amount of species in the Baltic sea at the salinity of $5-8^{0}/_{00}$ even surpasses the amount of species in slightly more fresh or salty waters. In the appropriate curve in the middle of the fall corresponding to "Artenminimum" there is a marked hillock (Fig. 2). This can be attributed to the fact that the total number of marine and freshwater species existing here is summed up and also properly brackish water species are added. However, it does not influence the general estimation of the $5-8^{0}/_{00}$ zone as a zone of junction or separation of freshwater and marine faunae.

Special studies about biological effects in the $5-8^{0}/_{00}$ zone (KHLEBOVICH 1962, 1965, 1968, 1969, 1974) proved

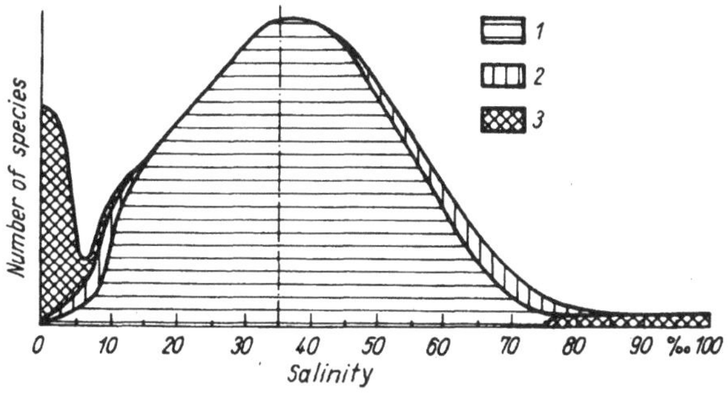

Fig. 1. Number of marine animal species (1), brackish water and euryhaline animals of marine origin (2) and freshwater animal species (of freshwater origin) (3). The left side is adopted from REMANE (1934), the right side constructed on the data published by HEDGPETH (1959). (After KHLEBOVICH 1962, 1969)

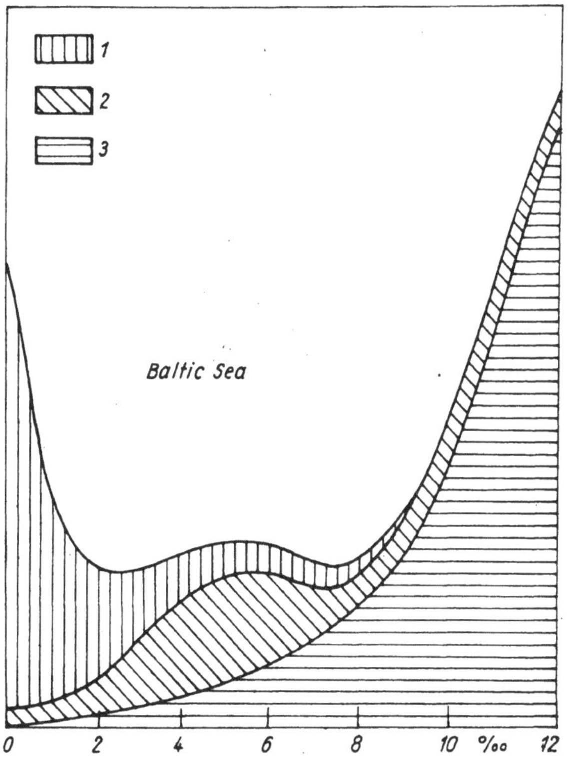

Fig. 2. "Artenminimum" in the Baltic sea. (After ZENKEVITSCH 1959). 1 — marine species; 2 — brackish water species; 3 — freshwater species

that events in this zone are more serious than a mere place of contact between the last marine and freshwater species. Even for those forms which cross this zone (widely euryhalineor migrating upstream-downstrcam) it occurs to be the upper or the lower limit of reproduction. Here the trend of physiological processes, especially the salt-water regulation, is radically changed. The use of the same criteria for the internal medium of organisms, their blood plasma and coelomic fluid, and for the external environment can be justified taking into account substantial similarity in the ionic composition of mentioned organic fluids and oceanic waters.

It has been found that the internal medium expressed in per mille ($^0/_{00}$), a unit familiar to hydrochemists, also sharply changes the biochemical reactions above or below $5^0/_{00}$. This has given ground to substantiate the concept of critical salinity in biology. This concept establishes the universal character of salinity of $5—8^0/_{00}$ as a barrier separating the main fauna complexes, ecological groups, various types of cellular and biochemical stability. Subsequently KINNE (1971) has introduced for the $5—8^0/_{00}$ salinity zone an apt term — horohalinicum (from greek "horeo" — divide). Apparently the notions "critical salinity" and "horohalinicum" are not completely synonymous. The first is more wide as it belongs to ecology and physiology, the second is attributed to ecology only.

In the open seas in which the ionic composition of water is near to ocean, horohalinicum also corresponds to the $5—8^0/_{00}$ salinity, similarly to the Baltic sea. Thus, even in semi-inland Azov sea, lying so far from the ocean, the salinity characteristics of horohalinicum are just the same. However, in the inland lake-type Caspian and Aral seas with untypical ionic ratio, the salinity values of the critical zone are different. According to ALADIN (1983), in the chain Ocean — Caspian sea — Aral sea, the horohaline zone expands and shifts towards higher salinity values (Fig. 3). It is interesting to note that chlorinity in these cases remains approximately unchanged — about $2.8—4.4^0/_{00}$.

The REMANE's curve is limited to a characteristic of changes in the fauna compositions in the range between the freshwater and the normal oceanic water. There was made an attempt to prolong this curve into regions of high salinity. For this purpose a list of species found in various water basins with the salinity above the oceanic (HEDGPETH 1959) was analyzed. For each species according to data from literature the higher salinity level of its natural distribution was established. On this basis a curve of the relation between the number of species and the salinity value was plotted (Fig. 4). Then the resulting curve was combined with the REMANE's curve. For the first time such a combined plot with appropriated substantiations and references was presented in the Soviet "Journal of general biology" in 1962 and repeatedly presented in the international journal "Marine Biology" (KHLEBOVICH 1962, 1969). Dates of our publications and,

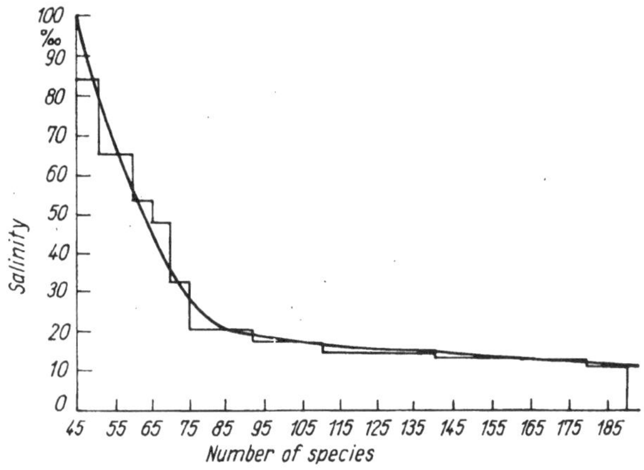

Fig. 4. Number of metazoan animal species (excluding of Insecta) in hypersaline waters. (On the data published by HEDGEPETH 1959, after KHLEBOVICH 1962)

which is of no less importance, the substantiation of the course of the right part of the curve by corresponding exact material give us ground to deny the originality to subsequent publications (HEDGPETH 1967; KINNE 1971).

In the whole the combined curve shows the non-linear character of variation of number of species and types of fauna within the whole practicable salinity range of development of life on earth.

It is interesting that the curve is symmetrical in a certain sense: the symmetry axis corresponds to salinities where the maximum numbers of species of hydrobionts and the bloom of purely marine fauna are observed. Off the symmetry axis the number of species gradually becames smaller (in the left side only to critical salinities) and the fauna acquires a more and more pronounced freshwater character. The inhabitants of ultrahaline waters are related not to the marine fauna but rather to the freshwater fauna which is more distant from it according to the salinity requirements.

Such a large-scale summary of data on the dependence of a number of species and of qualitative composition of organisms on the value of salinity of natural waters, referring to other large taxons of organic nature except animals, is not yet available. However, there is an impression that the distribution of bacteria ((RHEINHEIMER 1981), fungi (MANSUETI 1961) and brown algae (REMANE 1971) in the whole follows the pattern obtained for the animals.

The understanding of special ecological and evolutionary significance of the barrier of critical salinity allows to come to a final solution in the discussion which is being in progress for already three quaters of this century — concerning the role of dissolved organic matter (DOM) in extra-intestinal nutrition of water organisms.

Specialists in ecology of freshwater animals usually adhere to the position of KROGH who denied any importance of DOM in nutrition of aquatic organisms.

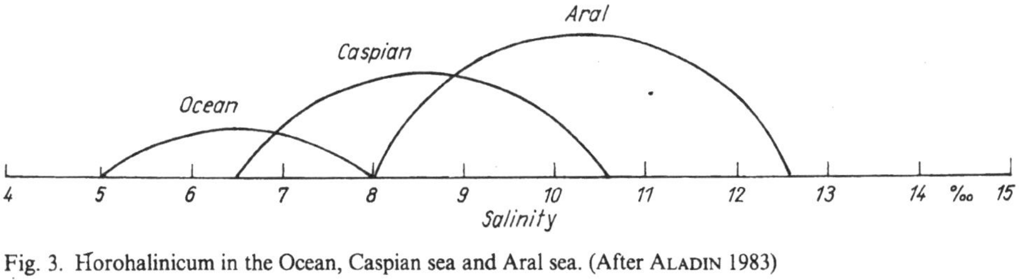

Fig. 3. Horohalinicum in the Ocean, Caspian sea and Aral sea. (After ALADIN 1983)

Marine biologists, especially after the works of STEPHENS who received interest in PÜTTER's "osmotic nutrition" hypothesis, have collected various facts on the important role of DOM in energy balance of various animals (JØRGENSEN 1976). In one of the early investigations of this sort performed on the polychaetous annelid *Nereis limicola*, it was shown that the intensity of absorption of DOM by integuments is a salinity-dependent process which totally discontinues below $6^0/_{00}$ (STEPHENS 1964). Later it was confirmed by experiments on the North American mollusc *Rangia cuneata* (ANDERSON & BEDFORD 1973; ANDERSON 1975). Investigations of such sort in which the process is studied in a wide salinity range over many points are very scarce. Therefore our laboratory had undertaken a special investigation of salinity dependence of lissolved glycine (DOM, most frequent in the sea water) by widely euryhaline bivalvia *Mytilus edulis, Mytilus edulis kussakini, Laternula limicola, Macoma balthica takahokoensis* and *Corbicula japonica* (KHLEBOVICH et al. 1982; KOMENDANTOV 1986; KHLEBOVICH & KOMENDANTOV 1986). In all cases at the critical salinity the absorption of aminoacid was hardly noticeable or totally discontinued below $5^0/_{00}$. It is evident that uptake of dissolved organic matter by integuments is exceptionally a problem of marine hydrobiology and may not be taken into account by freshwater hydrobiology. In natural conditions this phenomenon is separated by the critical salinity of 5—$8^0/_{00}$.

For biologists it would be tempting to co-ordinate the data on nonlinearity of biological processes in the salinity gradient with an appropriate picture of variation of hydrochemical processes. The problem may be formulated as follows: if freshwaters are characterized by diversity of chemical composition with the prevalence of bivalent ions and marine waters — by the constancy of ion ratio (KNUDSEN rule) and by the dominancy of univalent ions, one should like to know where the salinity boundaries of the applicability of the KNUDSEN formula lie.

According to tabulated data of WITTIG (1940) on the salinity dependence of the relative calcium content obtained from the actual mutually distant geographical points we have made a plot, Fig. 5 (KHLEBOVICH 1968).

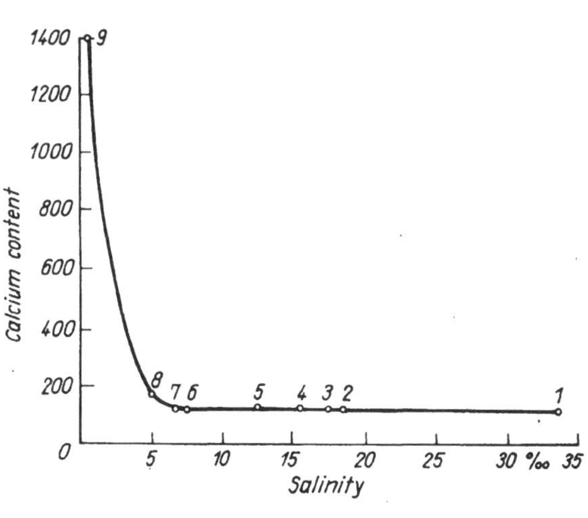

Fig. 5. Relative calcium content as a function of salinity at various locations: 1 — near Oslo; 2 — Kiel Bay; 3 — Bornholm Deep; 4 — east coast of the Baltic sea; 5 — Kiel Inlet; 6 — Arkona Deep; 7 — east coast of the Baltic sea; 8 — estuary of the Schwentine River; 9 — estuary of the Schwentine River. (Based on the tabulated data by WITTIG 1940, after KHLEBOVICH 1968). Calcium content in % from minimum

Actually the plot represents the relationship between the "freshwater" ion of calcium and the "marine" ion of chlorine. The curve obtained may be interpreted as a proof of violation of KNUDSEN law of ion ratio constancy for near-critical salinities.

The upper salinity boundary of the constancy of ion ratios lies above $42^0/_{00}$. Among other things, valency of ion of iron is changed here (HATZIKAKIDIS 1952). Thus, the law of constancy of ion ratios is valid in the salinity range between 5—8 and 42—$45^0/_{00}$.

Some researchers believe that estuary waters acquire marine features already at a salinity above $2^0/_{00}$ (SIMONOV 1969; MILLERO 1981). However, this is simply the result of logical analysis "on the river side". At the salinity of $2^0/_{00}$ water stops to be riverine, but stable hydrochemical character is acquired at salinities of approximately $5^0/_{00}$ and above. It is clearly seen that this stenoionic range coincides with the boundaries of existence of purely marine fauna.

It should be noted that the salinity dependence of the content of biogenic elements, microelements and dissolved organic matter (revue: GORDEEV 1983) is much more complicated. It may only be assumed, that there will be a strong influence of their biological absorption, or vice versa, biological excretion, which processes as already seen must themselves be salinity-dependent. Particularly, in places of mixture of fresh and marine waters, where zooplankton is largely represented by perished individuals (KOVAL 1984), there should be relatively many both DOM and biogens.

Of physical phenomena in the salinity gradient most prominent is the change of charge on particles drawn by water which occur evidently about $5^0/_{00}$ (PRAVDIČ 1970; PRAVDIČ et al. 1981). It is appropriate to mention here that in most methodic guides it is not recommended to determine the salinity from the electroconductivity at the salinity below $5^0/_{00}$. This change of charge would result in varying character of sediment formation on either side of this salinity barrier due to increased flocculation during the change of a surface charge. Thus clays (illite, kaolinite) most usually drawn by rivers quickly fall out into sediment at the salinity above $4^0/_{00}$ (WHITENHOUSE et al. 1960). Some examples of such kind are presented by GORDEEV (1983). Most interesting investigations of flocculation of various material are presented in the experiments by SHOLKOVITZ (1981).

There is no doubt that particle flocculation caused by the change of charge at these particles at near-critical salinities also changes other physical characteristics of water, e.g. transparency.

Owing to peculiar of sedimentations along the estuary axis two aspects of the problem should be paid special attention. First, it becomes possible to make a more precise estimation of the character of sedimentation in the river mouths on geological scale. Second, special attention should be paid to peculiarities of sedimentation of new materials and substances of anthropogenic nature in the river mouths from the point of view of their excessive concentration in the limited areas.

Thus, the most prominent feature of physical, chemical and biological processes in the salinity gradient of natural waters is their non-linearity and the special character of the barrier function of salinities in a narrow range of about $5^0/_{00}$. The zone of critical salinity or horohalinicum should be regarded as a nucleus of estuaries (KINNE 1971; STAROBOGATOV & KHLEBOVICH 1978).

This must be always considered in historical (organic evolution, geological processes) and ecological aspects (including of anthropogenous ones) of investigations of brackish waters.

References

ALADIN, N. V. (1983): On displacement of the critical salinity barrier in the Caspian and Aral seas, the Branchiopoda and Ostracoda taken as examples. Zool. Žurn. (Moscow) **62** (5), 689—694 (in Russian, English summary).

ANDERSON, J. W. (1975): The uptake and incorporation of glycine by the gills of *Rangia cuneata* (GRAY) (Mollusca, Bivalvia) in response to variations in salinity and sodium. In: F. J. VERNBERG (ed.), Physiological ecology of estuarine organisms, pp. 90—112, San Diego.

— & BEDFORD, W. B. (1973): The physiological response of the estuarine clam *Rangia cuneata* (GRAY) to salinity. II. Uptake of glycine. Biol. Bull. (Woods Hole). **144** (2), 229—247.

GORDEEV, V. V. (1983): Rechnoi stok v okean i cherty ego geokhimii (River inputs to Ocean and characters of their hydrochemistry). Moscow.

HATZIKAKIDIS, A. D. (1952): Research and extension of KNUDSEN's Tables, determining salinity in sea water. Praktica Hellenic Hydrobiol. Inst. (Athens) **6** (1), 55—71.

HEDGPETH, J. W. (1959): Some preliminary considerations of the biology of inland mineral waters. Archivio oceanogr. e limnol. (Venezia) **11**, Suppl., 111—141.

— (1967): Ecological aspects of the Laguna Madre, a hypersaline estuary. Estuaries, Amer. Ass. Adv. Sci. **83**, 408—419.

JÄRVEKÜLG, A. A. (1979): Bottom fauna of the east part of Baltic sea. Tallin (in Russian).

JØRGENSEN, C. B. (1976): AUGUST PÜTTER, AUGUST KROGH, and modern ideas on the use of dissolved organic matter in aquatic environments. Biol. Rev. (Cambridge) **51** (3), 291—328.

KHLEBOVICH, V. V. (1962): Osobenosti sostava vodnoi fauny v zavisimosti ot solenosti sredy (Peculiarities of the aquatic fauna composition in relation to the salinity of the medium). Žurn. ob. biol. (Moscow) **23** (2), 90—97 (in Russian, English summary).

— (1965): Physiologia eurigalinnosti: kritičeskaya soljonost vnešnej i vnutrennej sredy (Physiology of euruhalinity: critical salinity of the environment and inner medium). In: Voprosy hidrobiologii, p. 440—441. Moscow (in Russian).

— (1968): Some peculiar features of the hydrochemical regime and the fauna of mesohaline waters. Mar. Biol. **2** (1), 47—49.

— (1969): Aspects of animal evolution related to critical salinity and internal state. Mar. Biol. **2** (4), 338—345.

— (1974); Kritičeskaya solenost biologičeskikh processov (The critical salinity of biological processes). Leningrad (in Russian).

— (1986): K biologicheskoi tipologii estuariev Sovetskogo Soyuza (On biological typology of the estuaries of the USSR). Trudy Zoologicheskogo Instituta (Proc. Zool. Inst. Acad. Sci. USSR) **141**, p. 5—16. Leningrad (in Russian, English summary).

— & KOMENDANTOV, A. YU. (1986): O solenostnoi zavisimosti potrebleniya ROV vodnymi bespozvonočnymi (On the salinity dependence of DOM uptake by water invertebrates). In: The First Conference of Moldavian Hydrobiologists, pp. 147—148. Kishinev (in Russian).

— — & YAKOVISHINA L. A. (1982): Salinity dependence of the glycine uptake by *Mytilus edulis* of the White sea and Sea of Japan. Zool. Žurn. (Moscow) **61** (5), 783—786 (in Russian, English summary).

KINNE, O. (1971): Salinity — Invertebrates. In: O. KINNE (ed.), Marine Ecology **2**, Part 2, pp. 821—995. London.

KOMENDANTOV, A. YU. (1986): Solenostnye reakcii estuarnykh polikhet i dvustvorčatykh mollyuskov Yuzhnogo Primorja (Salinity reactions of the estuarine Polychaeta and Bivalvia from the south part of Primorje territory). Avtoref. kandidat. dissert. ZIN AN SSSR, Leningrad (in Russian).

KOVAL, L. G. (1984): Zoo- and necrozooplancton of the Black Sea. Kiev (in Russian).

MANSUETI, R. I. (1961): Effect of civilization on striped bass and other estuarine biota in Chesapeake bay Tributaries. In: Proc. Gulf and Carribean Fisheries Inst., 14th Annual Session, pp. 110—136.

MILIERO, F. J. (1981): Chemical speciation of ionic components in estuarine systems. In: River inputs to ocean system. UNEP and UNESCO, pp. 116—131. Switzerland.

PRAVDIČ, V. (1970): Surface charge characterization of sea sediments. Limnol. and Oceanogr. **15**, (2), 230—233.

— JEDNAČAK-BISCAN, J., & JURACIČ, M. (1981): Physico-chemical parameters discribing the role of particulate material in estuarine waters. In: River inputs to ocean system. UNEP and UNESCO, pp. 188—196. Switzerland.

REMANE, A. (1934): Die Brackwasserfauna. Zool. Anz. (Suppl.) **7**, 34—74.

— (1940): Einführung in die zoologische Ökologie der Nord- und Ostsee. Tierwelt der Nord- und Ostsee **34**. Leipzig.

— (1971): Ecology of brackish water. In: Biology of brackish water. Binnengewässer **25**, pp. 1—210. Stuttgart.

RHEINHEIMER, G. (1981): Estuarine bacterial populations and their role in the decomposition of organic material. In: River inputs to ocean systems. UNEP and UNESCO, pp. 283—290. Switzerland.

SHOLKOVITZ, E. R (1981): Experimental studies of interactions during the mixing. Ibid., pp. 149—151.

SIMONOV, A. I. (1969): Gidrologia i gidrokhimia ustjevogo vzmorja v moryakh bez prilivov (Hydrology and hydrochemistry of the mouth regions in the tideless seas). In: Trudy Gosudarstvennogo Oceanograficheskogo Instituta **92**. Moscow (in Russian).

STAROBOGATOV, YA. I., & KHLEBOVICH, V. V. (1978): Problemy tipologii solonovatykh vod (The problems of brackish water typology). Gidrobiol. Žurn. (Kiev) **14** (6), 3—6 (in Russian).

STEPHENS, G. P. (1967): Dissolved organic material as a nutritional source for marine and estuarine invertebrates. In: Estuaries. Amer. Ass. Adv. Sci. **83**, 367—373.

WHITEHOUSE, U. G., JEFFREY, L. M., & DEBRECHT, J. D. (1960): Differential settling tendencies of clay minerals in saline waters. Clay Min. **1** (1), 1—79.

WITTIG, H. (1940): Über die Verteilung des Kalziums und der Alkalinität in der Ostsee. Kieler Meeresforsch. **3** (2), 460—496.

ZENKEVITSCH, L. A. (1959): The classification of brackish water basins as examplified by the seas of the USSR. Archivio oceanogr. e limnol. (Venezia) **11** (Suppl.), 53—62.

Address of the author:
V. V. KHLEBOVICH
Zoological Institute
Academy of Sciences of the USSR
Universitetskaya nab. 1
199034 Leningrad
USSR

Limnologica (Berlin) **20** (1), 9—14, April 1990

Ernst-Moritz-Arndt-Universität, Sektion Geologische Wissenschaften, Greifswald, DDR, Akademie der Wissenschaften der DDR, Institut für Meereskunde, Rostock-Warnemünde, DDR

An Actualistic Model of Mud Deposition and Diagenesis for the Western Baltic Sea

By R.-O. NIEDERMEYER & D. LANGE

With 3 Figures and one Table

Abstract

The paper deals with results of a complex study on the deposition and diagenesis of modern Baltic Sea mud. The Baltic Sea as an intracontinental brack water sea which has a geological history only of a few millenniums, a limited water exchange capacity with the world ocean, strong thermohaline water stratification, high biological productivity and large river water input represents favourable conditions for studies on modern sedimentation processes from an actualistic point of view.

Using a characteristic type of Baltic sediments (mud/Schlick) which contains high amounts of organic matter, an attemption is made to establish a depositional and diagenetic model of the western area of the Baltic Sea which is part of the G.D.R. continental margin. With regard to this model both the mud as a product of sedimentation processes and the depositional conditions which are due to physical, chemical and biological features and interactions are dealt with. The supposed actualistic model gives the opportunity for general conclusions, for example, on the formation of organic carbon-rich sediments and black shales.

1. Introduction

Current sedimentological investigations of modern marine sediments deal with aspects of matter origin, migration and accumulation processes. Therefore, it is important to study both the sea water and the sediment. As an intracontinental brack water sea having a strongly limited water exchange capacity with the North Sea the Baltic Sea (semi-enclosed sea) gives a good opportunity for research work on modern mud sediments which occur within the basin areas. — The purpose of this paper is to attempt a complex characterization both of the depositional process and of the sedimentary product concerning the Baltic Sea mud (Schlick) which was dealt with in a detailed study (NIEDERMEYER 1987).

2. Location and Methods

The study areas (Mecklenburg Bight, Arkona Basin) are parts of the so-called transitional zone between North Sea and Baltic Sea water. Both areas represent shallow basins exarated by Weichselian glacial ice movements at various time intervals between 75,000 and 10,000 B.P. — The water depth of the Mecklenburg Bight is between 20 and 26 m, that of the Arkona Basin (site AB 113) is about 45 m. —

Sampling procedures and measurements were carried out in 1983, 1985 and 1986 by means of the Oceanographic Measuring Chain (OM 75) and a short gravity corer (cf. LANGE & NIEDERMEYER 1985). The following examination methods were used (in detail see NIEDERMEYER 1987): Hydrographic measurements (e.g. temperature, salinity, oxygen), sea water and pore water chemistry, quantitative and qualitative determination of suspended matter in sea water, grain size analysis of sediments, organic carbon and total carbonate of sediments, general and clay mineralogical composition of sediments.

3. Depositional Conditions

Mainly during the summer season there is in both areas a strong thermohaline water stratification which, depending on weather-induced water turbulences and mixing, partly disappears in the colder seasons. Especially, in the Mecklenburg Bight the depths of the thermocline and the halocline may coincide, whereas in the deeper Arkona Sea exists a permanent halocline which occurs in a water depth of about 30 m. The hydrographic measurements carried out during the sampling periods are in good correspondence with the general long-term trends of temperature and salinity of the Western Baltic Sea (cf. MATTHÄUS 1979, 1984).

The oxygen content of the Baltic Sea water is measured and interpreted since the last century at various places (MATTHÄUS 1978, 1985). According to these results generally a ± negative oxygen budget is assumed which, among other things, is due to oxygen depletion caused by the oxydation of high amounts of organic matter. The strong water stratification especially favours the oxygen depletion. But there is an oxic to suboxic environment which, in general, occurs in the Western Baltic Sea. The average content of dissolved oxygen decreased at the sea bottom of the Mecklenburg Bight to 0.1 ml O_2/l in which several sites show a total absence of dissolved oxygen. — In the Arkona Sea the mean oxygen content of the surface water layer (0—20 m) amounts to between 5 and 10 ml O_2/l (MATTHÄUS 1978). In the deep water layer sometimes extremely low contents (0.3—0 ml O_2/l) have been measured (NEHRING 1980).

Due to a permanent mixing of oceanic and fresh water masses brack water seas show a special composition. The reason are important differences in the average composition of river water and sea water regarding dissolved solids and the abundance of characteristic ions (cf. LEEDER 1983, p. 15). These general features can be applied to the anomalies of the ionic composition of the brackish Baltic water which earlier was established (NEHRING & ROHDE 1966, 1967; NEHRING 1980). — Using the ratios of the ionic concentrations (Na^+, K^+, Ca^{2+}, Mg^{2+}) to chlorinity characteristic seasonal variations could be detected. The ratios for magnesium, calcium and potassium are highest in December (late autumn) whereas in summer the ionic concentrations especially of magnesium and calcium are reduced which is due to the thermal and biological features of the Baltic water, i.e. to metabolic processes of organisms. — Moreover, differences of the ionic composition, especially regarding calcium and magnesium, between the surface water and the deep water well correspond with the results earlier obtained by NEHRING & ROHDE (1967). This fact could be in connection with the process of biogenic carbonate production by the phytoplankton during periods of bloom. It was established that in summer the boundary of the maximum phytoplankton biotop (about 10—12 m water depth) well correlates with the boundary between the surface water and the deep water which show varieties in the ionic composition. — The Mg/Ca-ratio, especially of the bottom water, is of great importance for processes of early diagenetic mineral formation, e.g. in the authigenesis of dolomite [$CaMg(CO_3)_2$]. The bottom water partly migrates within the upper horizons of the modern mud and represents a compound of the pore waters. Mg/Ca-ratios which were calculated for the deep water layers of the study areas amount to average values of 3.33 (Mecklenburg Bight) and 2.82 (Arkona Sea; Fig. 1). The oceanic ratio of 3.14 is between these Baltic Sea ratios. The decreasing ratio from the Mecklenburg Bight to the Arkona Sea is produced by the increasing calcium excess of the Baltic water from west to east which is due to calcium input, especially by large rivers.

Pore waters represent the link between sea water and sediments. Especially, they are of great importance in the early diagenesis and the formation of authigenic minerals (DAHMKE et al. 1986; NIEDERMEYER & LANGE 1989). Using the ratios of selected major pore water compounds (Na^+, K^+, Ca^{2+}, Mg^{2+}) to chlorinity (ŠIŠKINA 1972), the pore waters of the Western Baltic Sea can be classified as of the chloride-alkali-type, which seems to be typical for clayey and organic-rich oozes in marginal seas, coastal and shelf regions. The Mg/Ca-ratios of the investigated pore water samples are clearly > 3 and at some sites drastically increase at sediment depths > 30 cm (Fig. 1). According to FOLK & LAND (1975) positive salinity gradients and Mg/Ca-ratios > 3 favour the formation of authigenic dolomite.

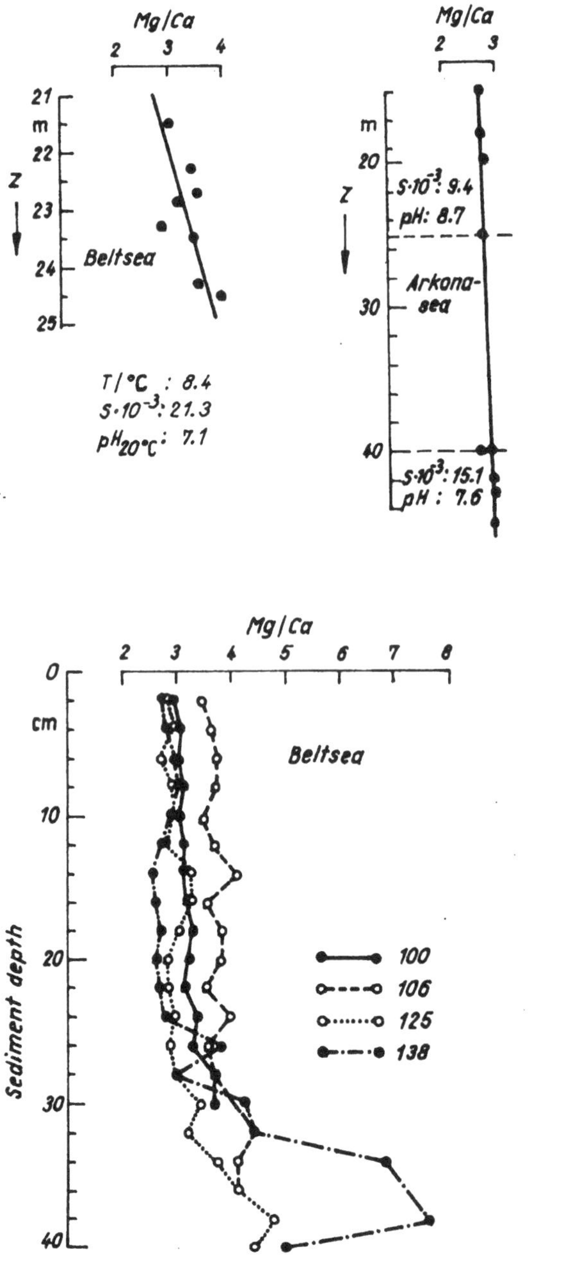

Fig. 1. Mg/Ca — ratios of bottom sea water and pore waters of the Western Baltic Sea

4. Suspended Matter (Seston)

Our results (LANGE & NIEDERMEYER 1985; NIEDERMEYER 1987) confirm the concentration values for suspended matter of the Western Baltic Sea which recently were published by GEORGI (1985). Accordingly, one can assume a concentration level of the surface water which amounts to between 2 and 3 mg (dry weight)/l. With increasing water depth the concentrations of suspended matter seem to decrease and then again to grow in the bottom water. Considering the results of PUSTELNIKOV (1984) the following aspects could explain the different vertical concentrations: The rapid development and extension of huge plankton populations beginning generally in the early spring (march) and which are due to increasing light supply and thermal radiation (photosynthesis; SCHULZ 1986) lead to higher concentrations of organic compounds in the surface water. The lower amount of suspended matter below the surface water, i.e. below the thermocline, may be caused by a temperature gradient of about 2—4 °C/m and a limitation of the euphotic zone (insufficient light supply). Due to an increased amount of clastic detritus in the deep water layer (below the halocline) which can be in connection with episodic bottom currents and re-suspension the concentration of suspended sediments grows again.

According to earlier results (ZEITZSCHEL 1965) the suspended sediments of the nearby Eckernförder Bay consist of 95% detritus (89.5% inorganic and 10.5% organic compounds) and 5% plankton organisms. Since there is a significant increase of the primary production rates in the Baltic water during the last two decades which is caused by a high input of nutrients (e.g. phosphorus, nitrogen) and eutrophication (BABENERD & ZEITZSCHEL 1985; NEHRING 1985) the amount of organic compounds of the suspended sediments increased as well. By ashing of seston samples which were extracted from bottom water of the Mecklenburg Bight we obtained up to 40% of organic compounds. This is nearly the same value reported by GEORGI (1986) as a minimum for the high-eutrophic Darß-Zingster Bodden (38%). By means of microscopical studies (scanning electron microscopy, optical microscopy) we identified 19 phytoplankton genera (39 species; mainly diatoms), pollen grains, plant relics (e.g. algal filaments) and bacterial mats as typical organic compounds. Among the inorganic compounds subangular micaceous minerals, lamellar shale grains, quartz-, feldspar- and carbonate particles dominate.

5. Mud Deposits

The dark coloured soft basin deposits can be classified as coarse mud (Grobschlick; Mecklenburg Bight) and as medium mud (Mittelschlick; Arkona Basin). These sediments mainly consist of particles which were suspended in the water column for a long time and generally contain <1% fine sand. Fig. 2 shows characteristic mud compositions of the Mecklenburg Bight (site MB 301) and the Arkona Basin (site AB 113). — Radiometric dating (lead-210 method) of a sediment core (MB) reveals a low sediment accumulation rate of 0.6 mm/yr.

The content of organic carbon in the mud, which ranges from 3.8—7.4% (mean value 5.7%) varies with

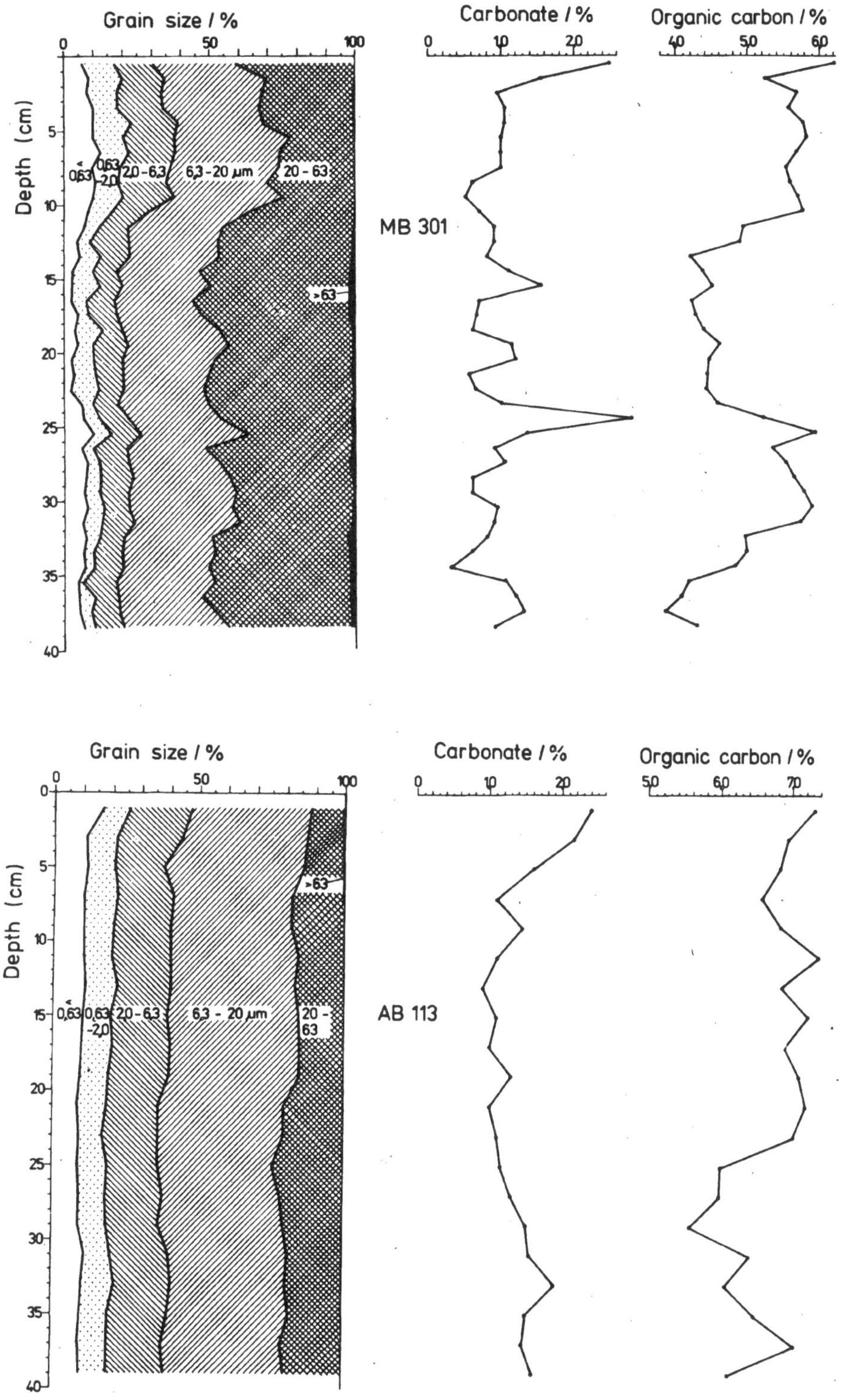

Fig. 2. Granulometrical and geochemical compositions of mud deposits of the Mecklenburg Bight (MB 301) and the Arkona Basin (AB 113)

depth (Fig. 2). There is an average C_{org}-value of 6.7% for sediments of the Arkona Basin. According to results obtained by the application of the computer programme OPTRIS (nonlinear regression; KOLYSCHKOW & PESCHEL 1985) to the sedimentological data, especially the size fractions fine silt and medium silt seem to contain the organic carbon compounds (correlation coefficient/r = 0.71—0.73). On the other hand, there is a negative correlation (r = —0.79) between coarse silt and organic carbon. — These high C_{org}-values of the Baltic Sea mud show similarities with C_{org}-contents mentioned by RULLKÖTTER (1985) and STEIN et al. (1986) for ancient Atlantic black shales which have been studied with regard to hydrocarbon formation.

On an average the carbonate content of the mud ranges from 0.8—1.4%. Since no calcite but dolomite is established by means of X-ray studies it can be assumed that the carbonate mainly consists of magnesium-rich phases (e.g. dolomite). The variations of the carbonate contents of the mud could be interpreted as varves and the formation of which may be due to an increased organic carbonate production in summer and the deposition and dolomitization of dissolution relics.

The results of clay mineralogical investigations show a qualitatively and quantitatively homogeneous composition of the clay fraction (< 2 μm). On an average the clay fraction consists of 42.6% illite, 12.8% kaolinite, 4% chlorite and 40.6% mixed layer. The source of the clay minerals within the modern mud is related to terrigenous and submarine erosion of Weichselian boulder clays.

6. The Model of Deposition and Diagenesis

The sedimentological investigations which were carried out yielded some clues both to deposition and to diagenesis of mud in the marine-brackish Baltic Sea. On the basis of a detailed study (NIEDERMEYER 1987) a general sedimentary model of the mud-forming depositional environment of the Western Baltic Sea has been established (Fig. 3). There are two sedimentary basins (Meck-

Table 1. Diagenetic zones for marine mud successions. ΔT °C — increase in temperature with depth below sediment/water interface due to a gradient of 27.5 °C/km (after CURTIS 1977)

Depth (km)	ΔT (°C)	Porosity (%)	Diagenetic zones (minerals formed)
			1 **oxidation**
0.0005			
			2 **sulphate reduction** pyrite, calcite, dolomite (low Fe-carbonates ^{12}C-enriched), kaolinite?, phosphate?
0.01	0.2	80	
			3 **fermentation** high Fe-carbonates, calcite, dolomite, ankerite, siderite, ^{13}C-enriched
1.00	28	31	
			4 **decarboxylation** siderite
2.50	69	21	
			5 **hydrocarbon formation** (a) wet – oil (b) dry – methane montmorillonite → illite (a) disordered, (b) ordered
7.00	192	9	
			6 **metamorphism** (a) 200 °C chlorite (b) 300 °C mica, feldspar, epidote?

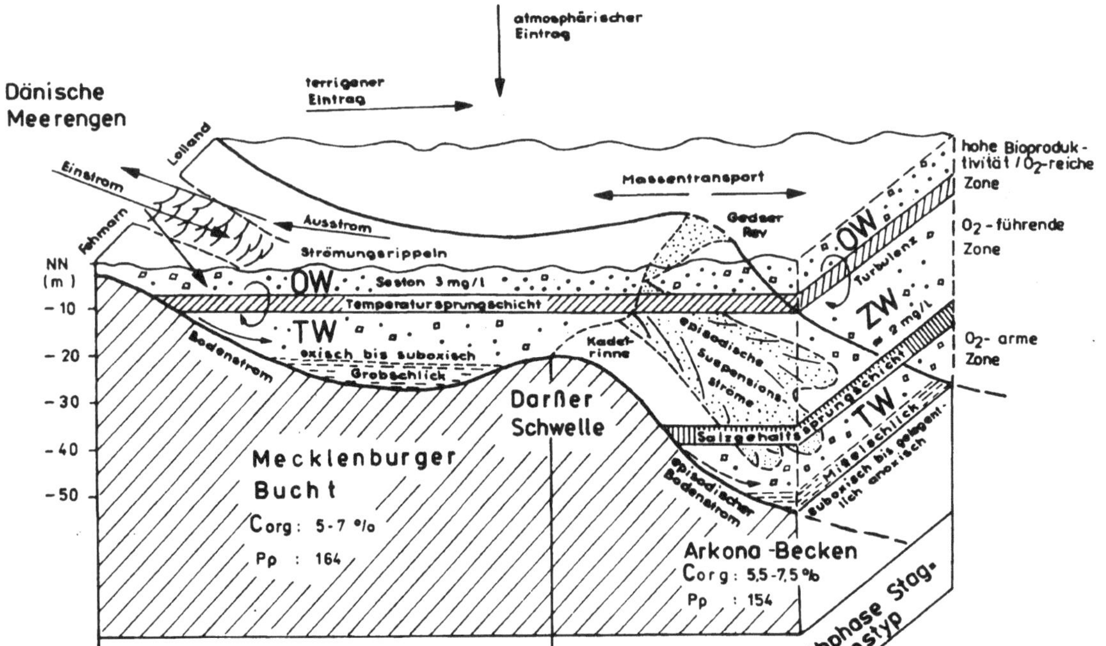

Fig. 3. Actualistic model of the mud-forming depositional environment of the Western Baltic Sea

lenburg Bight, Arkona Basin) which morphologically are separated from each other by a submarine sill region (Darßer Schwelle). Moreover, a submarine channel (Kadetrinne) which was formed during the Late Weichselian glaciation occurs at the sill and links the basins. Both the sill and the channel are important elements concerning the intensity of different transport processes (mass-, salt- and sediment transport). The model shows a progressive trend of mud-finering from west to east. The permanent halocline of the Arkona Sea which, in the shallower Mecklenburg Bight, does not display the same thickness and stability may be one reason for the deposition of finer mud within the eastern basin.
There are several similarities to models of the formation of organic carbon-rich sediments (THUNELL et al. 1984; STEIN et al. 1986). Considering these criteria, the Western Baltic Sea can be described as a productivity type with regard to the formation or organic carbon-rich sediments, whereas transitions to slightly stagnant conditions in the Arkona Basin may modify this productivity model. There are nearly the same rates of primary organic production (Pp/Fig. 3) calculated for both basins (164 and 154 g C/m^2/yr; SCHULZ 1986). Moreover, it can be said that the investigated mud sediments are in good agreement with the diagenetic zones 1 (oxydation) and 2 (sulphate reduction) which have been established by CURTIS (1977) for marine mud successions (Table 1). According to this diagenetic model the mud-depositing environment of the Western Baltic Sea can be an actualistic example both for the formation of organic carbon-rich sediments and for aspects of black shale formation.

Zusammenfassung

Die Arbeit enthält kurzgefaßt Ergebnisse komplexer Untersuchungen zur Schlicksedimentation and -diagenese im Gebiet der westlichen sowie des Übergangs zur mittleren Ostsee. Als ein geologisch sehr junges intrakontinentales Brackwassermeer mit einem stark eingeschränkten Wasseraustausch zur Nordsee, ausgeprägter thermohaliner Wasserschichtung, hoher Bioproduktivität sowie beträchtlicher Flußwasserzufuhr bietet die Ostsee ausgezeichnete Bedingungen für das Studium rezenter Sedimentationsprozesse unter aktualistischem Blickwinkel.

Am Beispiel eines charakteristischen und an organischen Bestandteilen reichen Sediments, des Schlicks, wird versucht, für den im Bereich des DDR-Festlandssockels liegenden Teil des Ostseegebietes ein Ablagerungs- und Diagenesemodell zu entwerfen und zu diskutieren. Dabei werden neben dem Schlick als Produkt von Sedimentationsprozessen insbesondere auch seine Ablagerungsbedingungen betrachtet, die Ausdruck physikalischer, chemischer und biologischer Faktoren und Wechselwirkungen sind. Das vorgestellte aktualistische Modell bietet die Möglichkeit zur Ableitung allgemeiner Erkenntnisse, so z. B. zu Fragen der Bildung C_{org}-reicher Sedimente und von Schwarzschiefern.

Резюме

В работе содержатся коротко изложенные резултаты комплексных исследований отложений и диагенеза шлика в районах западного и переходного к среднему Балтийского моря. Это море как геологически очень молодое внутриконтинентальное море смешенной воды с ограниченным обменом с Северным морем, с выраженным температурным и солевым наслоением, с высокой биоактивностью и значительным притоком воды из рек представляет собой хороший пример длю изучению современных процессов отложения седиментов с точки зрения актуализма. На примере характерного и багатого органическими составляющими седементе шлике делается попытка развития и обсуждения модели отложений и диагенеза в области шельфа ГДР. На ряду со шликом как продукта процесса отложения рассматриваются условия его отложения, которые являются выражением физических, химических и биологических факторов и их взаимодествия. Представленная актуалистическая модель позволяет получить общие представления об образовании богатых органическим углеводом осадков и черных сланцев.

References

BABENERD, B., & ZEITZSCHEL, B. (1985): Trends für eintragsrelevante Faktoren und für die Nährsalzkonzentrationen im Wasser der Kieler Bucht. Ein Beitrag zur Erforschung der Eutrophierung der Nord- und Ostsee. Ber. Inst. Meereskd. Univ. Kiel, Nr. 148, 45 S.

CURTIS, C. D. (1977): Sedimentary Geochemistry: Environments and Processes dominated by involvement of an aqueous phase. Phil. Trans. Roy. Soc. London A, **286**, 353—372.

DAHMKE, A., SCHULZ, H. D., & WEBER, W. (1986): Mineralstabilitäten und Frühdiagenese in Ostseesedimenten. Meyniana **38**, 109—124.

FOLK, R. L., & LAND, L. S. (1975): Mg:Ca Ratio and Salinity: Two Controls over Crystallization of Dolomite. Am. Ass. Petroleum Geologists/AAPG, Bull. **59**, 60—68.

GEORGI, F. (1985): Verteilung und Beschaffenheit des Sestons in inneren Küstengewässern der DDR (Darß-Zingster Boddengewässer) sowie der westlichen und mittleren Ostsee unter besonderer Berücksichtigung der Wasseraustauschprozesse zwischen beiden Systemen. Beitr. Meereskd. **52**, 35—48.

— (1986): Die chemische Zusammensetzung des Sestons in inneren Küstengewässern der DDR (Darß-Zingster Boddengewässer, südliche Ostsee). Acta hydrochim. hydrobiol. **14** (6), 581—594.

KOLYSCHKOW, P., & PESCHEL, G. (1985): Untersuchung der Zusammenhänge zwischen den Eigenschaften geologischer Objekte mit Hilfe nichtlinearer Regressionsbeziehungen. Z. angew. Geol. **31** (7), 175—178.

LANGE, D., & NIEDERMEYER, R.-O. (1985): Zur Untersuchung von Sedimentationsprozessen in der westlichen Ostsee (Mecklenburger Bucht) — Methodik und erste Ergebnisse. Wiss. Z. Univ. Greifswald, Math.-Nat. R. **34** (4), 27—30.

LEEDER, M. R. (1983): Sedimentology. Process and Product. London.

MATTHÄUS, W. (1978): Zur mittleren jahreszeitlichen Veränderlichkeit im Sauerstoffgehalt der offenen Ostsee. Beitr. Meereskd. **41**, 61—94.

— (1979): Langzeitvariationen von Temperatur, Salzgehalt und Sauerstoffgehalt im Tiefenwasser der zentralen Ostsee. Beitr. Meereskd. **42**, 41—93.

— (1984): Zur mittleren jahreszeitlichen Veränderlichkeit von Temperatur und Salzgehalt in der Mecklenburger Bucht. Beitr. Meereskd. **50**, 9—23.

— (1985): Grundzüge der regionalen und physikalischen Ozeanographie der Ostsee. Geogr. Ber. **114** (1), 1—15.

NEHRING, D. (1980): Das Nährstoffregime der Ostsee — seine Veränderungen im Zeitraum 1969—1978 — Ursachen und Auswirkungen. Diss. B, Inst. Meereskd. AdW der DDR, Rostock-Warnemünde.

— (1985): Grundzüge der chemischen und biologischen Ozeanographie der Ostsee. Geogr. Ber. **114** (1), 17—28.

— & ROHDE, K.-H. (1966): Anomale Ionenverhältnisse in der Ostsee. Monatsber. Dtsch. Akad. Wiss. Berlin **8** (5), 385—390.

NEHRING, D., & ROHDE, K.-H. (1967): Weitere Untersuchungen über anomale Ionenverhältnisse in der Ostsee. Beitr. Meereskd. **20**, 10—23.

NIEDERMEYER, R.-O. (1987): Beiträge zur komplexen Untersuchung von rezenten Sedimentationsprozessen in Schlickgebieten der westlichen und mittleren Ostsee. Diss. B, Univ. Greifswald, Math.-Nat. Fak., Greifswald.

— & LANGE, D.: Modern mud deposits of the Western Baltic Sea (Mecklenburg Bight). Sedimentary environment and diagenesis. Beitr. Meereskd. **60**, 5—20.

PUSTELNIKOV, O. S. (1984): Vzvešennye veščestva (suspended matter). In: D. E. GERŠANOVIC (ed.), Očerki po biologičeskoj produktivnosti Baltijskogo Morja. Koordinierungszentrum RGW. Thema „Weltozean", Bd. 1, S. 313—334. Moskau (in Russisch).

RULLKÖTTER, J. (1985): Schwarzschiefer in der Tiefsee — Entstehung und Kohlenwasserstoffpotential. Kernforschungsanlage Jülich GmbH, Institut für Erdöl u. Organische Geochemie (ICH-5), 49—54.

SCHULZ, S. (1986): Ergebnisse ökologischer Untersuchungen im pelagischen Ökosystem der Ostsee. Diss. B, Inst. Meereskd. AdW der DDR, Rostock-Warnemünde.

ŠIŠKINA, O. V. (1972): Geochimija morskich i okeanskich ilovych vod (Geochemistry of marine and oceanic pore waters). Moskau (in Russisch).

STEIN, R., RULLKÖTTER, J., & WELTE, D. H. (1986): Accumulation of organic carbon-rich sediments in the Late Jurassic and Cretaceous Atlantic Ocean — A Synthesis. Chem. Geol. **56**, 1—32.

THUNELL, R. C., WILLIAMS, D. F., & BELYEA, P. R. (1984): Anoxic events in the Mediterranian in relation to the evolution of Late Neogene climates. Mar. Geol., **59**, 105—134.

ZEITZSCHEL, B. (1965): Zur Sedimentation von Seston, eine produktionsbiologische Untersuchung von Sinkstoffen und Sedimenten der westlichen und mittleren Ostsee. Kieler Meeresforsch. **21**, 55—80.

Addresses of the authors:
Dr. RALF-OTTO NIEDERMEYER
Sektion Geologische Wissenschaften
Ernst-Moritz-Arndt-Universität
Friedrich-Ludwig-Jahn-Str. 17a
Greifswald
DDR-2200

Prof. Dr. sc. DIETER LANGE
Akademie der Wissenschaften der DDR
Institut für Meereskunde
Seestr. 15
Rostock-Warnemünde
DDR-2530

Limnologica (Berlin) **20** (1), 15—28, April 1990

Akademie der Wissenschaften der DDR, Institut für Meereskunde, Rostock-Warnemünde, DDR

Metal Distribution in Sediments of the Baltic Sea

By L. Brügmann & D. Lange

With 4 Figures and 3 Tables

Key words: Trace metals: Sediments; Baltic Sea; Contamination.

Abstract

In 1981 and 1983 more than 20 cores of silty and clayey sediments were taken from representative net-sedimentation areas of the Baltic Sea. The 20 or 25 cm long cores were dissected into 1 cm slices and analysed on the easily leachable fraction and on the "total" contents of metals (Mn, Cr, Zn, Pb, Ni, Cu, Co, Cd, Hg) together with background determinants as rates of linear sedimentation/mass accumulation, grain size, organic/inorganic carbon, Al, Fe, P, N. Several metals (Cd, Zn, Cu, Cr, Ni) showed maximum contents in the very fine grained mud sediments of the centre of deep basins, e.g. in the Gotland Deep, which are remote from possible land-based emission sources. Maximum lead contents were found more near-shore (Lübeck Bight, Gdansk Basin). For mercury, the Arkona Basin seems to act as a preferred "trap". This is possibly due to a co-precipitation with organic (humic) macromolecules which flocculate in the southwestern Baltic Sea owing to their first contact with more saline waters.

The vertical profiles of most metal contents in the sediment cores seem to be influenced strongly by the background conditions (redox-potential, pH, salinity, organic carbon content) prevailing in the water column during the deposition and in the sediments where they cause re-mobilization/-distribution of metals in the course of the early diagenesis. Thereby, the supposed geochronological registration of the contamination history of the Baltic Sea with metals is seriously disturbed.

Estimates were made on the mass accumulation of metals, organic carbon, phosphorus and nitrogen in Baltic sediments between 1900 and 1983 (1981) and on the maximum net-fluxes towards and in the sediments. The results of uni- and multivariate statistical calculations were used to draw conclusions on the main forms in which the metals could occur.

Introduction

Innumerable natural and anthropogenically influenced processes release increasing amounts of partly toxicological relevant metals into the working and living environment of the modern man. If presently, due to diffuse input pathways or because of a lack of necessary technologies a desirable recycling or depositing in immobilized form is still impossible, the only final sink of those metals are supposed to be the sediments of the lakes, seas and oceans. Even today, very often the metal accumulation in sediments is still oversimplified considered as an one-way process resulting in vertical concentration profiles which reflect the geochronology of the metal contamination (Pheiffer Madsen & Larsen 1986). However, there is increasing awareness of the fact now, that depending on their physico-chemical form and binding characteristics the deposited metals may be re-mobilized and redistributed in the sediments or even released into the adjacent bottom waters.

In the Baltic Sea the background conditions for the deposition of particulate matter are very complex and may differ from one basin with net-sedimentation to another. Therefore, reliable assessments on the state of the Baltic sediments in relation to their metal contamination must be based on two types of information, firstly on the vertical and horizontal distribution patterns of the metals and secondly on many sedimentological and geochemical properties of the same samples. The present paper is aimed to contribute to this task.

Materials and Methods

Sediment sampling was performed in several major basins and subregions of the Baltic Sea for which net-accumulation is assumed. This included the Bay of Bothnia, the Bothnian Sea, the Northern Baltic Proper, the Landsort Deep, the Gotland, Gdansk, Bornholm and Arkona Basins, the Mecklenburg and Lübeck Bights, the Kattegat and Skagerrak. Where possible, the central and deepest parts of those areas with high accumulation rates of fine grained material were chosen. Often, this was identically with stations of the Baltic Monitoring Programme (BMP) of the Helsinki Commission (HELCOM 1983).

In 1981, 10 sediment cores of 25 cm length were taken from the GDR rv "A. v. Humboldt" and from the Finnish rv "Aranda". In 1983, further 14 cores of 20 cm length were sampled (Fig. 1). The corer (diameter 2.5 inches) of the Niemistö type (Niemistö 1974) was equipped inside with a piacryl tube. Immediately after sampling the cores were dissected into 1 cm slices. These slices were sealed in plastic petri dishes and stored deep-frozen until analysis.

Parallel cores were taken for dating, pore water sampling and other sedimentological investigations. The pore water was separated by centrifugation, pressure filtered (0.45 μm membrane) and acidified with 0.1 ml HNO_3 per 10 ml sample.

In the land laboratory the 1 cm thick sub-samples were freeze-dried, sieved (0.1 mm) and homogenized. The "slightly extractable" metal fractions were leached with 0.5 N HCl (25 ml/0.5 g dry sediment; Chester & Voutsinou 1981). For "total" analyses two different digestion procedures were applied:

a) The samples taken 1981 were treated with a mixture of HCl/HNO_3/$HClO_4$/HF in PTFE "pressure bombs" at 150 °C, evaporated to dryness, taken up in diluted HCl and centrifuged.

b) The samples from 1983 were digested in the pressure bombs with a 1:3 mixture of HCl and HNO_3 (Breder 1982), decanted and centrifuged.

In the extracts and digestions the metals were preferably determined by flame AAS (Zn, Mg, Ca, Pb, Ni, Co, Cu, Fe, Mn, Cr, Al). The cadmium analyses and part of the lead determinations were carried out by flameless AAS with matrix modification and platform techniques.

Mercury was determined using the cold vapour reduction/aeration method (Hatch & Ott 1968). Initially, the samples were subjected a pressure digestion with HNO_3 and an after-treatment with potassium permanganate solution. The mercury was first collected on silica wool coated with gold, then released by heating through a resistance wire and finally recorded by AAS at 253.7 nm.

Dating of the sediment samples was performed with the 210-Pb technique (Häsänen 1977; Pheiffer Madsen & Sorensen 1979).

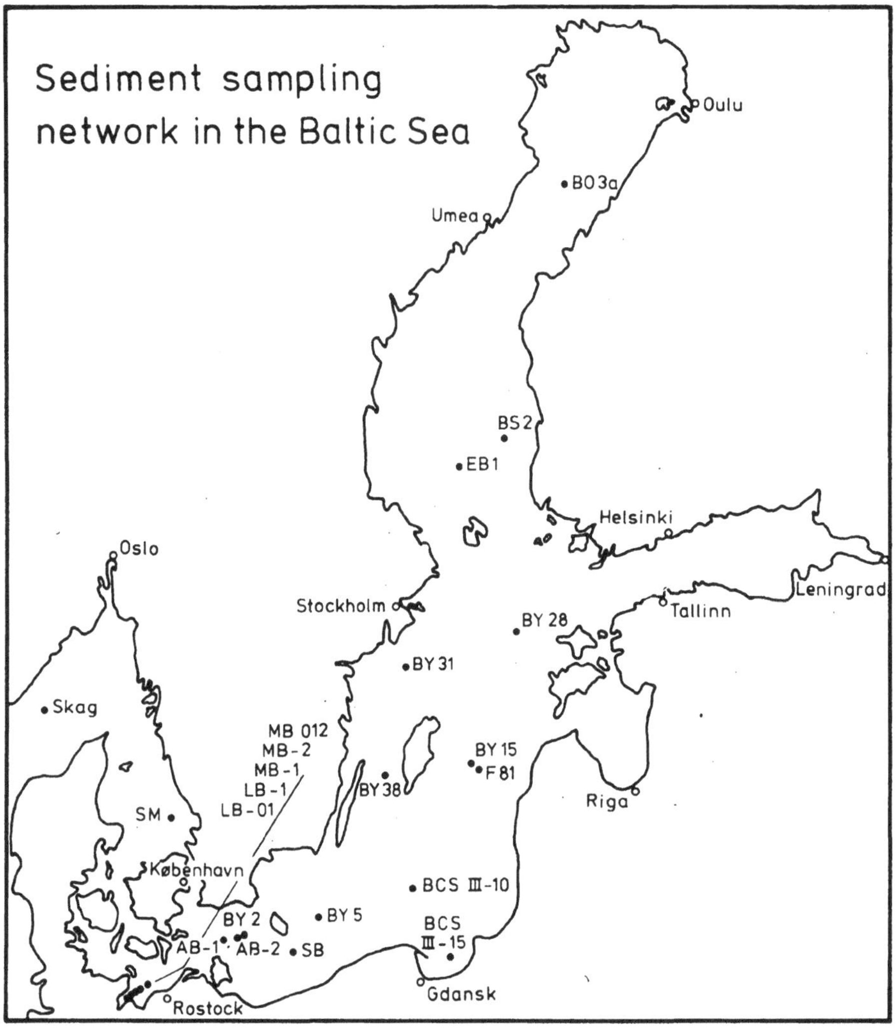

Fig. 1. Sampling network

By the use of reference materials with certified metal contents and other properties the performance of the applied methods was ensured.

For details on the methods applied for the determination of background parameters see BRÜGMANN & LANGE (1983).

Results and Discussion

For predictions on possible impacts of certain metals in sediments it is necessary to get an idea on the levels of their portions which are potentially available for the biota or other remobilization processes. The percentage of these fractions from the total contents which are extractable by 0.5 N HCl was below 50% for elements which exist predominantly in geochemically inert detritus material (Al, Cr, Fe). Frequently, other elements (Pb, Zn, Mn) are leached nearly quantitatively by HCl whereas Cu, Ni, Co and Cd occupy a medium position with respect to their leachable fractions. As a rule, the percentage of the HCl soluble fraction decreases with increasing sediment depth. This could be due to increasing binding strength, e.g. by transformation of amorphous phases into more crystalline one. Another explanation would be, that this behaviour could reflect the decrease of anthropogenic and relatively mobile fractions with depth. Leaching of samples of a sediment core from the Mecklenburg Bight with several different agents has shown this phenomenon already earlier, especially for cadmium (BRÜGMANN & HENNINGS 1982). In 1981, at the FIMR/Helsinki metal analyses by AAS were performed on cores taken in parallel to ours. These measurements followed a nitric acid digestion. For the investigated elements, most satisfactorily for copper, the nitric acid values were always between our results from HCl leaching and total digestions (BRÜGMANN et al. 1982). The vertical profiles of the metal contents were very similar. Even minima or maxima which were restricted to only some 1 cm slices appeared in the same depth of parallel cores.

This could be an indication, that no significant "patchiness of sediments" exists for the different deep basins of the Baltic Sea. This is true at least for the magnitude of the distance between coring locations which sometimes was extended to several hundred metres.

Using reference materials it could be shown that the digestion with HNO_3/HCl (3:1) resulted in significantly lower metal contents (Fe, Cr, Ni, Cu, Co) in relation to the HF treatment and to the certified values. Therefore, both data sets for the field samples were considered separately. In Figs. 2a—2d are presented some examples of the vertical profiles for background parameters (dry matter — TM, organic carbon, carbonate, nitrogen, phosphorus, grain size, age) and for metals which were measured partly by three methods of different leaching strengths. In addition, the dissolved (< 0.45 μm) concentrations of Fe, Mn, Zn, Cu, Ni, Pb, Cd, phosphate and the salinity in pore waters are shown.

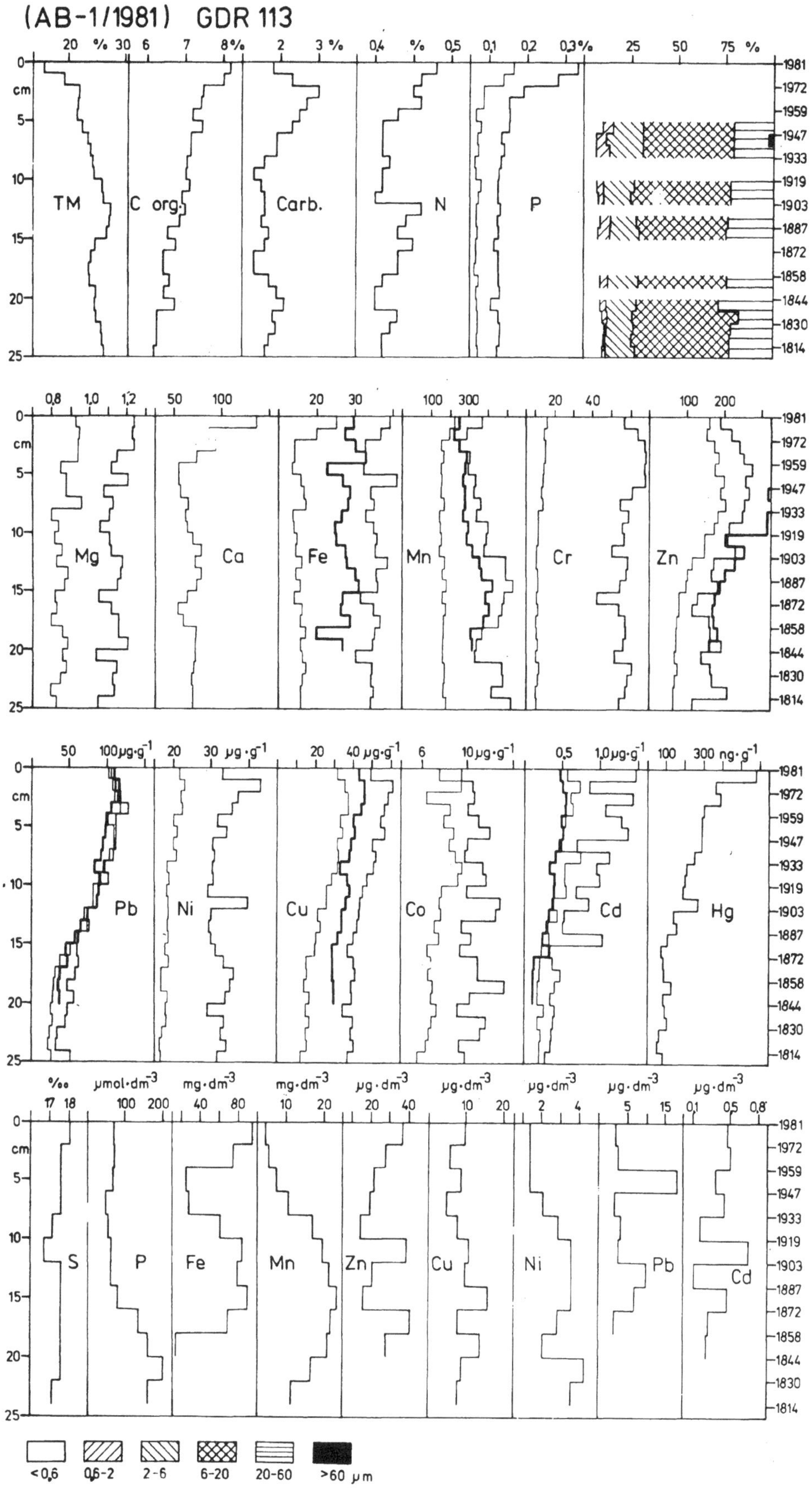

Fig. 2. Examples of dated profiles of trace metals and background parameters in sediment and pore water samples from the Baltic Sea 1981 and 1983
a) station "AB-1"/1981

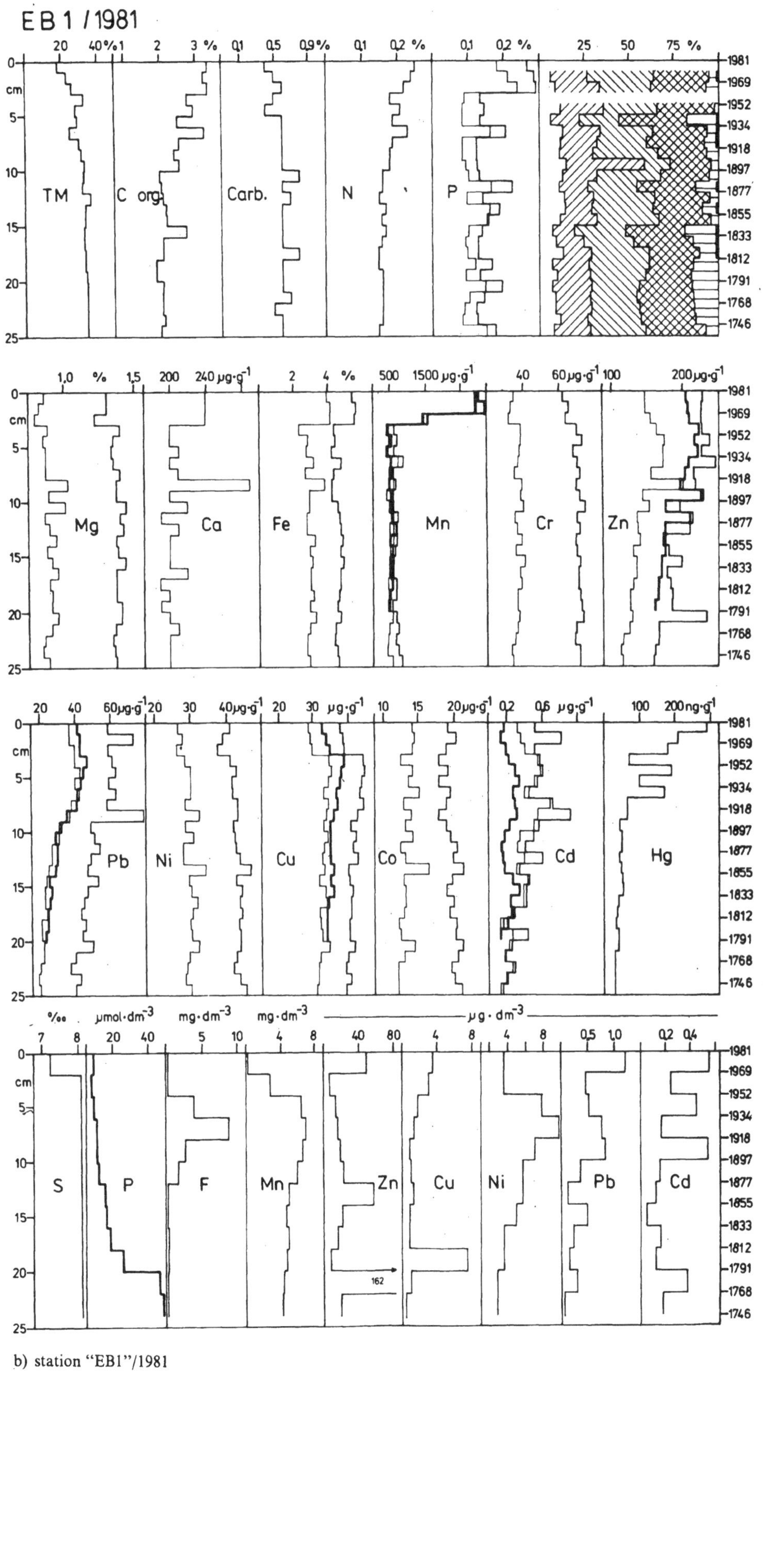

b) station "EB1"/1981

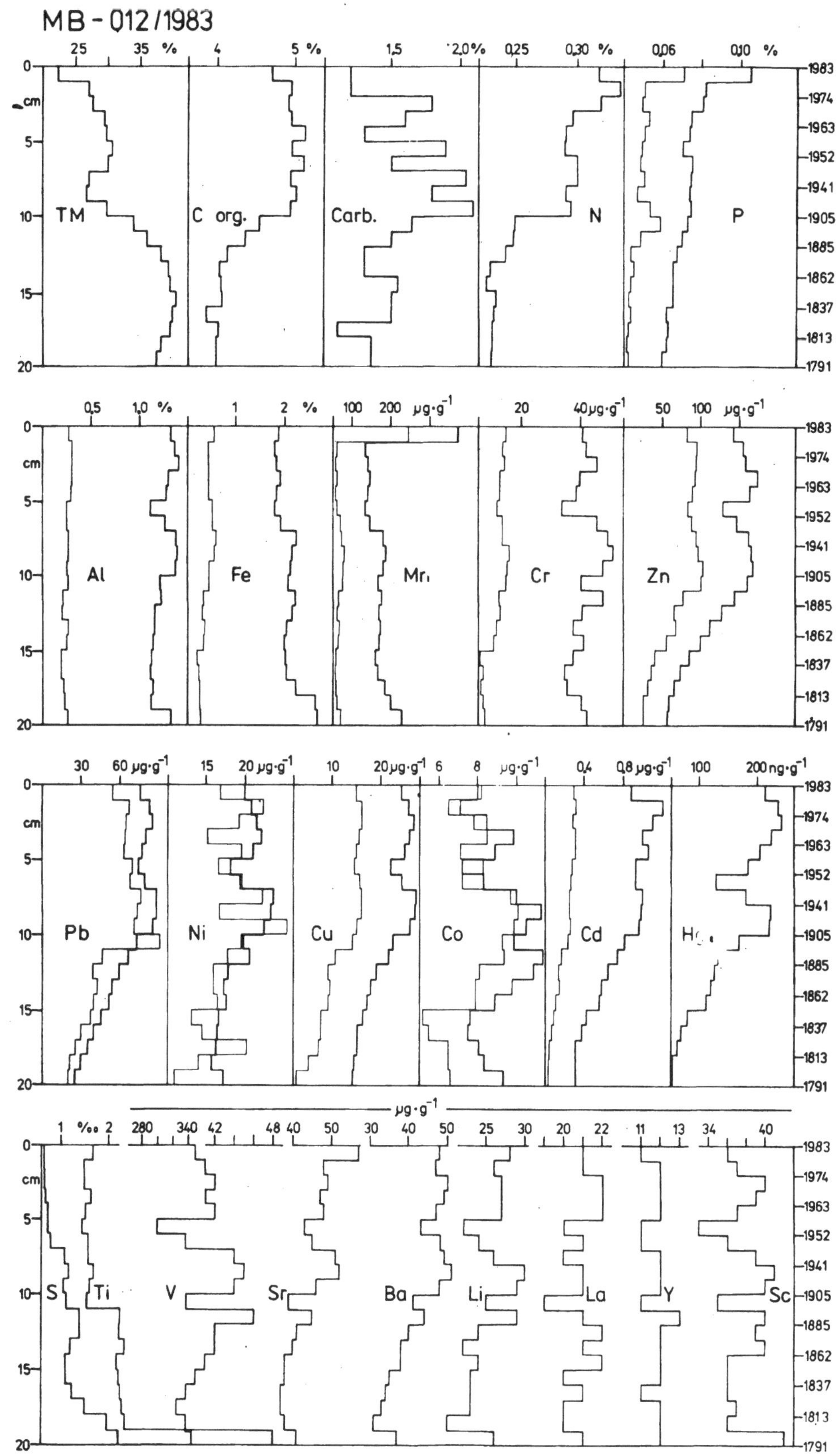

c) station "MBO12"/1983

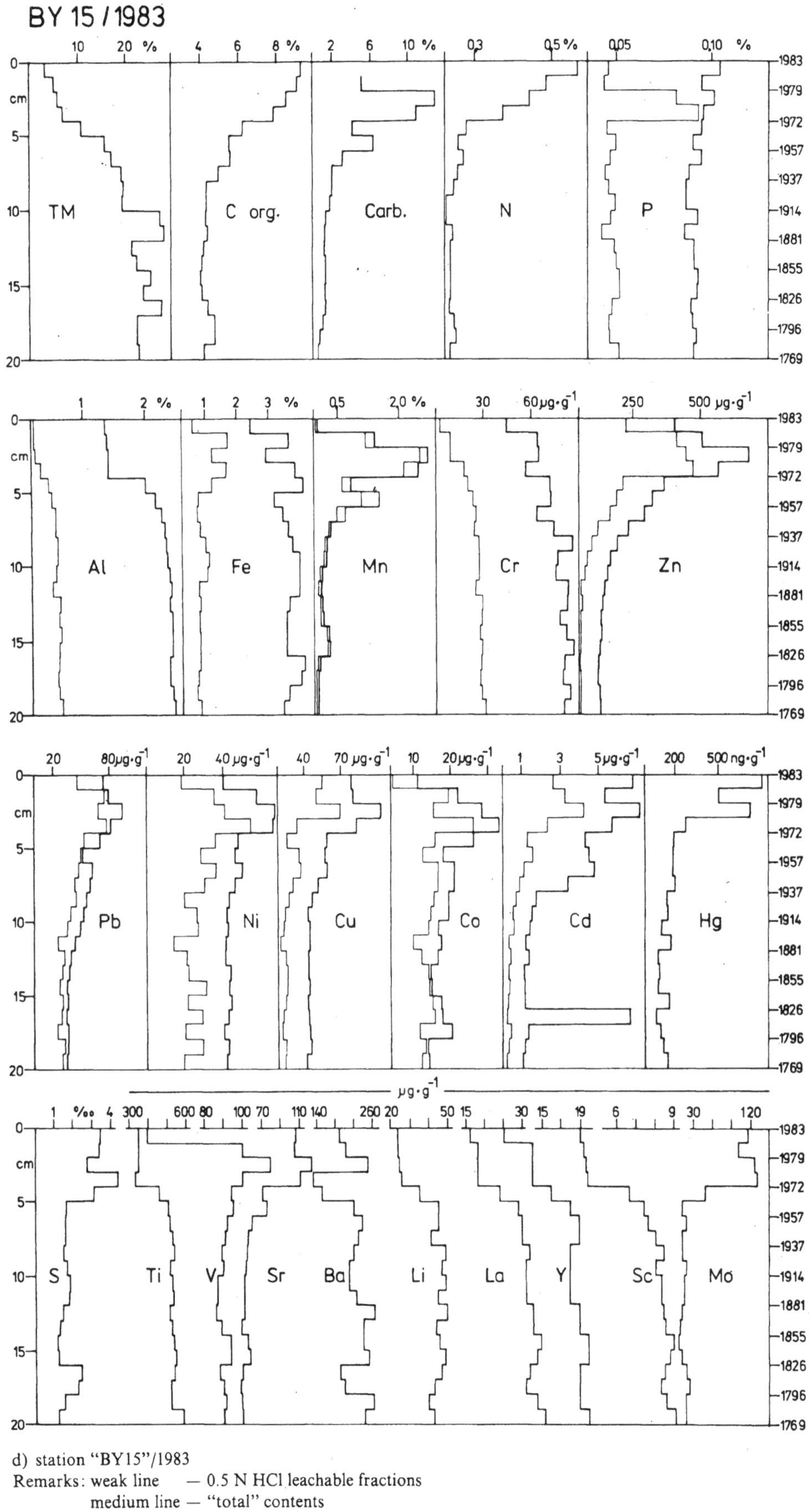

d) station "BY15"/1983

Remarks: weak line — 0.5 N HCl leachable fractions
medium line — "total" contents
strong line — nitric acid digestions

More than 99% of the investigated material had grain sizes below 63 μm and must mainly categorized as silt or muddy silt. At stations "F81" and "BY31" the clay fractions further increase and the sediment type changes to silty mud. Most of the trace metals (Zn, Hg, Cd, Pb, Cu) increase towards the water/sediment interface. For other elements this feature is not so pronounced (Co, Ni) and is entirely absent for Cr and Fe. For manganese, very often an inverse relationship between the contents in the total sediment and in pore water was observed. For instance, an increase of Mn in the core "LB-1" with depth was accompanied by a gradual decrease in pore waters from 0.5 mg/dm^3 to about one tenth of this concentration in 25 cm.

The difference between the total contents and those leached by 0.5 N HCl seems to reflect the detritical part. This part is of different magnitude, e.g. very low for Pb or Zn and dominating for Cr and Fe. Nevertheless, in most cases for all these elements this detritical fraction is nearly constant throughout the whole core. It is obviously, that the vertical gradients for most of the elements are mainly caused by the differences in the more mobile portion. This portion includes those metals adhered by physical and chemical sorption and those present in carbonates, hydroxides/oxides and in organic detritus. This behaviour was observed also for phosphate with a more or less constant inert fraction around 400/500 μg/g. The phosphate fraction which is leached by

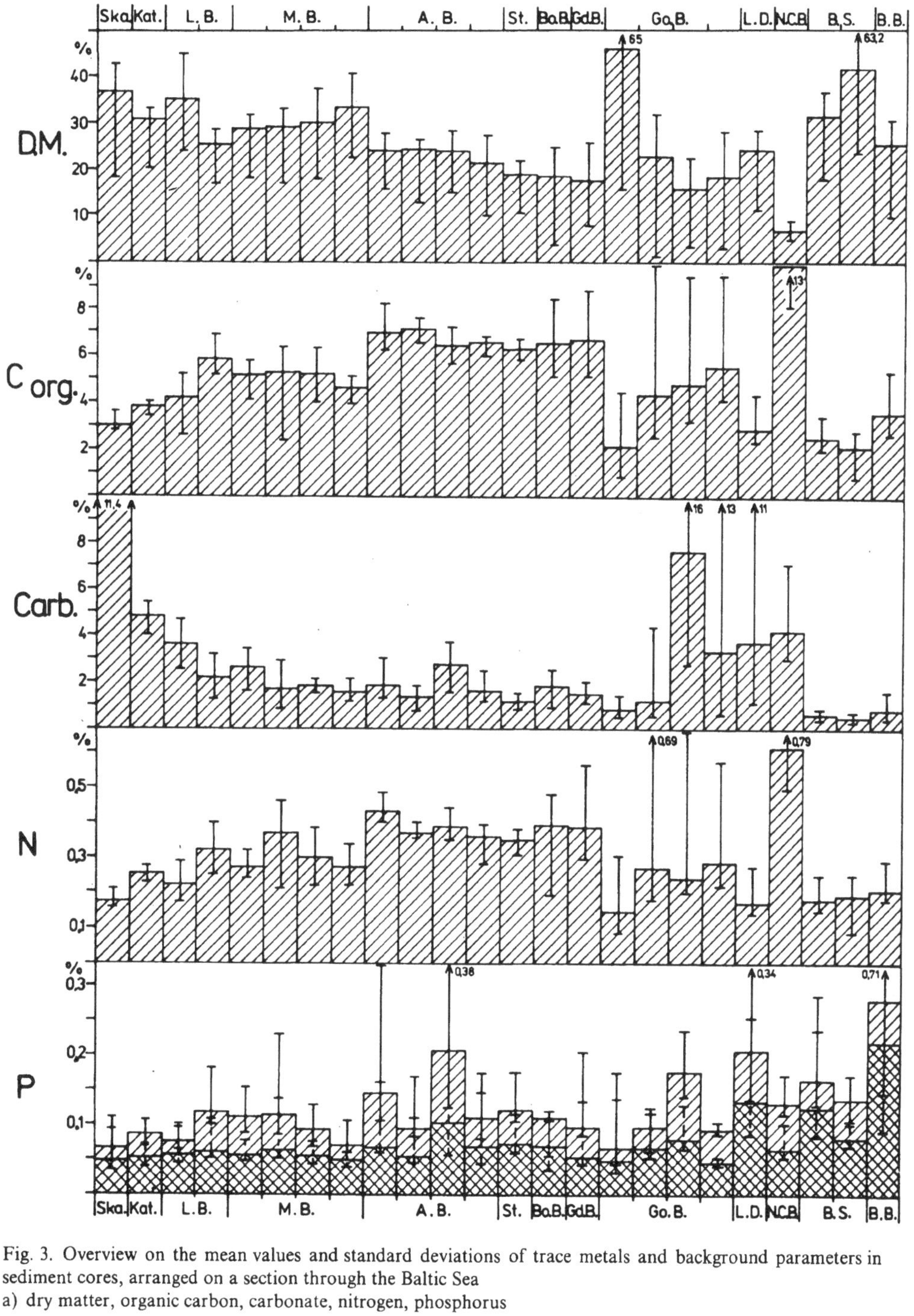

Fig. 3. Overview on the mean values and standard deviations of trace metals and background parameters in sediment cores, arranged on a section through the Baltic Sea
a) dry matter, organic carbon, carbonate, nitrogen, phosphorus

HCl decreases together with the redox-potential whereas parallel the concentrations in pore waters increase. The inert fraction seems not to be slightly available for a remobilization under anoxic conditions.

For manganese, extremely high contents — up to 4.5% — were observed for the Gotland Basin and for the Landsort Deep (Fig. 3b). Frequently, pronounced maxima appeared at the sediment/water interface or just below. This great excess of Mn above a small decritical fraction was soluble in 0.5 N HCl. Maximum zinc contents were met in the central Baltic Sea, e.g. at stations "BY15", "BY31" and "BY28". Generally, to the north (Bothnian Bay) and south (Kattegat) the contents decreased. However, this was accompanied by a significant increase towards land-based sources, e.g. from the central Mecklenburg Bight to the Lübeck Bight.

For lead, the expected anthropogenically governed distribution patterns were met, with maxima in the Lübeck Bight and in the Gdansk Deep. For most cores, at the first 10 or 15 cm sections the difference between the "total" contents and the HCl extracted fractions is minor. However, for the Landsort Deep and the Gulf of Bothnia a significant detritical/inert fraction was noted. Probably, due to diagenetic re-distribution, several times the maximum Pb contents were observed one or two cm below the water/sediment interface. For copper, at all samples a detritical component of about 10 to 25 µg/g was present. The Cu profiles are similar to those of lead whereas the appearance of regional maxima resembles those of zinc, that means that the highest contents were found in the central basins and a decrease to the north and south was visible. It is interesting to note, that the

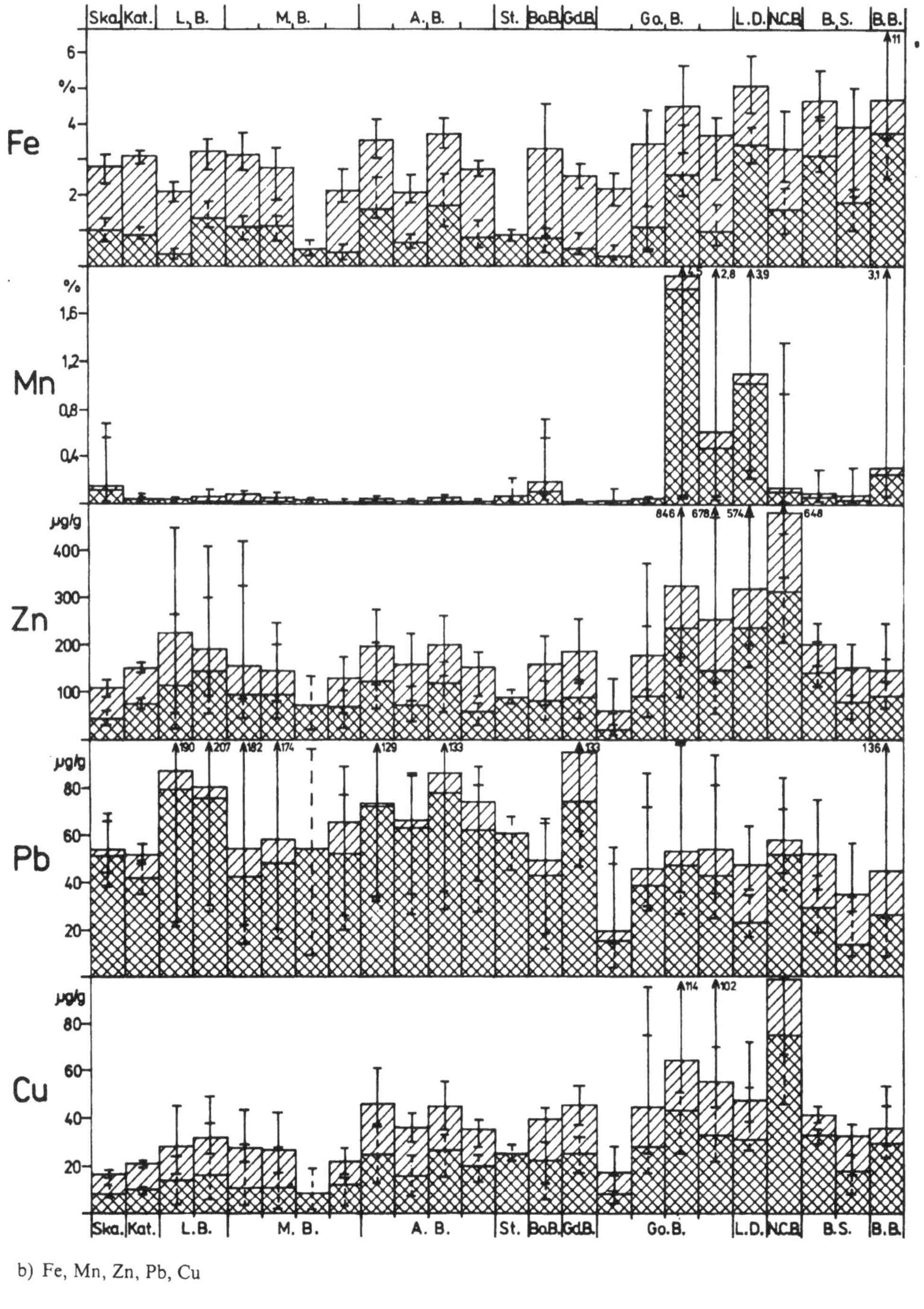

b) Fe, Mn, Zn, Pb, Cu

percent "chemical availability" for Cu seems to be relatively high in the Gulf of Bothnia. This is in agreement with observations on the existence (pK values between 11 and 12) and stability of Cu complexes with humic materials of different origin (SOHN & WEESE 1986) and on the strength of 1 N HCl leaching which recovers Cu associated with organic carriers and with the most important inorganic phases as amorphous Mn and Fe oxides/hydroxides (LUOMA 1986). These types of binding should represent the main forms in which Cu exists in oxygenated sediments of areas with high contents of iron, manganese and humic compounds as they are typically for the Gulf of Bothnia. On the other hand, in the deep central basins with anoxic waters and sediments the HCl insoluble copper sulphides must be taken into consideration.

For chromium, a relatively even distribution of the low HCl soluble fraction throughout the cores together with an absence of higher contents at the sediment/water interface seems to be typically. Again, maximum contents appear in the central basins. The distribution patterns of the nickel contents resemble those of copper with a maximum in the Gotland Basin, without indication of near-shore "hot-spots" and with an even inert fraction superimposed by more variable HCl extractable phases.

For cadmium, anthropogenic influences from land runoff seemed to be reflected by higher contents in cores from the Lübeck Bight and from the Gdansk Deep. Opposite to lead, however, maximum contents were found in the central basins. Mean values up to 12 µg/g are extraordinary for the Baltic Sea because they fall into the range of contamined lakes or rivers. Minimum values

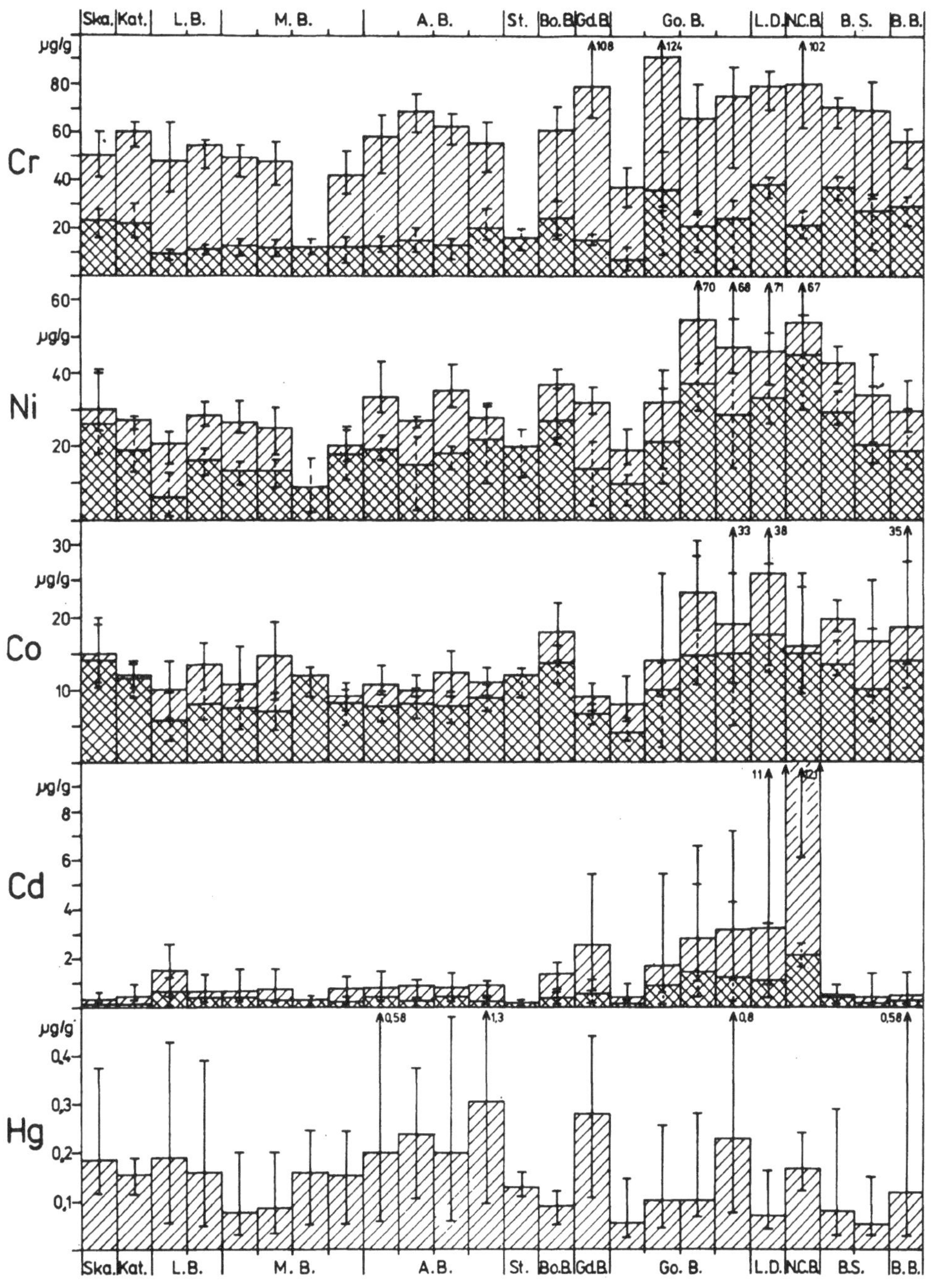

c) Cr, Ni, Co, Cd, Hg

were observed in the sufficiently ventilated Gulf of Bothnia. With increasing probability of the occurrence of anoxic conditions in bottom waters and sediments the cadmium contents seemed likewise to increase. The Cd "excess" above CLARKE or clay background values (0.2 to 0.3 μg/g) was only partly extractable by 0.5 N HCl treatment. This behaviour would correspond with the existence of sulphidic forms of Cd. Therefore, an extensive inflow of saline waters into the Baltic Sea would probably cause a remarkable remobilization of such

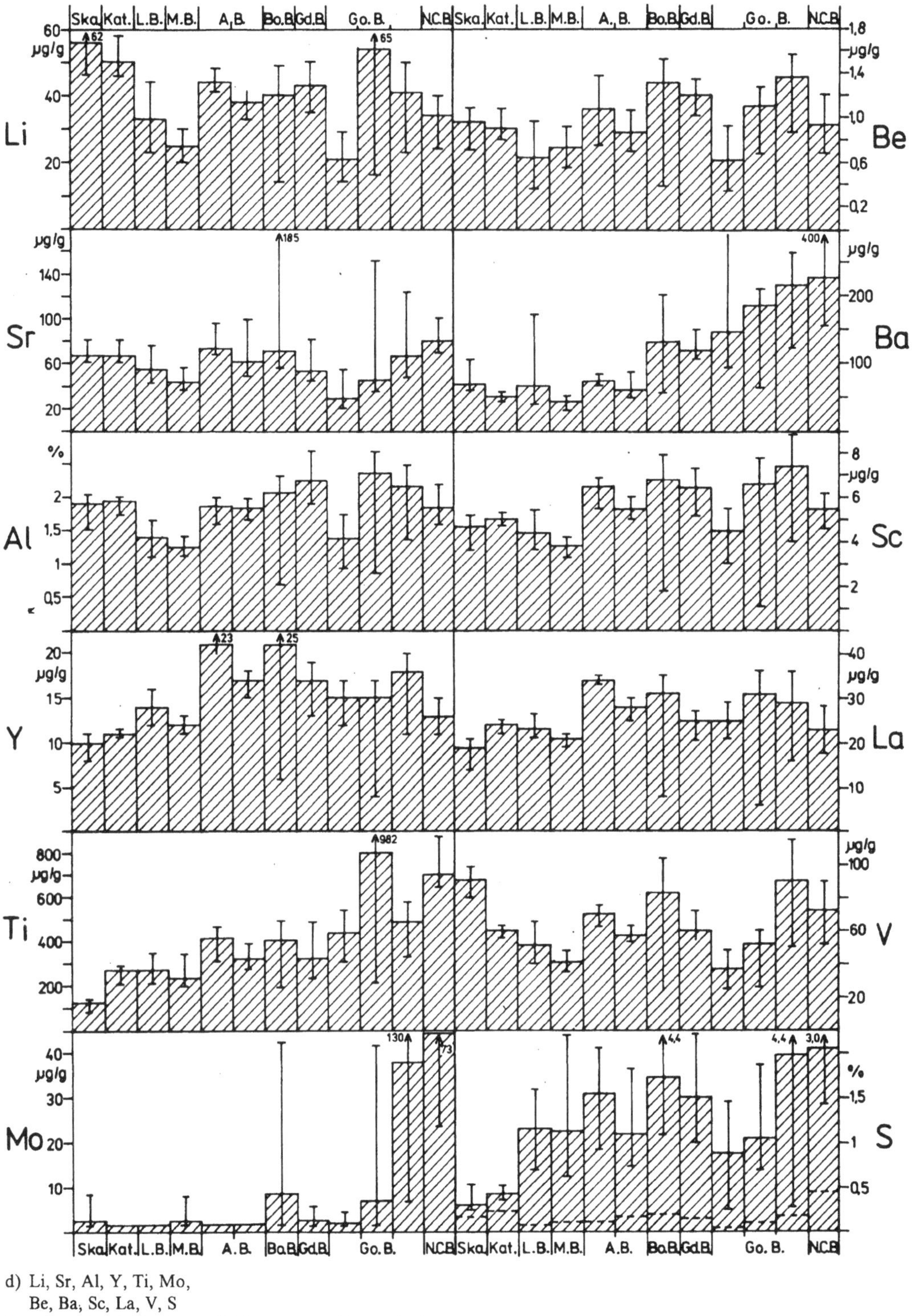

d) Li, Sr, Al, Y, Ti, Mo,
Be, Ba, Sc, La, V, S

Remarks: hatched columns/full lines — "total" contents
double hatched columns/broken lines — 0.5 N HCl fraction
"Ska." — Skagerrak; "Kat." — Kattegat;
"L.B." — Lübeck Bight; "M.B." — Mecklenburg Bight;
"A.B." — Arkona Basin; "St." — Stolp Trench;
"Bo.B." — Bornholm Basin; "Gd.B." — Gdansk Basin;
"Go.B." — Gotland Basin; "L.D." — Landsort Deep;
"N.C.B." — Northern Central Baltic;
"B.S." — Bothnian Sea; "B.B." — Bothnian Bay

sulphidic cadmium deposits due to oxidation processes (Calmano & Förstner 1985). The extreme enrichment of Cd in anoxic sediments is not unique for the Baltic Sea. We found more than 50 µg Cd/g dry material in cores of diatomaceous oozes off Namibia/Walvis Bay (Brügmann 1980). Holmes (1986) explained the extreme enrichment of Cd (up to 130 µg/g), Zn (up to 1.1%) and Hg (up to 10 µg/g) in Texas marine sediments by the precipitation of the corresponding metal sulphides.

The distribution patterns of the mercury contents are different from those of the other metals. A preferred place for the accumulation of this element seems to be the Arkona Basin. This is interpreted as follows: The concentrations of colloidal and dissolved humic acids in Baltic waters decrease sharply in the western Baltic Sea due to flocculation of the organic macromolecules. Mercury forms with humic compounds complexes with formation constants up to the order of pK 20. Therefore, the metal could precipitate together with the organic carrier. Likewise, this would explain the relatively low mercury content of sediments from the Gulf of Bothnia. The low salinity and high humic concentration of these waters prevent the flocculation and enables the Hg transfer to other parts of the Baltic Sea.

In pore waters, the elements Cd, Pb and Zn showed up to 100 times higher concentrations in relation to the adjacent water layer. The Cu and Ni values were only slightly elevated. The shape of the nickel profiles is very similar to those of manganese. This suggests similar redox behaviour of both elements and/or adsorption of Ni species onto particulate Mn carriers and their release from those carriers following re-dissolution under anoxic conditions. Frequently, if the redox regime switches from "oxic" to "sub-oxic", the increase of soluble Mn(II) is accompanied by significant peaks in the Pb, Zn or Cd profiles. Apparently, the soluble Fe(II) has only a very limited range of existence in the sub-oxic region of the profiles. In the presence of oxygen, Fe(II) becomes quickly re-oxidized. Under anoxic conditions, it is precipitated as sulphide. The Mn(II) pore water concentrations in anoxic sediments are controlled mainly by formation of rhodochrosite ($MnCO_3$ with Fe, Zn, Cd, Ni, Pb, Cu and/or Co admixtures — e.g., up to 12% Zn —; see Hartmann 1964).

To get an impression on mass fluxes of trace metals in Baltic sediments the gradients of their concentrations in pore waters were used supposing full control by diffusion. The results of calculations (Table 1) show that the maximum fluxes in the sediments are in the same order of magnitude as the mean annual metal accumulation averaged over the last ten or more than 80 years. It is unknown, how much of these potentially mobile metals will actually return to the bottom waters.

Table 1. Annual maximum fluxes of trace metals in sediments compared to mean accumulation rates (µg/cm²/yr)

	Maximum fluxes in sediments	Mean accumulation rates	
		1971—1981	1900—1983
Fe	3500	2400	—
Zn	22	16	4.9
Pb	10	4.3	1.4
Cu	1.0	3.0	1.0
Ni	0.6	2.4	0.8
Cd	0.3	0.12	0.05

The profiles of metal contents in dated sediment cores were used for crude estimates on the metal masses accumulated in the time period from 1900 until sampling. This was done as follows:

a) Re-calculation of measured metal contents (µg/g) into metal masses for the different 1 cm segments (µg/cm³) applying salinity-corrected dry masses.

b) Retrospective summarizing of the metal masses per segment until 1900 (µg/cm²).

c) Extrapolation of the accumulated metal masses from the investigated cores to the corresponding net-sedimentation areas.

Only representative cores without serious indication of disturbances including erosion were selected. It was roughly estimated that from the total area of the different basins (Ehlin et al. 1974) about one third is characterized by net-sedimentation (Brügmann 1986; Larsen &

Table 2. Masses of metals, organic carbon, nitrogen and phosphorus accumulated between 1900 and 1983 (1981) at different stations of the Baltic Sea

Station	C-org	N	P	Zn	Cr	Pb	Ni	Cu	Cd	Hg
	mg/cm²			µg/cm²						
LB01	40	2.3	0.75	269	47	114	20	33	5.7	0.255
LB-1	192	10.8	3.90	884	161	387	93	118	3.0	0.659
MB-1	154	8.8	3.42	626	148	252	80	94	2.8	0.344
MB-2	172	11.9	3.75	594	153	279	80	99	2.7	0.405
MB-012	181	11.1	2.82	573	158	297	79	94	3.6	0.750
AB-1	209	12.2	4.64	676	173	308	96	149	3.0	0.826
AB-2	190	11.7	3.26	623	178	324	105	139	2.5	0.745
BY2	67	4.0	1.61	162	55	84	29	37	0.9	0.438
BY5	64	3.7	1.09	193	60	59	38	40	1.6	0.694
BCSIII-15	81	5.1	1.15	242	98	120	36	56	4.3	0.417
BCSIII-10	45	3.0	1.22	148	56	61	30	33	1.1	0.147
BY38	65	4.2	1.51	280	121	73	50	72	2.8	0.178
F81	22	1.2	0.70	185	15	26	18	26	1.4	0.073
BY15	91	4.6	1.52	455	123	100	80	92	5.5	0.352
BY31	123	7.5	9.67	1495	352	217	208	216	16.7	0.342
BY28	98	6.0	1.29	440	74	52	50	89	10.3	0.143
EB1	93	6.4	5.53	720	219	198	132	135	2.0	0.403
BS2	33	3.0	3.69	333	126	110	78	60	1.5	0.228
B03a	46	2.5	3.26	192	56	94	34	44	0.8	0.332

Table 3. Masses of metals, organic carbon, nitrogen and phosphorus accumulated between 1900 and 1983 (1981) in different sub-areas of the Baltic Sea (EHLIN et al. 1974); one third of the sea bottom considered as "net-sedimentation area"

Net-sedimentation area (km^2)	C-org	N	P	Zn	Cr	Pb	Ni	Cu	Cd	Hg
mass/area mass, absolute	mg/cm^2 10^6 t			$\mu g/cm^2$ 10^3 t						
Bothnian Bay	46	2.5	3.3	192	56	94	34	44	0.8	0.33
(12268)	5.6	0.31	0.41	24	6.9	11.5	4.2	5.4	0.10	0.041
Bothnian Sea	63	4.7	4.6	527	173	154	105	98	1.8	0.32
(26358)	16.6	1.24	1.21	139	45.6	40.6	27.7	25.8	0.47	0.083
Northern Baltic Proper	98	6.0	1.3	440	74	52	50	89	10.3	0.14
(15770)	15.5	0.95	0.21	69	11.7	8.2	7.9	14.0	1.62	0.023
Gotland Deep	91	4.6	1.5	455	123	100	80	92	5.5	0.35
(14797)	13.5	0.68	0.22	67	18.2	14.8	11.8	13.6	0.81	0.052
Gdansk Basin	81	5.1	1.2	242	98	120	36	56	4.3	0.42
(8533)	6.9	0.44	0.10	21	8.4	10.2	3.1	4.8	0.37	0.036
Landsort Deep	123	7.5	9.7	1495	352	217	208	216	16.7	0.34
(2382)	2.9	0.18	0.23	36	8.4	5.2	5.0	5.1	0.40	0.008
Western Gotland Basin	65	4.2	1.5	280	121	73	50	72	2.8	0.18
(9025)	5.9	0.38	0.14	25	10.9	6.6	4.5	6.5	0.25	0.016
Bornholm Basin	64	3.7	1.1	193	60	59	38	40	1.6	0.69
(12997)	8.3	0.48	0.14	25	7.8	7.7	4.9	5.2	0.21	0.090
Arkona Basin	155	9.3	3.2	487	135	239	77	108	2.1	0.67
(6224)	9.6	0.58	0.20	30	8.4	14.9	4.8	6.7	0.13	0.042
Sound/Belt	157	9.5	3.1	589	133	266	70	88	3.6	0.48
Sea (6784)	10.7	0.64	0.21	40	9.0	18.0	4.7	6.0	0.24	0.033
Kattegat	52	3.4	1.3	197	77	70	35	27	0.7	0.23
(7348)	3.8	0.25	0.10	15	15.7	5.1	2.6	2.0	0.05	0.017
Accumulation rate (weighted mean, $mass/cm^2/yr$)	0.99	0.061	0.031	4.9	1.4	1.4	0.81	0.95	0.046	0.0044
Sum of masses (122486 km^2) (in 10^6 t or 10^3 t, respectively)	99.3	6.13	3.16	491	141	143	81	95	4.7	0.44

PHEIFFER MADSEN 1986; SALO et al. 1986). It was decided, to cover the time period back to the turn of the century. This was due to the fact that even in the case of high linear sedimentation rates with 20 or 25 cm core sections the last 80 years were entirely enclosed. In this time period the most extensive/intensive industrial development took place in the countries bordering the Baltic Sea.

The resulting accumulated metal masses are presented in Tables 2 and 3. In relation to the observed metal contents they reflect a different picture. Whereas, for instance, in sediments of the Gotland Basin and of the northern Baltic Proper maximum zinc contents were recorded, the maximum zinc mass accumulation took place at stations of the western Baltic Sea. Even in the Bothnian Sea (station "EB-1") the zinc accumulation of the last 80 years surpassed that of the Gotland Deep up to two times. Regional differences found before for the copper or nickel contents are now balanced if the accumulated masses are considered. This means, despite of the fact, that the very fine grained mud sediments of the central Baltic Basins exhibit high metal contents, the annually deposited metal masses are not extraordinary due to the low accumulation rate. An exception must be made from this statement for cadmium which — deposited probably as sulphide — shows in the central basins both types of maxima, for the contents and for the accumulated masses.

Statistical treatment of the data was performed on three different levels of abstraction:

Firstly, by the calculation of a correlation matrix for all variables was aspired to detect possible relations between the trace metals and how they correlate with background parameters as organic carbon, carbonate, phosphorus, nitrogen, iron, manganese, aluminium and/or scandium.

Secondly, based on significant (> 99% probability) positive correlations, a selection was made for pairs of one "dependent" and several "independent variables" for which multiple regression equations were computed.

Thirdly, previously computed data, as e.g., (a) mean values of relevant parameters for the different cores, (b) accumulated masses of metals and other constituents of the cores or (c) through the contents of organic carbon and/or scandium "normalized" mean values, were subjected to a principal component factor analysis (PCA) aimed to evaluate and group the 22 cores.

Obviously, the investigated materials constituted a mixture of

- geochemically only slightly altered weathering products of the earth curst,
- hydrolysed weathering products (clay minerals),
- sapropelitic material, deposited under reducing conditions, and
- (co-)precipitates (carbonates, phosphates, sulphides, oxides/hydroxides...).

From the correlation matrices which were calculated for each core separately some conclusions could be drawn with regard to the possible binding forms of the metals. However, it should always be kept in mind, that the main forms in which the trace metals exist may be subjected to strong fluctuations from one net-sedimentation area to another.

Mercury seems to be combined predominantly with organic matter, especially with compounds rich in nitrogen. However, in cores from the northern parts of the Baltic Sea (Gotland Basin to Bothnian Sea) mercury (co-)precipitation with manganese oxihydrates seems to compete. In the anoxic mud deposits at stations "BY28", "BY15" and "BY38" sulphidic forms could play a significant role, too.

For cadmium, the organic binding forms prevail. In the Gotland Basin and in the Gulf of Bothnia an association with Mn/Fe-rich materials (oxides/hydroxides, phosphates) could be remarkable. (Co-) precipitation as carbonates must be considered for the Arkona and Gotland Basins. The occurrence of Cd sulphides is attributed to the cores "BY38", "BY28" and "BY15". The speciation of zinc in the sediments is similar to cadmium. The main forms seem to include organic associates, carbonates and sulphides (only "BY15"). Significant positive correlations with Al and/or Sc as indicator of noteworthy detritic fractions were limited to the western Baltic Sea and to the southern Gotland Basin. In relation to Zn and Cd, the organically associated lead forms are of minor importance. Probably, lead phosphates, e.g. of the pyromorphite type ($Pb_5Cl(PO_4)_3$), must be taken into consideration. Sulphidic forms can not be excluded at cores "BY15" and "BY28", in the latter case as co-precipitates with FeS_x. Significant detritic Pb fractions are indicated for stations in the Lübeck Bight, Kattegat and southern Gotland Basin.

For copper, again the organically associated forms dominate. At core "BY15" carbonatic and sulphidic fractions must be considered. Binding to inert material seems to exist at those stations mentioned for lead. At the expense of organic forms for nickel the detritic binding becomes more important. The probability of the occurrence of carbonate phases must be valued similarly as done before for Cd and Zn. A nickel accumulation by Fe/Mn precipitates seems to be documented, too. Only minor indicated are sulphidic forms. Only for a few cores, chromium was weakly correlated with the organic carbon. The same statement must be made with regard to possible relations to the contents of sulphur, manganese, phosphorus, nitrogen or carbonate. Otherwise, always a high significant correlation with detritic components, as e.g. Al, Sc and Fe, was observed. Concerning the relative importance of different binding forms, cobalt could be placed between Ni and Cr. Organic Co phases are present, detritic forms dominate and for some cores carbonate and/or sulphide (co-) precipitates are indicated.

The "total" digestions of samples taken in 1983 were analyzed by IC(A)P on additional elements including S, Ti, V, Sr, Ba, Li, La, Y, Sc and Mo. Whereas metals like Ti, La, Li, Y and Sc are typical representatives of the lithogenic sediment fraction, the distribution patterns of molybdenum seem to be worth for more in-depth interpretation in the context of this paper. Unfortunately, only in eight out of twelve cores taken in 1983 the Mo contents were significantly higher than the limit of determination and/or showed reasonable vertical gradients. The geochemistry of this element in marine sediments was sometimes discussed together with that of manganese (SHIMMIELD & PRICE 1986). This is, because both elements are influenced easily by changes of the redox regime. However, only at two cores this inter-element relationship was indicated. Otherwise, there is no doubt, that Mo enters the sediment together with organic material (HOLLAND et al. 1986). Possibly, there it is accumulated finally as sulphide. A significant positive correlation between Mo and "excess" sulphur was not only seen at sediments formed below basins with frequently anoxic bottom waters but likewise in areas with well-oxygenated deep waters. Therefore, it can be concluded, that the MoS_2 precipitation occurs only in those sediment zones where the most intensive sulphate reduction just takes places. If this is true, the Mo may hardly be used as a paleoredox indicator of anoxia in deep bottom waters as proposed by HALLBERG (1974). In addition, a maximum of Mo which was observed in several dated cores from different locations in the Gotland Basin did not fall into the same time span. Consequently, these maxima seem to reflect the accumulation conditions in the sediment and not the deposition regime in the water column above.

Because of differences in the accumulation conditions for trace metals in sediments between coring sites, the calculated multiregression equations must differ significantly, too. Some examples shall illustrate this for 4 elements:

Core 'LB-01'	R	B
Zn = −700 + 635 * Al + 28 * Fe − 4.7 * oC	.88	.77
Ni = .44 + 17 * Al − .98 * Fe + .33 * oC − 31 * P	.97	.94
Cu = −39 + 64 * Al − 3.3 * Fe − 1.0 * oC − 147 * P	.95	.90
Hg = −607 + 753 * Al − 42 * Fe − 39 * oC	.89	.79
Core 'MB-012'		
Zn = −234 + 15 * Al + 58 * oC + 764 * P + 16 * aC	.88	.78
Ni = −10 + 13 * Al + 2.1 * oC + 13 * P + 2.4 * aC	.93	.87
Cu = −28 + 7.5 * Al + 5.6 * oC + 167 * P + 1.9 * aC	.97	.95
Hg = −304 + 61 * Al + 1340 * N + 255 * P	.91	.82
Core 'BCS III-10'		
Zn = −42 + 31 * Al + 488 * N − 115 * P − 71 * Mn	.98	.96
Ni = 1.5 + 9.0 * Al + 1.2 * oC + 8.8 * P − 1.5 * Mn + .73 * Fe	.96	.93
Cu = −6.7 + 7.6 * Al + 2.1 * oC + 175 * P − 133 * Mn	.97	.94
Hg = −70 + 9.3 * Al + 394 * N + 1070 * P − 785 * Mn	.97	.95
Core 'BY 38'		
Zn = −214 + 16 * oC + 2830 * P + 16 * aC + 29 * S	.95	.90
Ni = −4.4−3.6 * Sc + 191 * P + 159 * Mn + 9.9 * Fe	.93	.86
Cu = −66 + 4.2 * oC + 808 * P + 3.8 * aC + 8.1 * S	.96	.93
Hg = −60 + 23 * oC + 483 * P + 17 * S	.92	.85
Core 'BY15'		
Zn = −48 + 56 * oC − 633 * P + 8.0 * aC + 80 * Mn − 11 * S	.99	.98
Ni = 18 + 1.0 * oC + 219 * P − 2.7 * aC + 18 * Mn + .53 * S	.98	.95
Cu = −9.7 + 5.9 * oC + 333 * P + 1.6 * aC + 3.4 * Mn − 2.4 * S	.99	.98
Hg = −562 + 2150 * N + 3055 * P + 30 * Mn − 64 * S	.97	.93

Remarks: — Zn, Ni, Sc and Cu in µg/g; Hg in ng/g
— oC, aC, Al, Mn, S, P and N in %
— oC = organic carbon; aC = carbonate
— R = correlation coefficient
— B = measure of exactness

It should be kept in mind that some of the "independent" variables are interrelated. Therefore, despite only those variables were selected for the multiple regression calculations which were significantly positive (univariate) correlated with the trace metals, some of them "contributed" now "negatively" to the construction of the measured metal contents.

In Fig. 4 part of the results of the principal component factor analysis is shown. The "individuals" are 21 cores taken from the Baltic Sea in 1981 and 1983. The masses of organic carbon, nitrogen, phosphorus, Cd, Cr Cu, Hg, Ni, Pb and Zn which accumulated between 1900 and 1983 (1981), were used as similarity measure for the grouping of the cores. The results reflect — inter alia — the unique position of the Landsort Deep and the close relationship between groups of cores taken from the western Baltic Sea, from the Kattegat/Skagerrak or from the Arkona and Bornholm Basins. The "variables" organic carbon, nitrogen and lead are closely interrelated.

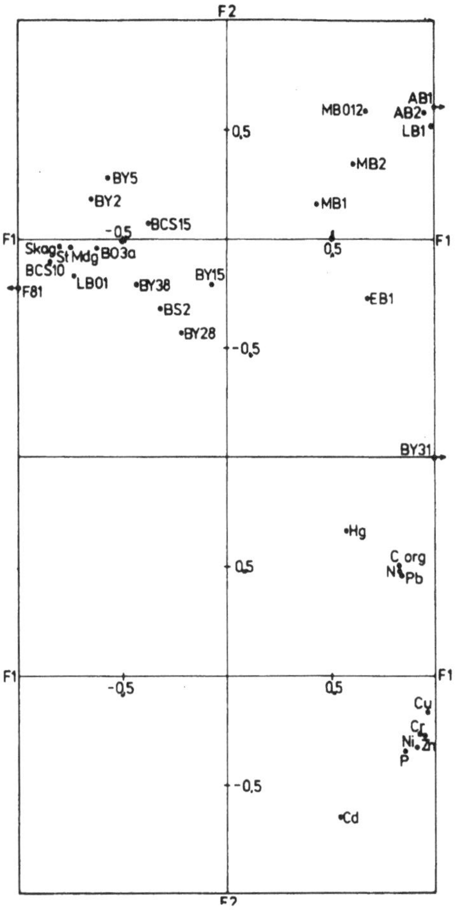

Fig. 4. Examples of results of the Principal Component Factor Analysis (PCA)
"individuals" — 21 sediment cores from the Baltic Sea
"variables" — between 1900 and 1983 (1981) accumulated masses of C-org, N, P, Cd, Cr, Cu, Hg, Ni, Pb and Zn
F1 = 69.8%, F2 = 19.8% (of variance)

A second group comprises Cu, Cr, Zn, Ni and P whereas Hg and Cd are more separated from those.

Acknowledgements

The authors are indebted to Dr. NIEMISTÖ from the Finnish Institute of Marine Research (FIMR), Helsinki/Finland who gave valuable advice for proper coring and for the interpretation of the field data. In addition, he provided us with several dissected sediment cores and dating results from parallel cores. Other 210-Pb dating results were kindly submitted by Dr. GELLERMANN (BAF Freiberg) and D. WEISS (SAAS Berlin). On part of the samples Dr. P. VOGLER (ZIPE Potsdam) carried out supplementary and repetition analyses for several metals, phosphorus and sulfur by atomic emission spectroscopy (IC-A-P). The bulk of those data will be presented in a separate publication.

References

BREDER, R. (1982): Optimization studies for reliable trace metal analysis in sediments by atomic absorption spectrometric methods. Fresenius Z. Anal. Chem. **313**, 395—402.

BRÜGMANN, L. (1980): Spurenmetalle in marinen Proben — Formen des Auftretens und ihre analytische Bestimmung. Diss., Rostock-Warnemünde.

— (1986): Influence of coastal zone processes on mass balances for trace metals in the Baltic Sea. Rapp. P.-v. Reun. Cons. int. Explor. Mer **186**, 323—336.

— & HENNINGS, U. (1982): Stratigraphie und Spezifizierung von Metallen in Sedimenten der westlichen Ostsee. Acta hydrochim. hydrobiol. **10**, 41—53.

— & LANGE, D. (1983): Geochemische und sedimentologische Untersuchungen an einem Sedimentkern aus dem Schlickgebiet der Lübecker Bucht. Gerlands Beitr. Geophys. **92**, 241—268.

— LANGE, D., & NIEMISTÖ, L. (1982): Comparative geochemical studies on sediment cores from different areas of the Baltic Sea. Proc. 13th Conference Baltic Oceanographers, Helsinki, 198—205.

CALMANO, W., & FÖRSTNER, U. (1985): Schwermetallbindungsformen in Küstensedimenten — Standardisierung von Extraktionsmethoden. Report, Hamburg, 183 pp.

CHESTER, R., & VOUTSINOU, F. G. (1981): The initial assessment of trace metal pollution in coastal sediments. Mar. Poll. Bull. **12**, 84—91.

EHLIN, U., MATTISON, I., & ZACHRISSON, G. (1974): Computer based calculations of volumens of the Baltic Sea area. Proc. 9th Conference Baltic Oceanographers, Kiel, 114—128.

HALLBERG, R. O. (1974): Paleoredox conditions in the Eastern Gotland Basin during the recent centuries. Merentutkimuslait. Julk./Havsforskningsinst. Skr. No. **238**, 3—16.

HARTMANN, M. (1964): Zur Geochemie von Mangan und Eisen in der Ostsee. Meyniana **14**, 3—20.

HÄSÄNEN, E. (1977): Dating of sediments, based on 210-Po measurement. Radiochem. Radioanal. Letters **31**, 207—214.

HATCH, W. R., & OTT, W. L. (1968): Determination of submicrogram quantities of mercury by atomic absorption spectrophotometry. Anal. Chem. **40**, 2085—2087.

HELCOM — Baltic Marine Environment Protection Commission (1983): Guidelines for the Baltic Monitoring Programme for the Second Stage, Helsinki, December 1983.

HOLMES, C. W. (1986): Trace metal seasonal variations in Texas marine sediments. Mar. Chem. **20**, 13—27.

HOLLAND, H. D., LAZAR, B. & MCCAFFREY, M. (1986): Evolution of the atmosphere and oceans. Nature **320**, 27—33.

LARSEN, B., & PHEIFFER MADSEN, P. (1986): Heavy metal budgets for selected Danish sea areas. Rapp. P.-v. Reun. Cons. int. Explor. Mer. **186**, 244—250.

LUOMA, S. N. (1986): A comparison of two methods for determining copper partitioning in oxidized sediments. Mar. Chem. **20**, 45—59.

NIEMISTÖ, L. (1974): A gravity corer for soft sediments. Merentutkimuslait. Julk./Havsforskningsinst. Skr. No. **238**, 33—38.

PHEIFFER MADSEN, P., & LARSEN, B. (1986): Accumulation of mud sediments and trace metals in the Kattegat and the Belt Sea. Report of the Marine Pollution Laboratory **10**.

— & SORENSEN, J. (1979): Validation of the lead-210 dating method. J. Radioanal. Chem. **54**, 39—48.

SALO, A., TUOMAINEN, K., & VOIPIO, A. (1986): Inventories for some longlived radionuclides in the Baltic Sea. The Science of the Total Environment **54**, 247—260.

SHIMMIELD, G. B., & PRICE, N. B. (1986): The behaviour of molybdenum and manganese during early sediment diagenesis — offshore Baja California, Mexico. Mar. Chem. **19**, 261—280.

SOHN, M., & WEESE, D. (1986): 13-C NMR spectra and Cu(II) formation constants for humic acids from fluvial, estuarine and marine sediments. Mar. Chem. **20**, 61—72.

Address of the authors:
Prof. Dr. sc. L. BRÜGMANN
Prof. Dr. sc. D. LANGE
Akademie der Wissenschaften der DDR
Institut für Meereskunde
Seestr. 15
Rostock-Warnemünde
DDR-2500

Limnologica (Berlin) **20** (1), 29—32, April 1990

Wasserwirtschaftsdirektion Küste, Stralsund, DDR

On the Variations of Some Chemical Criteria in the Unterwarnow — an Estuary of the Western Baltic Sea in 1977—1986

By ALEXANDER BACHOR

With 3 Figures and one Table

Key words: Estuary; Hydrography; Salinity; Dissolved oxygen; BOD_2; Dissolved nutrients.

Abstract

The estuary Unterwarnow has been investigated periodically since 1977. A short description is given of the favourable hydrological conditions and the high anthropogenic load.

Spatial and temporal variations in salinity, dissolved oxygen, BOD_2 and dissolved inorganic nutrient levels are discussed. Diagrams show the spatial variations in all of these criteria in the surface layer and the temporal variations in nitrate, orthophosphate and dissolved oxygen levels at one station in the estuary from 1977 to 1986. The relation between nitrate level and surface runoff is shown.

1. Introduction

Due to its use as a shipping route and watercourse, the estuary Unterwarnow represents a very important water for the county town Rostock.

The Unterwarnow extends from the town port of Rostock in the south to the Baltic Sea in the north. Its total length is 11 km. The Unterwarnow, 0.2 to 0.9 km wide, is very narrow. In the northeast, just before entering the Baltic, it broadens into the Breitling, which is separated from the Baltic Sea by a barrier-beach with dunes. Together, the Unterwarnow and Breitling are one of the GDR's smallest inland coastal waters with an area of 12.5 km² and about 50 million m³ volume. The mean water depth is 4 m.

Hydrological conditions in the Unterwarnow are characterized by water exchange with the nearby Baltic Sea and land runoff, and these are reflected in variations in physical, chemical and biological quality criteria. Results of the 10-year analysis (1977—1986) concerning the distribution of salinity, dissolved inorganic nutrients, dissolved oxygen and BOD_2 are presented in the following.

2. Material and Methods

The grid consists of 6 stations; 4 in the Unterwarnow, 1 in the mouth into the Baltic Sea and 1 in the Breitling (Fig. 1). All stations are located in the navigational channel, so that, beside water-surface samples, samples were also taken from the bottom-near water at a depth of 4 m (stations 1 and 6), 7.5 m (stations 2, 3 and 4) and 13.5 m (station 5) except during periods of ice cover. One to two samples were taken a month. Altogether, during the period investigated, 178 sample series were analysed for the Unterwarnow and 168 sample series for the Breitling. Physical and chemical quality criteria were determined according to "Selected methods of water analysis" (1986) and "Selected methods for determining solutes in marine and brackish water" (RHODE & NEHRING 1979). RUTTNER samplers were used to collect water samples from different depths.

3. Results and Discussion

3.1. Hydrological conditions and anthropogenic loads

The water balance of the water body governs the water quality. In terms of water inputs to the Unterwarnow the inflow of Baltic water dominates with a 10-year mean with 857 million $m^3 \cdot a^{-1}$. Freshwater discharge, mainly from the upper Warnow, with a 10-year mean of 633 million $m^3 \cdot a^{-1}$, also constitutes a major item in the water balance of the Unterwarnow. The total mean water inflow is sufficient for about 30 turnovers a year in the Unterwarnow and Breitling, resulting in a mean residence time of only 12 days. Waters with such intensive water renewal can withstand high loads.

In contrast to these favourable hydrological conditions, the anthropogenic load on the waters is high. The ratio of catchment area (3.209 km²) to area of the estuary (12.5 km²) is sufficient to expect a high loading, and the organic and nutrient loads in the Unterwarnow proved to be very high (BACHOR 1989). The main sources of these are sewage effluent from the town Rostock and loads imported from the upper Warnow.

3.2. Variations in salinity

The Unterwarnow is characterized by the major salinity variations in time and space typical of an estuary. The greatest variations in salinity are to be noted in the southern part of the waters (Table 1). From station 1 to station 2, a distance of only about 2 km, the salinity of the water surface almost doubles (Fig. 1). The salinity increases progressively up to the mouth into the Baltic Sea. The greatest vertical differences in salinity are found in the southern Unterwarnow as well. In this stretch the discharged freshwater and inflowing Baltic water collide directly. At a depth of 4 m at station 1, the mean salinity was twice that measured at the surface. Due to increasing intermixing of salt- and freshwater, the vertical salinity differences decrease northwards.

Seasonal variations in salinity are influenced considerably by the distribution of inflow and outflow, depending on hydrological and meteorological conditions. At station 5, annual fluctuations are relatively harmonic with maxima in October and November and a more or less steady low salinity from February to August. This corresponds approximately to the mean annual salinity fluctuations in the surface layer of the Mecklenburg Bay as calculated by MATTHÄUS (1984). The amplitude of the mean annual salinity in the Unterwarnow increases with increasing effect of land-runoff. As a rule, high land-runoff between February and April results in a clear drop in the salinity of the surface water in the southern Unterwarnow. In contrast, near the bottom, a decrease of salinity is observed only from June/July on. Therefore, vertical salinity differences are greatest from March to April.

3.3. Variations in dissolved oxygen and BOD_2

Results of analysis for dissolved oxygen and BOD_2 in these waters indicate the great degradable organic load. In the period studied, the mean oxygen concentration at the surface of the southern Unterwarnow decreased from station 1 to station 3 (Fig. 1). This was accompanied by increase in the biochemical oxygen demand (BOD_2). The lowest mean dissolved oxygen concentration and the highest mean BOD were measured at station 3. This station is located about 800 m north of the outfall from the Central Sewage Treatment Works of Rostock. The great demand on the oxygen budget, particularly in the southern Unterwarnow, is also shown by measurements of the oxygen concentration in the near-bottom layer (Table 1). At a depth of 7.5 m at station 2, about one fourth of all measured oxygen concentrations were below the limit of 4 mg $\cdot$ l^{-1} O_2. This is due not only to the sewage load, but partly also to the high seston load of the Warnow. Much of the particulate substances and sediments carried with land-runoff into the Unterwarnow and degraded at the bottom of the waters, a process in which oxygen is consumed. Oxygen conditions improve only north of station 3, where the influence of relatively clear Baltic water becomes increasingly apparent.

In the course of the year, dissolved oxygen levels vary inversely to the water temperature. Intra-annual variations result from different heating and cooling of the water surface, different advective processes and different levels of primary production by phytoplankton. The oxygen minimum in August coincides with the temperature maximum. However, the oxygen maximum is governed less by temperature than by the phytoplankton-bloom in spring. Sometimes, very high oxygen supersaturations from May to July indicate the eutrophic status of these waters. According to HÜBENER et al. (1984), the phytoplankton are of major importance for the oxygen input to the Unterwarnow, and also for its selfpurification potential, for most of year. On the other hand, high primary production leads to a high rate of sedimentation of autochthonous organic material, the biochemical degradation of which in turn has negative consequences for the oxygen budget of the waters. This biochemical degradation may be responsible for the distinct seasonal variation in dissolved oxygen levels in the near-bottom layer, where amplitudes are also greater than those at the water surface.

3.4. Variations in dissolved nutrients

Nitrate and orthosilicate levels vary inversely to the horizontal and vertical salinity distribution (Table 1). Both nutrients are imported mainly with freshwater discharge

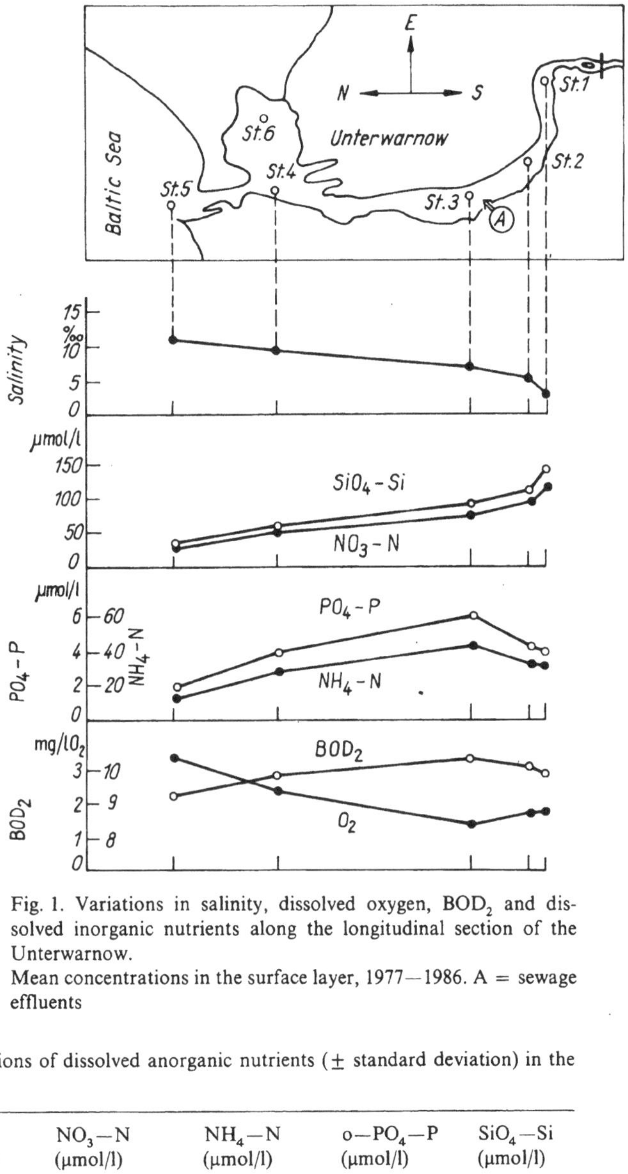

Fig. 1. Variations in salinity, dissolved oxygen, BOD_2 and dissolved inorganic nutrients along the longitudinal section of the Unterwarnow.
Mean concentrations in the surface layer, 1977—1986. A = sewage effluents

Table 1. Mean salinity, dissolved oxygen, BOD_2 and mean concentrations of dissolved anorganic nutrients (± standard deviation) in the Unterwarnow, 1977—1986

Station	Measurement Depth (m)	Salinity (‰)	O_2 (mg/l)	BOD_2 (mg/l O_2)	NO_3—N (µmol/l)	NH_4—N (µmol/l)	o—PO_4—P (µmol/l)	SiO_4—Si (µmol/l)
1	0.5	3.3 ± 3.1	8.8 ± 2.8	2.9 ± 1.6	112.6 ± 116	30.5 ± 24	3.7 ± 3.3	140.3 ± 89
	4.0	6.7 ± 4.2	7.3 ± 3.4	2.5 ± 1.5	77.0 ± 95	33.6 ± 22	3.9 ± 3.0	105.9 ± 67
2	0.5	5.6 ± 3.2	8.7 ± 2.7	3.1 ± 1.6	91.2 ± 99	31.7 ± 23	4.2 ± 3.2	109.9 ± 73
	7.5	11.4 ± 2.6	6.8 ± 3.1	2.3 ± 1.3	26.7 ± 33	30.1 ± 21	4.6 ± 4.1	53.4 ± 37
3	0.5	7.4 ± 2.7	8.3 ± 2.9	3.4 ± 1.9	72.4 ± 83	43.2 ± 31	6.1 ± 5.1	88.3 ± 57
	7.5	11.8 ± 2.6	7.8 ± 2.8	2.2 ± 1.3	30.7 ± 39	25.9 ± 23	4.3 ± 4.4	46.4 ± 31
4	0.5	9.6 ± 2.5	9.4 ± 2.0	2.9 ± 1.6	52.5 ± 65	27.7 ± 24	4.0 ± 2.4	52.9 ± 41
	7.5	12.1 ± 2.3	9.3 ± 2.2	1.7 ± 1.3	25.5 ± 32	16.6 ± 21	2.6 ± 2.0	34.7 ± 26
5	0.5	11.4 ± 2.3	10.4 ± 1.7	2.2 ± 1.2	26.1 ± 38	11.1 ± 12	1.7 ± 1.2	29.8 ± 39
	13.5	14.3 ± 2.7	9.6 ± 2.3	1.4 ± 1.1	8.1 ± 16	6.0 ± 8	1.3 ± 0.9	16.9 ± 12
6	0.5	10.5 ± 2.3	9.8 ± 2.1	2.3 ± 1.5	72.1 ± 106	31.8 ± 51	3.6 ± 2.5	49.5 ± 49
	4.0	11.3 ± 2.5	9.4 ± 1.8	1.8 ± 1.4	57.4 ± 92	26.7 ± 41	3.5 ± 2.2	47.5 ± 45

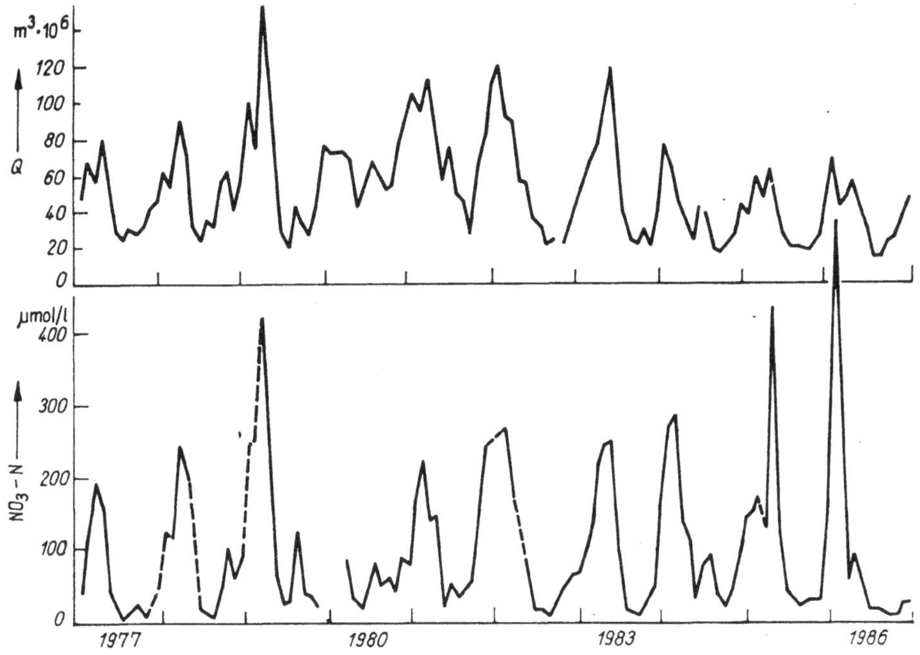

Fig. 2. Variations in nitrate levels in the surface layer of the Unterwarnow (St. 2) as a function of land-runoff. Mean per month

from the upper Warnow and show similar concentration gradients in the direction of the Baltic Sea (Fig. 1). Fig. 2 illustrates the close link between land-runoff and nitrate content in the Unterwarnow. High discharges from January to April result in nitrate peaks in the waters. It is worth mentioning that the highest mean monthly nitrate values were recorded in relatively flow-rich months of the flow-poor years 1985 and 1986. The close nitrate-flow relation is caused by rainwash and leaching in the Warnow catchment area. Besides depending on flow, the distinct seasonal variation with high amplitude also reflects nitrate assimilation by phytoplankton.

Orthosilicate is subject to strong seasonal variations, too, owing to its inclusion in the biochemical cycling of matter, particularly in diatoms. Orthosilicate concentrations may be decreased to the limit of detectability during the diatom bloom in May. The vertical differences of concentration also vanish at this time. However,

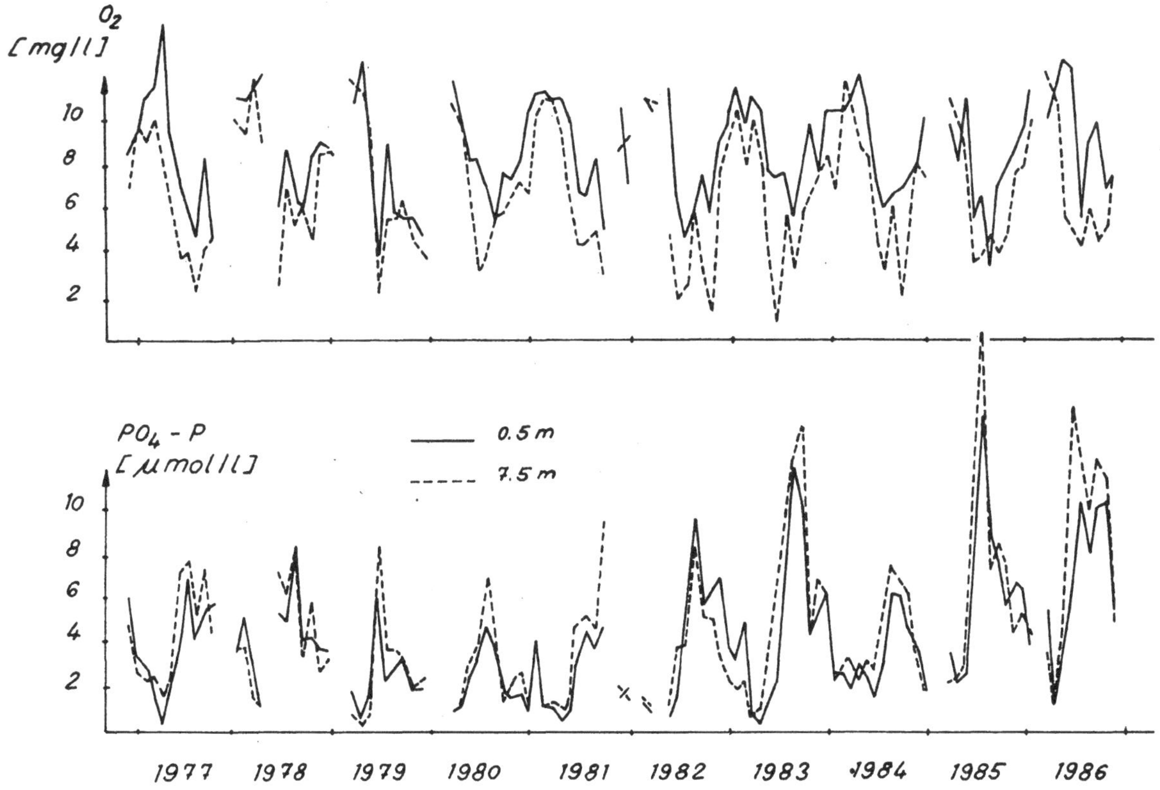

Fig. 3. Variations in dissolved oxygen and orthophosphate content at different depths in the Unterwarnow (St. 2), Mean per month

silicate concentrations in the open water body increase again drastically following the minimum in May.

Ammonium and orthophosphate, unlike nitrate and orthosilicate, show a significant increase along the longitudinal section of the Unterwarnow (Fig. 1). Mean concentrations at station 2 and, especially, at station 3 were higher than at station 1. This is largely due to the sewage effluent from the town Rostock. Most of this effluent reaches the Unterwarnow about 800 m south of station 3 via the Central Sewage Treatment Works. The development of Rostock and the increased phosphorus content of sewage has resulted in increasing phosphorus loading of the Unterwarnow. This has manifested itself as increased orthophosphate concentrations in the waters (Fig. 3). Seasonal variations in orthophosphate show a contrary behaviour to the oxygen content of the waters. Orthophosphate concentrations are lowest in spring, the minimum coinciding with the spring peak of phytoplankton. They start to increase again in late May and early June. As a rule, the maximum is measured in August. The summer concentrations are distinctly from those recorded in winter. Beside the year-round loading from the town, concentrations of allochthonous and autochthonous algal biomass in the estuary are higher in summer. Mineralization of these large amounts of organic matter results in an increase of orthophosphate concentrations in the waters. Moreover, processes linked to nutrient return from the sediment play an increasingly important role as the oxygen concentration decreases (BALZER 1978; NEHRING 1981, 1982). During oxic conditions, mineralization causes leads mainly to a phosphate increase, whereas under anoxic conditions the exchange at the interface water-sediment plays a dominant role.

4. Summary

The water quality of the estuary Unterwarnow has been investigated periodically since 1977. Measured chemical water quality criteria show major variations in time and space which depend on the geographical-hydrological conditions of the waters, the meteorological situation, anthropogenic loads and the dynamics of metabolism.

Variations in salinity are governed mainly by distribution of inflow, outflow and land-runoff, but also depending on hydrological-meteorological conditions.

Spatial variations in dissolved oxygen are strongly influenced by anthropogenic impacts. Oxygen conditions at the water surface are most unfavourable near the sewage outfall of the town Rostock. High oxygen supersaturations in spring and summer point to eutrophic status of this estuary. On the one hand, the phytoplankton is of great importance as an oxygen source, but, on the other, the oxygen budget of the waters is loaded additionally by biochemical degradation of the resulting algal biomass and that from the River Warnow.

Concerning their distribution in the waters, nitrate and orthosilicate levels vary inversely to salinity. Both compounds are imported mainly with land-runoff. There is a close correlation between the nitrate content of Unterwarnow water and River Warnow discharge. High land-runoff is linked to high nitrate concentrations. Nitrate and orthosilicate concentrations decrease as the fresh water mixes increasingly with Baltic Sea water. In contrast, ammonium and orthophosphate increase along the longitudinal section of the estuary. Concentrations of these two nutrients are highest near the sewage outfall. Levels of all dissolved inorganic nutrients show considerable seasonal variation depending on the development of phytoplankton. They are lowest at the time of maximum biomass in spring. In further course of the orthophosphate, orthosilicate and ammonium concentrations are governed by mineralization and nutrient recycling from the sediment.

The estuary Unterwarnow makes a most important contribution to the degradation of organic matter from the town Rostock and the rest of its catchment area. Beyond that, it acts as a "biological filter" in which the most of the nutrients input is bound and deposited in the sediment. This very important barrier function of these waters for the Baltic should definitely be maintained.

References

Autorenkollektiv (1986): Ausgewählte Methoden der Wasseruntersuchung, Bd. 1: Chemische, physikalisch-chemische und physikalische Methoden. Jena.

BACHOR, A. (1989): Untersuchungen zum Salz- und Nährstoffhaushalt der Unterwarnow — einem Ästuar der westlichen Ostsee. Beitr. Meereskd. **59**, 3—18.

BALZER, W. (1978): Untersuchung über Abbau organischer Materie und Nährstoff-Freisetzung am Boden der Kieler Bucht beim Übergang vom oxischen zum anoxischen Milieu. Rep. SFB 95 Wechselwirkung Meer — Meeresboden Nr. 36. 1—126 (nach NEHRING 1981).

CORRENS, M. (1976): Beitrag zur Hydrographie der Unterwarnow. Acta Hydrophys. **21** (3/4), 183—226.

HÜBENER, T., GASE, G., GENCIOVA, A., KRÄMER, A., MEHLKOPF, T., PAULICKOVA, Z., KELL, V., & SCHIEWER, U. (1984): Biomasse, Primärproduktion und Sukzessionsfolge in der Unterwarnow. — Wiss. Z. W.-Pieck-Univ. Rostock, Math.-nat. R. **33** (6), 43—52.

MATTHÄUS, W. (1984): Zur mittleren jahreszeitlichen Veränderlichkeit von Temperatur und Salzgehalt in der Mecklenburger Bucht. Beitr. Meereskd. **50**, 9—23.

NEHRING, D. (1982): Langzeittrends des Phosphat- und Nitratgehalts in der Ostsee. Beitr. Meereskd. **47**, 61—86.

— & FRANCKE, E. (1981): Hydrographisch-chemische Untersuchungen in der Ostsee von 1969—1978. Geod. Geoph. Veröff. R. IV, **35**, 39—220.

RHODE, K.-H., & NEHRING, D. (1979): Ausgewählte Methoden zur Bestimmung von Inhaltsstoffen im Meer- und Brackwasser. Geod. Geoph. Veröff. R. IV, **27**, 1—68.

Address of the author:

Dipl.-Chem. ALEXANDER BACHOR
Wasserwirtschaftsdirektion Küste
Badenstr. 18
Stralsund
DDR-2300

Limnologica (Berlin) **20** (1), 33—36, April 1990

Akademie der Wissenschaften der DDR, Institut für Meereskunde, Rostock-Warnemünde, DDR

Ecological Investigations in the Onshore Pelagic Zone Near Warnemünde from March 1985 to March 1986

By W. Kaiser, A. Irmisch, D. Nehring, F. Georgi & G. Breuel

With 3 Figures

Abstract

From March 1985 to March 1986 ecological investigations were carried out in the pelagic zone near Warnemünde with the aim of analysing the annual variations in some oceanological parameters and ascertaining the productivity and eutrophication of this region. Main results are as follows:

— All parameters studied show annual variations, which are greatest for temperature, nutrients and biological parameters.
— Short term variations (few days) were strong, except in the case of temperature.
— Close connections exist between the biological and chemical parameters.
— The influence of the discharge from the River Warnow on the ecological conditions could not be clarified.
— As comparisons of chlorophyll and seston concentrations with those of the Arkona Basin and the Lübeck Bay show, the area studied can be characterized as a slightly eutrophic region of the Mecklenburg Bay.

Introduction

Ecological investigations were carried out in the onshore region off Warnemünde (~3—4 sea miles from the coast; water depth: 13 to 15 m) from 11th March 1985 to 3rd March 1986. Samples (about 60 l each) of surface water were taken and the surface temperature was measured on a total of 69 dates. The samples were analyzed for the following parameters 3—4 h after collection (maximum delay: 12 h):

— salinity,
— inorganic nutrient concentrations (PO_4^{3-}, NO_3^-, NO_2^-, NH_4^+),
— dissolved organic compounds (urea, carbohydrates),
— chlorophyll and phaeopigment concentrations,
— potential primary production (P_{pot}) and
— seston dry weight (seston-dw).

Standard methods were used for all analyses (Rohde & Nehring 1979; Jeffrey & Humphrey 1975; Edler 1979; Georgi 1985). Meteorological data such as wind force and direction as well as cloudiness were taken from the daily weather report issued by the Meteorological Service of the GDR.

Results and Discussion

The area investigated was characterized by large salinity fluctuations, but these were not related in any way to local wind conditions. According to Lange (1975) wind conditions over the Kattegat, Kiel Bay and Mecklenburg Bay do not affect the current through Fehmarn Belt, this being influenced mainly by differences between the water levels of the North Sea and the Baltic which in turn are caused by the general weather situation over the North East Atlantic (Lass, pers. comm.). The eastward or westward currents in the Western Baltic can extend upwards to the surface and thus, together with the discharge from the River Warnow, lead to these significant salinity variations.

Minor short term temperature variations were observed, these slight changes being superimposed on the pronounced seasonal variation with the temperature maximum at about the end of July (Fig. 1).

Distinct seasonal variations were found in the biological parameters, inorganic nutrients and dissolved organic compounds. The spring phytoplankton bloom started in the second half of March, reaching its peak on 24th March 1985. The onset of mass algal development in this region therefore coincides with that shown by seasonal variations found throughout the Western Baltic (Von Bodungen 1975; Smetaček 1975, 1977; Steemann-Nielsen 1964; Gargas et al. 1978; Edler 1977; Kaiser & Schulz 1976; Kaiser et al. 1981). The chlorophyll concentration then decreased, reaching very low values during the last 10 days of April. It can therefore be assumed that the spring phytoplankton population is sedimented almost completely (Smetaček 1977; Von Bröckel 1975). Brief, but considerable, increases in potential primary production and slight increases in biomass were observed three times up to the end of July. The low nutrient concentrations measured at the same times did not permit increases of such a magnitude, and it must therefore be assumed that urea was used as a nitrogen source in addition to nitrate and ammonium. In addition, the collapse of the spring diatom bloom was followed by the rapid development of peridineans with a heterotrophic capability and greater use of organic compounds could therefore be made quite generally. Since this group of algae has a fairly low mean chlorophyll content (Hagmeier 1961) an increase in biomass (plankton volume or numbers of individuals) need not be accompanied by a corresponding rise in chlorophyll concentration. Advection may also have contributed to nutrient renewal because the periods of maximum production in late March/early April and June coincided with periods of strong winds. Moreover, the peridineans are motile and can therefore move actively to deeper layers to obtain nutrients.

Contrary to the observations reported by Kell et al. (1982) the summer production maximum was not reached until the second half of September. Although nitrate and phosphate concentrations were somewhat higher after July than from May to late July, the nutrients needed for this summer production maximum must have resulted from the more rapid remineral-

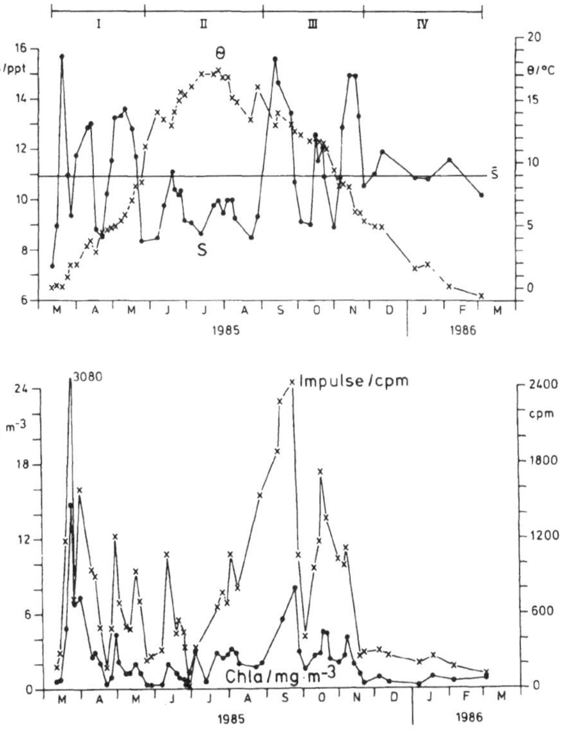

Fig. 1. Salinity, temperature, chlorophyll a and potential primary production (Impulse/cpm)

zation of organic matter. The NH_4 curve in Fig. 2 shows that ammonium might have been used more intensively as a nitrogen source because it remains almost at a low level until well into September although it should actually rise owing to increasing excretion by zooplankton. Considerable amounts of nitrogen can also enter the water in summer owing to the exudation of amino acids by nitrogen fixing cyanophyceans.

The summer phytoplankton bloom collapsed abruptly within a few days and was followed only a little later by the development of the autumn bloom in early to mid-October. The greater availability of inorganic nutrients owing to more intensive mixing and the decline of the zooplankton population (reduced grazing) were responsible for these maxima. Such autumn blooms are observed regularly in the Western Baltic (KELL et al. 1982; SMETAČEK 1975, 1977; VON BODUNGEN 1975), but in the Baltic proper they may fail to appear if meteorological conditions are unfavourable (SCHULZ & KAISER 1976).

There is little resemblance between the curves for phytoplankton biomass equivalent (chlorophyll a) and seston dry weight. It can be seen from Fig. 2 that these curves are most similar when the phytoplankton standing crop is high and therefore the portion of phytoplankton in the seston is large. Assuming the chlorophyll concentration to be nearly 2.1% of the phytoplankton dry weight (HAGMEIER 1961), the annual mean of 2.5 mg Chl a corresponds to a phytoplankton dry weight of 0.11 g and is therefore one order of magnitude lower than the abioseston dry weight. Although a more favourable ratio of 0.70 g plankton biomass to 3 g seston dry weight was found on 24th March 1985 when the plankton standing crop was highest, the amount of abioseston was still very much higher than that of phytoplankton. In other words although seasonal variations in phytoplankton production can contribute to the variations in seston dry weight, other factors have a considerably greater influence. These doubtless include resuspension following periods of strong wind (up to early July) and the influx of suspended matter from the River Warnow. The lower values recorded from June to the end of August show that some of the seston obviously sedimented to the bottom when there was little wind. The seston and chlorophyll values both increased from the end of August to the end of September, then the chlorophyll values dropped again abruptly. Slower sedimentation owing to more intensive water motion may have been the reason why there was no corresponding decrease in seston content. The mean seston dry weight of $1.72\ g \cdot m^{-3}$ calculated for the whole duration of the investigation confirms that only moderate values $< 3\ g \cdot m^{-3}$ occur in the coastal waters of the GDR (GEORGI 1985).

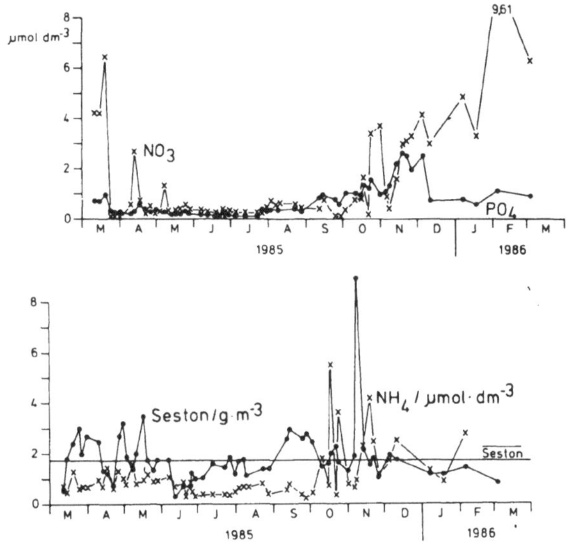

Fig. 2. Concentrations of nitrate, orthophosphate, ammonia and seston dry weight

The seasonal variations in inorganic nutrient concentrations in the area studied are similar to those recorded for other parts of the Western Baltic (see for example STIENEN 1986; VON BODUNGEN 1975; NEHRING 1981), and this is also reflected in the ratios between nutrients ($NO_3^- + NO_2^- + NH_4^+/PO_4^{3-}$; NO_3^-/PO_4^{3-}). Although the Redfield ratio for the Baltic differs considerably from that for the oceans (NEHRING 1981), in the productive season it is considerably lower even than in the well mixed layer in winter owing to the more rapid consumption of inorganic nitrogenous compounds and recirculation of phosphorus.

Urea enters sea water as an end product of protein metabolism in animals, particularly zooplankton (CONOVER 1978; WHITLEDGE 1981), and through bacterial decomposition of purine and pyrimidine (MAITA et al. 1973, cited in MC CARTHY 1980; SATOH et al. 1980). In coastal regions the possibility of import with surface run-off must also be considered.

Urea concentrations in the investigation area (Fig. 3) were substantially higher than those recorded for the Baltic proper. The five-year mean for the Arkona Sea in the period 1980 to 1984, for instance, was $0.46\ \mu mol \cdot dm^{-3}$ with values varying between 0.08 and $1.13\ \mu mol \cdot dm^{-3}$ (IRMISCH 1986).

STEINMANN (1976) reported similar concentrations in Kiel Bay to those recorded off Warnemünde, attributing them to anthropogenic influences which may also be responsible for the high values found in our investi-

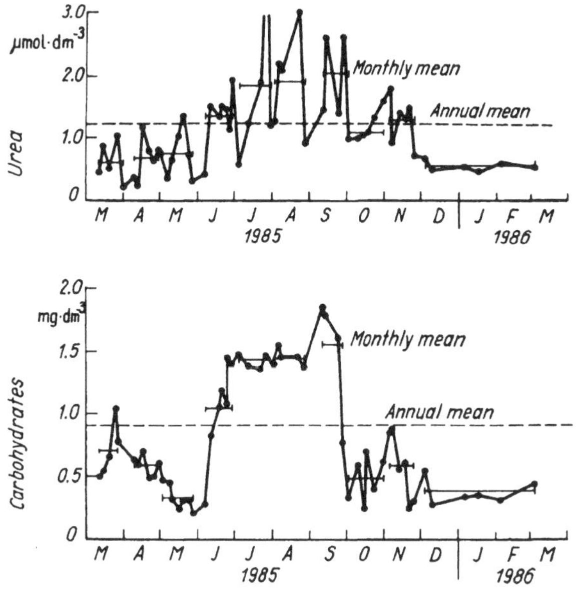

Fig. 3. Concentrations of urea and carbohydrates

gation area. BACKHAUS (1983) noted that the River Warnow carries a considerable urea load, and the fact that the period of high urea concentrations coincided with that of low salinities from May to August suggests that urea levels in the investigation area may be influenced by water discharged from the river. This explanation, however, is not fully consistent with the different short-term variations in both parameters (see Figs. 1 and 3).

According to KAISER & SCHULZ (1986) zooplankton biomass development in Mecklenburg Bay in 1985 followed the pattern given in Table 1.

Table 1. Zooplankton biomass development in Mecklenburg Bay in 1985

	14 March	21 March	25–27 April	4–5 June	1–2 August	13–14 November
Zooplankton biomass (ml · m^{-3}, 200 µ net, 0–25 m)	0.17	0.22	0.28	0.43	0.45	0.22

If it is assumed that zooplankton development in our investigation area off Warnemünde does not differ appreciably from that in the other parts of the Southern Baltic, it can be concluded that the high urea concentrations observed there in summer could also be due to excretion by zooplankton. The summer urea maximum in the Arkona Sea shows that this is quite possible (IRMISCH 1986).

It is impossible to say on the basis of our data to what degree the autochthonous and allochthonous fractions respectively are responsible for the much higher urea concentrations observed off Warnemünde than in the Baltic proper, but they are definitely attributable directly or indirectly to run-off.

Phytoplankton is able to use urea as a nitrogen source particularly if concentrations of inorganic nitrogen compounds are low (CARPENTER et al. 1972; HARVEY & CAPERON 1976; BUTLER et al. 1979). Moreover, urea is decomposed relatively quickly by heterotrophic microorganisms, STEINMANN (1976) reporting turnover times of 3.4 to 60 h in Kiel Bay, and the ammonia liberated in the process can also replace nutrients consumed by algae. In other words owing to these two pathways through which it can serve as a nutrient source for phytoplankton in the Baltic, urea perhaps represents a nitrogen potential that contributes particularly in summer to sustaining high rates of primary production when inorganic nutrient levels are low. This is supported by enclosure experiments with plankton communities under spring conditions (SCHULZ et al. 1985). The urea and algal nutrient concentrations in the enclosures decreased at almost equal rates while the primary production and the phytoplankton biomass increased. Our data do not reflect this phenomenon, and it therefore seems possible that in our investigation area urea consumption was balanced by zooplankton excretion and the urea influx from the River Warnow.

Carbohydrate concentrations in the sea area off Warnemünde are only slightly higher than in the Baltic proper. Dissolved carbohydrate concentrations measured in the Arkona Sea from 1980 to 1984, for instance, varied between 0.40 and 1.20 mg · dm^{-3}, the five-year mean being 0.75 mg · dm^{-3} (IRMISCH 1987).

Autochthonously formed dissolved carbohydrates consist of phytoplankton exudations and decomposition products (HELLEBUST 1974; BOLZE & SOEDER 1978) and can also enter the water when feeding zooplankton injure phytoplankton cells (LAMPERT 1978). Excretion by zooplankton represents another source (BURNEY et al. 1979). It is therefore not surprising that seasonal variations in carbohydrate concentration resemble the seasonal variations in primary production and chlorophyll concentration, although short term variations may differ. The relationship between these parameters during our investigations are also revealed statistically by the correlation coefficients:

	Potential primary production	Chlorophyll concentration
Dissolved carbohydrate concentration	$r = 0.43$ $p = 0.01$ $n = 52$	$r = 0.33$ $p = 0.05$ $n = 57$

A dependence between carbohydrate concentration and phytoplankton development has also been found in the Baltic proper (IRMISCH 1987), and LIEBEZEIT (1980), studying the free dissolved glucose and fructose concentrations in Kiel Bay, noted that their seasonal variations resembled those of primary production.

The carbohydrate concentrations found in the summer in our investigation area off Warnemünde are higher relative to the spring values than would be expected from comparison of the corresponding values for primary production or chlorophyll concentration. This may be due to the fact that in spring a large proportion of the phytoplankton biomass can settle to the bottom before stratification has taken place (SMETAČEK 1975) or grazing by zooplankton has started. The situation is different in summer: the warm surface layer is separated from the deeper cold water by a thermocline which interrupts the flux of matter between the two water bodies, and most of the phytoplankton, like the dissolved carbohydrates contained in their exudate and decomposition products, remains in the surface layer. In addition the amounts of carbohydrate entering the water from phytoplankton cells injured by zooplankton while grazing and excretion from zooplankton is very probably higher in summer than in spring owing to the greater biomasses involved.

The results of our investigations can be summarized briefly as follows:

a) All parameters show seasonal variations which, as expected, were most pronounced in the case of temperature, the biological parameters and parameters affected directly by these two. Seasonal variations in seston content are distinct, but less pronounced than those in other parameters owing the large proportion of abioseston.

b) Short term variations with a duration of only a few days were slight in the case of temperature but considerable for all other parameters.

c) The relationships between the biological and chemical parameters are closer than those between the meteorological/hydrological and chemical/biological parameters. The degree to which ecological conditions are influenced by the River Warnow could not be ascertained.

d) Comparison of our chlorophyll and seston data with corresponding data from the Arkona Sea and Lübeck Bay shows that the sea area off Warnemünde can be regarded as a part of Mecklenburg Bay that is only slightly eutrophic, if at all, owing to the unrestricted exchange of water in this open sea area.

References

BACKHAUS, B. (1983): Untersuchungen zum Vorkommen von Harnstoff in Oberflächengewässern des Nordens der DDR. Dipl.-Arb., Wilhelm-Pieck-Universität Rostock.

BODUNGEN, B. VON (1975): Der Jahresgang der Nährsalze und der Primärproduktion des Planktons in der Kieler Bucht unter Berücksichtigung der Hydrographie. Diss., Christian-Albrechts-Universität Kiel.

BOLZE, A., & SOEDER, J. (1978): Konzentrationsänderungen der extrazellulären organischen Substanz bei Synchronkulturen von *Scenedesmus acutus* var. *alternans* HORTOB. Arch. Hydrobiol. **82**, 142–154.

BRÖCKEL, K. VON (1975): Der Energiefluß im pelagischen Ökosystem vor Bocknis Eck (westliche Ostsee). Diss., Christian-Albrechts-Universität Kiel.

BURNEY, C. M., KENNETH, M. J., LAVOIE, D. M., & SIEBURTH, J. McN. (1979): Dissolved organic carbohydrates and microbial ATP in the north Atlantic: concentrations and interactions. Deep-Sea Res. **26**, 1267–1290.

BUTLER, E. I., KNOX, S., & LIDDICOAT, M. I. (1979): The relationship between inorganic and organic nutrients in sea water. J. mar. Biol. Ass. U.K. **59**, 239–250.

CARPENTER, E. J., REMSEN, C. C., & WATSON, S. W. (1972): Utilization of urea by some marine phytoplankters. Limnol. & Oceanogr. **17**, 265–269.

CONOVER, R. J. (1978): Transformation of organic matter. In: KINNE, O. (ed.), Marine Ecology. Vol. IV, p. 221–449. Chichester.

EDLER, L. (1977): Phytoplankton and primary production in the Sound. PhD Thesis, Univ. Göteborg, 82 pp.

— (1979): Recommendations on methods for marine biological studies in the Baltic sea. Phytoplankton and chlorophyll. The Baltic Marine Biologists (BMB), Publ. No. **5**, 25–38.

GARGAS, E., AERTEBERG NIELSEN, G., & MORTENSEN, S. (1978): Phytoplankton production, chlorophyll a and nutrients in the open Danish waters 1975–1977. The Belt Project, 63 pp.

GEORGI, F. (1985): Verteilung und Beschaffenheit des Sestons in inneren Küstengewässern der DDR (Darß-Zingster-Boddengewässer) sowie der westlichen und mittleren Ostsee unter Berücksichtigung der Wasseraustauschprozesse zwischen beiden Systemen. Beitr. Meereskd. **52**, 35–48.

HAGMEIER, E. (1961): Plankton-Äquivalente. Kieler Meeresforsch. **17**, 32–47.

HARVEY, W. A., & CAPERON, J. (1976): The rate of utilization of urea, ammonium and nitrate by natural populations of marine phytoplankton in an eutrophic environment. Pacific Sci. **30**, 329–340.

HELLEBUST, J. A. (1974): Extracellular products. In: STEWART, W. D. P. (ed.), Algal physiology and biochemistry, Vol. 10, pp. 839–863. Oxford.

IRMISCH, A. (1986): Untersuchungen über den gelösten Harnstoff in der Ostsee. Beitr. Meereskd. **55**, 29–37.

— (1987): Untersuchungen über die gelösten Kohlenhydrate in der Ostsee. Beitr. Meereskd. **56**, 19–26.

JEFFREY, S. W., & HUMPHREY, G. F. (1975): New spectrophotometric equations for determining chlorophylls a, b_1, c_1 and c_2 in higher plants, algae and natural phytoplankton. Biochem. Physiol. Pflanzen (BPP) **167**, 191–194.

KAISER, W., & SCHULZ, S. (1976): Zur Ursache der zeitlichen und räumlichen Differenzen des Beginns der Phytoplanktonblüte in der Ostsee. Fischerei-Forsch. **14**, 77–81.

— RENK, H. & SCHULZ, S. (1981): Die Primärproduktion der Ostsee. Geod. Geoph. Veröff. R. IV, **33**, 27–52.

— & SCHULZ, S. (1985): Die produktionsbiologischen Bedingungen und ihre Veränderungen in der Ostsee im Jahre 1985. Unveröff. Ber. Inst. Meereskd., Warnemünde.

KEIL, V., BLOSS, W., BÖRNER, R., HAGEMANN, M., & VÖKLER, T. (1982): Die Sukzession des Phytoplanktons in der Mecklenburger Bucht (südliche Ostsee) in den Jahren von 1977–1980. Wiss. Z. W.-Pieck-Univ. Rostock, Naturwiss. R. **31**, 27–30.

LAMPERT, W. (1978): Release of dissolved organic carbon by grazing zooplankton. Limnol. & Oceanogr. **23**, 831–834.

LANGE, W. (1975): Zu den Ursachen langperiodischer Strömungsänderungen im Fehmarnbelt. Kieler Meeresforsch. **31**, 65–81.

LIEBEZEIT, G. (1980): Aminosäuren und Zucker im marinen Milieu — neuere analytische Methoden und ihre Anwendung. Report SFB Nr. 53, Kiel.

Mc CARTHY, J. J. (1980): Nitrogen. In: MORRIS, I. (ed.), The physiological ecology of phytoplankton, pp. 191–234. Oxford.

NEHRING, D. (1981): Hydrographisch-chemische Untersuchungen in der Ostsee von 1969–1978. II. Die chemischen Bedingungen und ihre Veränderungen unter besonderer Berücksichtigung des Nährstoffregimes. Geod. Geoph. Veröff. R. IV, **35**, 39–219.

ROHDE, K.-H., & NEHRING, D. (1979): Ausgewählte Methoden zur Bestimmung von Inhaltsstoffen im Meer- und Brackwasser. Geod. Geoph. Veröff. R. IV, 27, 1–68.

SATOH, Y., OKINO, T., & AOYAMA, K. (1980): Correlation between urea and other chemical and biological parameters in waters of Lake Suwa, Japan. Int. Revue ges. Hydrobiol. **65**, 445–454.

SCHULZ, S., & KAISER, W. (1976): Produktionsbiologische Untersuchungen in der Ostsee 1975 und einige spezielle Ergebnisse aus dem Jahre 1974. Fischerei-Forsch. **14**, 53–63.

— BREUEL, G., IRMISCH, A., JOST, G. & SIEGEL, H. (1985): Ergebnisse ökologischer Untersuchungen an eingeschlossenen Planktongemeinschaften der Arkonasee im Frühjahr 1981. Geod. Geoph. Veröff. R. IV, **41**, 1–66.

SMETAČEK, V. (1975): Die Sukzession des Phytoplanktons in der westlichen Kieler Bucht. Diss., Christian-Albrechts-Universität, Kiel.

— (1977): Der Jahreszyklus des Phytoplanktons: Ursache und Bedeutung für das marine Ökosystem. Verh. Ges. Ökologie, Kiel.

STEEMANN-NIELSEN, E. (1964): Investigations of the rate of primary production at two Danish light ships in the transition area between the North Sea and the Baltic. Medd. Danmarks Fisk. Havundersögelser **4**, 31–70.

STEINMANN, J. (1976): Untersuchungen über den bakteriellen Abbau von Harnstoff und Harnsäure in der westlichen Ostsee. Bot. Mar. **19**, 47–58.

STIENEN, C. (1986): Phytoplanktonentwicklung in Abhängigkeit von der Nährsalzkonzentration. Ber. Inst. Meereskd. Kiel, Nr. 152.

WHITLEDGE, T. E. (1981): Nitrogen recycling and biological populations in upwelling ecosystems. In: RICHARDS, F. A. (ed.), Coastal upwelling, American Geophysical Union, 257–273. Washington.

Address of the authors:

Dr. WOLFGANG KAISER
Dr. ANDREAS IRMISCH
Prof. Dr. sc. DIETWART NEHRING
Dr. FRANK GEORGI
Dipl.-Biol. GÜNTHER BREUEL
Akademie der Wissenschaften der DDR
Institute für Meereskunde
Seestr. 15
Rostock-Warnemünde
DDR-2530

Limnologica (Berlin) **20** (1), 37, April 1990

Ernst-Moritz-Arndt-Universität Greifswald, Sektion Biologie, Biologische Station Hiddensee, Kloster/Hiddensee, DDR

Cyanobacterial Blooms and Nitrogen Fixation in Coastal Waters of the Arkona Sea, 1974—1986

By HELMUT HÜBEL & MARIANNE HÜBEL

Key words: Cyanobacteria; Water bloom; Nitrogen fixation; N-input; Baltic

Abstract

Using the acetylene reduction technique nitrogen fixation rates are determined at the shore region of the western Arkona Sea during blue-green algal vegetation period in 1974—1986. Measurements pointed out at short intervals are demonstrating the dynamics in space and time. At the shore region the annual N-input varies from 219—1395 mg $N \cdot m^{-2}$. Resulting from 13 years an average value of 680 mg $N \cdot m^{-2} \cdot a^{-1}$ is given. At the off-shore region corresponding dates are 243—1536 mg $N \cdot m^{-2} \cdot a^{-1}$ and 787 mg $N\ m^{-2} \cdot a^{-1}$. The amounts of N-input and the water temperature are positive correlated. The annual N-input of 680 mg $N \cdot m^{-2}$ at the shore region gives an increase of 40.5% of the inorganic N-content of the surface water and respectively that of 787 mg $N \cdot m^{-2} \cdot a^{-1}$ at the off-shore region an increase of 46.9%.

Doz. Dr. sc. HELMUT HÜBEL
Biologische Station Hiddensee
Kloster/Hiddensee
DDR-2346

Limnologica (Berlin) **20** (1), 39—45, April 1990

Institute of Oceanography, Gdańsk University, Gdynia, Poland

Important Ecological Features of the Polish Coastal Zone of the Baltic Sea

By MARCIN PLINSKI

With 7 Figures

Introduction

The Polish zone of the coastal Baltic waters bordered to the West, by the Pomeranian Bay, with the GDR, and to the East by the Gulf of Gdańsk, with the USSR, has a great economical importance and it has been the subject of numerous investigations for a long time. The coastal waters reaching the 20 m isobath were accepted as the coastal zone. It covers therefore the Szczecin Lagoon, the Pomeranian (Pomorze) Bay, the strip of the coastal waters of the open coast from several to a dozen nautical miles wide, and in the Gulf of Gdańsk the narrow coastal zone including Puck Bay. In the Gulf of Gdańsk the direct influence of the land factors considerably projects beyond the isobath of 20 m, so the deeper regions of the Gulf will be also considered here. All over this coastal zone it is susceptible to the inflow of various types of pollution from the land. The result is the systematic deterioration of the state of cleanliness of the sea water. The first signals of growing pollution of Polish coastal waters were noted in the '50s, when increased concentrations of nitrogen and phosphorus compounds were observed; this followed by the intensive production of organic matter enhancing the eutrophication of the water.

As from the mid 1970s, a new phenomenon was noted — particularly in the region of the Gulf of Gdańsk — that of high bacterial contamination, which frequently prevented enjoyment of the use of the beaches and sea bathing. In recent years, in particular, the catastrophic sanitary contamination and run-off of toxic substances (heavy metals, chlorinated hydrocarbons, oil-originated hydrocarbons, etc.) to the sea are accompanied by disease in fish, which periodically takes on an epidemic character. This, in consequence, creates a new ecological situation, which is highly specific for the Gulf of Gdańsk.

Environmental characteristic

As regards the environmental factors, the most comprehensive data concern temperature and salinity. Basing on data covering many years the highest mean values of water temperature are typical for the lagoons, average — for the gulfs, the lowest for the coastal zone of the open sea. The lowest temperature occurs in the lagoons generally in January, but in the open part of the coast mainly in February, especially in the eastern part. The highest water temperature is reached earlier in the lagoons (in July), than in the gulfs and along the

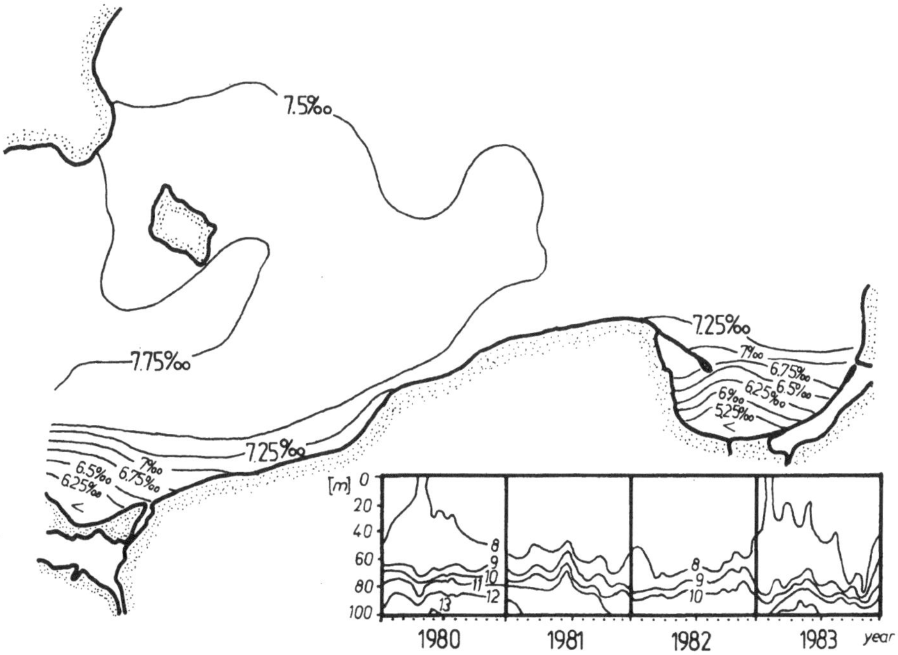

Fig. 1. Mean annual (1950—1970) salinity in the surface layer and in cross-section in the Gdańsk Deep (after CYBERSKA, in print)

open coast (in August). The thermocline develops only in the Gulf of Gdańsk, because of the depth (more than 40 m). The temperature in the lower layers oscillates around the values below 4 °C, round the year. These layers consist of a water mass with a higher salinity than that of the surface almost always being above $10^0/_{00}$. The surface water salinity is around $6-7^0/_{00}$ with a clearly defined increase away from the Vistula estuary (Fig. 1). At the central coast the extent of the rivers does not exceed several nautical miles so it is rather of local importance. The riverain outflow in the Pomeranian Bay occurs clearly in the southern part (TRZOSIŃSKA et al. 1989).

Shallow waters of the Bay of Pomerania and of the coastal zone are good oxygenated, but there were noticed some transitory and local shifting of the oxygen equilibrium occurring mainly in the regions of rivers mouth. Only in the Gdańsk Deep, caused by the stratification, there was formed the anaerobic region containing hydrogen sulphide in some years (Fig. 2).

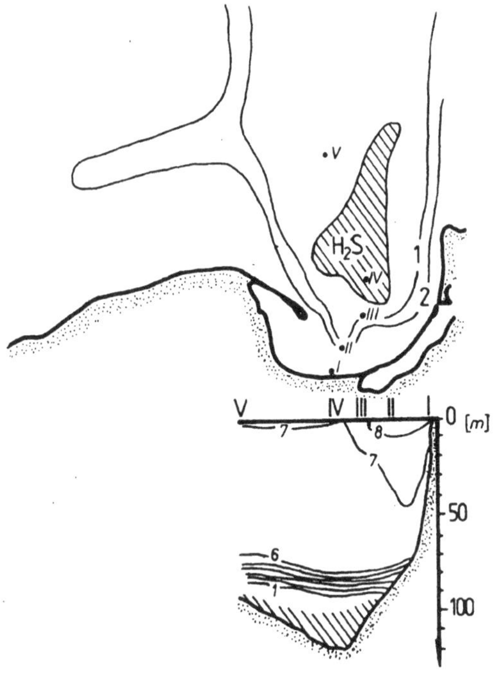

Fig. 2. Concentration of oxygen (ml/l) and occurrence of hydrogen sulphide in the bottom water of the Gulf of Gdańsk region (after MAJEWSKI, in print) and in cross-section in summer 1975 and 1981 (after MILEWSKA & ANDRULEWICZ 1982)

In the shallow zone of the central coast the mean concentration of the phosphorus compounds remains at the level similar to that in the open sea, while the concentration of all the nitrogen compounds is approximately 50—100 per cent higher than those in the open sea. The discharge of nutrients from the land-based sources is marked here periodically by the local increase of the maximal concentrations up to about 3.0 $\mu mol/dm^3$ of phosphates, 20.0 $\mu mol/dm^3$ of nitrates, 1.1 $\mu mol/dm^3$ of nitrites, 20 $\mu mol/dm^3$ of ammonia, 55 $\mu mol/dm^3$ of organic nitrogen compounds and 2 $\mu mol/dm^3$ of organic phosphorus compounds. According to the data from period 1979—1983 the Pomeranian Bay is the most eutrophicated region of the Polish coast. It is expressed by the very high concentration of nutrients. Nitrates constitute there the main fraction (69—77 per cent) of the inorganic nitrogen compounds in almost all over the year without summer, when the strong predominance of the ammonia (79 per cent) takes place. The average concentration from 1979—83 of nitrogen forms (NO_3 + NO_2 + NH_4) was 10 $\mu mol/dm^3$ (6.0 $\mu mol/dm^3$ in summer; 20.7 $\mu mol/dm^3$ in winter). The highest concentration of nutrients has been noticed around the mouths of Swina and Dziwna. The main amount of nutrients in Pomerania Bay comes from the Szczecin Lagoon, where their concentration is also very high because of the Odra inflow (TRZOSIŃSKA et al., in print).

In the distribution of nutrients in the Gulf of Gdańsk, there is a marked influence of point sources, such as: the Vistula and outflow from sewage-treatment plants. In the period 1979—1983 the average value of phosphates in the surface waters was 0.46 $\mu mol/dm^3$ and of nitrates 4.54 $\mu mol/dm^3$, while at the same time for the central coast the content of phosphates was 0.38 $\mu mol/dm^3$ and nitrates 2.64 $\mu mol/dm^3$. The concentration of all nutrient salts in the euphotic layer is much higher in the coastal zone than in the remaining parts of the bay (Fig. 3). In view of the very large quantities of phosphorus and

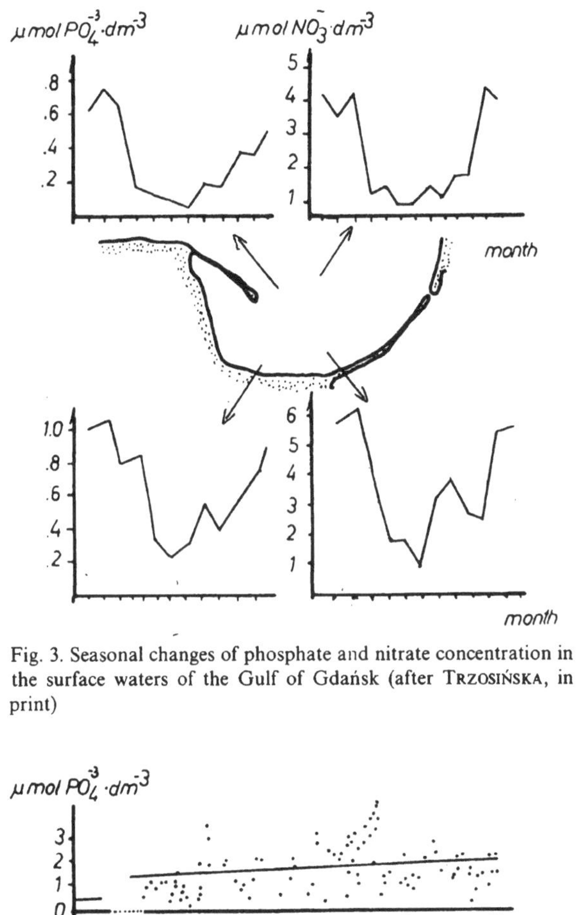

Fig. 3. Seasonal changes of phosphate and nitrate concentration in the surface waters of the Gulf of Gdańsk (after TRZOSIŃSKA, in print)

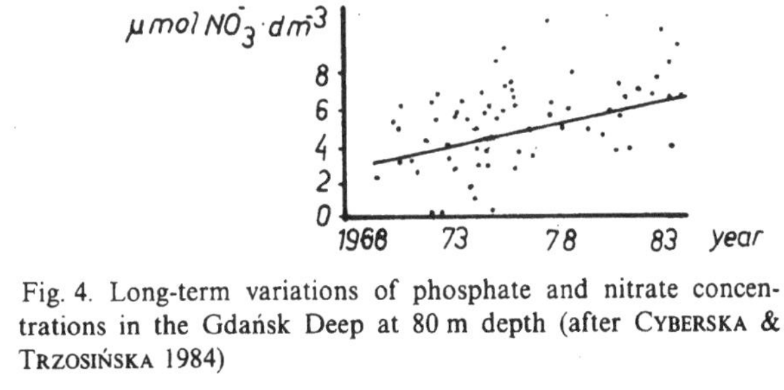

Fig. 4. Long-term variations of phosphate and nitrate concentrations in the Gdańsk Deep at 80 m depth (after CYBERSKA & TRZOSIŃSKA 1984)

nitrogen transported by the Vistula in the warm seasons (run-off of artifical fertilizers), these differences increase during the vegetative period. Despite of the high primary production, phosphates and nitrates are not usually fully exhausted in the inshore zone. Below the halocline the distribution of nutrients is almost even in the whole of the Gdańsk Basin and their occurrence is ruled by the same laws as in the central Baltic. Based on investigations over a period of many years, the essential statistical accumulation was found, which at a depth of 80 m is expressed by the coefficients: 0.03 μmol PO_4/dm^3/year and 0.24 μmol NO_3/dm^3/year (Fig. 4) (TRZOSIŃSKA, in print).

Biological features

Plankton

Plankton of the coastal zone is most of all stressed by the fresh water flowing from the land, that results in the formation of the specific brackish water community, which is additionally modified due to the increased eutrophication. This modification is strongly marked in the composition of the groups as well as in the number of the species. This statement considers both phyto- and zooplankton.

The species from the following divisions: Cyanophyta, Euglenophyta, Pyrrhophyta, Chrysophyta with two classes Chrysophyceae and Bacillariophyceae as well as Chlorophyta are found in phytoplankton. Diatoms are the most numerous group; independently of the region they appear in the open waters and in the coastal zone. The increased participation of blue-green algae is strongly marked in the coastal zone while Pyrrhophyta appear here in smaller amounts than in the open waters.

The seasonal appearence of particular groups of algae in phytoplankton is characterized by a clear pattern. In early spring the bloom of diatoms takes place, while the centric species of genus *Chaetoceros*, *Melosira* and *Thalassiosira* as well as *Achnanthes taeniata* outnumber the others. The number of these taxons decreases in May while in this period *Sceletonema costatum* appears in greater quantities. The bloom of *Heterocapsa triquetra* has been found for several years to take place in June. Summer is characterized by the predominance of blue-green algae with dominating *Aphanizomenon flos-aquae*. In autumn diatoms, especially *Coscinodiscus granii*, dominate the phytoplankton. This outline has been noted for many years (PLIŃSKI et al. 1985). The maximum of the quantity of the given species can be a subject of some temporal shift during the annual cycle, what is caused by the differences in the climate conditions of the particular years, however the above mentioned sequence of species is always kept (Fig. 5).

In the last period, the expansion of the species of the genus *Microcystis*, mainly *Microcystis aeruginosa* has been noted in many areas, especially in the Szczecin Lagoon. The qualitative and semi-quantitative investigations carried out in the last ten years period confirmed the tendency for the increasing intensity of the bloom of Cyanophyta. The appearance of the large amounts of Cyanophyta mainly *Microcystis aeruginosa* has been observed to start earlier each year, even in June, and the culmination of their bloom occurs in August—September, when the amount of Cyanophyta in the Szczecin Lagoon exceeds sometimes 98 per cent of phytoplankton

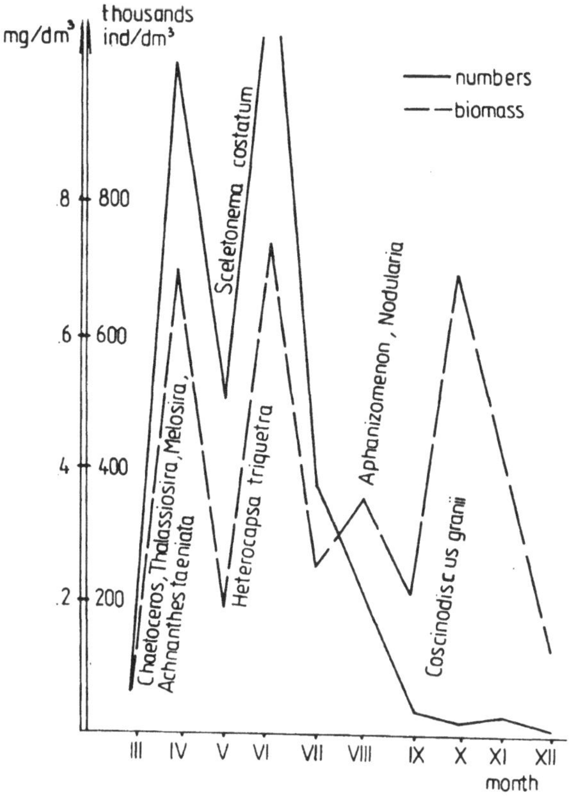

Fig. 5. Seasonal occurrence and biomass of main taxa of phytoplankton in the years 1977 and 1978 in the Gulf of Gdańsk

organisms. However after the period of maximal development there is no typical, sudden breakdown of the bloom connected with the depletion of the food reserves. The decrease of the water temperature seems to be the only factor with a restaining effect on the bloom (TRZOSIŃSKA et al. 1989).

The spatial differences as well as the seasonal fluctuations in the concentration of chlorophyll, an important indicator of the plant biomass, reflect the relations typical for phytoplankton. The highest chlorophyll concentrations are found in the zone of the big rivers' estuaries, while the farther it is from the coast the clearer is the downward trend. In the Polish area of the Baltic Sea the highest values have been detected in the Szczecin Lagoon. For example, the concentration of chlorophyll "a" amounted there to 184 mg/m^3 in April 1981 while in August 1983 to 314 mg/m^3. The annual mean values calculated for the period 1975—1985 kept on the quite equal level of 60.5 mg/m^3 with the standard deviation ± 17.5 mg/m^3. The trend of increase has not been discerned, only some variations in particular years, however the content of chlorophyll in the Szczecin Lagoon is elevated, indicating the high degree of eutrophication of this aquatorium (TRZOSIŃSKA et al. 1989). The content of chlorophyll is lower in the Gulf of Gdańsk; for chlorophyll "a" it ranges from 1 to 90 mg/m^3 and for chlorophyll "c" from 1 to not much more over 40 mg/m^3. The concentrations of carotenoids fluctuate between 1 and 35 mg/m^3, while the concentrations of chlorophyll "b" are considerably lower and don't exceed 2 mg/m^3 (LATAŁA 1982). These data consider the coastal waters; much lower values have been found in the area of the Gdańsk Deep, for example the maximum concentration of chlorophyll "a" was equal to 15 mg/m^3 (RENK 1973), and in the open waters of the southern Baltic these values are even lower (Fig. 6).

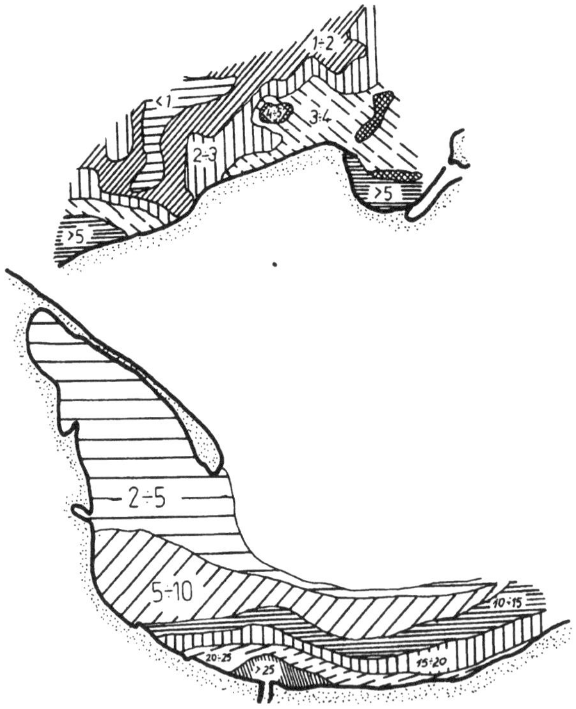

Fig. 6. Mean chlorophyll "a" concentration in the top water layer (mg/m^3) (after RENK et al. 1986, LATAŁA 1985)

The increase in the production of phytoplankton created the more advantageous trophic conditions for zooplankton. The basis of zooplankton biomass and its quantity in the open waters of the Baltic Sea is constituted during the last decade by *Pseudocalanus elongatus*, *Temora longicornis*, *Acartia* spp., *Centropages hamatus*, *Evadne nordmanii*, *Bosmina maritima*, *Podon* spp., *Synchaete* spp., *Fritillaria borealis* and *Aurelia aurita*. During the last twenty years the clear increase of the percentage of the number of herbivorous crustaceans has been noticed. It has to be stressed that in the years 1951—1983 the wet biomass of mesozooplankton increased on average in the open waters of the central coast from 7.8 to about 20 g/m^2 (TRZOSIŃSKA et al., in print). A statistically significant trend was discerned in the long-term variations of the occurrence of mesozooplankton in the southern Baltic during the period 1951—1974. The increase of the mesozooplankton biomass was expressed by the annual coefficient of 0.38 g/m^2 (CISZEWSKI 1985).

The above mentioned species maintain also in the gulf areas, however while closer to the coast the composition shows some specific character resulting from the influence of the land. In general the same species dominate in the whole area of the Gulf of Gdańsk. The following ones belong to them: *Synchaeta baltica*, *S. monopus*, *Keratella quadrata* as well as two species from the genus *Acartia* — *A. bifilosa* and *A. tonsa*. The number and participation of *Temora longicornis*, *Pseudocalanus elongatus*, *Fritillaria borealis* and *Bosmina maritima* increases together with the enlarging distance from the coast and with the increase of the water depth (WIKTOR et al. 1982).

The region of the Vistula river's estuary differs strongly in its character from the rest of the Gulf of Gdańsk area. Together with the Baltic species the fresh water species can be found here, for example the following species from the class crustacea: *Daphnia cucullata*, *Chydorus sphaericus*, *Alona quadrangularis* and *Eudiaptomus* sp. They appear in small numbers, and their extent is limited only to the area of the strongest influence of fresh waters near the Vistula mouth. The rotifers especially typical for the eutrophic waters appear in greater numbers as far as the number of species and the density of the population are considered. The rotifers reach wider than crustaceans. Species from the genus *Brachionus*, mainly *B. calyciflorus* and *B. angularis* have especially high numbers as well as wide extent (WIKTOR & ZMIJEWSKA 1985).

During the whole year the following species are detected in the Pomeranian Bay: *Acartia bifilosa*, *Temora longicornis*, *Pseudocalanus elongatus* and *Centropages hamatus*, while *Cyclops viridis*, *Acartia longiremis*, *Eurytemora hirundoides*, *Acartia tonsa*, *Evadne nordmanni*, *Podon* spp., *Bosmina maritima* and veliger of *Mytilus edulis* appear seasonaly. The seasonal changes in the structure of the dominants have been observed — in winter *Acartia bifilosa* and *A. longiremis* constitute the dominating species, in spring — *Acartia longiremis*, *A. bifilosa* and *Pseudocalanus elongatus* are the dominants, in summer — *Evadne nordmanni*, *Eurytemora* spp., then *Acartia tonsa*, larvas of *Mytilus edulis*, *Acatria bifilosa* and *Bosmina maritima*, in autumn — *Acartia tonsa* and *A. bifilosa* (TRZOSIŃSKA et al. 1989).

Rotifers are the important element of the zooplankton in the Szczecin Lagoon beside the crustaceans, whose composition has not changed considerably during the last twenty years. Some trend in the changes of the number of the rotifers has been noticed. In the fifties it was found, that crustaceans prevailed rotifers only in the summer and early-autumn plankton, while in 1979—1980 the domination of the crustaceans was noted (TRZOSIŃSKA et al., in print).

Benthos

The investigations on the sea sedentary flora were concentrated in the area of the Gulf of Gdańsk and precisely in the Puck Bay. Species from the following divisions: Chlorophyta, Phaeophyta, Rhodophyta and from Spermatophyta are found in the phytobenthos of this region, as it is typical for the litoral of all the seas. The number of the algae species found in the Gulf of Gdańsk is low in comparison with the oceanic coasts, because of the low water salinity. Actually there are not much more than 20 algae species and several vascular plant species noted in the phytobenthos of the southern Baltic (PLIŃSKI, in print).

During the last years considerably changes in the composition and distribution of the Gulf of Gdańsk phytobenthos were noted. The number of species diminished by 50 per cent during eighty years, it can be assumed, however, that the droping out of species took place in the last years. Green algae from genus *Enteromorpha* and *Cladophora* as well as Phaeophyta — two species from family Ectocarpaceae: *Pilayella litoralis* and *Ectocarpus siliculosus* are the most numerous species. The presence of several species of *Ceramium* and Spermatophyta (*Zostera marina*, *Ruppia rostellata*, *Zannichellia palustris*) has been still found but very seldom. However the progressive reduction of the ranges of *Zostera marina*, *Furcellaria fastigiata* and *Fucus vesiculosus* has been observed for several years. Since 1984 these two last species have disappeared completely. The species from family Ectocarpaceae, which develop in masses occupy the place of *Furcellaria fastigiata* and

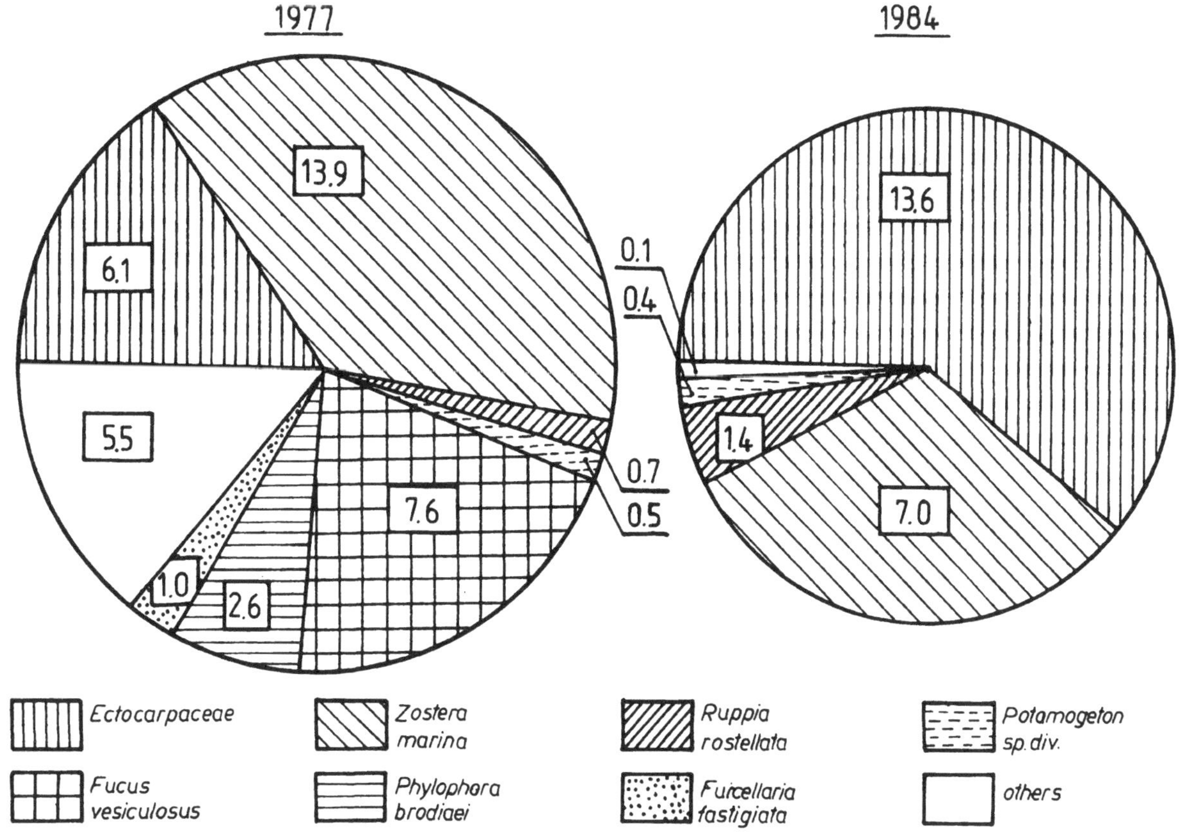

Fig. 7. Relationship between main components of phytobenthos in the inner part of Puck Bay in two compared years (dry weight, g/m²)

Fucus vesiculosus. This fact is reflected also in the biomass. The dry weight of *Pilayella litoralis* and *Ectocarpus siliculosus* in the Puck Bay increased from 6.1 g/m² in 1977 to 13.6 g/m² in 1984 (Fig. 7). The limitation of the bathymetric range of the macroflora is the other important symptom of biocoenotic changes in the Gulf of Gdańsk. At the beginning of the present century benthic algae appeared to the depth of 25 m, in the years 1979—1980 they were found at the 9 m depth and in 1985 the maximum range was found to be 6 m (PLIŃSKI & FLORCZYK 1984).

Some changes which are directly connected with the deterioration of the physico-chemical conditions at the bottom have been also detected in the zoobenthos. It considers most of all the deeps. The occurrence of *Macoma calcarea—Pontoporeia femorata* community, which wet weight was evaluated to 25.6 g/m² was noted in the Gdańsk Deep in 1925 (HAGEMEIER 1926). In 1951 the presence of *Scoloplos armiger—Macoma baltica* community with the average biomass of about 3.0 g/m² was detected (DEMEL & MULICKI 1954). However in 1980 in the same area of the Gdańsk Deep the presence of only *Harmothoe sarsi* with the average biomass of 2.11 g/m² was proved, while in 1981 the complete destruction of macrozoobenthos was stated; only *Harmothoe sarsi* was present at the depth of 80—90 m, producing the biomass of 0.07—0.14 g/m² (OKOŁOTOWICZ 1985). A quite different situation can be seen in the coastal zone, where *Nereis diversicolor*, *Macoma baltica*, *Mytilus edulis*, *Mya arenaria*, *Hydrobia ulvae* and *H. ventrosa* belong to the most numerous species. The considerable number is also characteristic for Oligochaeta. There are many species which appear seldom and only in some areas.

Great differentiation can be seen also in the distribution of the biomass which varies from 1 to over 300 g/m² (WENNE & WIKTOR 1982; HERRA & WIKTOR 1985). Polychaeta, Crustacea, Bivalvia and Gastropoda are the main constituents of the macrozoobenthos biomass in the coastal zone of the Gulf of Gdańsk, others do not play an important role. The trend to impoverish the bottom of the shallow zone has been clearly seen in the biomass distribution, while areas of the greater productivity were shifted to the deeper regions. The maximum of the zoobenthos biomass has been detected at the depth of about 20 m. The biomass composition has also changed, viz. the participation of crustaceans (*Idotea*, Gammaridae) has been reduced while that of the molluscs (*Mytilus*, *Macoma*, *Cardium*) has increased (LEGEŻYŃSKA & WIKTOR 1981). These transformations can be connected with the increase of the trophy of the coastal zone, especially evident in the area of the Vistula estuary. Similar situation, but not so intensive, has been observed in the Pomeranian Bay.

Turbellaria, Oligochaeta, *Nereis diversicolor*, *Pygospio elegans*, *Mya arenaria*, *Hydrobia* spp., *Macoma baltica*, *Cardium glaucum* belong to the most often appearing organisms in the macrobenthos of the Pomeranian Bay. Furthermore, *Manayunkia aestuarina* and *Corophium volutator* are common but only in the western part of the bay. Many years' data concerning the distribution of zoobenthos indicate the fertilizing effect of the Szczecin Lagoon waters discharged to the lagoon of the Świna River. The density of zoobenthos in the areas to the north of Międzyzdroje and far to Ławica Orla was on average 5—10 times higher than in the areas where there was no influence of the Odra River (TRZOSIŃSKA et al. 1989).

The bottom fauna of the Szczecin Lagoon is of fresh water character and is constituted, especially at the muddy bottom, mainly of representatives of Oligochaete and Chironomidae (WIKTOR 1954; WOLNOMIEJSKI et al. 1984; DZYCIMSKI 1985, cit. after TRZOSIŃSKA et al. 1989). The quantitative relations among those most important components differ in time and space, but even at the quantitative domination of Oligochaeta, Chironomidae and mainly *Chironomus plumosus* reach the biggest biomass. Besides, Ostracoda play the important role in biomass. Despite the above mentioned groups of macrofauna *Dreissena polymorpha*, occurring in the form of close agglomerations plays an important role in the coastal zone on the slopes of the shallows. In the fifties it covered the bottom area of about 46 km^2. Also in the eighties the presence of *Dreissena polymorpha* in the considerable amounts was noticed (TRZOSIŃSKA et al. 1989).

Ichthyofauna

Herring, cod, sprat and flat-fishes belong to the fish species which are being caught in the greatest number in the coastal zone of the southern Baltic, while roach, pike-perch, bream, seatrout, eel, pike and perch are the most often found in the gulfs; among the nonindustrial the gobid family of fishes and stickleback appear in considerable amounts. The strong trend of the increasing fertility has been noted for many years in the southern Baltic, for example in the period of twenty years since 1950 to 1970 the fishing increased three times and presently the fishing yield in the southern Baltic reaches 50—60 kg from 1 hectare; it concerns such sea fishes as herring, sprat and cod (ZMUDZIŃSKI 1986). However in the case of other species which have been until now caught in great quantities in the gulfs, the clear structural revaluation took place. The catches of seatrout, salmon and eel have decreased in the Gulf of Gdańsk while the catches of roach increased. The so called "ulcerous disease" was the cause of eel's death in masses on the turn of the seventies and the eighties. The drastic drop in the catches of pike-perch and lavaret was also observed. There are surely many causes, which act complementary. In the last period the catches of seatrout have increased a little, thanks to the intensive stocking with fry. The sea species which periodically appear in the coastal zone are not so strongly influenced by the increasing pollution of this zone. The fluctuations of their participation in the boat-fishing catches are mainly dependent on the variable fertility of their generations in the successive years. However the state of the stock of the southern Baltic cod is alarming, that was reflected by the break of its catches in spring 1986. Herring, cod, flounder and sprat, moreover several species having strong biological connections with the Szczecin Lagoon and inland waters, among others pike-perch, seatrout, lavaret, eel, bream, roach belong to the main species, which are fished industrially in the Pomeranian Bay. Herring has become the basic species in the fishing since the beginning of the seventies while the percentage of the flat-fishes in the catches decreased. Nearly complete disappearance of flat-fishes has been noted in the west part of the Bay. The considerable decrease of the cod population has been observed in the whole Pomeranian Bay (TRZOSIŃSKA et al. 1989).

Problems of the fish illness in the Polish zone of the Baltic fishery arose not earlier than at the end of seventies, when the amount of fishes with the pathological symptoms increased considerably. This problem took an extremely sharp form when the eels died in masses in 1979 and in the following years. The most often noted symptoms are ulcerations mainly at demersal fish (cod, eelpout, flounder) and from pelagical fish at sprat; extravasations and congestions at cod and herring; fin defect at cod and herring. The percentage of the ill fishes has oscillated considerably in the particular years. Eelpout, cod, flounder and sprat suffer the most often. The frequency of the occurrence of the ulcerous disease at cods fished in the Gdańsk Basin is similar to that in the Bornholm Deep, in the Pomeranian Bay and on the fishing grounds of Kołobrzeg-Darłowo but much higher than of Ustka-Łeba or Władyslawowo (TRZOSIŃSKA et al. 1989).

Conclusion

The Polish coastal zone of the Baltic Sea consisting of the Gulf of Gdańsk, narrow water stripe of the central coast and the Pomeranian Bay is under the strong influence of the land factors, among others such big rivers as the Vistula and the Odra. At present the total loads of nitrogen and phosphorus discharged from the land within the borders of Poland are estimated to reach $209.6 \cdot 10^3$ tons TOT-N/year and $19.1 \cdot 10^3$ tons TOT-P/year. 90.2 per cent of total nitrogen comes from the riverine input, 8.2 per cent from the municipal wastes and 1.6 per cent from the industry. The proportions in the total phosphorus load are different: 60.2 per cent is discharged through rivers, 15.7 per cent originates from the municipal wastes while 24.1 per cent comes from the industrial wastes. The essential changes in the species composition, distribution and numbers, i.e. the biomass of the particular species have been found in the biocoenosis of the coastal waters of the southern Baltic. They result in the shifts of the dominates. Particularly intensive changes were noted in the gulfs and nearby of the river mouths. The significant changes have been found in the benthos, particularly in the phytobenthos. The nearly complete dominant exchange in phytobenthos occurred in the Puck Bay — the spawning ground of many fish species. *Pilayella litoralis* and *Ectocarpus siliculosus*, developing in masses, occupy the place of *Furcellaria fastigiata* and *Fucus vesiculosus*. The reduction of the bathymetric range of macroflora to 6 m constitutes the other important symptom of biocoenotic transformations in the Gulf of Gdańsk. The present biocoenosis differs structurally from those described 40—50 years ago. The question remains how deep the changes are and if they are yet reversible.

References

CISZEWSKI, P. (1985): Long-term trends in mesozooplankton biomass development in the southern Baltic. Oceanologia **22**, 63—70.

CYBERSKA, B. (in print): Atlas zasobów, walorów i zagrożeń środowiska geograficznego Polski (Atlas of the resources, values and emergencies of the geographical environment of Poland), ed. ST. LESZCZYCKI. Instytut Meteorologii i Gospodarki Wodnej w Gdyni, Gdynia.

— & TRZOSIŃSKA, A. (1984): Environmental conditions in the Gulf of Gdańsk during the last decade, 1974—1983. Proceedings of the XIV Conf. Baltic Oceanogr. **2**, 490—509.

DEMEL, K., & MULICKI, Z. (1954): Studia ilościowe nad wydajnością dna południowego Bałtyku (Quantitative studies on the biological yield of the southern Baltic bottom). Prace MIR w Gdyni **7**.

HAGEMEIER, A. (1926): Die Arbeiten mit dem Petersenschen Bodengreifer auf der Ostsee, April 1925. Ber. Deutsch. Komm. Meeresforsch. N. F. Bd. **2—4**, 4.

HERRA, T., & WIKTOR, K. (1985): Skład i rozmieszczenie fauny dennej w strefie przybrzeżnej Zatoki Gdańskiej Właściwej (Composition and distribution of bottom fauna in coastal zone of the Gulf of Gdańsk Proper). Studia i Materiały Oceanol. Gdańsk **46**, 115—142.

LATAŁA, A. (1982): Zawartość chlorofilu w wodach powierzchniowych zachodniej części Zatoki Gdańskiej (The chlorophyll content in the surface waters of the western part of Gdańsk Bay). Studia i Materiały Oceanolog. Gdańsk **39**, 15—33.

— (1985): Zawartość chlorofilu w wodach powierzchniowych Zatoki Gdańskiej Właściwej w roku 1981 (The chlorophyll content in the surface water of the Gulf of Gdańsk Proper in 1981). Studia i Materiały Oceanolog. Gdańsk **46**, 11—22.

LEGEŻYŃSKA, A., & WIKTOR, K. (1981): Fauna denna Zatoki Puckiej Właś ciwej (Bottom fauna of Puck Bay Proper). Zesz. Nauk. BiNoZ Uniwersytet Gdańsk. Oceanografia **8**, 63—78.

OKOŁOTOWICZ, G. (1985): Biomasa makrozoobentosu polskiej strefy Bałtyku wskaźnikiem jej zanieczyszczenia (Biomass of the macrozoobenthos of the Polish Baltic Zone as the indicątor of its pollution). Biuletyn MIR **5/6**, 27—40.

MAJEWSKI, A. (in print): Atlas zasobów, walorów i zagrożeń środowiska geograficznego Polski (Atlas of the resources, values and emergencies of the geographical environment of Poland), ed. ST. LESZCZYCKI. Instytut Meteorologii i Gospodarki Wodnej w Gdyni, Gdynia.

MILEWSKA, E., & ANDRULEWICZ, E. (1982): An assessment of chemical parameters measured in the Polish sector of the Baltic Sea in 1979—81. Proceedings of the XIII Conference of the Baltic Oceanographers, Vol. II, 818—844.

PLIŃSKI, M. (in print): Roślinność. W: Monografia Bałtyku południowego (Flora. In: Monography of the southern Baltic). Gdańskie Tow. Naukowe.

— & FLORCZYK, I. (1984): Changes in the phytobenthos resulting from the eutrophication of the Puck Bay. *Limnologica* (Berlin) **15** (2), 325—327.

— — & PICIŃSKA, J. (1985): Skład i liczebność fitoplanktonu Zatoki Gdańskiej Właściwej (Composition and abundance of phytoplankton in the Gulf of Gdańsk Proper). Studia i Materiały Oceanolog. Gdańsk **46**, 23—64.

RENK, H. (1973): Primary production and chlorophyll content in the Baltic Sea. Part II. Chlorophyll „a" distribution. Pol. Arch. Hydrobiol. **20**, 237—255.

— NAKONIECZNY, J., OCHOCKI, S., GROMISZ, S., & WIKTOR, J. (1986): Primary production and chlorophyll "a" concentration in the southern Baltic in 1982 and 1983. Annales Biol. ICES **40**, 81—83

TRZOSIŃSKA, A., PLIŃSKI, M., & RYBIŃSKI, J. (1989): Charakterystyka hydrochemiczna i biologiczna polskiej strefy przybrzeżnej Bałtyku (Hydrochemical and biological characteristics of the Polish coastal zone). Studia i Materiały Oceanol. Gdańsk **54**, 5—70.

— (in print): Związki azotu i fosforu w Basenie Gdańskim (Nitrogen and phosphorus compounds in the Gdańsk Basin). In: Procesy sedymentacyjne w Basenie Gdańskim (Sedimentation processes in the Gdańsk Basin). Instytut Meteorologii i Gospodarki Wodnej w Gdyni, Gdynia.

WENNE, R., & WIKTOR, K. (1982): Fauna denna przybrzeżnych wód Zatoki Gdańskiej (Benthic fauna of the inshore waters of Gdańsk Bay). Studia i Materiały Oceanol. Gdańsk **39**, 137—171.

WIKTOR, K., & ZMIJEWSKA, M. I. (1985): Skład i rozmieszczenie zooplanktonu przybrzeżnych wód Zatoki Gdańskiej Właściwej (Composition and distribution of the coastal waters of the Gulf of Gdańsk Proper). Studia i Materialy Oceanol. Gdańsk **46**, 65—114.

— CYLKOWSKA, U., & OSTROWSKA, K. (1982): Zooplankton przybrzeżnych wód Zatoki Gdańskiej (Zooplankton in the coastal waters of Gdańsk Bay). Studia i Materiały Oceanol. Gdańsk **39**: 77—136.

ZMUDZIŃSKI, L. (1986): Zanieczyszczenia Morza Bałtyckiego (Pollution of the Baltic Sea). Kosmos **35**, 79—93.

Address of the author:

MARCIN PLIŃSKI
Institute of Oceanography
Gdańsk University
Gdynia
Poland

Limnologica (Berlin) **20** (1), 47—51, April 1990

Wilhelm-Pieck-Universität Rostock, Sektion Biologie, Rostock, DDR

Characteristics of Phytoplankton in Brackish Waters of Different Trophic Levels

By N. Wasmund

With 5 Figures and one Table

Key words: Trophic level; Primary production; Phytoplankton biomass; Diversity; Succession.

Abstract

Investigations on phytoplankton have been carried out in the waters south of the Darss-Zingst Peninsula. These consist of a chain of several water bodies (boddens), differing in their trophic level. Annual phytoplankton production and a mean biovolume of 100—200 g C · m^{-2} and 2—10 $mm^3 \cdot l^{-1}$ respectively were found for the mesotrophic level, of 200—400 g C · m^{-2} and 10—20 $mm^3 \cdot l^{-1}$ for the eutrophic level, and of 400—600 g C · m^{-2} and 20 (—35) $mm^3 \cdot l^{-1}$ for the polytrophic level.

Since the water becomes more turbid as the trophic level increases (for instance in polytrophic waters the top metre of the water column accounts for 85—95% of planktic primary production), the higher the trophic level, the smaller is the contribution of the phytobenthos to the total primary production. It is difficult to assign specific indicator species to different trophic levels because the relatively slight differences in species composition are masked by the much greater seasonal and annual variations.

The Department of Biology of the Wilhelm Pieck University Rostock has been studying the phytoplankton population of the chain of shallow waters south of the Darss-Zingst peninsula without interruption for almost 20 years. The peninsula separates these waters from the open water of the Baltic Sea, with which they are connected only by a narrow channel at their eastern end. Known as "boddens", they are separated from each other by several islands and peninsulas and thus form a continuous chain stretching from west to east south of the Darss-Zingst Peninsula.

Since the boddens are interconnected only by relatively narrow stretches of water, the exchange of the total water body takes place very slowly. Consequently, as the river water entering the boddens mainly at the western end mixes with the water entering from the Baltic Sea, the chain of boddens does not act as a uniform water body, each bodden having its own typical salinity ranging from almost zero in the west to nearly 14‰ in the east. This increase in salinity from west to east is accompanied by corresponding decreases in nutrient concentration, turbidity, algal biomass and primary production.

Fig. 1 presents the long term means for primary

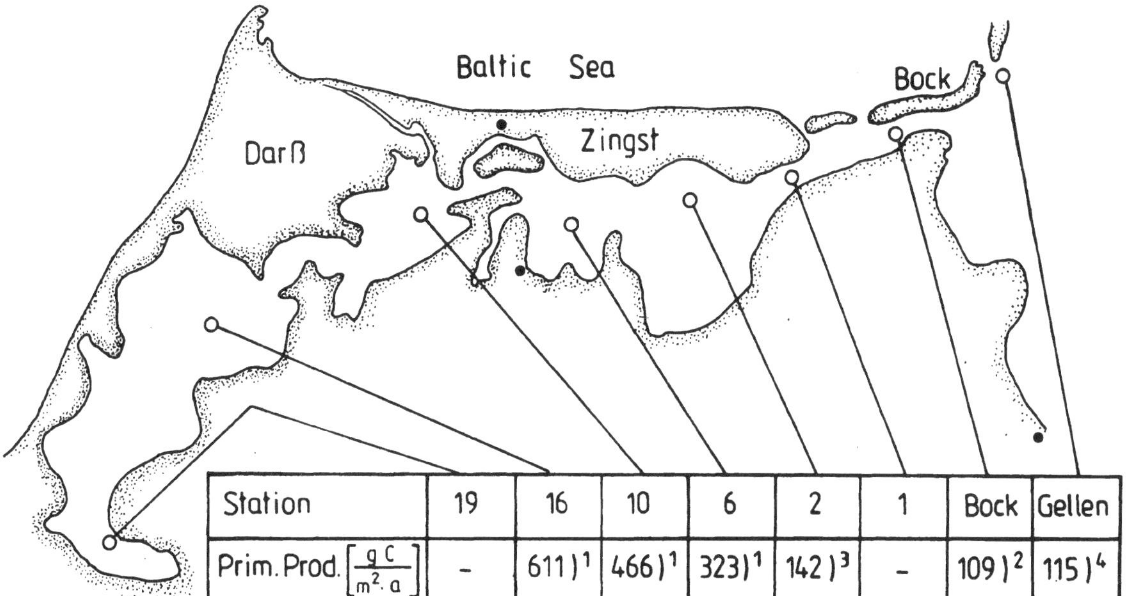

Station	19	16	10	6	2	1	Bock	Gellen
Prim. Prod. $\left[\frac{g\,C}{m^2 \cdot a}\right]$	-	611)[1]	466)[1]	323)[1]	142)[3]	-	109)[2]	115)[4]
Biovol.)[5] $\left[\frac{mm^3}{l}\right]$	27.9	23.8	21.1	15.6	7.4	3.6	-	-

Fig. 1. Long term primary production means (after Hübel, unpubl.) and biovolume (after Nasev, unpubl.) for phytoplankton at various stations in the boddens south of the Darss-Zingst Peninsula and in the region of the Baltic Sea close to the bodden outlet.
Investigation period:)[1] 1964—1968, 1971—1974, 1976—1978
)[2] 1971—1978
)[3] 1972—1978
)[4] 1976—1978
)[5] 1971—1980

Table 1. Assignment of primary production and biovolume values of the phytoplankton to different trophic levels and specification of the waters with the corresponding trophic levels (after SCHLUNGBAUM et al. 1987, modified)

Trophic level	Water	Annual primary production $g\,C \cdot m^{-2} \cdot a^{-1}$	Daily prim. prod. in the maximum production zone (April—Sept.) $mg\,C \cdot m^{-3} \cdot d^{-1}$	Phytoplanktonbiovolume (annual mean) $mm^3 \cdot l^{-1}$
mesotroph	Grabow	100—200	80—300	2—10
eutroph	Barther Bodden	200—400	300—900	10—20
polytroph	Bodstedter Bodden	400—600	900—1800	> 20
hypertroph	Saaler Bodden	> 600	>1800	

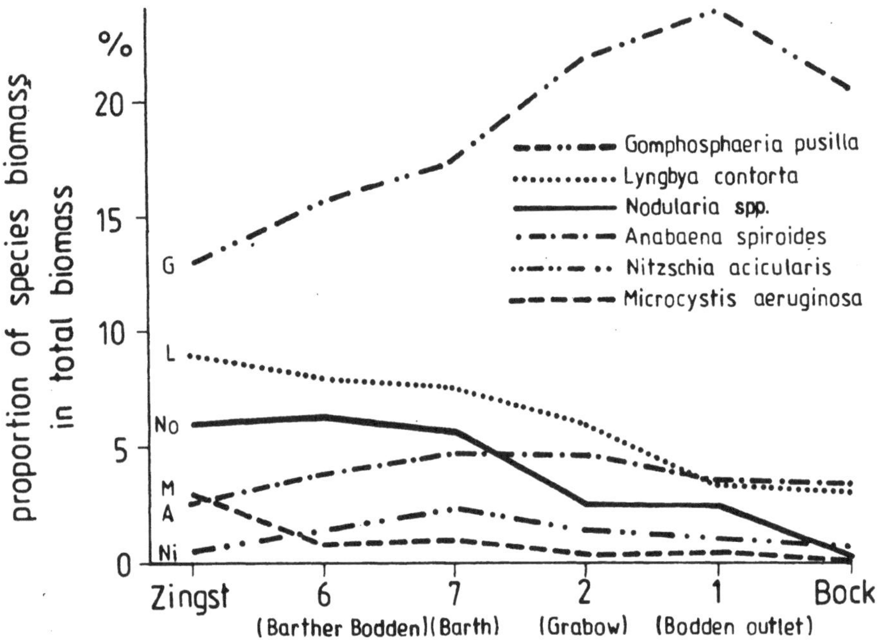

Fig. 2. Distribution of 6 selected species of algae in the eastern part of the bodden chain south of the Darss-Zingst Peninsula (means for the years 1984—1986)

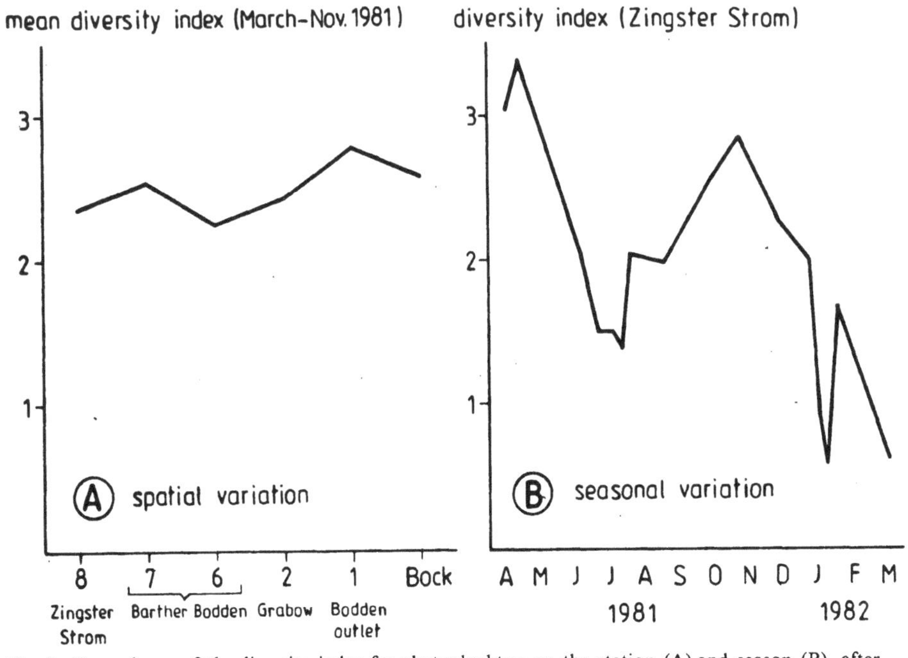

Fig. 3. Dependence of the diversity index for phytoplankton on the station (A) and season (B), after LOBODA (1982)

production and phytoplankton biovolume in tabular form. The total range of the values is quite considerable owing to the great difference between the quality of the water of the different boddens.

The unit studying these waters has now linked its data with the corresponding trophic levels ranging from mesotrophic in the Grabow Bodden (stat. 2) to hypertrophic in parts of the Saaler Bodden (stat. 16 and 19).

Table 1 presents only a few of the biological parameters that are closely linked to the nutritive status of waters and are therefore useful for estimating trophic levels. Our assignment of production and biomass data to different trophic levels is not only based on data collected in the boddens south of the Darss-Zingst Peninsula, but also on those of the boddens around the Isle of Rügen. The system we propose here is certainly valid also for other coastal waters of the Baltic. If it is accepted generally, it might help to achieve some degree of homogeneity among the estimates given of the trophic levels of coastal waters because production and biomass (or biovolume) values are objective measures of trophic levels.

Several other criteria can also naturally be used to assess trophic levels, including for instance the primary production vs. depth profile. In mesotrophic waters the light penetrates to a great depth, and the euphotic zone is consequently several metres deep: the metre-thick (1.0 to 2.0 m depth) zone of maximum production accounts for only 15—45% of total production. In polytrophic waters the maximum production zone (from 0 to 1.0 m depth in this case) is responsible for 85—95% of the total production of the whole water column; photosynthesis is virtually impossible at depths greater than only 2 m in such waters even on sunny days, and for benthic algae even the light intensity at a depth of 1 m is insufficient for survival.

This explains why benthic production is low in eutrophic to hypertrophic waters even if they are shallow. In the Barther Bodden for instance, which has an average depth of only 1.8 m, the phytoplankton accounts for no less than 93% of the total primary production (WASMUND 1986). In a comparable mesotrophic coastal water (Lake Grevelingen, The Netherlands), in contrast, phytoplankton was responsible for only 60% of total primary

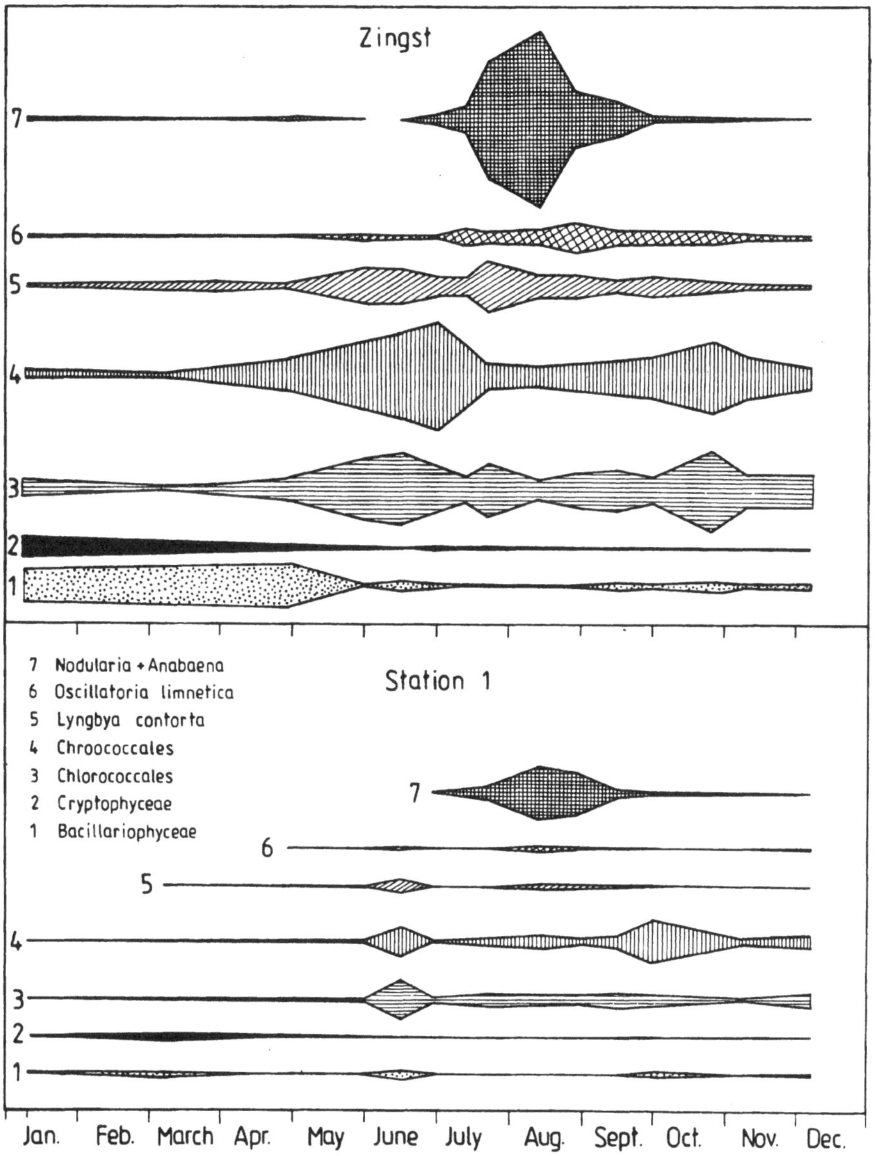

Fig. 4. Comparison of seasonal development of the most important groups of algae (biovolume) at a eutrophic station (Zingst) and a mesotrophic station (outlet from boddens) in 1984

production despite the greater mean depth (5.3 m) (DE VRIES 1984). The phytobenthos was able to flourish there owing to the high transparency of the water. Under oligotrophic conditions in Chukchi Sea near Barrow, Alaska, where the water is also 5 m deep, the contribution of phytoplankton to total primary production was only 31% (MATHEKE & HORNER 1974), and in an ultraoligotrophic water (Lake Char) it even sank to a mere 20% (WELCH & KALFF 1974).

In view of this it seems reasonable to assume that there is some link between trophic level and the relative proportions of benthic and planktic primary production. This, however, would be a very cumbersome way of measuring trophic levels and is actually merely an expression of the turbidity of the water. In other words it should be sufficient to measure only the turbidity or Secchi depth. These physical methods are much simpler than biological methods. On the other hand, however, they are also much more susceptible to a variety of nuisance factors such as wind-induced resuspension of particulate matter which reduce transparency and thus make the trophic level of the water appear higher than it actually is.

I do not wish to follow this train of thought here, however, but to concentrate on biological parameters. It is well known that species composition varies with water quality. *Microcystis* and *Merismopedia* species, for instance, are more prevalent in hypertrophic regions, whereas *Anabaena spiroides* and *Nodularia* spp. seem to prefer the eutrophic to polytrophic areas. The dominant green algae such as *Scenedesmus quadricauda*, *Tetrastrum komarekii* and *Monoraphidium contortum* are relatively uniformly distributed and are therefore unsuitable for use as indicator species.

The distributions of a few species of algae in the mesotrophic to eutrophic regions of the boddens south of the Darss-Zingst Peninsula are shown in Fig. 2. The species concerned were chosen because they obviously prefer certain areas of the boddens. On the basis of our data, however, it is impossible to decide whether this preference for certain water bodies is a result of the decreasing trophic level or the increasing salinity as the distance to the Baltic Sea decreases.

There can be little doubt that the decreasing numbers of species from west to east is a salinity effect because many fresh water species are unable to tolerate salinities even as low as $5^0/_{00}$. PANKOW & MARTENS (1973), for instance, found 343 species of planktic and benthic algae in the Saaler Bodden compared with only 268 in the Bodstedter Bodden, 249 in Zingster Strom and 230 in the Grabow.

This is reflected in the diversity index, which is significantly higher for the Saaler Bodden than for the more eastern boddens (NASEV 1976). Bodstedter Bodden with a salinity of scarcely $5^0/_{00}$ is a transitional region, representing the limits of existence for some fresh water species. Hence, the theory that the diversity of fauna and flora decreases as the nutrient load of a water becomes greater does not apply here on account of the salinity effect. In our coastal waters it is not possible to judge water quality from the diversity of the algal population.

This was also shown by LOBODA's (1982) studies of the eastern part of the bodden chain. There, too, the diversity index does not differ from one station to another, although there are clear seasonal variations (Fig. 3). In other words, in the eastern part of the bodden chain species composition and diversity vary considerably in time, but only slightly in space.

To illustrate this we shall consider the development of the different algae at the Zingst station, where the water is eutrophic, and at station 1, which is situated in the mesotrophic water at the outlet of the boddens (Fig. 4). The biovolumes at the mesotrophic station are naturally much lower, and the "blooms" there are much shorter than at the eutrophic station, but the basic trends in seasonal development are the same: The year starts with high cryptophycean and bacillariophycean biomasses. These are followed by the development of the chlorophyceans, coccal cyanobacteria (Chroococcales) and, subsequently, *Lyngbya contorta*. These populations de-

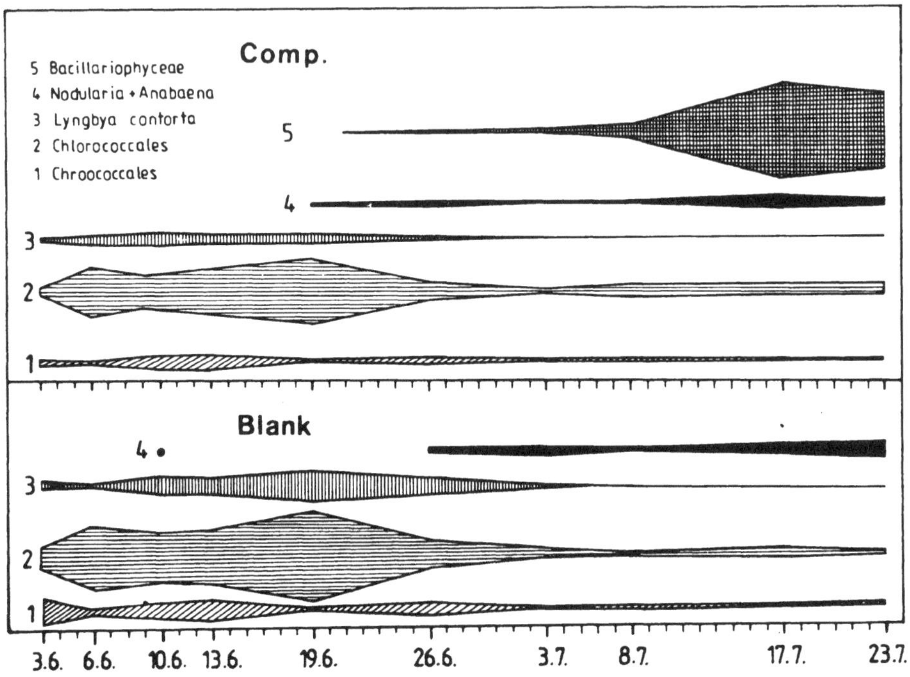

Fig. 5. Comparison of the development of the most important groups of algae (biovolume) in the course of an experiment with (Comp.) and without (Blank) enclosure and nutrient enrichment

cline rapidly in early July with the onset of a nitrogen-fixing cyanobacteria bloom, which is accompanied by the development of *Oscillatoria limnetica* and *Lyngbya contorta*. The collapse of this bloom marks the onset of renewed development of the coccal cyanobacteria, chlorophyceans and, to a lesser degree, the bacillariophyceans. The biomasses then decrease slowly as the end of the year approaches. It is therefore clear that trophic level generally affects rather the magnitude of primary production and the biomass of the algae rather than their succession.

In experiments involving artificial nutrient enrichment (cf. SCHIEWER et al. 1988) we were able to show that very drastic intervention under very specific conditions is necessary to disrupt the natural algal community. Fig. 5 shows a particularly striking example in which the biomass development of the five most important functional groups was followed during an eight-week experiment. The top half of the figure shows the situation in a compartment with a volume of 2.3 m^3 that was held under field conditions. The bottom half shows the course of events under natural conditions outside of the compartment (blank). During the first three weeks, 6.35 g of phosphorus and 24.85 g of nitrogen were added to the water in the compartment without the species composition being influenced. A diatom (*Synedra*) bloom was later observed in the compartment, but not in the blank experiment, but only after the N:P ratio had been increased from 4 to 20 by reducing the phosphorus input.

Our observations of the natural situation in the boddens lead us to the following conclusions:
Although some species of algae show preferences for regions of a certain trophic level, these preferences are so weak that changes in trophic level do not result in major changes in species composition. Algal blooms are governed by seasonal (succession) rather than spatial (differences in trophic level) factors. Although trophic level does influence the duration of such blooms, this is reflected not in the species composition, but in the productivity and biomass levels of the populations concerned. The system we propose for judging the trophic levels of coastal waters is based on the assignment of certain ranges of primary production and algal biomass values to the different trophic levels.

References

DE VRIES, I. (1984): The carbon balance of a saline lake (Lake Grevelingen, The Netherlands). Neth. J. Sea Res. **18**, 511—528.

LOBODA, S. (1982): Phytoplanktonuntersuchungen in der Darß-Zingster Boddenkette (südliche Ostsee). Dipl.-Arb., Wilhelm-Pieck-Univ. Rostock.

MATHEKE, G. E. M., & HORNER, R. (1974): Primary production of the benthic microalgae in the Chukchi Sea near Barrow (Alaska). J. Fish. Res. Board Canada **31**, 1779—1786.

NASEV, S. (1976): Ökologische Phytoplanktonuntersuchungen in den Boddengewässern südlich des Darß und des Zingst. Diss., Wilhelm-Pieck-Univ. Rostock.

PANKOW, H., & MARTENS, B. (1973): Die Beziehung der Kleinalgenflora der Darßer Boddenkette (südliche Ostsee) zur Qualität dieser Gewässer. Wiss. Z. Univ. Rostock, Math.-nat. R. **22**, 1147—1151.

SCHIEWER, U., BÖRNER, R., & WASMUND, N. (1988): Deterministic and stochastic influence of nutrients on phytoplankton function and structure in coastal waters. Kieler Meeresforsch., Sonderheft 6, 173—183

SCHLUNGBAUM, G., NAUSCH, G., ARNDT, E.-A., BAUDLER, H., BIESTER, E., HÜBEL, H., KELL, V., KRÜGER, G., PANKOW, H., SCHIEWER, U., WASMUND, N., & WINKLER, H. (1987): Entwurf einer Küstengewässer-TGL. W.-Pieck-Univ. Rostock (unpubl.).

WASMUND, N. (1986): Die Größe der Primärproduktion im Barther Bodden (südliche Ostsee) unter besonderer Berücksichtigung des Mikrophytobenthos. Wiss. Z. Wilhelm-Pieck-Univ. Rostock, Math.-nat. R. **35** (5), 22—27.

WELCH, H. E., & KALFF, J. (1974): Benthic photosynthesis and respiration in Char Lake. J. Fish. Res. Board Canada **31**, 609—620.

Address of the author:

Dr. NORBERT WASMUND
Wilhelm-Pieck-Universität Rostock
Sektion Biologie
Doberaner Str. 143
Rostock
DDR-2500

Limnologica (Berlin) **20** (2), 53—56, April 1990

AdW der DDR, Institut für Geographie und Geoökologie, Bereich Hydrologie, Berlin, DDR

Comparison of Methods for Estimation of Integral Primary Production in Shallow Aquatic Ecosystems with Special Regard to Turbulent Mixing

By BRIGITTE NIXDORF, HORST BEHRENDT & RITA STELLMACHER

With 3 Figures and 3 Tables

Abstract

Three methods for estimation of integral primary production were compared: a) traditionally in-situ incubation of several samples along a vertical profile, b) exposition of samples in the optimum production horizon, only, and c) besides these static methods production samples were vertically moved in a mixed layer by a lift. In general, there is a good agreement between the measured and calculated results of integral primary production estimated with different methods. Static in-situ incubation of O_2-samples yields to an underestimation of O_2-primary production in mixed layers. The quantification of these results guarantees a reliable estimation of integral primary production and offer the possibility to calculate this integral from just one measurement. It is an approach to make the measurements more rapid and economical and more real by simulation of vertical movement of algae. The problems of application of the represented methods are discussed.

Introduction

The measurement of phytoplankton primary production is one of the most important prerequisites for a better understanding of circulation of matter and energy flow, and succession of phytoplankton, respectively, in aquatic ecosystems. The magnitude of primary production depends, besides on biological parameters like chlorophyll content, mainly on the light conditions in the water. Because light climate is changing with depth and time the estimation of carbon input and phytoplankton balance processes is possible on the basis of integral primary production (P_{tot}) only. In general, the valuation of P_{tot} is realized by means of exposition of numerous samples along a vertical depth profile and subsequent calculation of the integral. Independent of the used method for estimation of integral primary production (O_2, ^{14}C) the procedure of measurement has substantial disadvantages:

- A relatively great measurement expense. The estimation of one value of integral primary production is laborious and time consuming. It requires the exposition of at least four samples as parallels in different horizons of the euphotic layer.
- The static exposition of samples in fixed depths is problematic in well mixed eutrophic water bodies (shallow lakes, estuaries, rivers). The whole possible spectrum of natural algal movement induced by turbulence or active migration is suppressed in the closed sample volume during incubation under constant light conditions for several hours (see also BEHRENDT 1989).

Different attempts to overcome these disadvantages are the subject of this paper.

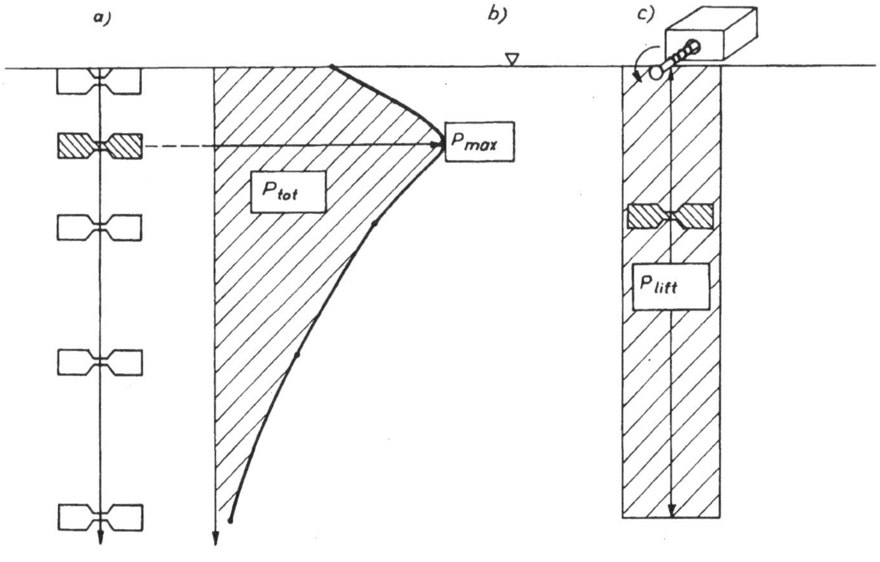

Fig. 1. Methods for estimation of integral primary production:
a) traditional incubation
b) incubation of samples in the optimum production layer
c) movement of samples in the production layer by a lift

Methods

The investigations of primary production were carried out in Lake Müggelsee, a eutrophic, polymictic shallow lake with an area of 7.2 km^2 and a mean depth of about 4.9 m. Samples from the euphotic zone above the deepest point (7 m) and mixed samples, resp., have been incubated at intervals of one week in the vegetation periods since 1978.

The procedure for estimating O_2- and ^{14}C-primary production is described in NIXDORF (1985) and NIXDORF & FULDA (1989). After filling samples as parallels into 120 ml glass bottles and inoculation with $NaH^{14}CO_3$ the samples were exposed in different ways (Fig. 1): a) as a vertical series in different horizons of the euphotic zone depending on Secchi-depth (P_{tot}, see NIXDORF 1982), b) only in the optimum layer P_{max}, STELLMACHER & NIXDORF 1984), c) the samples were moved up and down in the mixed layer by a lift (P_{lift}, BEHRENDT 1989). While the first method (a) is the basis for estimation of integral primary production, the last two methods (b, c) were chosen as a test making the measurements more rapid and economical. The third method, additionally, is a further approach to measure a more real primary production considering the vertical movement of algae in a mixed layer.

Results

1. Relationships between P_{tot} and P_{max} by application of ^{14}C-method using different statistical statements

Some proposals to estimate the area integral of primary production (P_{tot}) in the Lake Großer Müggelsee as a function of maximum value (P_{max}) measured in the optimum horizon have been presented in the paper of STELLMACHER & NIXDORF (1984). Here, the unknown coefficients have been estimated by samples measured in 1979/80. Now, these coefficients have been reestimated for the period 1979/83. The model validation was realized using an independent sample, that of 1986. The results are summed up in Table 1.

Table 1. Comparison of the coefficients estimated for different samples of primary production measurements in Lake Müggelsee (P_{tot} in mgC $\cdot$ m^{-2} $\cdot$ h^{-1}, P_{max} in mgC $\cdot$ m^{-3} $\cdot$ h^{-1}, Secchi depth SD in m)

Model	Sample size	Coefficient "a" estimated for the sample 1979/83	Model fitting
$P_{tot} = a * P_{max}$			
SD $\leqq$ 1 m	57	1.06	77%
SD > 1 m	134	1.18	64%
1 < SD $\leqq$ 2 m	89	1.14	77%
SD > 2 m	45	1.5	90%
$P_{tot} = a * P_{max}\ SD^{0.7}$			
SD $\leqq$ 1 m	57	1.3	64%
SD > 1 m	134	0.89	79%

If the P_{tot} values were drawn as a function of Secchi depth, it was found that it can be approximated by three parts of straight lines. That is the reason why a further division in depth steps ($\leqq$1 m, 1 < SD $\leqq$ 2 m, >2 m) was introduced. The used models explain only between 64 and 79% of the variability. One can increase the model fitting only by testing the experimental conditions, the availability of model assumptions and the introduction of other main influence factors, but this is not the aim of the present paper.

The model validation has been realized using an independent sample, that of 1986. The application of linear regression model with auxillary conditions gave the best fitted results. This is the same conclusion as found in the paper cited before.

2. Relationships between P_{tot} and P_{max} by application of O_2-technique

The O_2-production per unit area (P_{tot} in mg $O_2 \cdot m^{-2} \cdot d^{-1}$) has been correlated to the production in the optimum production layer (P_{max}). This correlation is significant:

$$P_{tot} = 1.01\ P_{max}^{0.96}, \quad r = 0.89, \quad n = 77\,. \tag{1}$$

For calculation of P_{tot} we considered the water column up to the compensation depth only, i.e. the depth where the community oxygen production and respiration are equal. The relationship between the two production terms is demonstrated in Fig. 2.

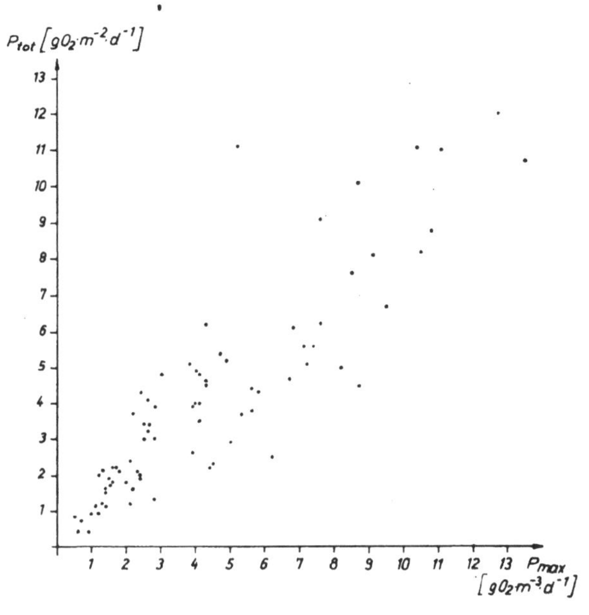

Fig. 2. Dependence of integral oxygen primary production (P_{tot} in g $O_2 \cdot m^{-2} \cdot d^{-1}$) on the maximum oxygen production in the optimum layer (P_{max} in g $O_2 \cdot m^{-3} \cdot d^{-1}$) in Lake Müggelsee

3. Comparison of integral primary production results investigated by common static in-situ incubation and by vertical movement of samples in the mixed layer

On the understanding of partial or complete mixing of shallow aquatic ecosystems throughout the year the influence of turbulent mixing on the primary production of phytoplankton was investigated by application of a bottle lift from 1984—1986. For valuation of changes of primary production induced by vertical movement in a light gradient in comparison to the common static incubation the production per unit area (P_{tot}) has been related to the production obtained by lift measurements (P_{lift}). Contrary to P_{max} the value of P_{lift} is an integral too, describing the average of integral primary production within the zone where the samples were moved by the lift. For calculation of the production sum per unit area the average value of P_{lift} is multiplied with the depth of passed lift length.

There is a good agreement between P_{tot} and P_{lift} for ^{14}C-technique (Fig. 3):

$$P_{tot} = 0.93 \pm 0.03\, P_{lift}, \quad n = 129, \quad \alpha = 0.95\,. \qquad (2)$$

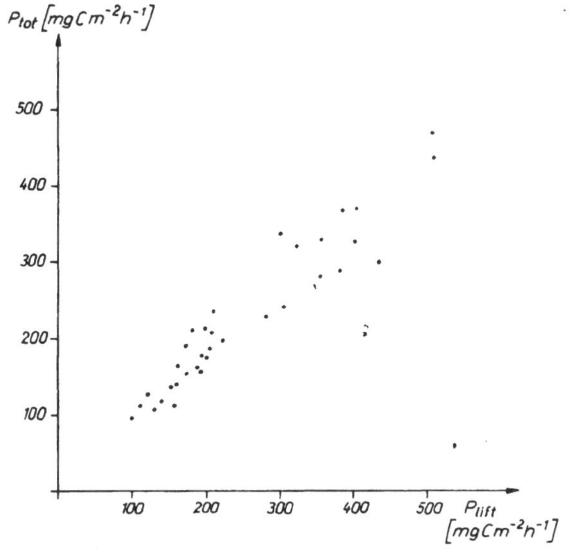

Fig. 3. Relationship between integral primary production estimated by static in-situ incubation (P_{tot} in mg C · m^{-2} · h^{-1}) and by moving samples in the mixed layer of Lake Müggelsee by a lift (P_{lift} in mg C · m^{-2} · h^{-1}) by application of ^{14}C method in the vegetation period 1986

In contrast to this there are remarkable discrepancies between P_{tot} and P_{lift} estimated by O_2-technique. On an average, P_{tot} measured by static vertical incubation is only a little more than half the production integral calculated from lift measurements:

$$P_{tot} = 0.59 \pm 0.15\, P_{lift}, \quad n = 38, \quad \alpha = 0.95\,. \qquad (3)$$

Description and interpretation of this phenomenon are more complex and, therefore, the subject of another paper (NIXDORF & BEHRENDT 1990).

Discussion

Generally, there is a good agreement between the results of the different in-situ expositions for application of ^{14}C-technique. These close relationships offer the possibility of calculating the depth production integral from just one measurement. They guarantee a reliable estimation of integral primary production. In that way, radioactive material and filters as well as time for preparation of samples can be economized by 60—70%.

Nevertheless, there are some problems associated with the application of the represented methods:

1. A single static measurement for calculating integral primary production supposes the knowledge of the optimum production layer. This zone was estimated in dependence on Secchi-depth and incident radiation for Lake Müggelsee in previous investigations (NIXDORF 1982). In the case of higher radiation intensity (> 10 MJ · m^{-2} · d^{-1}) the optimum layer was determined in half the Secchi-depth if the transparency of water was lower than one meter. If Secchi-depth exceeded 1 m, the optimum layer was found in 0.5 m, approximately. The relationship is valid for almost all dominating phytoplankton groups with the exception of colony forming blue-green algae. These algae are protected from high light intensity by pigment composition, the ability of buoyancy and forming of colonies (PAERL & USTACH 1982). This results, among other things, in a displacement of the production maximum from a depth of about 0.5 m to the surface, where other algal groups show a distinct inhibition of primary production (HORSTMANN 1975; NIXDORF & STELLMACHER 1987).

In the case of low radiation (< 10 MJ · m^{-2} · d^{-1}) there is no photoinhibition of primary production in the surface samples for all algal groups. Therefore, P_{max} is measured in the surface independent of Secchi-depth. Contrary to BARANOV (1979), ROMANENKO (1975), ERNST & REINHARDT (1980), and BUL'ON (1983) we obtained no better relationships between P_{tot} and P_{max} by consideration of Secchi-depth (see Table 1) or other optical parameters. Similar unsatisfactory results were obtained by application of photosynthesis-light models (PATTEN 1968) for calculation of P_{tot} in Lake Müggelsee in 1979/80 (NIXDORF 1982). Our results correspond to those of DOKULIL (1984), who described the relationship between P_{tot} and P_{max} of phytoplankton oxygen production in Lake Neusiedlersee by a hyperbolic function.

2. In contrast to ^{14}C-measurements of integral primary production the vertical distribution of O_2-samples yields also negative values below the compensation depth, i.e. the depth where O_2-community production and -respiration are balanced. The relationship between P_{max} and P_{tot} depends, therefore, on the depth of the water column which was considered for estimation of P_{tot}. In our measurements, only the water depth up to the compensation depth was taken into account. Considering different water layers (average, mixing, or maximum depth) exceeding the compensation depth the represented relationship (equ. 1) is unvailable. Additionally, the depth of compensation and the dark respiration in the aphotic layer must be considered for these cases.

3. A good agreement between vertically moved and statically incubated ^{14}C-samples is also reported by JEWSON & WOOD (1975), DOKULIL et al. (1978), MARRA (1978) and YODER & BISHOP (1985). One of the unsolved problems for application of lifts or similar devices is the estimation of the actual velocity and the kind of algal movement in a mixed layer and, therefore, the subsequent simulation of form and velocity of this transport. A spectrum of velocities (0.08—3 cm · s^{-1}) was tested by means of lifts in Lake Müggelsee, where a weakly marked maximum was found in periods of 10—20 min. This pointed to the existence of an optimum period, which was also proved by HARRIS (1978) for the photosynthetic oxygen evaluation of algal cultures.

Generally, the application of lifts or similar devices for estimation of P_{tot} seems to possess most advantages in mixed water bodies. It is possible to vary the length and velocity of the vertical movement by simple means. In this way, values of P_{tot} are obtained for different water depth, (e.g. mixed or euphotic layer) by the attempt to simulate the natural mixing conditions. For estimating of ^{14}C-production integral it is necessary to reach or to exceed the lower production depth (i.e. usually the depth of 1% incident radiation level estimated by previous measurement of extinction).

Further, our studies prove the hypothesis that during static incubations primary production in well mixed waters is neither systematically under- nor over-estimated by traditional incubation method that ignores mixing induced irradiance fluctuations by application of ^{14}C-technique, but not of O_2-technique.

Acknowledgement

We wish to thank INGRID HOFFMANN for accurate assistance.

Addendum

Our investigations have been continued during the last two vegetation periods (1987/88). The combination of our previous results, which are presented in this paper, and the current results requires the following correction: Turbulence induced mixing enhances the integral primary production by application of the ^{14}C-technique by about 18 %. The details of the influences on primary production and discrepancies between the applied methods are described in NIXDORF & BEHRENDT (1990).

References

BARANOV, S. A. (1979): Durchlässigkeit des Wassers, Biomasse und Produktion der Planktonalgen. Gidrobiol. Žurn. **15** (4), 18—23 (in Russ.).

BEHRENDT, H. (1989): Devices for turbulence simulation during in situ measurements of biological processes in aquatic ecosystems. II. A bottle lift for simulation of vertical movement in natural water. Acta hydrophys. **33** (1), 9—14.

BUL'ON, V. V. (1983): Primärproduktiondes Planktons in Binnengewässern. Trudi zool. Inst. Leningrad Tom **98**, 13—45 (in Russ.).

DOKULIL, M. (1984): Assessment of components controlling phytoplankton photosynthesis and bacterioplankton production in a shallow, alkaline, turbid lake (Neusiedlersee, Austria). Int. Rev. ges. Hydrobiol. **69** (5), 679—727.

— HAMMER, L., & JEWSON, D. H. (1978): Vergleichende Untersuchungen zur Primärproduktion im Neusiedlersee. O_2, 14-C und Experimente mit künstlicher Zirkulation. Biol. Forsch. Inst. Burgenland **29**, 60—73.

ERNST, D., & REINHARDT, D. (1980): Primary productivity measurements and carbon metabolism in Steinhuder Meer and Lake Dümmer. In: M. DOKULIL, H. METZ & D. JEWSON (eds.), Shallow Lakes. Contributions to their limnology (Developm. in Hydrobiol. **3**), pp. 11-16. The Hague, Boston, London.

HARRIS, G. P. (1978): Photosynthesis, productivity and growth: The physiological ecology of phytoplankton. Arch. Hydrobiol., Beih. **10** (I–IV), 1—171.

HORSTMANN, U. (1975): Eutrophication and mass development of blue-green algae in the Baltic. Merentut Kimuslait. Julk./Havsforskningsinst. Skr. No. **239**, 83—90.

JEWSON, D. H., & WOOD, R. B. (1975): Some effects on integral photosynthesis of artificial circulation of phytoplankton through light gradients. Verh. Int. Verein. Limnol. **19**, 1037—1044.

MARRA, J. (1978): Phytoplankton photosynthetic response to vertical movement in a mixed layer. Mar. Biol. **26** (3), 203—208.

NIXDORF, B. (1982): Die Sichttiefe als ein Maß zur Charakterisierung der Vertikalverteilung der Primärproduktion im Müggelsee. Acta hydrophys. **27** (3/4), 223—228.

— (1985): Photosynthetic activity of dominant algal species in eutrophic shallow lake (Großer Müggelsee, Berlin) investigated by microautoradiography. Int. Rev. ges. Hydrobiol. **70**, 237 to 246.

— & FULDA, S. (1989): A method for estimation of phytoplankton dark losses by application of ^{14}C-technique. Arch. Hydrobiol., Beih. Ergebn. Limnol. **30**.

— & BEHRENDT, H. (1990): Discrepancies between O_2- and ^{14}C-method resulted from measurements of primary production under consideration of vertical mixing in a shallow eutrophic lake. Verh. Int. Verein. Limnol. **24** (in press)

— & STELLMACHER, R. (1987): Long-term variations of primary production in a eutrophic shallow lake with special regard to photosynthetic efficiency of blue-green algae. Acta hydrophys. **31** (3/4), 181—190.

PAERL, H. W., & USTACH, J. T. (1982): Blue-green algal scums: An explanation for their occurrence during freshwater blooms. Limnol. Oceanogr. **27** (2), 212—217.

PATTEN, B. C. (1968): Mathematical models of plankton production. Int. Rev. ges. Hydrobiol. **53**, 357—408.

ROMANENKO, W. J. (1975): Zusammenhang zwischen der Photosyntheseintensität bei gleichmäßiger Algenentwicklung im Wasserkörper und der Durchsichtigkeit nach Secchi. Biol. Binnengew. Inf. Bull. **19**, 15—18 (in Russ.).

STELLMACHER, R., & NIXDORF, B. (1984): Ein Beitrag zur Anwendung der Regressionsanalyse im Zusammenhang mit der Schätzung der Primärproduktion im Großen Müggelsee (Berlin) für den Zeitraum 1979 . . . 1982. Acta hydrochim. et hydrobiol. **12** (6), 653—657.

YODER, J. A., & BISHOP, S. S. (1985): Effects of mixing-induced irradiance fluctuations on photosynthesis of natural assemblages of coastal phytoplankton. Mar. Biol. **90**, 87—93.

Address of the authors:

Dr. BRIGITTE NIXDORF
Dr. HORST BEHRENDT
Dr. RITA STELLMACHER
AdW der DDR
Institut für Geographie und Geoökologie
Bereich Hydrologie
Müggelseedamm 260
Berlin
DDR-1162

Limnologica (Berlin), **20** (1), 57—60, April 1990

Wilhelm-Pieck-Universität Rostock, Sektion Biologie, Rostock, DDR

Basis of Salt Adaptation in Cyanobacteria

By N. Erdmann, M. Hagemann, C. Berg & S. Fulda

With 5 Figures

Key words: Dark; Excretion; Glucosylglycerol; Light; Salt adaptation.

Abstract

A hypothesis regarding the course of salt adaptation and metabolism of glucosylglycerol in cyanobacteria is represented. This model includes the different behaviour of salt-shocked cells under light and dark conditions and some features of the synthesis of glucosylglycerol, its regulation, and its release into the surrounding medium after a hypoosmotic shock.

1. Introduction

Organisms living in estuaries are confronted with fluctuating salinities. For survival the process of salt adaptation gets fundamental importance and, according to Kirst (1979), seems to have priority over all other vital processes.

When cells are shocked by higher salinities, water loss and uncontrolled uptake of ions are the first responses. It happens within a few minutes until a new osmotic parity is reached. This new situation is unfavourable for the whole metabolism and triggers the adaptation process. It includes mainly:

a) defence of the enhanced inflow of ions which have poisonous effects on metabolism and membranes,
b) restoration of the non-stress volume and turgor,
c) avoidance of the deactivation of biopolymers caused by the lowered water potential.

The main, but not the most economic strategy of becoming salt adapted consists in pumping out the additional ions and substitute them by organic solutes. They are accumulated in appreciable amounts and are called compatible solutes because they replace ions with respect to their osmotic potential. But they have an additional function, too: based on their excellent hydrophilic nature they protect macromolecules like enzymes against inactivation by the lowered water potential. Both of these functions are paraphrased by the name osmoprotective compound. At present about 15 osmoprotective compounds are known. Cyanobacteria of low salt tolerance accumulate mainly sucrose or trehalose, species of intermediate tolerance — glucosylglycerol (GG), high salt-tolerant species — glycinebetaine (Mackay et al. 1984; Reed et al. 1984).

Microcystis firma and *Synechocystis* PCC 6714 belong to the intermediate group and have been choosen as model organisms to understand the course of salt adaptation by studying the biosynthetic route of GG, its severe accumulation under stress, and disappearance after salt discharge in both light and dark.

2. Material and Methods

The methods are described in Hagemann et al. (1987) and Erdmann et al. (1989).

Abbreviations: GG — glucosylglycerol; G6PDH — glucose-6-phosphate dehydrogenase.

3. Results and Discussion

3.1. Salt adaptation in the light

Carbon dioxide fixed during photosynthesis is the precursor of GG in the light (Fig. 1). UDP-glucose and glycerol-1-phosphate are important intermediates along this route. For the increased synthesis of glycerol-1-phosphate additional reducing power is needed. It may be delivered by the pentose phosphate pathway which is activated under the conditions of a salt stress. G6PDH, the key enzyme of this pathway, shows enhanced activity in dependence of raising NaCl concentrations up to 500 mM (Fig. 2). The higher activity of G6PDH may also be necessary for the active exclusion of Na^+ via cytochrome oxidase of the respiratory chain.

In comparison to the biosynthetic pathway of the related osmoprotective compound galactosylglycerol (=isofloridoside, Kauss 1979) it seems likely that the two intermediates UDP-glucose and glycerol-1-phosphate were combined to GG phosphate. After dephosphorylation GG is accumulated in the cells according to the level of salt stress (Fig. 2). In adapted cells a linear relationship exists between NaCl concentration and GG content per dry weight. At the highest NaCl concentration tested (771 mM) more than 20% GG per dry weight were found.

The synthase reaction (Fig. 1) is probably the rate-limiting step along the whole pathway. Non-stressed cells do not accumulate GG, but its synthesis starts immediately after a salt shock. No lag phase is obvious indicating that no de-novo synthesis of the enzyme is necessary. With the exception of proline this is common for all other osmoprotective compounds, too. Furthermore, the initial rate of GG synthesis, that means the GG amount synthesized per time during the first 3 hours after salt shock, is independent of the NaCl concentration and is of the highest order at all (about 60 nmol GG per hour and per mm^3 total cell volume — Fig. 3).

On the contrary adapted cells show a different behaviour. In these cells the GG concentration is constant during time (e.g. 0.76 mol per litre of cell volume at 684 mM NaCl) and no turnover of GG was found (Hagemann et al. 1987). Therefore the biosynthetic apparatus has to supply GG only in such an amount as the cells do growth. This is the continuous rate of GG synthesis which clearly depends on the NaCl concentration (Fig. 3).

The different behaviour of the rate of GG synthesis could be explained by the assumption that, under fresh-

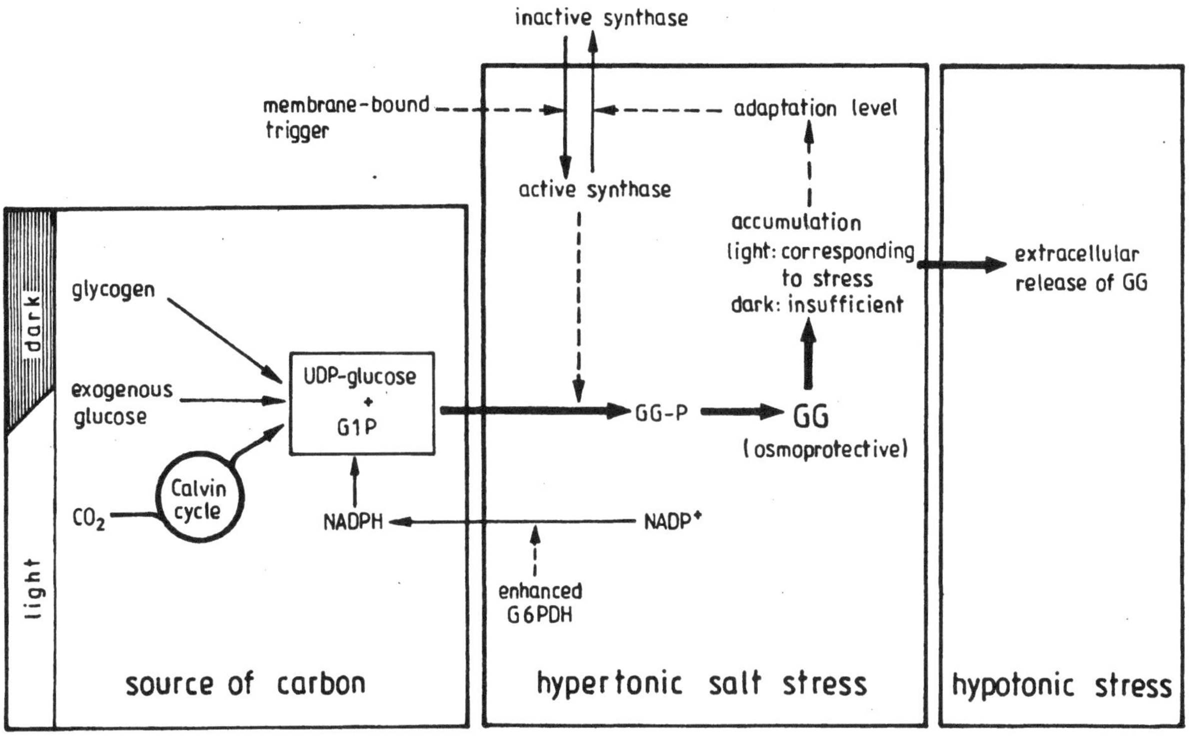

Fig. 1. Main reactions and adaptation behaviour of glucosylglycerol-accumulating cyanobacteria after hyper- and hypo-osmotic shock. (➡ main reactions, → additional reactions, ---→ catalytic or control activities; GG — glucosylglycerol, GG-P — glucosylglycerol phosphate, G1P — glycerol-l-phosphate, G6PDH — glucose-6-phosphate dehydrogenase)

water conditions, the GG phosphate synthase is present in a preformed, inactive stage; a salt shock activates a trigger which may be situated in or at the cytoplasmic membrane and transforms the whole enzyme pool into its active stage. Therefore accumulation of GG starts with its highest speed. When the GG content of adapted cells is reached, the active synthase is reversed partly into its inactive form probably under the control of osmotic equilibrium between inside and outside of the cells. This concept includes that, under freshwater conditions, this inactive enzyme represents a kind of metabolic ballast. But after a salt shock these cells are able to start immediately the adaptation without any lag. Another idea is that, under freshwater conditions, this enzyme has a different function (or catalytic activity) which is changed after an osmotic shock.

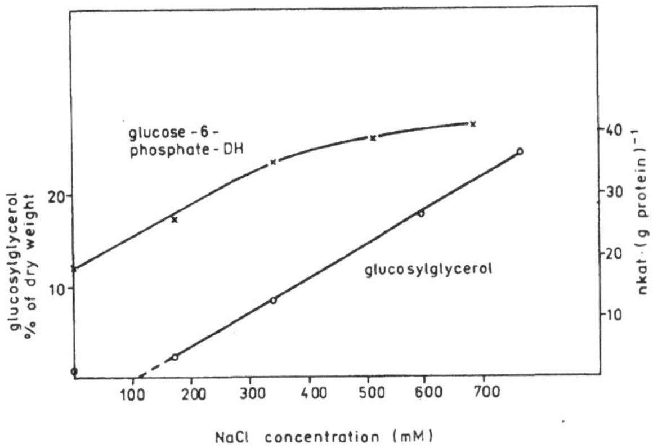

Fig. 2. Accumulation of glucosylglycerol and activity of glucose-6-phosphate dehydrogenase of salt-adapted cells of *M. firma*.
(Salt adaptation was performed by incubating cells in media with appropriate NaCl concentrations for 7 days)

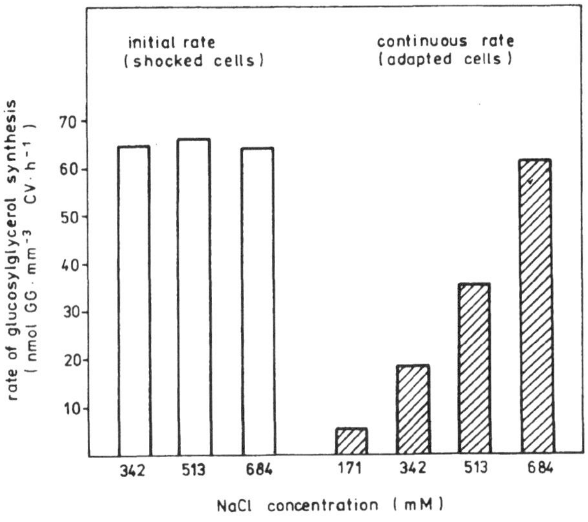

Fig. 3. Rates of synthesis of glucosylglycerol in shocked and adapted cells of *M. firma*.
[Initial rates represent average amounts of glucosylglycerol per hour and per total cell volume (CV) after 1, 2, 3 hours of stress; continuous rates were calculated from amounts of glucosylglycerol and doubling time of adapted cells]

3.2. Missing salt adaptation in the dark

Recently SCHIEWER et al. (1978) have shown that *M. firma* is unable to adapt to higher salt concentrations in the dark. Therefore we looked for the capacity of darkened cells of *M. firma* to produce GG. As in the light darkened cells do not synthesize GG under freshwater conditions; in Fig. 4 small amounts of GG are even indicated, but this is not GG because its position is unoccupied on the

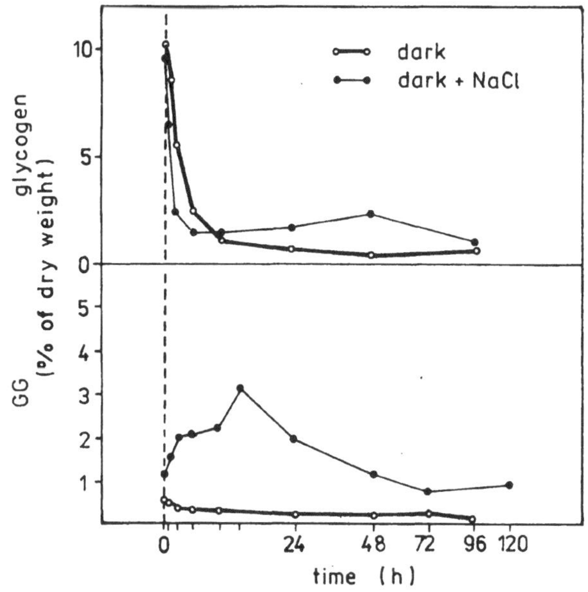

Fig. 4. Effect of NaCl (771 mM) on glycogen consumption and glucosylglycerol accumulation in the dark. (The first strokes of the time scale represent 0.5, 1.5, 3.5, and 7.5 hours; zero time indicates transfer of cells into dark and dark + NaCl, resp.)

appropriate thin-layer chromatogramms (not shown here). When the cells are submitted to a simultaneous salt and dark stress they start immediately with the GG synthesis. But it durates only a few hours and ceases thereafter (Fig. 4). The reason for the cease of the GG synthesis is the early depletion of the glycogen pool which is the only carbon source available in the dark. Glycogen is also rapidly consumed in non-stressed cells in the dark (90% during the first 8 hours). This seems to be a common feature of cyanobacteria, e.g. free-living cells of *M. aeruginosa* completely mobilize its glycogen pool in the dark within 48 hours (REYNOLDS et al. 1981).

The NaCl-triggered GG accumulation in the dark amounts to about 3% of dry weight what is insufficient for the protective and osmotic function, in comparison to the high amounts accumulated in the light at the same NaCl-concentration (see Fig. 2). Furthermore, after depletion of the glycogen pool GG seems to substitute for glycogen as an energy source and becomes respired. The protection level already attained gets lost. Additionally disappearing GG has to be replaced by toxic ions to effect osmotic balance. As a consequence, both processes cause the cell metabolism to abort.

Another possibility to prove GG synthesis in the dark is given by feeding ^{14}C-labelled glucose (for this experiment the cyanobacterium *Synechocystis* PCC 6714 was used). In the presence of NaCl labelled GG is synthesized but, in comparison to the same experiment in the light, GG synthesis does not dominate above all other metabolic pathways (ERDMANN et al. 1988).

Thus the inability of *M. firma* (cyanobacteria at all?) to adapt in the dark is caused by the inability of the cells to allocate to the GG synthesis the highest priority and, therefore, by the limited glycogen pool.

3.3. Hypotonic stress

When salt-adapted cells are transferred into freshwater the high intracellular osmotic potential causes water inflow which increases cell volume and turgor. Again

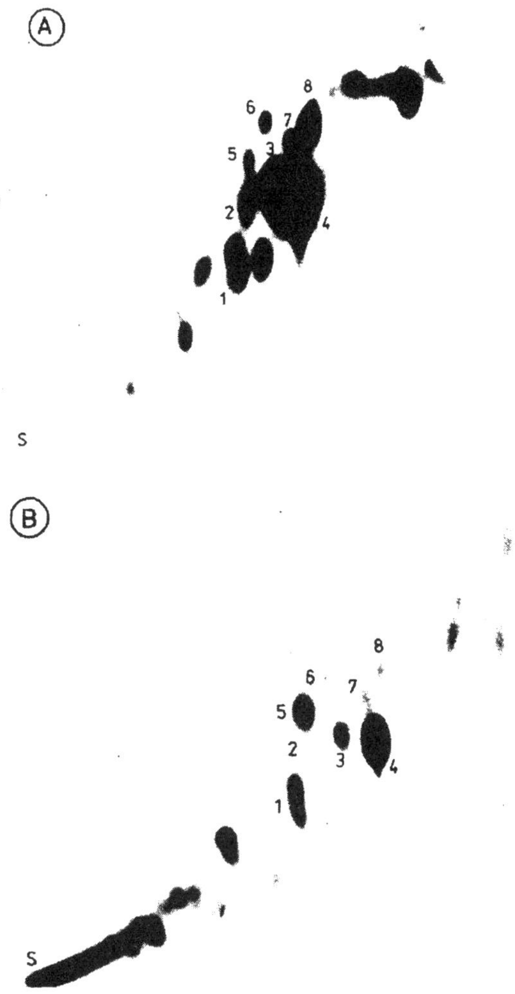

Fig. 5. Metabolite pattern of the external medium (A) and cell interiour (B) of *Synechocystis* PCC 6714 after transferring salt-adapted cells into salt-free medium.
[Cells were preloaded with ^{14}C; estimations were done 3 minutes after transfer by concentration (A)/extraction (B), thin-layer chromatography, and autoradiography; 1 = aspartic acid/sucrose, 2 = glycine/serine, 3 = glutamic acid, 4 = glucosylglycerol, 5 = glutamine, 6 = citrulline, 7 = threonine, 8 = alanine, S = start]

an adaptation process is initiated. It has mainly to execute the removal of all GG accumulated. This should occur by remetabolization to save fixed carbon. But this way is realized in a minor extent only. As it is demonstrated in Fig. 5 the bulk of GG is excreted into the medium. Fig. 5a shows the spectrum of labelled metabolites present after 3 minutes in the medium when salt-adapted, ^{14}C-loaded cells of *Synechocystis* PCC 6714 have been transferred from NaCl-containing (684 mM) into NaCl-free medium. This treatment did not cause the cells to burst because no sign of free phycocyanin could be observed in the supernatant after centrifugation.

Under the exudates GG dominates strikingly. Fig. 5b shows the spectrum of metabolites which have been remained in the cells; GG is present, too, but it is now in the concentration range of the other metabolites.

A comparison of both pictures reveals that the excretion concerns all low-molecular compounds. This is in accordance with the view of REED et al. (1985) that, after a hypotonic shock of cyanobacteria, an unspecific excretion takes place caused by a transient breakdown of the selective permeability of the cytoplasmic membrane.

3.4. Conclusions

It is obvious that *M. firma* and other cyanobacteria are well able to survive fluctuating salinities by the synthesis and removal of GG as the main adaptation prerequisite. But from the point of economy some shortcomings are involved: accumulation of an organic compound for osmotic parity is more expensive than ion accumulation (RAVEN 1985); it may be that there is a bulk of ballast protein under freshwater conditions; the strategy is non-functioning for dark periods exceeding 10 hours; the cells exhibit a through-away behaviour in the case of down stress. But these shortcomings should not be overestimated since in nature transitions are generally smooth and the adaptation process is challenged rarely with its full capacity.

4. Acknowledgement

The authors are greatly indepted to Chem.-Ing. K. SOMMEREY and Mrs. H. MENZEL for technical assistance.

References

ERDMANN, N., BERG, C., & HAGEMANN, M. (1989): Missing salt adaptation of *Microcystis firma* (Cyanobacterium) in the dark. Arch. Hydrobiol. **114**, 521—530.

HAGEMANN, M., ERDMANN, N., & WITTENBURG, E. (1987): Synthesis of glucosylglycerol in salt-stressed cells of the cyanobacterium *Microcystis firma*. Arch. Microbiol. **148**, 275—279.

KAUSS, H. (1979): Biochemie der osmotischen Regulation bei *Poterioochromonas malhamensis*. Ber. Deutsch. Bot. Ges. **92**, 11—22.

KIRST, G. O. (1979): Osmotische Adaptation der marinen Planktonalge *Platymonas subcordiformis*. Ber. Deutsch. Bot. Ges. **92**, 31—42.

MACKAY, M. A., NORTON, R. S., & BOROWITZKA, L. J. (1984): Organic osmoregulatory solutes in cyanobacteria. J. Gen. Microbiol. **130**, 2177—2191.

RAVEN, J. A. (1985): Regulation of pH and generation of osmolarity in vascular plants: a cost-benefit analysis in relation to efficiency of use of energy, nitrogen, and water. New. Phytol. **101**, 25—77.

REED, R. H., RICHARDSON, D. L., WARR, S. R. C., & STEWART, W. D. P. (1984): Carbohydrate accumulation and osmotic stress in cyanobacteria. J. Gen. Microbiol. 130, 1—4.

— RICHARDSON, D. L., & STEWART, W. D. P. (1985): Na^+ uptake and extrusion in the cyanobacterium *Synechocystis* PCC 6714 in response to hypersaline treatment. Evidence for transient changes in plasmalemma Na^+ permeability. Biochim. Biophys. Acta **814**, 347—355.

REYNOLDS, C. S., JAWORSKI, G. H. M., CMIECH, H. A., & LEEDALE, G. F. (1981): On the annual cycle of the blue-green alga *Microcystis aeruginosa* KÜTZ. emend ELENKIN. Phil. Trans. Roy. Soc. London, B, Biol. Sci. **293**, 419—477.

SCHIEWER, U., ERDMANN, N., & KUHNKE, K.-H. (1978): NaCl-Wirkung auf die Photosyntheseintensität von Blaualgen. Die Wirkung unterschiedlicher NaCl-Konzentrationen auf die Photosyntheseintensität der Blaualgen *Microcystis firma* und *Synechocystis aquatilis*. Biochem. Physiol. Pflanzen **172**, 351—368.

Address of the author:
Dr. sc. N. ERDMANN
Dr. M. HAGEMANN
C. BERG
S. FULDA
Wilhelm-Pieck-Universität Rostock
Sektion Biologie
Doberaner Str. 143
Rostock
DDR-2500

Limnologica (Berlin) **20** (1), 61, April 1990

Wilhelm-Pieck-Universität Rostock, Sektion Biologie, Rostock, DDR

Ecological, Physiological and Historical Aspects of the Brackish Water Fauna Distribution

By Ernst Albert Arndt

Abstract

The distribution pattern of marine euryhaline and genuine brackish water animals in the estuaries and perennial brackish waters (e.g. Baltic Sea) of boreal biotopes is described.

Colonization patterns in tideless eutrophic to hypertrophic internal coastal waters between meso- and oligohaline regions (= horohaline region) are used in particular to illustrate different ways of genetic adaptation as a prerequisite for the successful colonization of this extreme biotope.

Cordylophora caspia, *Neomysis integer*, *Hediste (Nereis) diversicolor* and the "small polychaetes" *Alkmaria romijni*, *Streblospio shrubsoli*, *Manayunkia aestuarina* and *Fabricia sabella* are used as examples to show that causal analysis of the colonization pattern can be based only on the simultaneous consideration of physiological, ecological and historical data. It is scarcely possible to formulate generalizable rules concerning some specific "key" factors. Our present knowledge of the example species mentioned above permits to draw conclusions regarding the historical background of the different pathaways leading to the colonization of this extreme biotope and, thus, to be biogeography of those species.

Address of the author:

Prof. Dr. Ernst Albert Arndt
Wilhelm-Pieck-Universität Rostock
Sektion Biologie
Freiligrathstr. 7/8
Rostock
DDR-2500

Das vollständige Manuskript ist für eine Publikation in der Zeitschrift „Marine Ecology" PROGRESS SERIES vorgesehen.

Limnologica (Berlin) **20** (1), 63, April 1990

Wilhelm-Pieck-Universität Rostock, Sektion Biologie, Rostock, DDR

Energy Requirement of Some Brackish Water Invertebrates

By J.-A. VON OERTZEN & E.-M. SCHARF

Abstract

Reliable data concerning ecologically relevant energy requirements of the most important species of an ecosystem are indispensible for understanding the complex energy flow.

A collection of oxygen consumption values — all from our laboratory — for some micro-, meio- and macrofauna species of the Baltic are intended to contribute to our stock of such data.

The validity of regression lines for weight specific respiration rates is discussed versus representative single oxygen consumption data.

Specific brackish water problems such as the influence of salinity and oxygen depletion on oxygen uptake are illustrated for two selected species.

Despite open questions regarding certain contradictions between laboratory data and ecosystem values, a prime notation sheet for energy requirement data of Baltic marine species — in continuation of the suggestion by WIESER & KANWISHER (1960) — will be proposed, after more thorough discussion among specialists of the "Baltic Marine Biologist's" (BMB).

References

WIESER, W., & KANWISHER, J. (1960): Vorschlag zum Anlegen einer „Stoffwechselkartei" für marine Tiergemeinschaften. Zool. Anz. **165**, 370—375.

Address of the authors:
Dr. J.-A. VON OERTZEN
Dr. E.-M. SCHARF
Wilhelm-Pieck-Universität Rostock
Sektion Biologie
Freiligrathstr. 7/8
Rostock
DDR-2500

Limnologica (Berlin) **20** (1), 65—69, April 1990

Sektion Biologie der Wilhelm-Pieck-Universität Rostock, Rostock, DDR

Secondary Production of Calanoids (Copepoda, Crustacea) in Brackish Waters of the Southern Baltic

By Reinhard Heerkloss, Ulrich Brenning & Mathias Ring

With 6 Figures and one Table

Key words: Secondary production; Calanoids; Brackish water; Feeding rate; Tracer methods.

Abstract

In shallow brackish waters of the Baltic, calanoids of the genus *Eurytemora* and *Acartia* are the main components of the zooplankton during spring and early summer. Biomass and production data are presented for two waters differing in productivity. To estimate secondary production, two methods (a volumetric and a tracer method) were used and compared.

1. Introduction

Calanoid copepods are an important component of the coastal zooplankton of the southern Baltic. Monitoring data from different stations on the GDR's coast show that they dominate during spring and early summer. Because of their importance as food for fish larvae, their production is of interest in connection with fisheries.

The aim of this paper is to compare calanoid productivity in two waters, the Darss-Zingst estuary and Greifswald Bay (Fig. 1). Biomass and production data will be presented. It simultaneously has a methodological aim, however: beside using the classical method of calculating productivity from the potential growth of all nauplius and copepodite stages, production was also estimated by a tracer technique.

Characteristics of the waters investigated:
The Darss-Zingst estuary is a chain of lake-like waters south of Darss-Zingst Peninsula. The salinity increases from about $1^0/_{00}$ (sometimes fresh water) in the inner part to about $10^0/_{00}$ at the eastern end, where it has a narrow connection to the Baltic Sea. The estuary is polytrophic. The pH-value increases to 9.5 in summer. Green and bluegreen algae are the principal phytoplankton components (Börner 1984). Greifswald Bay is more open to the Baltic Sea and less influenced by fresh water inflows. Its salinity fluctuates between 7 and $8^0/_{00}$. As a rule, its phytoplankton consists of diatoms in spring and bluegreens in summer (Kell 1986). In both waters calanoid copepods account for almost the whole zooplankton biomass in spring and early summer. In Darss-Zingst estuary only the genuine brackish water species *Eurytemora affinis* (Poppe) is present during this time, but in Greifswald Bay *Acartia* species (mainly *Acartia bifilosa* Giesbrecht) are found in addition. There is a correlation between the relative abundance of *Eurytemora* and *Acartia* and the quality of phytoplankton. If *Achnanthes taeniata* dominates in the algal spring bloom, *E. affinis* will be the main calanoid species, whereas *Acartia* spp. are numerous if *Sceletonema costatum* dominates the spring bloom (Brenning & Kell 1985).

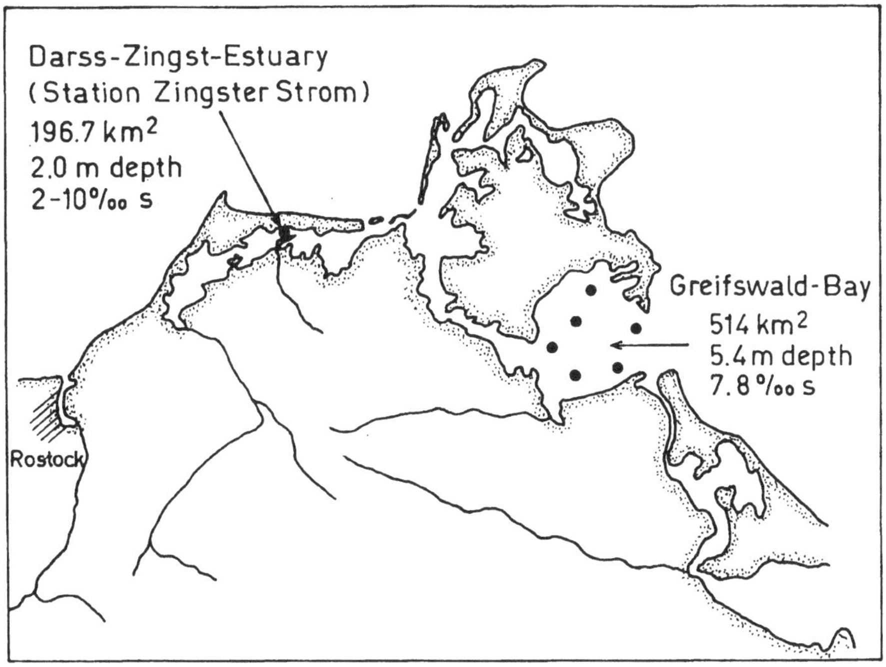

Fig. 1. Area of investigation

2. Methods

Sampling:
Zooplankton was sampled by standard methods (5 l-sampler, Hydrobios). The results from different stations in Greifswald Bay were averaged. In Darss-Zingst estuary the sampling point was in Zingster Strom about half-way along the chain (Fig. 1). The counts were performed with subsamples from a 50 l integrated sample. Production was estimated in two different ways.

1. The instantaneous production rate was calculated from the potential growth of all stages using CORCETT & MCLAREN's (1970) formula to correct for temperature influence.
2. The biomass values for each sample were multiplied by a coefficient for the weight specific feeding rate of adults (determined *in situ* with a ^{14}C-tracer method), a coefficient of assimilation efficiency (0.65 according to experimental tracer studies with ^{14}C and ^{51}Cr, Fig. 2) and a coefficient of production efficiency of

$K_2 = \frac{P}{A} = 0.4$ (VINOGRADOV & MENŠUTKIN 1977).

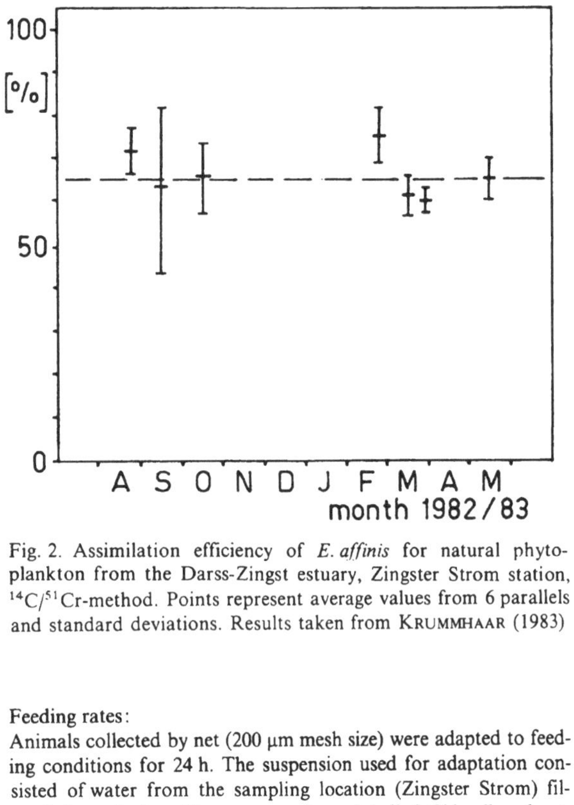

Fig. 2. Assimilation efficiency of *E. affinis* for natural phytoplankton from the Darss-Zingst estuary, Zingster Strom station, $^{14}C/^{51}Cr$-method. Points represent average values from 6 parallels and standard deviations. Results taken from KRUMMHAAR (1983)

Feeding rates:
Animals collected by net (200 μm mesh size) were adapted to feeding conditions for 24 h. The suspension used for adaptation consisted of water from the sampling location (Zingster Strom) filtered through glass fibre paper and non-labelled *Chlorella vulgaris* (10^6 cells/ml). 800—2000 individuals were adapted *in situ* in a 2000 ml container. To estimate the feeding rate, adapted animals were gently transferred by net into a suspension of ^{14}C-labelled *Chlorella* of the same concentration for 30 min. This incubation period is shorter than the gut passage time. To wash the animals they were transferred back to the suspension of non-labelled algae for 10 min and then killed in a 1% chloroform-in-water mixture. We killed the samples in this way because chloroform cleans the copepods of sessile ciliates which are abundant in the Darss-Zingst estuary (RING et al. 1986). After storage for 24 h, 20 animals (10 females and 10 males) were placed in an LSC counting vial and digested in NCS-tissue solubalizer. The radioactivity of this sample was then estimated by LSC-counting using the channel ratio method for quench correction. The animals lose a considerable amount of radioactivity during the first hours after fixation. Their radiation levels become stable only after 20 to 25 h of storage, when the measured level is 25 to 30% of that measured immediately after fixation. The dpm per animal where therefore corrected for these losses using the formula

$$y = 24.54 + 57.73\,x$$

where y = remaining activity (%) and x = dpm/cell of *Chlorella*. The relation between radioactivity and dry weight was used to calculate the weight specific feeding rate. The dry weight of the algae was determined on weighted membrane filters (Synpor 6). For the animals it was calculated from the body length according to the equation

$$\lg W = 2.33 \lg L - 6.12$$

(W = dry weight in μg; L = length of the cephalothorax in μm).

Estimation of the assimilation efficiency (according to CALOW & FLETSCHER 1972): Water from the natural location (Zingster Strom) was labelled either with ^{14}C-$NaHCO_3$ (3.7 MBq per litre) or ^{51}Cr-Na_2CrO_4 (37 MBq per litre) for 2 days under artificial light (3000—4000 Lux) while being agitated. The suspended matter was then extracted from the dissolved fraction by zentrifugation for 10 minutes (2000 rpm) and resuspended in diluted sea water with a salinity of 5‰. The radioactivity of each fraction was measured and the fractions were then mixed to obtain an isotope ratio of 1:1. After adaptation to the experimental temperature and salinity for 24 h (10 °C, 5‰), 30 to 60 animals were transferred by net to the experimental vessel containing 100 ml 30 μm-filtered suspension of double labelled food. After incubation for 3 h the animals were given a non-labelled yeast suspension (about 10^6 cells/ml, 30 μm-filtered) for 2 h. Fecal pellets were collected from both the algal and the yeast suspension by filtration through a 50 μm net, washed in 5‰-water, transferred to membrane filters and dried for 1 h at 60 °C. Radioactivity levels were measured with a Geiger-Müller-counter for ^{14}C and a NaJ-crystal scintillation counter for ^{51}Cr. The counts of the latter were corrected for the overlapping fraction of ^{14}C-radioactivity, which was about 2% of the cpm determined by the Geiger-Müller-counter. The radioactivity of the food was measured in aliquots on dried membrane filters at the beginning and the end of incubation. The efficiency of assimilation was calculated according to the formula

$$AE\,(\%) = 1 - \frac{A \cdot 0.9523 \cdot 100}{B}$$

A = relation of radioactivity ^{14}C:^{51}Cr of the fecal pellets; B = relation of radioactivity ^{14}C:^{51}Cr of the food; 0.9523 = constant to correct for the efficiency of ^{51}Cr assimilation (according to experiments performed separately = 5%).

3. Results and Discussion

The annual curve for zooplankton in the Darss-Zingst estuary is shown in Fig. 3. There was a pronounced increase in *E. affinis* during May, but almost the whole of this population had already disappeared in June. A second calanoid peak in late summer consisted of *Acartia tonsa* DANA, besides *E. affinis*.

In view of the quantitative and qualitative stability of the phytoplankton during the rise and decline of the copepods, the collapse of the population in June cannot have been caused by food shortage. Probably it was a consequency of high pH and temperature. In laboratory experiments RING (1987) demonstrated that feeding activity varies inversely with temperature between 10 and 20 °C and with pH at values between 8 and 10.

In Greifswald Bay the spring biomass values were only one fifth as high as those in the Darss-Zingst estuary (Fig. 4). *E. affinis* and *Acartia* sp.. appeared concurrently. This quantitative difference in zooplankton reflects the fact that spring primary production was comparatively low in Greifswald Bay.

Furthermore, the calanoid biomass reached its peak a month later than in the Darss-Zingst estuary. This may have been due to food shortage, because phytoplankton biomass figures were very low in mid-May. At this time the spring diatom peak had already declined, but the summer phytoplankton bloom had not yet occurred.

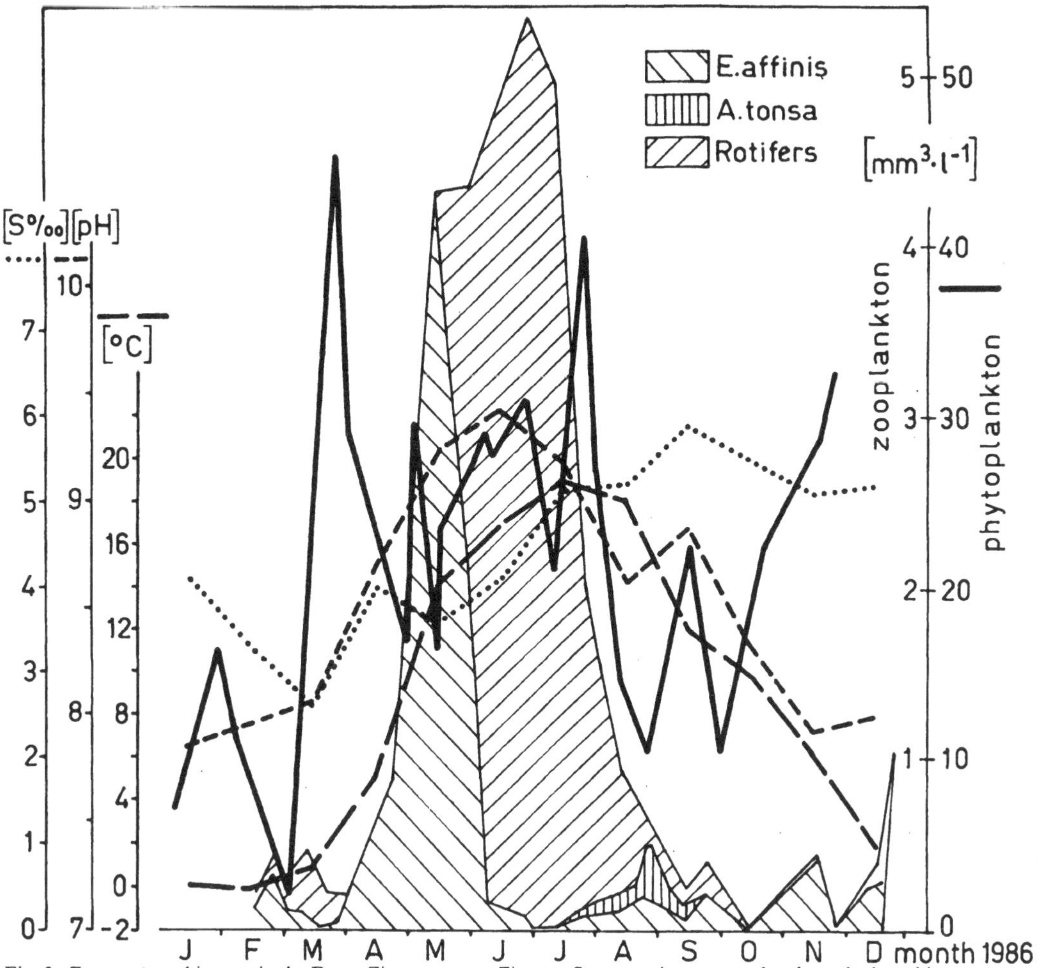

Fig. 3. Zooplankton biomass in the Darss-Zingst estuary, Zingster Strom station, versus the phytoplankton biomass, temperature, pH-value and salinity. Phytoplankton biomass from WASMUND (unpubl.)

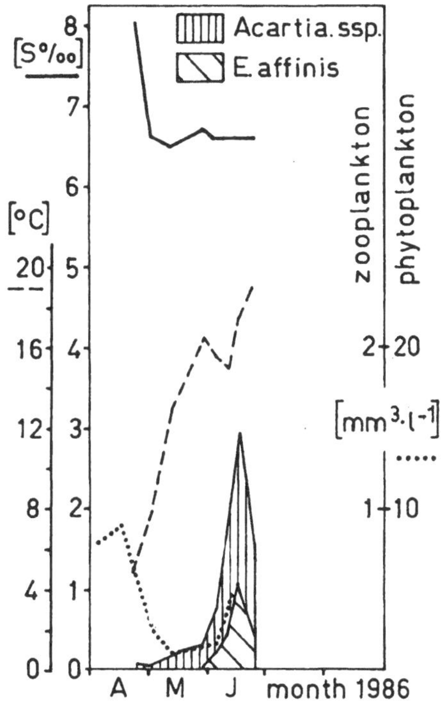

Fig. 4. Zooplankton biomass in Greifswald Bay from April to June versus phytoplankton biomass, temperature and salinity. The biomasses of zooplankters other than calanoids were very low and therefore have been ignored. Phytoplankton bio mass from KELL (unpubl.)

Primary production during this period was only about equal to calanoid secondary production when it reached its maximum level a month later (Table 1).

When *E. affinis* reached its biomass peak in Greifswald Bay in mid-June, the population in the Darss-Zingst estuary had already collapsed. Also, the sudden decline in population size observed almost every year in the Darss-Zingst estuary does not occur in Greifswald Bay. Data for the whole annual cycle in 1977, 1981 and 1984

Table 1. Daily calanoid production during the period of maximal biomass development. Figures are means for the given periods. Primary production figures were not measured directly, but were calculated from biomasses using P/B-coefficients typical of the periods and waters: Greifswald Bay: 1 mm³ ≙ 200 mgC · m⁻² · d⁻¹ (BRENNING & KELL 1985), Darss-Zingst estuary: 1 mm³ ≙ 105 mgC · m⁻² · d⁻¹ (HEERKLOSS et al. 1984). Factors used to convert biovolume to carbon according to HEERKLOSS & VIETINGHOFF (1981). A depth of 5.4 m (Greifswald Bay) and 1.8 m (Darss-Zingst estuary) is used to convert figures calculated per volume to figures per m².

	Period	Production (mgC · m⁻² · d⁻¹)		sp/pp
		Phytoplankton (pp)	Calanoids (sp)	
Greifswald Bay	1.–12. 5. 86	106	2.0	1:40.8
	10.–24. 6. 86	478	78.4	1: 6.1
Darss-Zingst estuary	9.–14. 5. 86	2515	117.3	1:21.4

show that *E. affinis* is present thoughout the season in Greifswald Bay. If productivity during the time of peak biomass is taken into account, calanoid production in the Darss-Zingst estuary is only about 1.5 times higher than in Greifwald Bay (Table 1). Primary production however is higher by a factor of 5. The food chain is therefore more efficient in Greifswald Bay. The calanoid production figures yielded by the tracer technique for the period of their spring peak are considerably higher than those obtained with the volumetric method. RING's (1987) weight-specific feeding rates for *E. affinis* give a curve similar to that for annual biomass (Fig. 5). A pronounced activity peak occurs in early May. It then drops again to very low values, but further peaks appear in June and August. Fig. 6 shows, that the tracer method yielded lower daily production figures than the volumetric method for almost all sampling dates. This situation was reversed only when the spring biomass peak reached its zenith in May. However, since production during May accounts for the main part of annual production, the annual production figure calculated by the tracer method was almost three times as high as that obtained with the volumetric method. At present it is difficult to decide which of the two methods yields the more realistic results. On the other hand, a calanoid production amounting to 3.2 or 9.1% of the annual primary production seems reasonable.

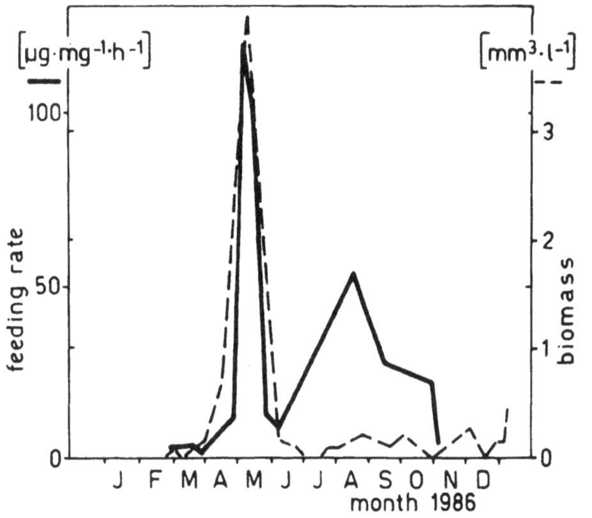

Fig. 5. Feeding rate per unit dry weight of *E. affinis* compared to its annual biomass curve. Darss-Zingst estuary, Zingster Strom station

As a rule, annual calanoid production in neritic and estuarine waters does not exceed 1 g C $\cdot$ m^{-2} $\cdot$ y^{-1}. The results for Greifswald Bay thus seem to be within the range of neritic productivity. In contrast, the high production values obtained for the Darss-Zingst estuary lie outside of the normal range and are only comparable to calanoid production in some highly productive lakes. For example, BOSSELMANN (1975) reported an annual calanoid production of 5 g C $\cdot$ m^{-2} $\cdot$ y^{-1} in Lake Esrom. Therefore, semi-enclosed estuaries with only a narrow connection to the open sea are probably more similar to lakes in terms of calanoid production. Despite their shallowness and low food chain efficiency, their potential for secondary production is considerable.

4. Zusammenfassung

Die Calaniden bilden einen wesentlichen Teil der Zooplanktonbiomasse in Boddengewässern der südlichen Ostsee. Ein Vergleich zwischen zwei morphometrisch verschiedenen Abschnitten, dem Zingster Strom und dem Greifswalder Bodden, zeigt charakteristische Unterschiede in der Artensukzession während des Frühjahrs. Während im Zingster Strom der Frühjahrs-peak in der Bio-

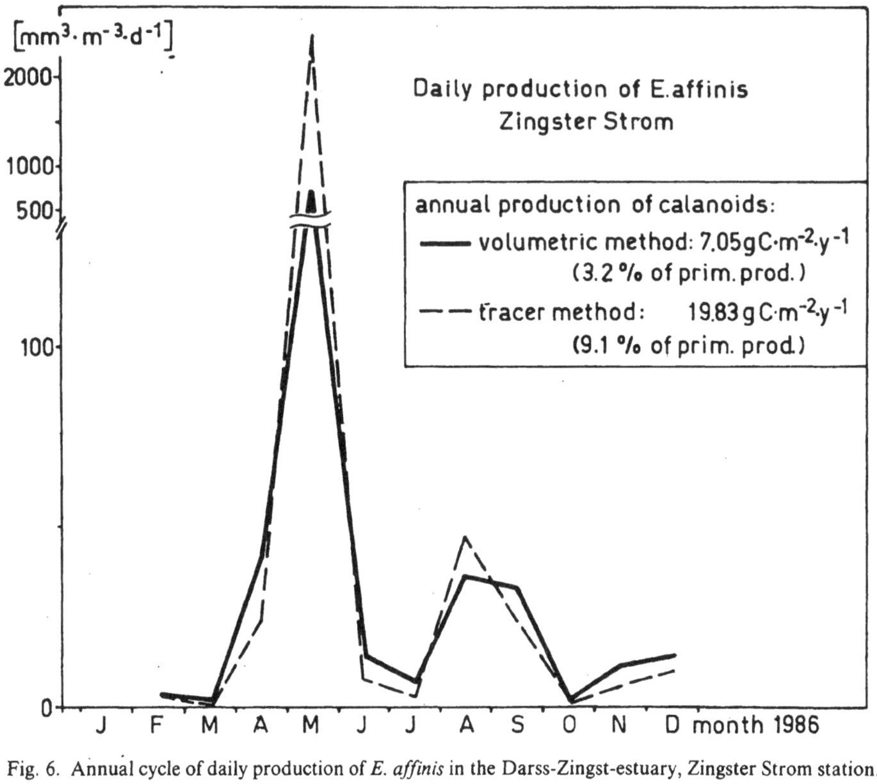

Fig. 6. Annual cycle of daily production of *E. affinis* in the Darss-Zingst-estuary, Zingster Strom station. Comparison between volumetric and tracer methods

masse fast ausschließlich durch *E. affinis* gebildet wird, kommen im Greifswalder Bodden neben dieser Art auch Vertreter der Gattung *Acartia* vor. Ihren Höhepunkt erreicht die Calanidenentwicklung im Greifswalder Bodden — bedingt durch einen Mangel im Nahrungsangebot des Phytoplanktons — einen Monat später als im Zingster Strom. Die Maximalwerte der Calaniden-Biomasse betragen 4,3 bzw. 1,5 mm^3/l. Mit Hilfe einer volumetrischen und einer Tracer-Methode wurde die Sekundärproduktion berechnet. Die Tracer-Methode lieferte dreifach höhere Werte. Das Verhältnis zwischen der Primärproduktion und der Sekundärproduktion der Calaniden weist darauf hin, daß das Nahrungsangebot im polytrophen Zingster Strom weniger effektiv ausgenutzt wird als im nur mäßig eutrophierten Greifswalder Bodden.

References

BÖRNER, R. (1984): Produktionsbiologisch-ökologische Untersuchungen am Phytoplankton des Zingster Stromes. Diss., Sektion Biologie, W.-P. Univ. Rostock.

BOSSELMANN, S. (1975): Production of *Eudiaptomus graciloides* in Lake Esrom 1970. Arch. Hydrobiol. **76**, 43—64.

BRENNING, U., & KELL, V. (1985): Abschlußbericht über Analyse der Phyto- und Zooplanktonverbreitung und deren Dynamik in Heringsaufwuchsgebieten und ihr Einfluß auf die Reproduktion von Nutzfischen. Forschungsber., Sektion Biologie, W.-P.-Univ. Rostock.

CALOW, P., & FLETCHER, C. R. (1972): A new radiotracer technique involving 14-C and 51-Cr for estimating the assimilation efficiencies of aquatic primary consumers. Oecologia (Berlin) **9**, 155—170.

CORCETT, C. J., & MCLAREN, J. A. (1970): Relationships between developmental rate of eggs and older stages of copepods. J. Mar. Biol. Ass. U. K. **50**, 161—168.

HEERKLOSS, R., & VIETINGHOFF, U. (1981): Biomasseäquivalente planktischer und benthischer Organismen in der Darss-Zingster Boddenkette. Wiss. Z. Univ. Rostock, Math.-nat. R. **30** (4/5), 31—36.

— ARNDT, H., HELLWIG, J., VIETINGHOFF, U., GEORGI, F., & WESSEL, B. (1984): Comsumption and assimilation by zooplankton related to primary production in the baltic coastal inlet Barther Bodden. *Limnologica* (Berlin) **15** (2), 387—394.

KELL, V. (1986): Untersuchungen zur Qualität und Quantität des Phytoplanktons des Greifswalder Boddens (südliche Ostsee) in den Jahren 1977—1985. II. Quantitativer Teil. Wiss. Z. Univ. Rostock, Math.-nat. R. **35** (5), 17—21.

KRUMMHAAR, B. (1983): Untersuchungen zur Nahrungsverwertung an Sekundärproduzenten. Dipl.-Arb., Sektion Biologie, W.-P.-Univ. Rostock.

RING, M., HEERKLOSS, R. & ALBRECHT, H. (1986): Saisonale Veränderungen in der Besiedlung von *Eurytemora affinis* (POPPE) (Calanoida, Copepoda) mit peritrichen Ciliaten. Wiss. Z. Univ. Rostock, Math.-nat. R. **35** (5), 64—68.

RING, M. (1987): Autökologische Untersuchungen an *Eurytemora affinis* (POPPE). Diss., Sektion Biologie, W.-P. Univ. Rostock.

VINOGRADOV, M. E., & MENSHUTKIN, V. V. (1977): The modelling of the open sea ecosystems. In: E. D. GOLDBERG (ed.), The sea: Ideas and observations on progress in the study of the sea, pp. 891—921. New York.

Address of the authors:
Dr. REINHARD HEERKLOSS
Prof. Dr. ULRICH BRENNING
Dr. MATHIAS RING
Sektion Biologie
Wilhelm-Pieck-Universität Rostock
Freiligrathstr. 7/8
Rostock
DDR-2500

Limnologica (Berlin) **20** (1), 71–73, April 1990

Department of Sanitary Hydrobiology, Olsztyn-Kortowo, Poland

Seasonal Variations of the Biomass of Copepoda and Rotatoria: Permanent Components of the Zooplankton in the Vistula Lagoon

By BOZENA ADAMKIEWICZ-CHOJNACKA & ZOFIA RÓŻAŃSKA

With 3 Figures

Abstract

The materials were collected in 1975, 1977, 1978, 1985 with a 5-l RUTTNER sampler. Weight standards for the rotifers were determined by measuring the length and breadth of the organisms and comparing the results with literature data or using the formulas of RUTTNER-KOLISKO (1972). Weight standards for Copepoda were determined from ČISLENKO's nomograms (1968).

Introduction

Zooplankton biomass in the Vistula Lagoon consists mostly of Copepoda and Rotatoria, Cladocera appear periodically only and are characterized by considerable variability in different years, both as regards their biomass and numbers. Hence, the present study focused on Copepoda and Rotatoria which are present in the zooplankton throughout the year and represent basic producers in the lagoon and determine its fishery productivity.

Material and Methods

Zooplankton samples were collected in 1975, 1977, 1978 and 1985 at 9 stations distributed all over the lagoon. Samples were collected at monthly intervals with a 5-l RUTTNER sampler. Due to the fact that the materials represented a long period of time it was possible to determine trends of the variations in the zooplankton biomass for the period from February till November.

All Rotatoria taxons present in the zooplankton were measured (body length and width) in order to work out their weight standards. Average values were calculated and compared with the data given in the literature (KOSOVA 1961; KOSTE 1978; KUTIKOVA 1970; MORDUCHAJ-BOLTOVSKOJ 1954; RUTTNER-KOLISKO 1972). In the case of the species not found in the available literature, weight standards were calculated from the formulas by RUTTNER-KOLISKO (1972). ČISLENKO's (1968) nomograms were used to determine weight standards for the Copepoda. Unit weights of all Copepoda species were obtained in this way.

Results and Discussion

The average seasonal biomass of Copepoda and Rotatoria varied seasonally from 1.2 to 4.9 $g \cdot m^{-3}$. A minimum was noted in winter. The biomass then increased rapidly to reach maximal values in May (Fig. 1). Another biomass peak was observed in August. In other months, i.e. in June, July and from September to November, the biomass remained on an average level, with only small variations between 2.1 and 2.4 $g \cdot m^{-3}$. Seasonal variations were mainly due to the Copepoda biomass. Rotatoria were of little significance in the zoo-

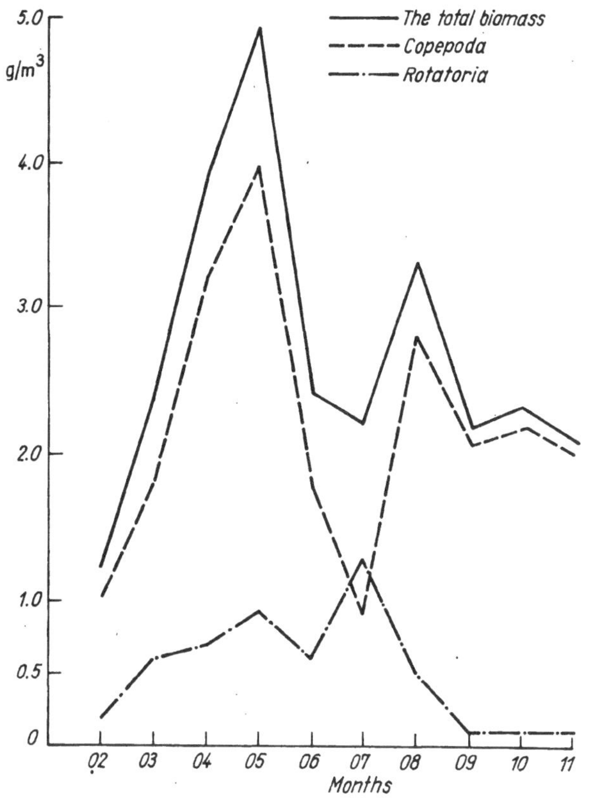

Fig. 1. Seasonal variations in the biomass of Rotatoria and Copepoda in the Vistula Lagoon

plankton biomass although their numbers were usually very high.

Only a few species of Copepoda were present in the Vistula Lagoon. *Eurytemora affinis* POPPE predominated all year round. *Acartia tonsa* (DANA) was fairly abundant in summer. Other, less important species were Cyclopoida and those entering the lagoon with Baltic waters, especially in winter.

Seasonal variations in the Copepoda biomass were characterized by two peaks (Fig. 2). The first occurred in spring and was composed of *Eurytemore affinis*. The biomass of this species increased from March until the maximum in May, when it reproduced intensively. Hence, all development stages of this species were present in the zooplankton in the latter month. The Copepoda biomass decreased to low levels in July, as did the abundance of *Eurytemora affinis*.

Another Copepoda biomass peak occurred in August. It was composed of: increasing levels of *Eurytemora affinis*, high biomass of *A. tonsa* (the highest in the season), and abundant nauplii of both species. In autumn, the warm-water *Acartia tonsa* decreased gradually, as did the number of nauplii.

Eurytemora affinis was most abundant in this period, when it reached its second peak. In winter, the Copepoda biomass was very low, its levels being similar to those in July. In this period, species entering the lagoon with Baltic water (such as *A. bifilosa* GIESBRECHT) became most important.

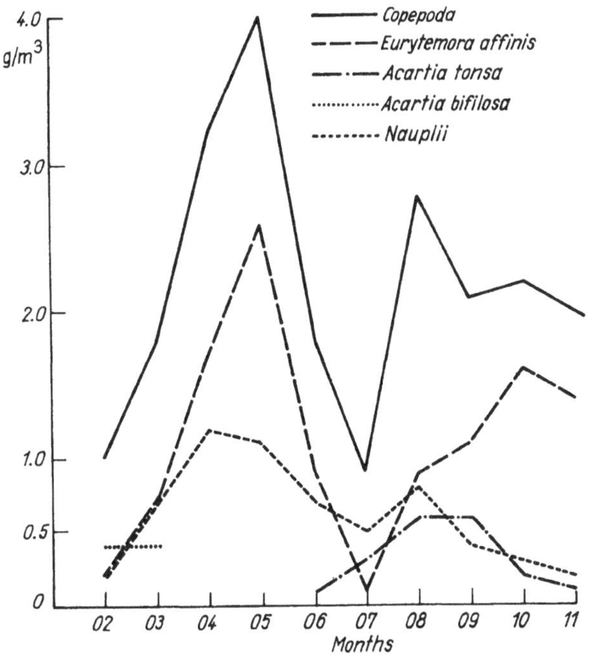

Fig. 2. Seasonal variations in the biomass of dominant Copepoda species

Studies carried out by ŽUDOVA (1978) in the Soviet part of the lagoon also revealed a spring maximum of Copepoda biomass. However, the values reported were very low, ranging from 0.735 to 2.362 $g \cdot m^{-3}$. Corresponding levels observed in the same year in the Polish part of the lagoon (ADAMKIEWICZ-CHOJNACKA & FAIT 1987) were 1.07 to 8.73 $g \cdot m^{-3}$. Moreover, the Soviet author stated that part of the lagoon near the Polish border should be regarded as characterized by the highest zooplankton biomass. These studies as well as studies carried out in Poland (RÓŻAŃSKA 1963; ADAMKIEWICZ-CHOJNACKA & FAIT 1987) suggest that *Eurytemora affinis* is most abundant in waters in which salinity is about $3^0/_{00}$, and that its abundance is responsible for the high Copepoda biomass in the Polish part of the Vistula Lagoon.

The July minimum of *Eurytemora affinis* may be caused by a variety of factors, including the high water temperature in this month (about 20 °C) which does not favour reproduction. Laboratory experiments (BRADLEY & KETZNER 1982) confirm that *Eurytemora affinis* prefers lower temperatures. At the same time, increased temperatures favour the development of *A. tonsa*. As stated by MCALICE (1981) this species is most abundant around 20 °C.

Thermal conditions induce "shifting" between the two species. Competition should also be taken into account. According to BRADLEY (1975), the competitive impact of *A. tonsa* on *E. affinis* was the main reason for the low abundance of the latter in summer. Moreover, blue-green algae develop in masses during summer (PLIŃSKI 1979). This may inhibit development of the zooplankton in the lagoon. HEERKLOSS et al. (1984) stated that the ability of Copepoda to assimilate the phytoplankton decreases in periods of abundant blue-green development. Under natural conditions, this might lead to a decrease in the biomass and abundance of *Eurytemora affinis* in the lagoon (HEERKLOSS et al. 1985).

As regards Rotatoria, 27 taxa were found in the Vistula Lagoon, but only a few species were important in terms of biomass (ADAMKIEWICZ-CHOJNACKA & LEŚNIAK 1985). These were: *Brachionus calyciflorus* PALLAS, *Brachionus angularis* GOSSE, *Filinia longiseta* (EHRENBERG), *Keratella cochlearis* (GOSSE), *Keratella quadrata* (MÜLLER), *Synchaeta* sp. and *Hexarthra fennica* (LEVANDER). The rotifers biomass showed considerable seasonal variation (Fig. 3). Minimal levels of 0.1—0.2 $g \cdot m^{-3}$ were observed in late autumn and winter. In February and March the biomass was composed mostly of *Synchaeta baltica*. Two rotifer biomass peaks were noted: in spring (either April or May, depending on the temperature) and in summer (July).

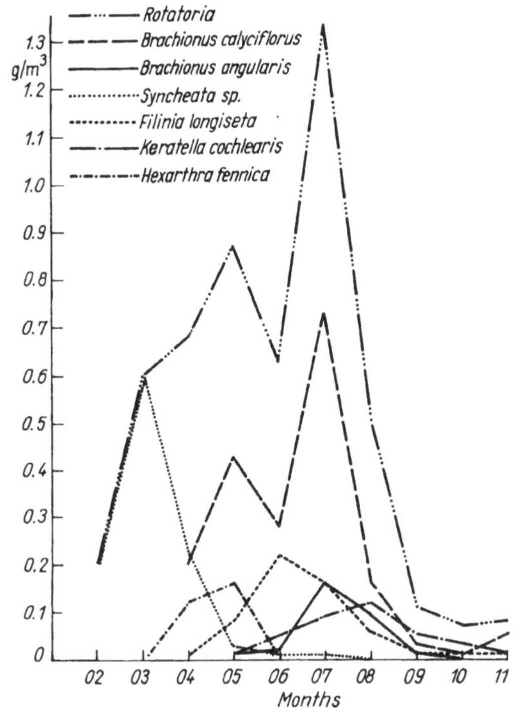

Fig. 3. Seasonal variations in the biomass of dominant Rotatoria species

In 1978 July was rather cool and the water temperature did not exceed 17 °C (ROŽANSKA & WIĘCŁAWSKI 1981). No characteristic July peak was observed in this year, and the rotifer biomass remained at a more or less the same level throughout the summer.

Spring and summer maxima of the rotifer biomass corresponded to maximal abundance of the main component i.e. *Brachionus calyciflorus*. Other species were also important in some months: *Hexarthra fennica* in May, *Filinia longiseta* in June and July, *Brachionus angularis* and *Keratella cochlearis* in July and August. After September, the biomass of Rotatoria decreased noticeably to about 0.1 $g \cdot m^{-3}$.

Seasonal variations in the zooplankton biomass suggest that Rotifers are most important in summer, especially in July. Similar results were obtained by ŽUDOVA

(1978) in the Soviet part of the lagoon. Considerable abundance and an increased proportion of Rotatoria in the zooplankton biomass in summer have also observed in other brackish water (WIKTOR 1964, 1972; KISELITE 1959).

Conclusions

1. Two peaks of the zooplankton biomass were observed in the season, in May and August. The lowest zooplankton biomasses were recorded in February and July.
2. Seasonal variations of the zooplankton biomass correspond to the variations in Copepoda biomass.
3. A significant proportion of Rotatoria was observed in the zooplankton in July. Higher levels of Rotatoria were also found in May.

References

ADAMKIEWICZ-CHOJNACKA, B., & FAIT, D. (1987): Ocena biomasy widłonogów (Copepoda) Zalewu Wiślanego w latach 1975, 1977 i 1978. Stud. i Mater. Oceanol. KBM PAN **51** (7), 283—312.

— & LEŚNIAK, D. (1985): Ocena biomasy wrotków (Rotatoria) słonawych wód Zalewu Wiślanego w latach 1975, 1977 i 1978. Stud. i Mater. Oceanol. KBM PAN **46**, 273—308.

BRADLEY, B. P. (1975): The anomalous influence of salinity on temperature tolerances of summer and winter populations of the Copepod *Eurytemora affinis*. Biol. Bull. **148**.

— & KETZNER, P. A. (1982): Genetic and non-genetic variability in temperature tolerance of Copepod *Eurytemora affinis* in five temperature regims. Biol. Bull. **162** (3), 233—245.

ČISLENKO, L. L. (1968): Nomogramy dlia opredelenia vesa vodnych organizmov po rozmerom i forme tela (morskoi mezobentos i plankton). Akad. Nauk SSSR Zoologicheskii Instytut, Leningrad.

HEERKLOSS, R., ADAMKIEWICZ-CHOJNACKA, B., & SCHNESE, W. (1985): Biomasse der planktischen Copepoden in inneren Küstengewässern der südlichen Ostsee. Wiss. Z. Univ. Rostock, Math.-nat. R. **34** (6).

— BÖRNER, R., KRUMMLAAR, B., ARNDT, H. & SCHNESE, W. (1984): Growth inhibition in estuarine zooplankters due to blue-green algae. Ophelia, Suppl. **3**, 45—49.

KISELITE, T. (1959): Zooplankton zaliva Kursju Mares. Akad. Nauk Litovskoj SSSR, Kursju Mares, 169—185.

KOSOVA, A. A. (1961): Vycislennye vesa nekotorych form zooplanktona nizovev delty Volgi. Trudy Astrach. Zapov. **5**, 151—159.

KOSTE, W. (1978): Rotatoria. Berlin, Stuttgart.

KUTIKOVA, L. A. (1970): Kolovratki fauny SSSR (Rotatoria). Leningrad.

MCALICE, B. J. (1981): On the Post-glacial History of *Acartia tonsa* (Copepoda: Calanoida) in the Gulf of Maine on the Gulf of St. Lawrance. Mar. Biol. **64**.

MORDUCHAJ-BOLTOVSKOJ, F. D. (1954): Materialy po srednemu vesu vodnych bespozvonocnych bassejna Dona. Problemy gidrobiologii vnutrennich vod **2**, 223—241.

PLIŃSKI, M. (1979): Kierunki zmian strukturalnych w fitoplanktonie estuariów Bałtyku Południowego. Zesz. Nauk. Uniw. Gdański. Rozprawy i Monografie **15**, 136.

RÓŻAŃSKA, Z. (1963): The Vistula firth zooplankton. Zesz. Nauk. WSR Olszt. **16**, 41—57.

— & WIĘCŁAWSKI, F. (1981): Zmiany czynników eutrofizacji wód Zalewu Wiślanego. Stud. i Mater. Oceanol. KBM PAN **34** (4), 5—32.

RUTTNER-KOLISKO, A. (1972: Rotatoria. Das Zooplankton der Binnengewässer. In: Die Binnengewässer, Bd. 26, S. 99—225. Stuttgart.

WIKTOR, K. (1964): Związki między produkcją zooplanktonu a odżywianiem się ryb planktonożernych. Stud. i Mater. MIR Gdynia **2**, 1—100.

— (1972): Zmiany w rozwoju planktonu Zalewu Szczecińskiego jako wynik zmian w warunkach fizyko-chemicznych środowiska, spowodowanych gospodarką człowieka. Ekosystemy morskie, Opracowanie MIR Gdynia **4**, 63—132.

ŽUDOVA, A. M. (1978): Zooplankton Vislinskogo Zaliva i ego dinamika. Gidrobiol. Ž. **14** (1), 65—70.

Address of the authors:
Prof. Dr. hab. ZOFIA RÓŻAŃSKA
Dr. BOŻENA ADAMKIEWICZ-CHOJNACKA
Department of Sanitary Hydrobiology
10-957 Olsztyn — Kortowo
Poland

Limnologica (Berlin) **20** (1), 75—78, April 1990

Institute of Oceanography, Gdańsk University, Gdynia, Poland

Biomass and Abundance of the Zooplankton from the Gulf of Gdańsk

By Krystyna Wiktor

With 4 Figures

Abstract

Investigations covered zooplankton distribution and composition in the coastal waters of the Gulf of Gdańsk. The biomass and abundance were compared on the basis of mean data for the period between May and October 1981. Materials were collected with an Apstein net (mesh size No. 25). The samples were taken vertically once a month from stations at depths of 5, 10, 20 and 35 m, located at four profiles

The standart weight of individuals according to the BMB Recommendations of Working Group 14 (1985) was used for biomass calculations. An exception to this was the biomass of Rotatoria, which was calculated according to the standard taken from Adamkiewicz-Chojnacka & Leśniak (1985).

The zooplankton in the study area is composed of Copepoda, Cladocera, velliger of Mollusca and Appendicularia. Copepoda have the highest biomass (59.5%) but Rotatoria are the most numerous (59.7%). It was found that the biomass increases with depth. The horizontal distribution of biomass is similar to the abundance distribution: the highest values are found in the middle of the study area.

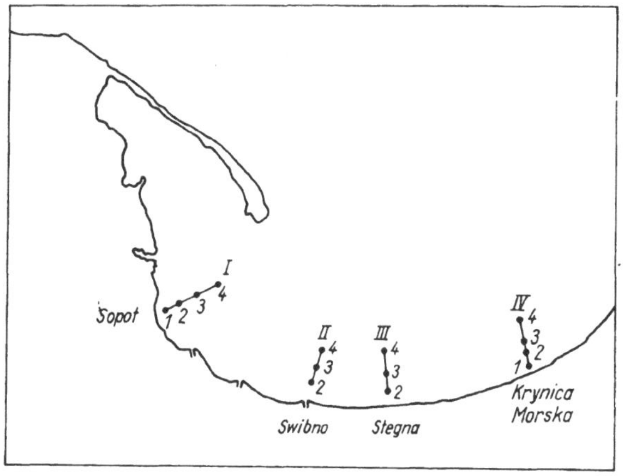

Fig. 1. Sampling station grid in the Gulf of Gdańsk

Introduction

The zooplankton of the Gulf of Gdańsk has been the objekt of numerous studies. Its species composition, distribution, abundance and seasonal changes are well know (Siudziński 1978; Wiktor et al. 1982; Wiktor & Żmijewska 1985). On the basis of the materials collected between May and October 1981, an attempt was also undertaken to estimate the biomass of zooplankton in that area.

The zooplankton of the Gulf of Gdańsk is composed of organisms of varying size: Crustaceans (Copepoda and

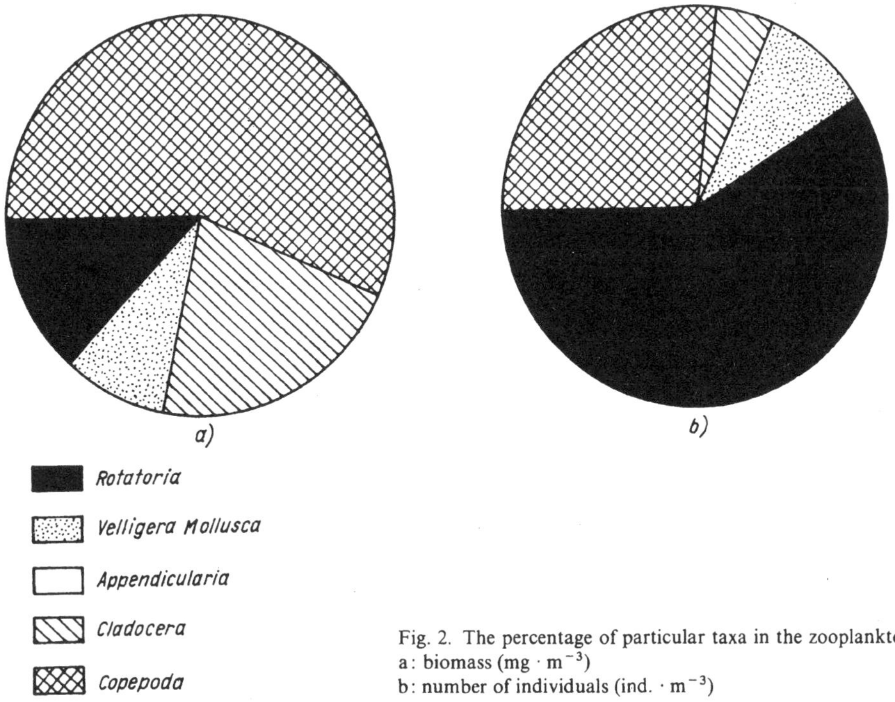

Fig. 2. The percentage of particular taxa in the zooplankton.
a: biomass (mg · m^{-3})
b: number of individuals (ind. · m^{-3})

Cladocera), Rotatoria (mostly small forms like *Keratella cruciformis*) as well as larvae of Molluscs (velligera). A question thus arise as to whether, and to what degree, the picture of the zooplankton obtained on the one hand by an analysis of abundance will differ from that derived from biomass analysis on the other.

The present paper is aimed at settling this issue.

Material and Methods

The basis for comparing the composition and distribution of zooplankton biomass and abundance in the Gulf of Gdańsk consisted of samples collected with an Apstein net with a mesh size No. 25. The samples were taken vertically between May and October 1981, once a month, along four transects, from points located on each transect at a depth of 10, 20 and 30—35 m (Fig. 1). Two points at a depth of 5 m were added at two transects. A detailed description of the method of sample collection and analysis has been given in previous publications dealing with the zooplankton of this area (WIKTOR & ŻMIJEWSKA 1985).

The paper mentioned above and unpublished results of analyses provided data on the abundance of zooplankton species in each period of material collection. These were averaged later for each point during the period May—October. Average data served for an estimate of mean biomass for each zooplankton component for a given point and water layer. Biomass estimates were based on the standard individual weights given for all species and, in the case of

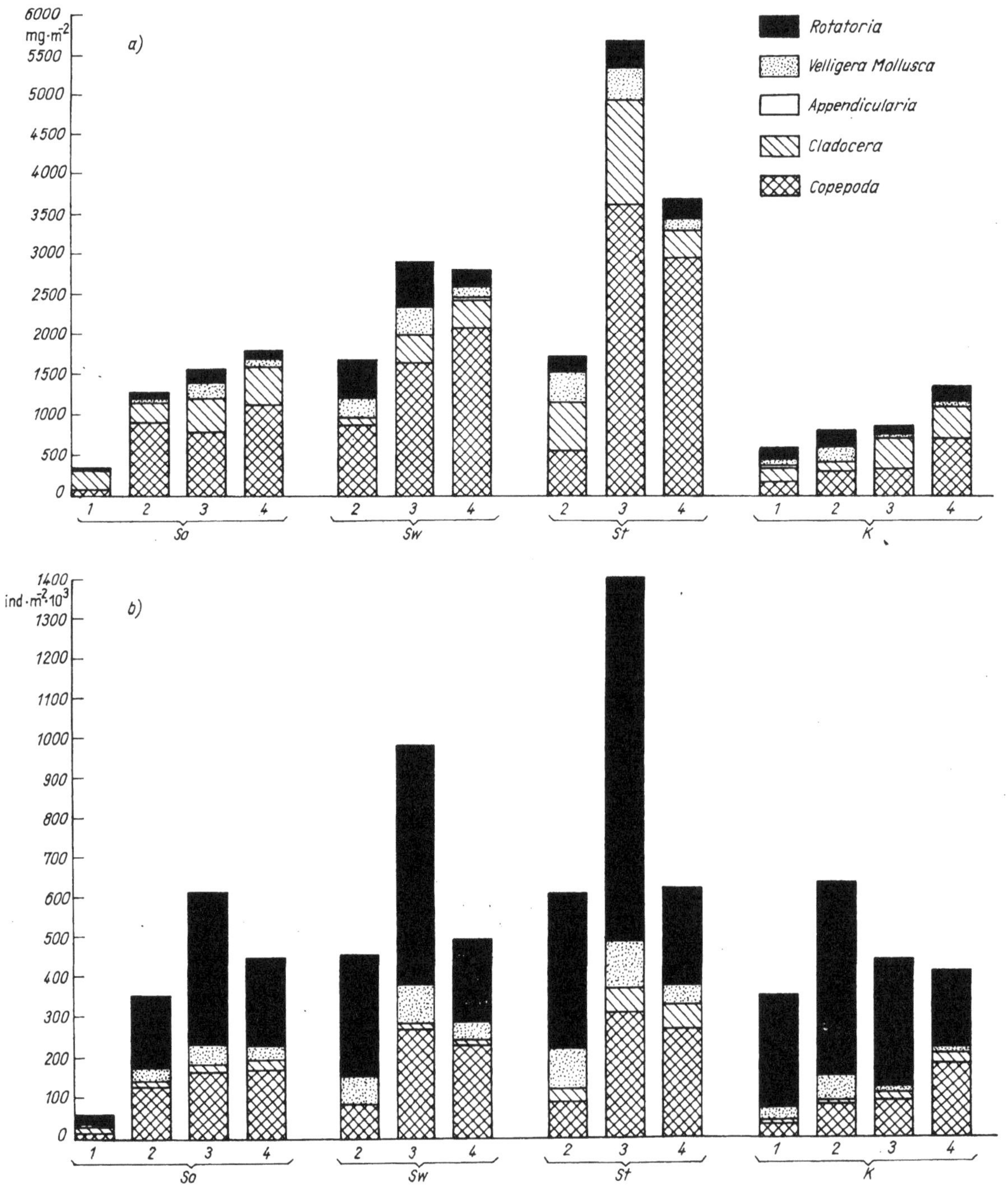

Fig. 3. The horizontal distribution of the zooplankton in coastal water in the Gulf of Gdańsk.
a: biomass (mg · m^{-2})
b: number of individuals (ind. · m^{-2})

Copepoda, for all developmental stages, in the BMB Publication No. 10 — Working Group 14 (1985). In addition, the standard given by ADAMKIEWICZ-CHOJNACKA & LEŚNIAK (1985) was used to determine the biomass of Rotatoria. Individual weights of mollusc larvae were determined from the nomograms of TSCHISLENKO (1968), after establishing the mean size of the larvae found during the study.

Results and Discussion

Zooplankton composition

As mentioned in the introduction, the zooplankton is mainly composed of Copepoda, Cladocera, Rotatoria and larval forms of Mollusca. In deeper waters, the only representative of Appendicularia, *Fritillaria borealis*, appears sporadically.

Crustaceans are dominant in the total zooplankton biomass — first of all Copepoda (59.5 %) and then Cladocera (21.4 %). Rotatoria constitute on the average 10.1 % and larval forms of Mollusca 8.9 % of the total biomass (Fig. 2a). The proportions of these components is quite different when total zooplankton abundance is analysed. Rotatoria predominate in this case represented mainly by forms of small size and individual weight (*Keratella, Brachionus, Synchaeta*). Their share in the total zooplankton abundance accounts on average for up to 59.7 %, whereas the biomass component, Crustacea, accounting for only 31.2 % (Copepoda 27.0 %, Cladocera 4.2 %) (Fig. 2b). The shares of mollusc larvae and *Fritillaria borealis* in total abundance is similar to their shares in terms of biomass (9.0 % and 0.1 %, respectively).

Zooplankton distribution

The general distribution of zooplankton in the water column under 1 m^2 exhibits similar tendencies, as regards both abundance and biomass: in both cases the highest values are found in the central part of the study area (up to 5713 $g \cdot m^{-2}$ and 1400 thous. $ind.m^{-2}$) and the lowest along the easternmost profile (564—1352 $g \cdot m^{-2}$ and 350—636 $thous.ind.m^{-2}$) (Figs. 3a and b).

Total zooplankton biomass in the water column under 1 m^2 increases with depth at each profile, while the highest zooplankton abundance is usually found at a depth of 20 m. In this water layer Rotatoria (*Keratella cruciformis* and *Synchaeta* sp.), i.e. organisms of small individual weight, usually have the greatest abundance; their great number therefore affects the biomass only to a small degree.

Like the horizontal distribution, vertical variations in zooplankton biomass and abundance observed at the four deepest points of each profile exhibit the same tendencies. Both parameters reach their highest values in the near-surface 10-m water layer. Zooplankton biomass and abundance decrease with depth, most clearly in the 10—20 m layer, while in the near-bottom layer they increase once again. Only at one point at the eastermost profile (point K-4), was a steady downward trend from surface to the bottom observed (Figs. 4a and b).

The share of individual components of zooplankton also changes with depth. As far as abundance is concerned, Rotatoria predominate in the near-surface layer; their share decreases with depth. The abundance of cladocerans and mollusc larvae also decreases, so that in the near-bottom layer Copepoda become the main component.

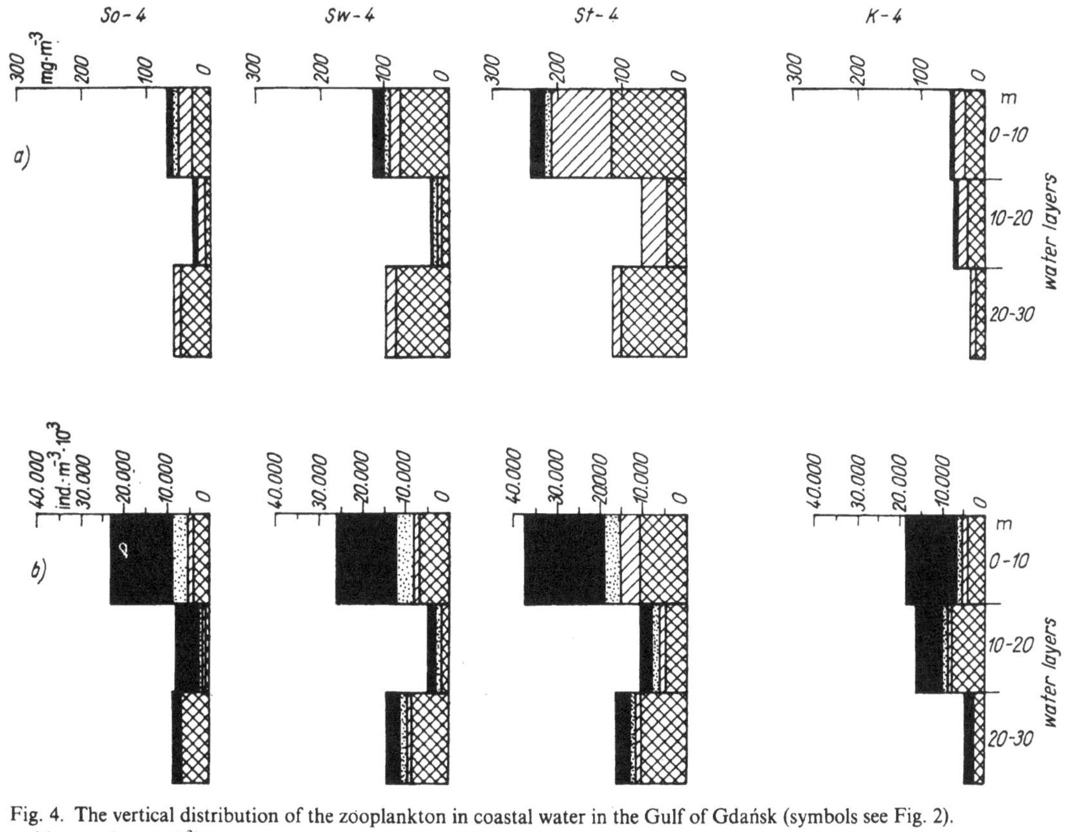

Fig. 4. The vertical distribution of the zooplankton in coastal water in the Gulf of Gdańsk (symbols see Fig. 2).
a: biomass ($mg \cdot m^{-3}$)
b: number of individuals ($ind. \cdot m^{-3}$)

As regard the biomass, the importance of Rotatoria is small even in the near-surface layer while cladocerans predominate next to copepods. Their biomass and share in the total biomass decrease with depth, like those of Rotatoria and mollusc larvae. In the near-bottom layer, Copepoda remain the main biomass component.

Conclusion

It appears from the comparative analysis that the use of the biomass of zooplanktonic organisms in place of their abundance is not particularly revealing since estimating the biomass of individual zooplankton components does not bring us closer to conclusions concerning their role in the biocenosis. The pictures of zooplankton distribution and developmental tendencies are similar, irrespective of whether they are based on abundance or biomass data. Conclusions concerning the role of individual components in the biocenosis are most accurate when based on production data and not biomass. Thus, in order to observe long term or seasonal changes in the zooplankton of a given area it is enough to investigate the abundance of individual components.

The use of the biomass of the zooplankton components may be of significance when comparing a single community inhabiting different areas that differ radically in species composition of higher taxa.

References

ADAMKIEWICZ-CHOJNACKA, B., & LEŚNIAK, D. (1985): Ocena biomasy wrotków (Rotatoria) słonawych wód Zalewu Wiślanego w latach 1975, 1977 i 1978 [Evaluation of biomass of the Rotifers (Rotatoria) in salty water of the Vistula Bay in 1975, 1977 and 1978). SIMO KBM PAN **47** (7), 273—308 (in Polish, with English summary).

Recommendations on methods for marine biological studies in the Baltic Sea-Mesozooplankton assessment, 1985. BMB Publication No 10 — Working Group 14, Ed. L. HERNROTH, pp. 1 to 32.

SIUDZIŃSKI, K. (1977): Zooplankton Zatoki Gdańskiej [Zooplankton from Gulf of Gdansk]. Studia i Mat. Morsk. Inst. Ryback. w Gdynia **A-18,** 1—111.

WIKTOR, K., CYLKOWSKA, U., & OSTROWSKA, K. (1982): Zooplankton przybrzeżnych wód Zatoki Gdańskiej (Zooplankton in the coastal waters of Gdańsk Bay). SIMO KBM PAN **39** (6), 77—136 (in Polish, with English summary).

— & ŻMIJEWSKA, M. (1985): Skład i rozmieszczenie zooplanktonu przybrzeżnych wód Zatoki Gdańskiej właściwej (Composition of the coastal water zooplankton of the Gulf of Gdańsk). SIMO KBM PAN **46** (7), 65—114 (in Polish, with English summary).

Address of the author:

KRYSTYNA WIKTOR
Institute of Oceanography
Gdańsk University
Czołgistow 46
Gdynia
Poland

Limnologica (Berlin) **20** (1), 79—81, April 1990

Akademie der Wissenschaften der DDR, Institut für Geographie und Geoökologie, Bereich Hydrologie, Berlin, DDR

A Method for the Separation of Zooplankton in High Eutrophic Water Bodies

By Horst Behrendt & Maria Krocker

With 2 Figures and ohne Table

Abstract

Methods of zooplankton separation from other particulate material based on net are not applicable for highly eutrophic lakes and reservoirs with a high content of filamentous algae.

A method for zooplankton isolation was developed for such water bodies especially. The method is based on two well known effects. At first the positive phototaxis of zooplankton individuals and, secondly, the phenomenon that these individuals migrate to the bottom of closed vessels. On the base of these phenomenona a zooplankton trap was constructed. Design, function of the trap and details of the procedure are described. Test of the trap showed different migration behaviour of several zooplankton groups. Regarding to the whole individual number, mesozooplankton (Cladocera and Copepoda) migrate in two hours in average to 80 per cent, but microzooplankton (nauplii and Rotatoria) only to 50 per cent. Results of separation give rise to the thesis that measurement of macrozooplankton influence on phytoplankton is possible.

Introduction

A better understanding of phytoplankton development requires the measurement of phytoplankton balance processes. At present methods for the in-situ estimation of phytoplankton growth and primary production are established but in general the problem of in-situ measurement of phytoplankton loss processes is not solved. The measurement of phytoplankton losses caused by zooplankton represents one example for this fact. First of all the in-situ estimation of this process is connected with the problem of zooplankton separation from phytoplankton and other particulate material before or after the exposition. Separation methods based on nets are applicable only under conditions of oligotrophic and up to a point mesotrophic water bodies, too. In cases of eutrophic or highly eutrophic lakes, reservoirs and estuaries these methods can not be used because a high content of filamentous algae hinders this separation procedure.

Especially for such water bodies, it was tried to develop a new method for zooplankton isolation. This method based on two well known effects. At first the positive phototaxis of zooplankton individuals and, secondly, the phenomenon that zooplankton individuals migrate to the bottom of a closed vessel.

Previously Franz (1911, 1912) investigated the influence of light on the behaviour of copepods and *Daphnia* species in detail. He pointed out that the effect of positive phototaxis occurred under conditions of a high zooplankton concentration in closed vessels and after a pre-darkening. He observed, additionally, a migration of the majority of zooplankton individuals to the bottom in such closed vessels. His conclusion that the phototaxis of zooplankton individuals is not an effect of adaptation but a short term reaction on abrupt irregular irritations was the base for the construction of a new quantitative separation method.

In the past different authors have tried to use the effect of positive phototaxis for the catch of zooplankton by means of light traps (Hungerford et al. 1955; Pienczynski 1962; Kowalczyk & Kowalik 1970; Heerkloss & Arndt 1981). Their experiences were used too.

The separation experiments were carried out with water samples of Lake Müggelsee mainly, a eutrophic, polymictic shallow lake with an area of 720 hectare and a mean depth of about 4.9 m.

Construction and function of the trap

According to the results of Franz (1911) and the experiences of the other authors two further limiting conditions were taken into account for light trap construction:

1. Minimal expenses for the trap elements and for construction
2. Application of the trap not only in laboratories but also under field conditions.

The new constructed light trap consists on three main sections and is shown in Fig. 1. The dark section is built up as a cylinder with a lid on the top and a tube piece on the bottom. The bottom of this section was arranged in a slanting direction that the deepest point of the dark section is at the begin of the tube piece and the passage to the light section. In contrast to the traps developed by Heerkloss & Arndt (1981) not glass but plastic was used as material for the dark section in order to prevent reflections of light on the border of this vessel.

The light section consists of a usual jar with a definite volume. Dark and light section are connected by a transparent or semitransparent plastic tube. At the tube is a clip to interrupt the connection between dark and light section before and after the separation procedure.

The function of the trap is described in the following:

1. Filtrated lake water is filled in the light section up to the top.
2. The plastic tube connected with the dark section is put into the full jar in such a kind that a tight connection is given.
3. Via dark section and tube filtrated lake water is filled further up to a little bit on the clip. Then the clip has to be closed and the overstanding water in the tube above the clip is putted away.
4. The water sample concentrated by a plankton net previously is filled into the dark section that air bubbles remain in the tube. The dark section is closed

by the lid and the whole light trap is placed in such a way that the dark section stands over the light section as presented in Fig. 1.

5. The clip is opened and the migration starts after some minutes for calming of water in the trap. During the separation the water temperature in the two sections of the trap should be the same to prevent a mixing of water by different density.
6. At the end of the separation procedure the clip has to be closed again and the two sections are separated.

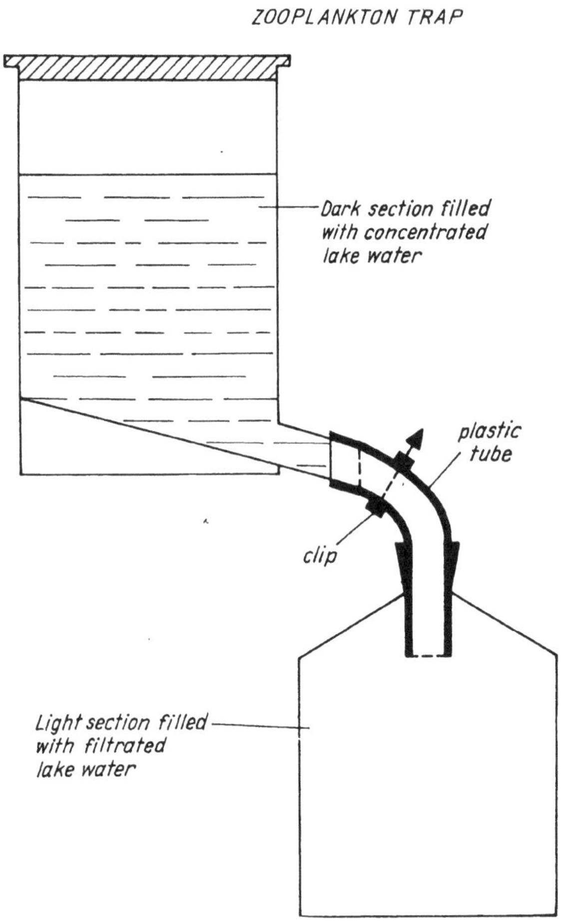

Fig. 1. Schematic representation of the zooplankton light trap

Results and Discussion

During the last three years more than 40 separation experiments were carried out. In all cases a very clear zooplankton sample could be obtained at the end of this procedure in the light section. In general the content of phytoplankton and other particulate material in the light section was smaller than 2 per cent in comparison to the isolated zooplankton biomass, what is in agreement with the investigation of HEERKLOSS & ARNDT (1981). It offers the possibility of using of the trap to get clear zooplankton samples for the estimation of chemical composition and for further grazing experiments.

The number of individuals of zooplankton species and the average length of the species were estimated in the light and in the dark section for the quantitative analysis. The effectiveness of the separation procedure was calculated as the capture rate (CR) defined as the ratio between the catched zooplankton in the light section to the total number or weight of zooplankton in the light and the dark section of the trap. The capture rate based on the total number of zooplankton individuals was in average:

$n = 33 \quad CR_{total} = 58 \pm 7\% \quad = 0.95\,.$

While for the microzooplankton (nauplii and Rotatoria) the capture rate relating to the number of individuals was only

$n = 33 \quad CR_{micro} = 48 \pm 8\% \quad = 0.95$

for the mesozooplankton (copepods and cladocerans) a capture rate was obtained to

$n = 33 \quad CR_{meso} = 82 \pm 4\% \quad = 0.95\,.$

It is obvious that the capture rate relating to the weight of zooplankton is located more at that of the mesozooplankton because these species are larger and possess a higher weight per individual

$n = 33 \quad CR_{weight} = 79 \pm 4\% \quad = 0.95\,.$

The same results could be observed for samples from other water bodies as Saidenbach reservoir, Lake Kaulsdorfer See and a small white water-lily pond.

Approximately, the same behaviour relating to capture rate is shown by different groups or species within the meso- and microzooplankton. A capture rate of 83 ± 4% and 85 ± 5% was found for cyclopoids and calanoids, respectively. *Daphnia* and *Bosmina* species seem to possess a little bit lower CR with 79 ± 6% and 70 ± 7%, respectively. The observed differences of catch effectivity were smaller for the two investigated groups of microzooplankton (nauplia and Rotatoria) with 49 ± 9% and 47 ± 8%, respectively.

The application of this zooplankton trap depends not only on the capture rate but also on the rate of repetition for several subsamples or parallel used traps. In all experiments two or more identical traps were applied. Subsamples for the different traps were produced very simple by means of several small and close connected tubes which are linked on the one side with the vessel of the total sample and on the other side with subsample jars. For the subsample procedure a mean error of subsamples was calculated to 20 per cent relating to the total individual number. The differences between the capture rates of different traps were smaller than 20% in all cases. Therefore, it may be assumed that the traps work with a higher precision than the subsample procedure.

Table 1 presents the change of capture rate of the different zooplankton groups during different periods of the year.

Two conclusions are derived from this behaviour (see Table 1). At first it may be assumed that according to FRANZ (1911) the positive phototaxis is mainly caused by short term irregular irritations. At second long term adaptation phenomenona seem to influence the catch effectivness of the trap, too. Dependences of capture rate

Tabl 1. Changes of catch effectiveness of zooplankton trap during different periods of the year.

Period	Capture rate of	
Period	Mesozooplankton	Microzooplankton
1. Half of May	84 ± 5%	58 ± 12%
2. Half of May	75 ± 10%	39 ± 19%
1. Half of June	70 ± 9%	41 ± 19%
September	85 ± 5%	48 ± 11%

on the temperature and the direct intensity of pre-light conditions could not be detected. While this independence on temperature is in agreement to the results of RUTTNER (1914), a contradiction exists in relation to light dependences. It seems to be possible that not the light intensity but the length of period between sunrise and sample collection determines the result of separation, because the collection of the water sample was carried out at the same time of the day over the whole investigation period. Fig. 2 represents this dependence. On the base of this relation may be assumed that the adaptation behaviour of the zooplankton species depends more on the time period between sunrise and sample collection if a certain

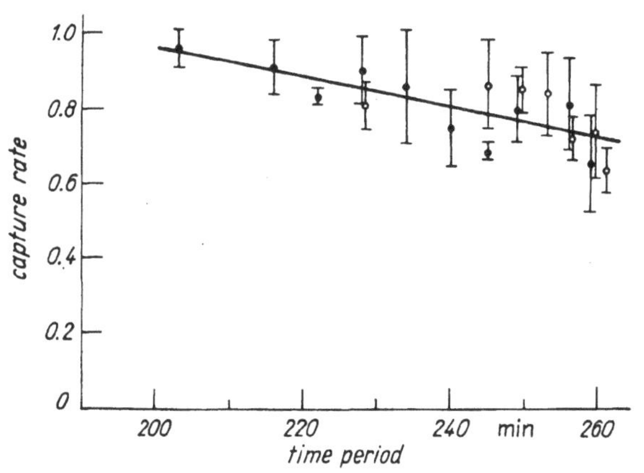

Fig. 2. Relation between the capture rate of mesozooplankton and the time period from sunrise to sample collection.
O measured capture rates during May/June 1986
× measured capture rates during May/June 1987
Results of regression:
n = 17
r = −0.738
CR = 1.747 − 0.0039 * t

light level is exceeded. Regarding to the correlation presented in Fig. 2 it has to be considered that the relation between capture rate and time is valid only for the given time range. Further experiments in this direction might give a possibility to get increasing effects for this separation method.

References

FRANZ, V. (1911): Beiträge zur Kenntnis der Phototaxis. Nach Versuchen an Süßwassertieren. Int. Rev. ges. Hydrobiol., Biol. Suppl., 2. Serie, **4**, 1–11.

— (1912): Weitere Phototaxisstudien. I. Zur Phototaxis bei Fischen. II. Phototaxis bei marinen Crustaceen. III. Phototaktische Lokomotionsperioden bei *Hemimysis*. Int. Rev. ges. Hydrobiol., Biol. Suppl., 3. Serie, **4**, 1–23.

HEERKLOSS, R., & ARNDT, H. (1981): Eine Lichtfalle zur Reinigung von Zooplanktonproben. Wiss. Z. W.-Pieck-Univ. Rostock, Math.-naturw. R. **30** (4/5), 49–52.

HUNGERFORD, H. B., SPANGLER, P. J., & WALKER, N. A. (1955): Subaquatic light trap for insects and other animals. Trans. Kans. Acad. Sci., Kansas **58**, 387–407.

KOWALCZYK, C., & KOWALIK, W. (1970): A new model of a plankton light trap. Wiad. Ecol. **16**, 68–76 (in Polish).

PIECZYNSKI, E. (1961): The trap method of capturing water mites (*Hydrocarina*). Ecol. Pol., Ser. B, **7**, 111–115.

RUTTNER, F. (1914): Bemerkungen zur Frage der vertikalen Planktonwanderung. Int. Rev. ges. Hydrobiol., Biol. Suppl., 4. Serie, **6**, 1–12.

Address of the authors:

HORST BEHRENDT
MARIA KROCKER
Akademie der Wissenschaften der DDR
Institut für Geographie und Geoökologie
Bereich Hydrologie
Müggelseedamm 260
Berlin
DDR-1162

Limnologica (Berlin) **20** (1), 83—88, April 1990

Institute of Fisheries Oceanography and Protection of Sea, Szczecin, Poland

Dynamics of Meiobenthic Communities in a Eutrophic and Polluted Estuary

By Teresa Radziejewska & Idzi Drzycimski

With 4 Figures and 2 Tables

Abstract

Seasonal phenomena were studied in meiobenthic communities sampled in 1985—1986 at 7 stations in the Szczecin Lagoon, a eutrophic and polluted link of the River Odra estuarine system. Both the abundance and composition of meiobenthos fluctuated widely at each station, no stable pattern of changes being detected. Changes in the total meiobenthos density as well as in densities of nematodes, ostracods, and harpacticoid copepods did, however, correlate with changes in bottom water salinity. Furthermore, ostracod densities correlated inversely with BOD_5 in bottom water, while harpacticoid abundances correlated directly with oxygen content.

The harpacticoid copepod subcommunity, consisting of a total of 10 species recorded in the Lagoon over the two years of study, showed signs of both seasonal succession and characteristic spatial distribution. Two brackish-water species. *Halectinosoma curticorne* and *Microarthridion littorale*, were the subcommunity dominants along the temporal and spatial axes. The overall distribution of harpacticoid copepods in the Lagoon reflected the salinity and pollution gradients.

Introduction

Estuaries are areas where environmental parameters fluctuate extensively, the fluctuations resulting in a stress situation the biota have to cope with (Levinton 1982). To quote Green (1969): "environmental variability is an important factor limiting the successional processes which lead to higher diversities and stability, and erratic, unpredictable variability is the most effective kind". One could interpret this statement as meaning that, when subjected to erratic environmental variability, biologically controlled animal communities give way to physically controlled assemblages.

In the present work we examine data on seasonal and spatial dynamics of meiobenthic communities in the Szczecin Lagoon, a eutrophic and polluted water body constituting a link in the River Odra estuarine system, in order to determine the kind and extent of controls operating on the communities studied.

Area of study

Meiobenthos was sampled in 1985—1986 at 7 stations located in various parts of the Szczecin Lagoon (Fig. 1). Fig. 2 presents changes in environmental parameters recorded over the period of study. Our previous paper (Radziejewska & Drzycimski (1988) describes the Lagoon's environment in more detail.

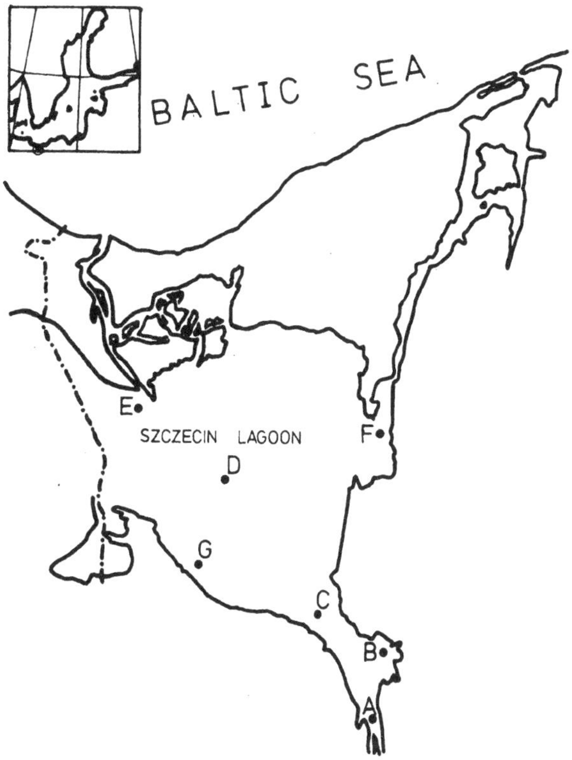

Fig. 1. The Szczecin Lagoon: area of study and sampling sites

Materials and Methods

Meiobenthic collections were made in 1985 and 1986, 4 times each year (April; June; September—October in 1985, August—September in 1986; November). Samples were taken with a 22.4 mm i.d. corer (Płocki & Radziejewska 1980), 3 samples per station being collected at most times. The samples were fixed with 4% formaldehyde and stained with Rose Bengal. After screening a sample through a 1 mm mesh size sieve, meiobenthic animals were extracted by decantation and sieving through a set of 3 sieves (0.200; 0.100; 0.040 mm mesh sizes). The animals extracted were identified to major taxa and enumerated, harpacticoid copepods being identified to species.

The following parameters determined for each station are used in the present analysis: mean densities of various taxa and of total meiobenthos; mean number of taxa; and mean taxon diversity index calculated according to the Shannon-Wiener formula (Pielou 1975): $H' = -\Sigma p_i \log p_i$
where p_i is the proportion of ith taxon. Statistical treatment of data involved testing for significance of Spearman's rank correlation coefficients (Sokal & Rohlf 1981) calculated for pairs of biotic and abiotic variables.

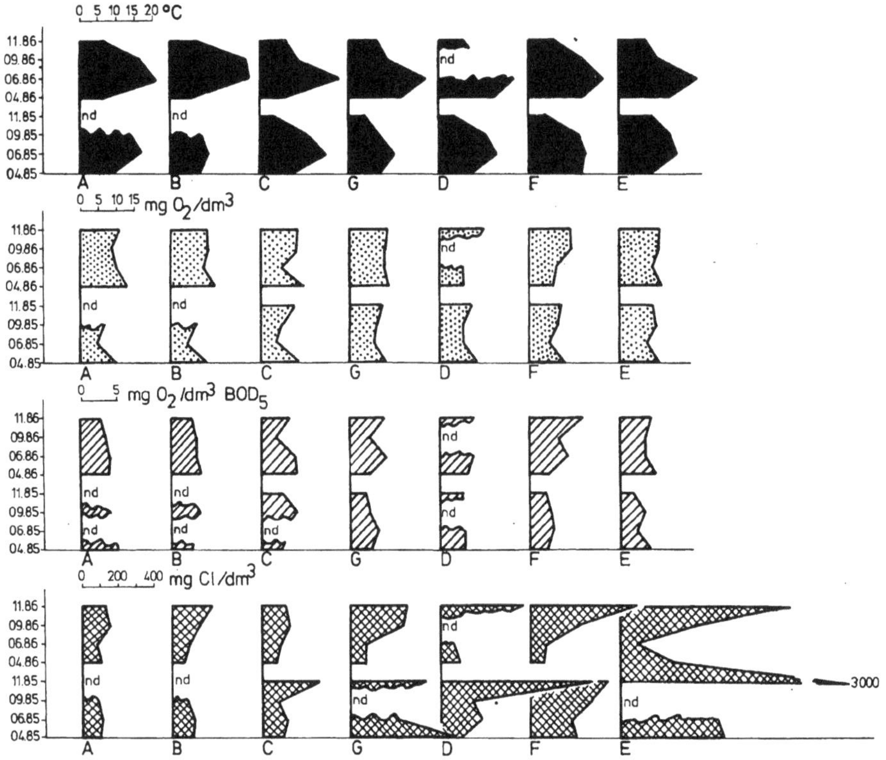

Fig. 2. Seasonal and spatial variability in environmental parameters in the Szczecin Lagoon within 1985—1986 (based on data in TADAJEWSKI et al., unpubl. report). nd — no data

Results and Discussion

Fig. 3 presents seasonal changes in the total density and diversity (number of taxa and diversity index) of meiobenthos at each station during the two years of study. As seen from the figure, densities and composition of meiobenthos at each station fluctuated considerably, no recognisable temporal pattern being detected, however. On the other hand, certain characteristics of spatial patterns can be discerned, involving a decrease in the overall density and reduction in diversity up the Lagoon toward the river mouth. This is treated in detail by RADZIEJEWSKA & DRZYCIMSKI (1988).

To check if any substantial biological accomodation, in the form of regular seasonal succession of major taxa such as that found by HEIP (1971) was present, we followed changes in abundance of three taxa, the Nematoda, Ostracoda, and Harpacticoida which made up the bulk of the Lagoon's meiofauna (RADZIEJEWSKA & DRZYCIMSKI (1988)) (Fig. 4). As opposed to HEIP (1971), we could find no consistent pattern of seasonal succession over the two years of study. We contend then that biological control in the meiobenthos of the Szczecin Lagoon is overruled by physical control in this highly variable environment.

In an attempt to identify factors, among those measured (TADAJEWSKI, BIAŁY, KNASIAK, STASZCZUK, SZUREK, PAWŁOWSKA, unpubl. report), most likely to affect the meiobenthos in the Lagoon, SPEARMAN's rank correlation coefficients (SOKAL & ROHLF 1981) were calculated for pairs of biological and environmental variables (Table 1). As shown in the table, the bottom water temperature, the most predictable, season-dependent variable, had no effect on the meiobenthos, producing no significant correlation.

Oxygen contents in the bottom water correlated significantly with densities of harpacticoids only, which is consistent with the commonly held notion that these crustaceans are very sensitive indicators of habitat deterioration manifested, in most cases, as reduced oxygen contents (RAFFAELLI 1982; WARWICK & GEE 1984).

Table 1. Significance of SPEARMAN's rank correlation coefficients for pairs of biotic and abiotic variables

Variable	Bottom water temperature	Bottom water oxygen content	Bottom water chloride content	Bottom water BOD_5
Total density	n.s.	n.s.	$p < 0.01$	n.s.
Density of Nematoda	n.s.	n.s.	$p < 0.05$	n.s.
Density of Ostracoda	n.s.	n.s.	$p < 0.01$	$p < 0.05$
Density of Harpacticoida	n.s.	$p < 0.05$	$p < 0.05$	n.s.
Number of taxa	n.s.	n.s.	n.s.	n.s.
Diversity index	n.s.	n.s.	n.s.	n.s.

n.s. = correlation non-significant;
$p < 0.05$ = correlation significant;
$p < 0.01$ = correlation highly significant;

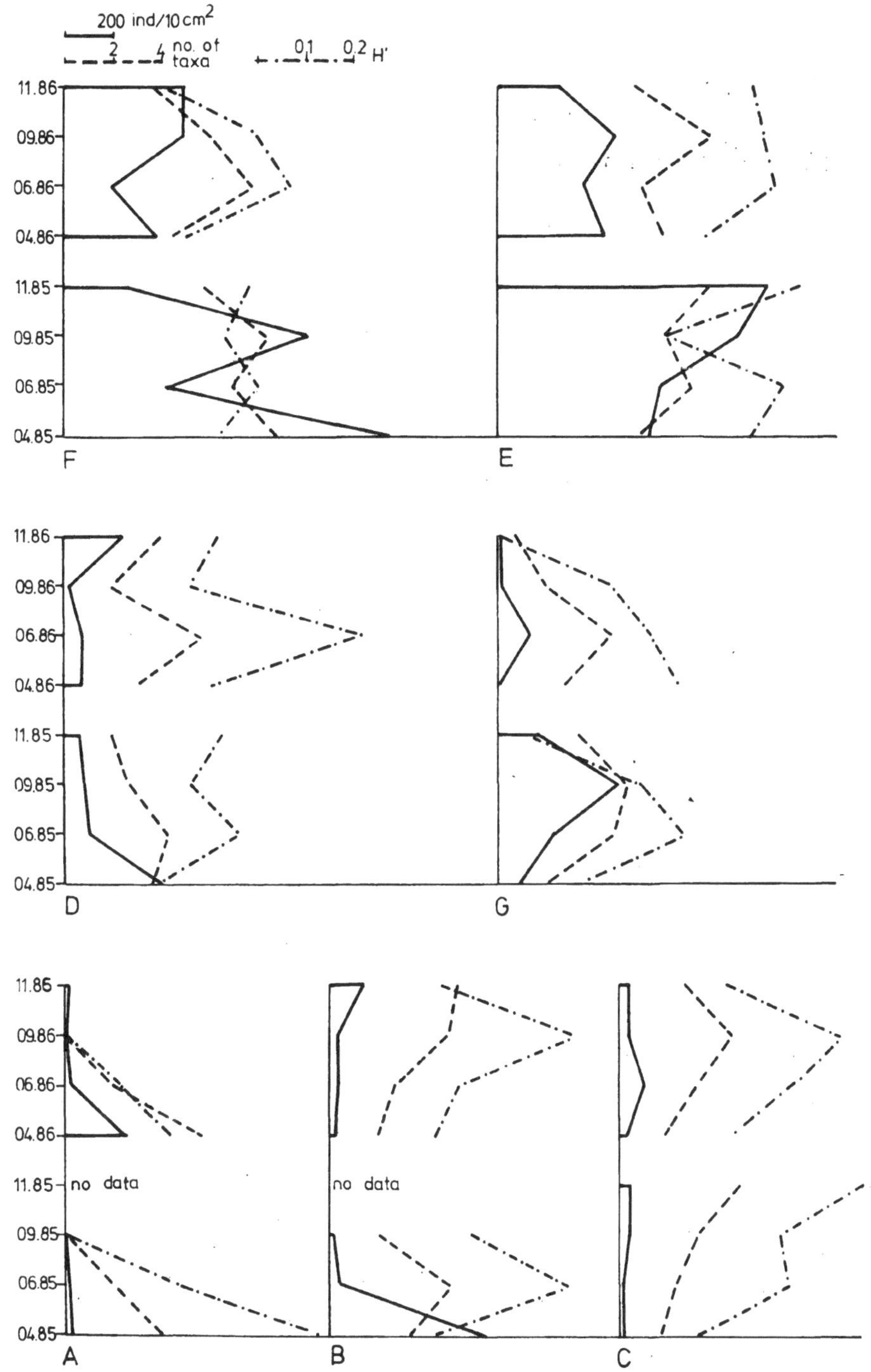

Fig. 3. Seasonal changes in mean total densities, mean numbers of taxa, and mean diversity indices in the Szczecin Lagoon meiobenthos within 1985—1986 (standard deviation bars omitted for clarity)

Chloride contents in the bottom water reflect the salinity changes in the Lagoon, resulting from inflows of the Baltic water. Chlorides (Table 1) are seen to correlate significantly with all density variables. Inflows of Baltic water into the Lagoon thus form one of the major factors enhancing the meiobenthos, possibly by favouring reproduction of brackish-water species, spreading them up the Lagoon and thus enabling colonisation, and by levelling off the pollution gradient to some extent.

The 5-day biochemical oxygen demand, a variable reflecting the amount of organic matter in the bottom water, was found to correlate significantly (and inversely) with densities of ostracods only.

Thus it may be concluded that the abundance of meiobenthos in the Szczecin Lagoon is related mainly to inflows of Baltic water into the Lagoon; the effects of inflows on the fauna may be both direct (enhancement of brackish-water species) and indirect (levelling off the pollution gradient, which results in a temporary improvement in habitat quality).

As stated earlier, and also in RADZIEJEWSKA & DRZYCIMSKI (1988), harpacticoid copepods proved 1988 to be the Lagoon meiobenthos component that is highly sensitive to environmental variability. Stepping down to the species level, the taxocene was found to consist of a total of 10 species recorded over the period

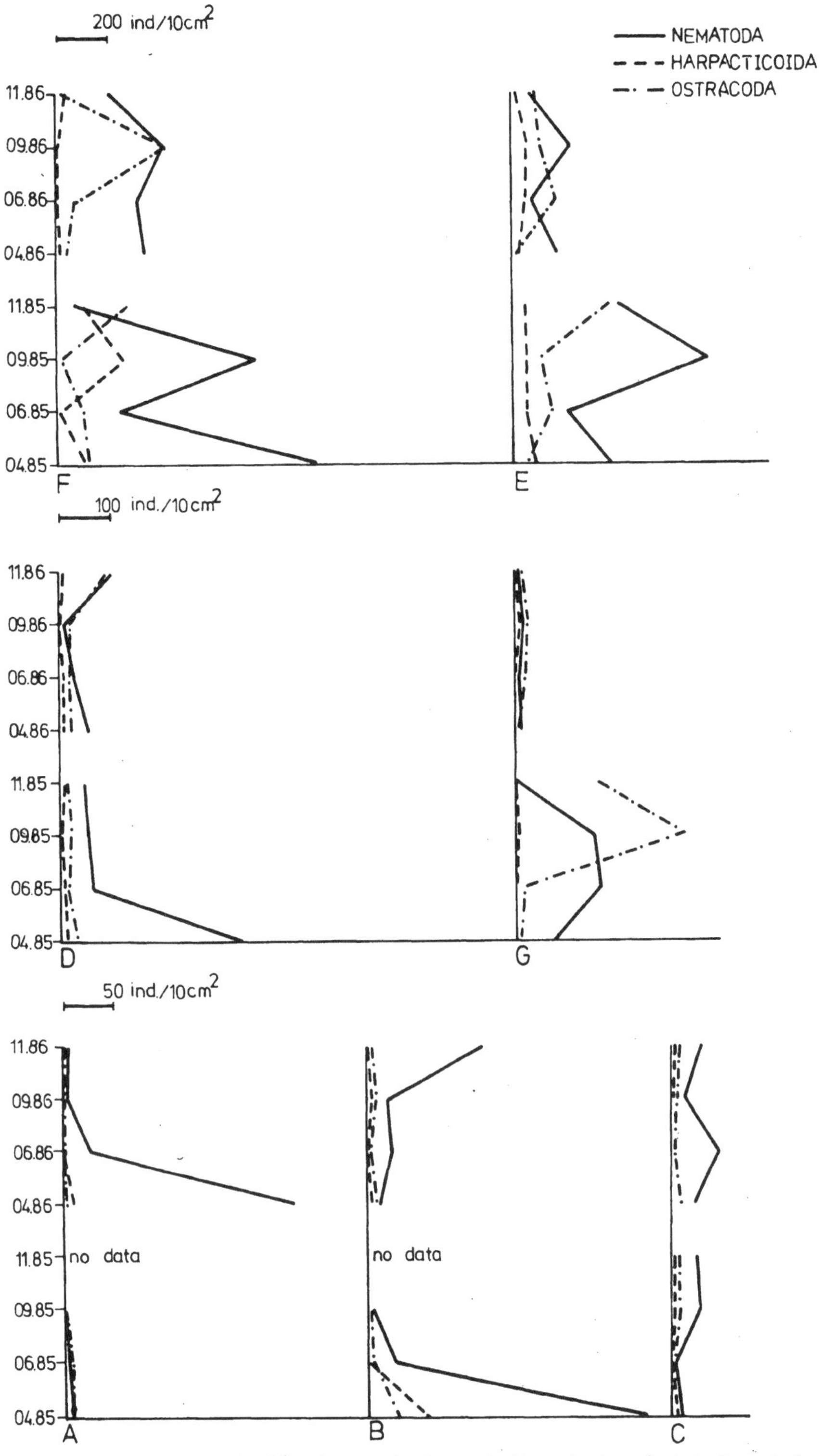

Fig. 4. Variation in mean densities of nematodes, harpacticoids, and ostracods at the Szczecin Lagoon stations in 1985—1986 (standard deviation bars omitted for clarity; note scale changes)

of study. Their occurrence along the spatial and temporal axes is presented in Table 2. The table shows certain species (*Attheyella crassa*, *Bryocamptus echinatus*, *Nitocra lacustris*, and *Paraleptastacus spinicauda*) to be exceedingly rare; when present, they appeared in very low densities only. Among the remaining, more common and abundant species, some signs of seasonal succession may be discerned: *Bryocamptus minutus*, *Canthocamptus staphylinus*, and *N. spinipes* were confined to colder seasons only (cf. SARVALA 1986).

However, the spatial axis in the harpacticoid subcommunity seems to be more pronounced than the temporal one, particularly with respect to the number of harpacticoid species per station and to *Halectinosoma curticorne* and *Microarthridion littorale*, the subcommunity dominants. As seen from Table 2, these two brackish-water species are most abundant at the most saline stations E and F. The species are known to live in the topmost layer of detritus-enriched sand (BELL & SHERMAN 1980; NOODT 1970). This, coupled with their ability to perform

Table 2. Occurrence of harpacticoid copepods in the Szczecin Lagoon

Species	Year	1985							1986						
	Station	A	B	C	G	D	F	E	A	B	C	G	D	F	E
	Month														
Halectinosoma curticorne	April		a			a	va	va				+	+		+
	June					+	+	+							a
	Sep.		+				+	a			+				a
	Nov.							a					+		+
Microarthridion littorale	April		+				+	+				+	+		
	June					+	a	a							a
	Sep.							+							
	Nov.							+							
Paraleptastacus spinicauda	April							+							+
	June														
	Sep.														
	Nov.														
Nitocra spinipes	April		+				+	+						+	
	June														
	Sep.													+	+
	Nov.														
N. lacustris	April													+	+
	June							+							
	Sep.														
	Nov.														
N. hibernica	April													+	
	June														
	Sep.						a								
	Nov.						a	a					+	+	
Canthocamptus staphylinus	April														
	June														
	Sep.									a					
	Nov.	a	a				a						a	a	
Bryocamptus minutus	April														
	June														
	Sep.						va								
	Nov.						va			+	+		a		
Br. echinatus	April														
	June												+		
	Sep.														
	Nov.														
Attheyella crassa	April								+						
	June														
	Sep.													+	
	Nov.														

\+ = present, very low density;
a = abundant;
va = very abundant;

active diurnal vertical migrations in the water column above the bottom (Arlt et al. 1980), makes them susceptible to passive horizontal transport with currents up the Lagoon (cf. Bell & Sherman 1980), which enables them to exploit suitable conditions created by the Baltic inflows.

Thus, the distribution of harpacticoid copepods in the Szczecin Lagoon reflects the environmental variability of the area, mainly its most unpredictable component governed by the inflows of the Baltic water.

Acknowledgements

We wish to thank Mr. Jerzy Masłowski for his assistance in the field, Ms. Ewa Kadziołka for help in sorting the samples, Ms. Halina Dworczak for preparing the figures, and Ms. Ewa Sobolewska for secretarial assistance.

References

Arlt, G., Radziejewska, T., & Rodbertus, L. (1980): Vorläufige Mitteilungen über Untersuchungen zur Vertikalwanderung der Meiofauna in einem Flachwassergebiet der Darß-Zingster Boddenkette. Wiss. Z. Univ. Rostock, Math.-nat. R. **29**, 123—125.

Bell, S. S., & Sherman, K. M. (1980): A field investigation of meiofaunal dispersal: tidal resuspension and implications. Mar. Ecol. Progr. Ser. **3**, 245—249.

Green, R. H. (1969): Population dynamics and environmental variability. Am. Zool. **9**, 393—398.

Heip, C. (1971): The succession of benthic micrometazoans in a brackish water habitat. Biol. Jaarb. Dodonaea **39**, 191—196.

Levinton, J. S. (1982): Marine Ecology. Englewood Cliffs.

Noodt, W. (1970): Zur Ökologie der Copepoda Harpacticoidea des Küstengebiets von Tvärminne (Finnland). Acta Zool. Fenn. **128**, 1—35.

Pielou, E. C. (1975): Ecological diversity. New York.

PŁOCKI, W., & RADZIEJEWSKA, T. (1980): A new meiofauna corer and its efficiency. Ophelia, Suppl. **1**, 213—233.

RADZIEJEWSKA, T., & DRZYCIMSKI, I. (1988): Meiobenthic communities in the Szczecin Lagoon. Kieler Meeresforsch. Sonderh. **6**, 162—172.

RAFFAELLI, D. (1982): An assessment of the potential of major meiofauna groups for monitoring organic pollution. Mar. Environ. Res. **7**, 151—154.

SARVALA, J. (1986): Patterns of benthic copepod assemblages in an oligotrophic lake. Ann. Zool. Fenn. **23**, 101—130.

SOKAL, R. R., & ROHLF, F. (1981): Biometry. San Francisco.

WARWICK, R. M., & GEE, J. M. (1984): Community structure of estuarine meiobenthos. Mar. Ecol. Progr. Ser. **18**, 97—111.

Address of the authors:

TERESA RADZIEJEWSKA
IDZI DRZYCIMSKI
Institute of Fisheries Oceanography and Protection of Sea
ul. Kazimierza Królewicza 4
71-550 Szczecin
Poland

Limnologica (Berlin) **20** (1), 89—94, April 1990

Akademie der Wissenschaften der DDR, Institut für Meereskunde, Rostock-Warnemünde, DDR

Results of Ecological Enclosure Experiments with Pelagic Communities of the Arkona Sea (Baltic Sea)

By SIGURD SCHULZ, GÜNTHER BREUEL & ANDREAS IRMISCH

With 4 Figures and one Table

Abstract

In spring, summer and autumn enclosure experiments have been carried out in the central part of the Arkona Sea. The natural surface water was enriched with nutrients up to the winter level and changes of the community structure (addition and removal of zooplankton) were conducted. In spring and summer the phytoplankton responded on the addition with a rapid uptake of nutrients and an increase in primary production and biomass. In autumn the uptake of nutrients was also quick, the productivity however did not answer and a clear increase of the biomass was only in diatoms visible.

The addition of zooplankton influenced the productivity of the phytoplankton only under summer conditions whereas in spring and autumn the effects were negligible.

As a conclusion the attempt is made to apply some of the results to possible consequences for the pelagic system of the Baltic Sea.

1. Introduction

The study of ecological processes running small scales in time and space (e.g. phytoplankton growth, nutrient uptake, grazing) are hardly to investigate using the normal oceanographical measuring strategies. Therefore the introduction of micro- and mesocosm-experiments was the most promising approach, although even not without restrictions (e.g. lack of water- and gas exchange etc.) in comparison with in situ or laboratory experiments (BOYD 1981).

The present paper summarizes some results of the enclosure experiments carried out by the Institute of Marine Research during different seasons and years in the Central Arkona Sea.

The aim of these experiments was to study the response of the phytoplankton community on

— nutrient additions and
— changes in the grazing pressure by addition or removal of mesozooplankton.

2. Material and Methods

The three experiments were carried out at the GDR-station 113 (54° 55,5′N, 13° 30,0′E) in the Arkona Sea, which is also a site in the frame of the Baltic Monitoring Programme of the HELCOM. The cylindrical enclosures with a volume of 25.4 l floated in 2.5 m water depth, further details on the containers and the performance of the experiments are described in SCHULZ et al. (1985).

The experiments were carried out to the following times:

ÖKEX' 81 8.—28. 4. 1981
ÖKEX' 78 7.—27. 7. 1978
ÖKEX' 83 13.—30. 9. 1983.

The chemical and biological methods used, follow the standards accepted and recommended for the Baltic Sea (ROHDE & NEHRING 1979; DYBERN et al. 1976). Zooplankton removal was done by sieving the water through a net with the mesh size of 90 µm, while in the experiments with zooplankton addition natural zooplankton was added.

3. Results

The experiments in April started just before the onset of the spring bloom, therefore in the first and second experiment (Fig. 1) the nutrient concentration showed the high winter level. About one week later the surface water was nearly exhausted from nitrogenous nutrients and the experiments with addition of nutrients started (exp. III and IV). The nitrate and ammonia were taken up rapidly, differences between the four experiments did not occur. The phytoplankton was dominated by diatoms like *Thalassiosira rotula*, *Achnanthes taeniata*, *Skeletonema costatum* and *Chaetoceros* sp. The amount of µ-algae was relatively small although they also increased during the experiments. Dinoflagellates occurred only in the last two experiments. The zooplankton community remained according to the season at a low developmental stage, nauplii and copepodids of calanoid copepods dominated the zooplankton. As could be seen from Fig. 1 additions of zooplankton as well as its removal did not influence the development of the algae in general.

In summer the surface water was completely exhausted from nitrogenous nutrients as it is typical for this season and area (Fig. 2). In these experiments deep water from the station (sampling depth 43 m) was added to the surface water in relation 2:1 to simulate a high ammonia input in preference to nitrate. The added nutrients were also in these experiments taken up rapidly down to zero.

The phytoplankton consisted mainly of cyanobacteria like *Nodularia spumigena* and *Aphanizomenon flos-aquae* and µ-algae. Dinoflagellates became more numerous during the experiments as well as diatoms. As could be seen from all experiments with deep water additions the µ-algae responded in the beginning of these experiments fast. The diatoms reacted similar with an increase in biomass. After a few days, however, the increase in phytoplankton biomass was only caused by cyanobacteria, pointing beside other effects to their ability of fixing molecular nitrogen (HÜBEL & HÜBEL 1980).

By this action also the control bags showed a clear increase in biomass, in contrast to the spring experiments.

Acartia spp. dominated the zooplankton, effects of the zooplankton activity could be observed in uptake of ammonia by the phytoplankton and in the phytoplankton biomass development itself. In the experiments with zooplankton addition the doubled zooplankton biomass

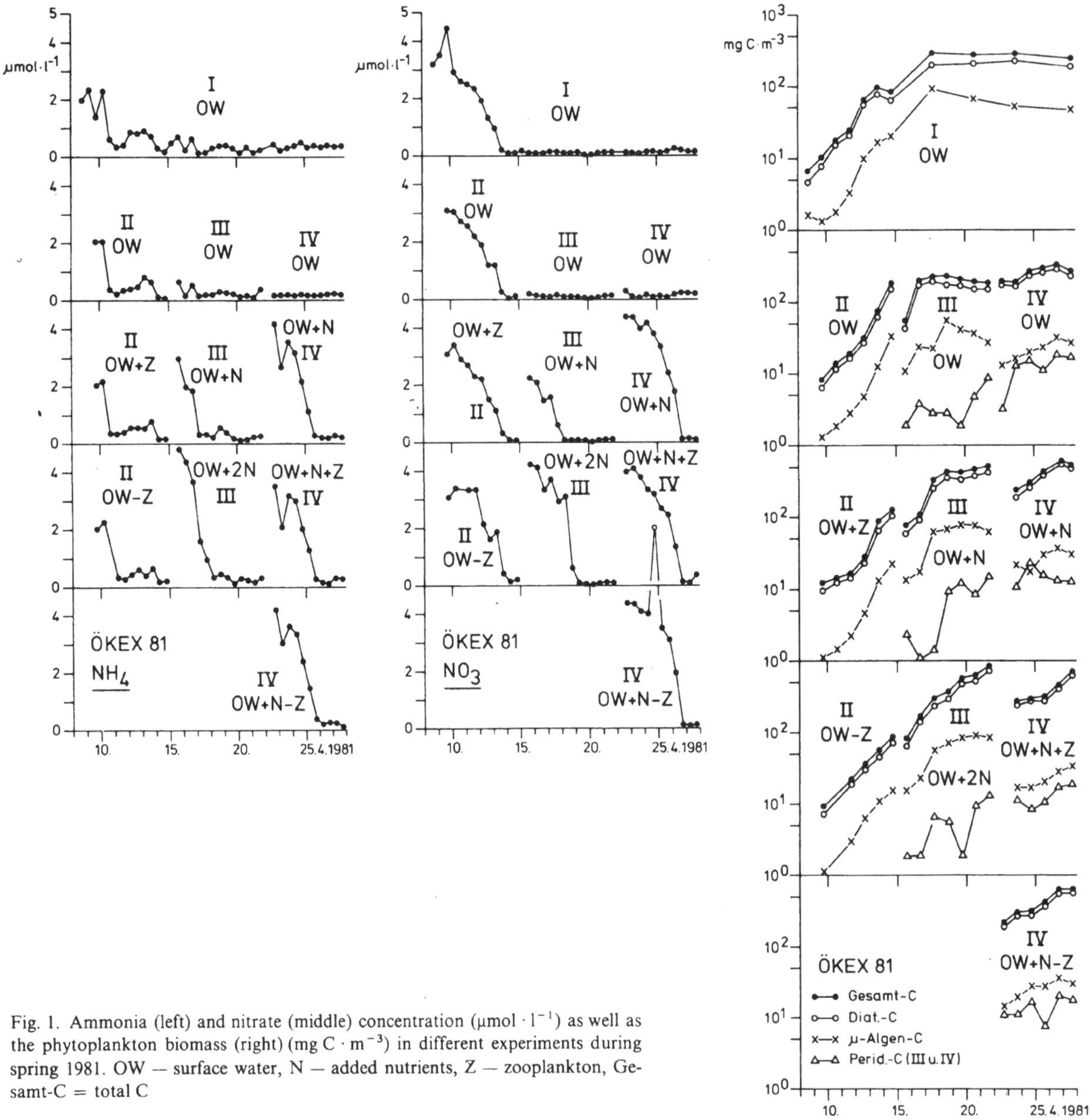

Fig. 1. Ammonia (left) and nitrate (middle) concentration (µmol · l^{-1}) as well as the phytoplankton biomass (right) (mg C · m^{-3}) in different experiments during spring 1981. OW — surface water, N — added nutrients, Z — zooplankton, Gesamt-C = total C

was reached and the ammonia was later diminished (Fig. 2) than in the parallel experiments. Diatoms, especially *Skeletonema costatum*, were preferably grazed in comparison with the other phytoplankton groups and the other batches.

Although the phosphate concentrations in the surface water in September showed a slight increase the limiting nitrogen was furtheron nearly depleted in the surface. After additions of nutrients to about the winter concentrations (Fig. 3) only in the third experiment both ammonia and nitrate were taken up, whereas in the second experiment preferably ammonia was used.

The phytoplankton was dominated by *Rhodomonas minuta*, diatoms and dinoflagellates played only a minor role and cyanobacteria were absent, because of the season. As could be seen from Fig. 3 the nutrient additions had only small effects on the phytoplankton biomass altogether. This is caused mainly by the inactive μ-algae. Diatoms on the other hand showed a remarkable increase in biomass in the third experiment.

According to the season the zooplankton activity and biomass dropped in the surface water, which was used in the experiments. As in summer furtheron *Acartia* spp. were the dominating part. Again the growing diatom population showed a sink in the experiments with added zooplankton in comparison with the parallel experiments.

4. Discussion

The phytoplankton took up the natural nutrient stock in spring experiment and the additional supplied nutrients during all three seasons very rapidly as could be seen in Figs 1—3.

The winter concentrations of phosphate, nitrate and ammonia were 0.61, 3.20 and 2.98 µmol · l^{-1} and daily absolute uptake rates of 0.16 µmol · l^{-1} for phosphate, 0.65 µmol · l^{-1} for nitrate and 0.65 µmol · l^{-1} for ammonia were calculated. In the other seasons the relative uptake rates were rather similar although the absolute rates differed in dependency on the initial concentrations (SCHULZ et al. 1988). The phytoplankton quickly responded to the nutrient input in spring and summer,

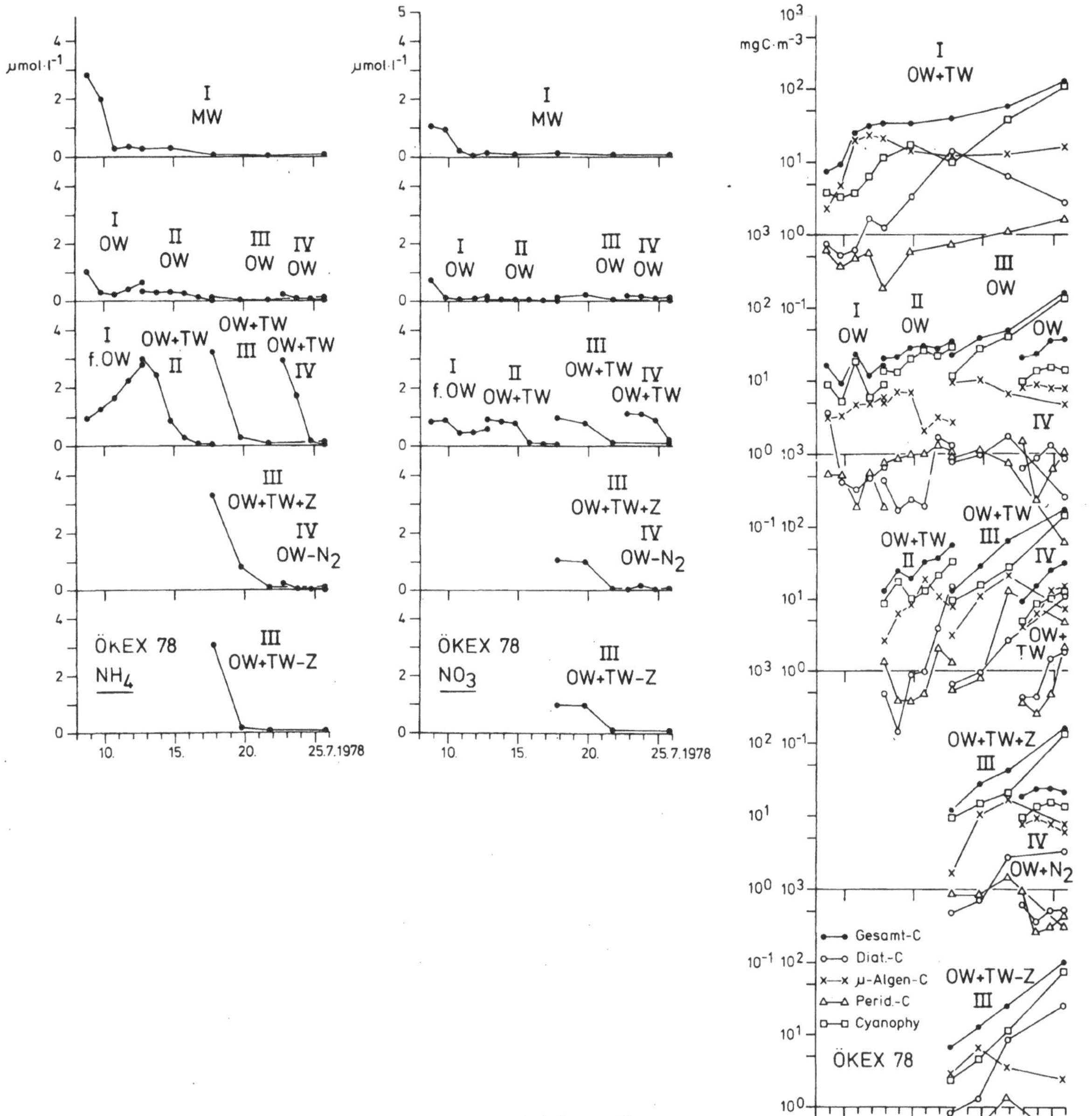

Fig. 2. Ammonia (left) and nitrate (middle) concentration (µmol · l^{-1}) as well as the phytoplankton biomass (mg C · m^{-3}) (right) in different experiments during summer 1978. TW — deep water

in autumn only diatoms and some small unidentified dinoflagellates grew, whereas growth of the dominating µ-algae could not be observed.

In all three seasons the species composition remained unchanged when nutrients were added. In dependency on the differing relation of phytoplankton species, however, changes in the dominance occur.

The mesozooplankton activity on the system is reflected in lowering the phytoplankton stock and in increasing the remineralisation. In dependency on the wellknown seasonal differences of mesozooplankton biomass and metabolic activity (Ackefors & Lindahl 1979) different results could be expected in the experiments with zooplankton addition or removal.

In April no reaction of the dominating diatoms could be found. The µ-algae, however, showed some grazing pressure. This assumption is based on the alterations in the percentual share of this group during the experiments (Table 1).

Table 1. Percentual portion ($\bar{x}$), standard deviation (s), minimal (min) and maximal (max) percentual portion of the µ-algae (%) of the total phytoplankton biomass and number of observations (n) during the second experiment in April 1981

	$\bar{x}$	s	min	max	n
OW	15.9	1.7	13.5	18.3	6
OW + Z	13.8	3.4	9.1	17.5	6
OW − Z	15.6	3.4	10.7	18.9	5

OW — surface water; Z — zooplankton.

It seems due to the composition of the zooplankton during the time, the dominating nauplii and young copepodids were not able to use the bulk of diatoms and feed because of the size mainly on the µ-algae.

In summer high zooplankton biomass and activity are typical (Schulz & Breuel 1981, 1984) and therefore

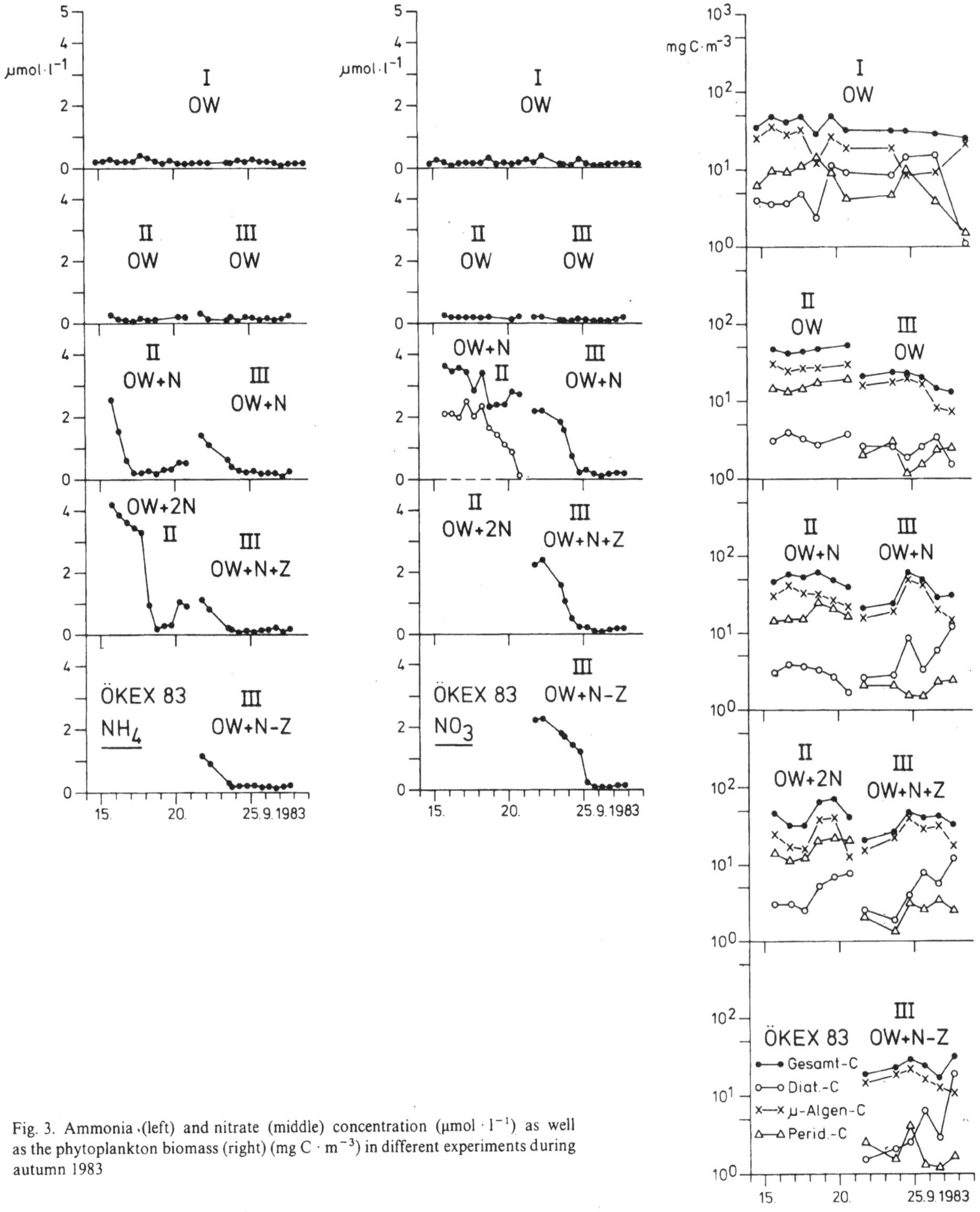

Fig. 3. Ammonia (left) and nitrate (middle) concentration ($\mu mol \cdot l^{-1}$) as well as the phytoplankton biomass (right) ($mg\ C \cdot m^{-3}$) in different experiments during autumn 1983

clear differences between batches with zooplankton addition and removal in respect to biomass development of the phytoplankton could be observed. The zooplankton fed mainly on the fast growing *Skeletonema costatum* population (Fig. 4), similar observations made BROCKMANN et al. (1977) and GRANELI (1981).

The primary production shows the highest values also in the + Z-experiment with the largest grazing pressure (Fig. 4). This was also found by MCKELLAR & HOBRO (1976). The reason is a higher remineralisation in the batch with added zooplankton and our results support the idea that the productivity of an ecosystem is to a great extent determined by its turnover rates (GOLDMAN 1985).

In autumn again the just developing *Skeletonema costatum* population was grazed preferably, but no effect could be observed on the ammonia concentration.

Different to spring and summer the zooplankton concentration in the + Z-bags was in autumn only somewhat higher than in the unchanged bags and therefore no clear differences between the bags could be obtained.

As the experiments have shown the input of nutrients leads not only to higher phytoplankton activity (primary production, biomass, chlorophyll) but also to a prolon-

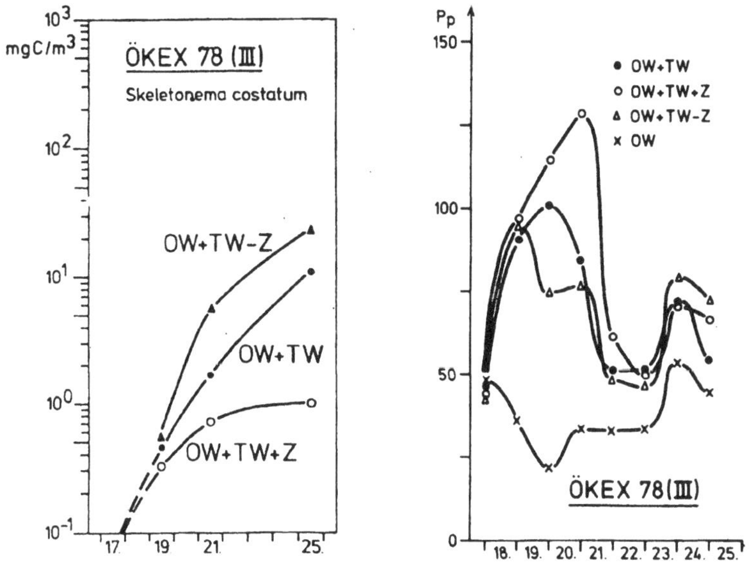

Fig. 4. Biomass of *Skeletonema costatum* (mg C · m^{-3}) and primary production (mg C · m^{-3} · d^{-1}) during ÖKEX' 78 (third experiment)

gation of the productive period (SCHULZ et al. 1985). This seems to be especially in spring conditions important because the gap in the time scales of phyto- and zooplankton development becomes smaller. That means that a larger share of the spring phytoplankton biomass might be used by the just developing zooplankton, increases the stock themselves and retains a larger part of chemical energy in the pelagic system. On this way the summer production might be influenced positively.

From the experiments however it turned also out that the input of nutrients in the different seasons does not lead to changes in the species composition. Only the dominance relations change as a response. This is also the actual result of most of the phytoplankton studies until now, which conclude that although the winter nutrient level increased significantly, changes in species composition could not be proved (WULFF et al. 1986).

Nutrient inputs to the pelagic system in different seasons have for the ecosystem a different meaning. In spring the phytoplankton production responds fastest to the impact and because of the early stage of the zooplankton development also the largest part of the addition is sedimented and acts as an oxygen sink in the deepwater. Only a small share increases the productivity over the whole year, as it was mentioned above.

The summer situation in the Baltic is characterized by a well balanced pelagic community (SCHULZ & BREUEL 1984), a low grazed down phytoplankton stock and a highly developed zooplankton community. This system reacts to an input of nutrients also very rapidly. In contrast to the spring conditions, however, the additional energy leads to an enhancement of the productivity of all compartments of the pelagic system. The organic substance is retained in the euphotic layer and circulated several times, sedimentation is negigible during this time (PRANDKE 1980; PEINERT et al. 1982). If the addition is not completely transferred to fish biomass or higher predators than later on it threatened the oxygen in the deep water again.

The autumn situation equals the spring conditions because again by the decrease of the zooplankton stock the pelagic system tends to a misbalance leading to a higher sedimentation of newly produced algal biomass.

References

ACKEFORS, H., & LINDAHL, O. (1979): Primary phytoplankton production in the Baltic proper, 1973—1976, in relation to secondary production. ICES paper C.M. **1979/L: 32**, 1—13.

BOYD, C. M. (1981): Microcosms and experimental planktonic food chains. In: LONGHURST, A. R. (ed.), Analysis of Marine Ecosystems, pp. 627—650. London.

BROCKMANN, U. H., EBERLEIN, K., HENTZSCHEL, G., SCHONE, H. K., SIEBERS, D., WANDSCHNEIDER, K., & WEBER, A. (1977): Parallel plastic tank experiments with cultures of marine diatoms. Helgol. Wiss. Meeresunters. **30**, 201—216.

DYBERN, B. I., ACKEFORS, H., & ELMGREN, R. (1976): Recommendations on methods for marine biological studies in the Baltic Sea. BMB Publication No. **1**, 1—98.

GOLDMAN, J. C. (1985): Oceanic nutrient cycles. In: FASHAM, M. J. R. (ed.), Flows of energy and materials in marine ecosystems. Theory and practice, pp. 137—170. London.

GRANELI, E. (1981): Experimental investigations of limiting nutrients for phytoplankton production in the brackish-water Oresund, SW Sweden. Thesis, Lund, 1—18.

HÜBEL, H., & HÜBEL, M. (1980): Nitrogen fixation during blooms of *Nodularia* in coastal waters and backwaters of Arkona Sea (Baltic Sea) in 1974. Int. Rev. ges. Hydrobiol. **65**, 793—808.

MC KELLAR, H., & HOBRO, R. (1976): Phytoplankton — zooplankton relationships in 100-liter plastic bags. Contributions from the Askö Laboratory, University of Stockholm, No. **13**, 1—83.

PEINERT, R., SAURE, A., STEGMANN, P., STIENEN, C., HAARDT, H., & STEMACEK, V. (1982): Dynamics of primary production and sedimentation in a coastal ecosystem. Neth. J. Sea Res. **16**, 276—289.

PRANDKE, H. (1980): Einige Ergebnisse der Lichtstreuuntersuchungen in der mittleren Ostsee. Beitr. Meereskd. **44/45**, 43—54.

ROHDE, K.-H., & NEHRING, D. (1979): Ausgewählte Methoden zur Bestimmung von Inhaltsstoffen im Meer- und Brackwasser. Geod. Geoph. Veröff. R. IV, H. **27**, 1—68.

SCHULZ, S., & BREUEL, G. (1981): Response of plastic-bag enclosed summer phytoplankton to a nutrient input under in situ conditions in the Arkona Sea (Baltic Sea). Kieler Meeresforsch., Sonderh. **5**, 91—100.

— — (1984): Variations in primary production, chlorophyll a, phytoplankton and zooplankton during an anchor station in the central Arkona Sea in July 1978. Part **4**, pp. 45—61. Leningrad.

— — & IRMISCH, A. (1988): Simulated eutrophication in enclosure experiments in the Arkona Sea. Kieler Meeresforsch. (in press).
— — — JOST, G., & SIEGEL, H. (1985): Ergebnisse ökologischer Untersuchungen an eingeschlossenen Planktongemeinschaften der Arkonasee im Frühjahr 1981. Geod. Geoph. Veröff. R. IV, H. **41**, 1—66.
WULFF, F., AERTEBJERG, G., NICOLAUS, G., NIEMI, Å., CISZEWSKI, P., SCHULZ, S., & KAISER, W. (1986): The changing pelagic ecosystem of the Baltic Sea. Ophelia, Suppl. **4**, 299—319.

Address of the authors:

Dr. sc. SIGURD SCHULZ
Dipl.-Biol. GÜNTHER BREUEL
Dr. ANDREAS IRMISCH
Akademie der Wissenschaften der DDR
Institut für Meereskunde
Seestr. 15
Rostock-Warnemünde
DDR-2530

Limnologica (Berlin) **20** (1), 95—100, April 1990

Wilhelm-Pieck-Universität Rostock, Sektion Biologie, Rostock, DDR

Carbon Flux Dynamics in a Shallow Eutrophic Estuary

By U. SCHIEWER, H. ARNDT, G. JOST, R. HEERKLOSS & N. WASMUND

With 3 Figures and 3 Tables

Key words: Eutrophic estuary; Mesocosm; Carbon flux; Microbial loop.

Abstract

Joint enclosure experiments were performed in Kirr Bucht, a shallow eutrophic bay in the Darss-Zingst estuary. The pelagic community played a major role in the main carbon flux of this ecosystem. Despite constant changes, various states of dynamic equilibrium were identified which were related to nutrient loadings and seasonal periodicity.

The results presented refer to the (N + P)-loaded enclosure from the 1st to 2nd and from the 5th to 6th week of the experiment. The carbon values describing phytoplankton biomass and activity during these two periods are nearly equal, but considerably more detritus was consumed during the second period than during the first. The links and dynamics of the carbon flux are discussed in terms of the principle organism groups or species involved.

Introduction

The chief characteristic of estuarine ecosystems is the great variability of their structural and functional parameters in space and time. Seasonal rhythms triggered by abiotic factors are also one of their typical features. To understand these rhythms it is essential to know the magnitudes of the different pools and to identify the fluxes between them. The Coastal Research Group of the Department of Biology at Rostock has used the advantages of mesocosm experiments (MENZEL & STEELE 1978; TAKAHASHI et al. 1975) to study the carbon, nitrogen and nutrient fluxes in the course of joint field experiments (SCHIEWER & JOST 1986). The experiments focused on the problem of controlling community structure and function "from below" (MULLIN 1982) by nutrient loading in "bottom-up" experiments. Nutrient loads directly affect the phytoplankton, the main biotic component of the ecosystem we studied. The ultimate aim is to identify possible ways to manipulate this particular estuarine ecosystem and to establish a scientific basis on which biological control technology or ecotechnology can be established.

This paper uses our investigations into the interactions between carbon flux and various ecosystem components to describe our approach. As examples we have selected two quasi-stationary states observed in the (N + P)-loaded enclosure during a long-term experiment (FLAK 86). These two states are characterized by approximately equal phytoplankton and detritus pool sizes, and identical, increasing phytoplankton productivity. They differ with respect to detritus pool size, however, which remained stable during the first period but decreased significantly during the second. We investigated the details of the structural and functional changes associated with these states.

Material and Methods

The experiments were performed in Kirr Bucht, a shallow region of the boddens south of the Darss-Zingst peninsula (Fig. 1A). Enclosures were used to isolate shallow bodies of water with a mean depth of 75 cm and a surface area of about 3 m^2 from the surrounding water (Fig. 1B). The enclosed water body had a volume of about 2.4 m^3. The enclosures (FLAK series) consisted of metal rings with a diameter of 2 m and a height of 40 cm which were open at the top and bottom. They were driven 35 cm into the sediment to prevent water exchange at the bottom. A polyethylene cylinder with a height of 95 cm was fastened to each metal ring. Floats at the top of each cylinder provided the necessary buoyancy and also prevented outside water from entering the enclosure during rough weather. The whole structure was protected against wind and wave action by a secondary structure (Fig. 1B).

The experiments began on 30 May 1986 when the enclosures were set up. They ended on 24 July 1986. The water in the enclosure considered here was enriched with NH_4Cl and NaH_2PO_4 from 03 June to 21 July 1986 to prevent possible nutrient limitation of phytoplankton development.

The following parameters were measured on at least 2.5 days per week throughout the experiment:

- chlorophyll a concentration by methanol extraction (RIEMANN 1978),
- phytoplankton composition and biomass (UTERMÖHL 1958),
- in situ phytoplankton production by the ^{14}C-method (SCHINDLER et al. 1972),
- number of bacteria by epifluorescence count (JOST & BALLIN 1980),
- bacterial secondary production by (^3H)thymidine incorporation into DNA (FUHRMAN & AZAM 1980, 1982),
- ciliate number by discrete counts of live organisms (individuals sorted into size classes for biovolume estimation),
- numbers and feeding rates of rotifers and crustaceans (SCHNESE & HEERKLOSS 1978), and
- detritus content (GEORGI et al. 1980).

The values used for the Figures are means of selected parameters in the (N + P)-loaded enclosure during the 1st—2nd and 5th—6th week. Table 1 presents the changes in phytoplankton biomass and detritus content during these periods.

Table 1. Changes in the phytoplankton and detritus pools (mg · m^{-3} · C) during the selected periods (1st—2nd and 5th—6th week) of the FLAK-experiment. (N + P)-loaded enclosure

Period	Phytoplankton	Detritus
1. period		
02. 06. 1986	9747	19985
14. 06. 1986	18199	18199
mean	13937	
2. period		
30. 06. 1986	5572	19291
11. 07. 1986	23891	12364
mean	14732	

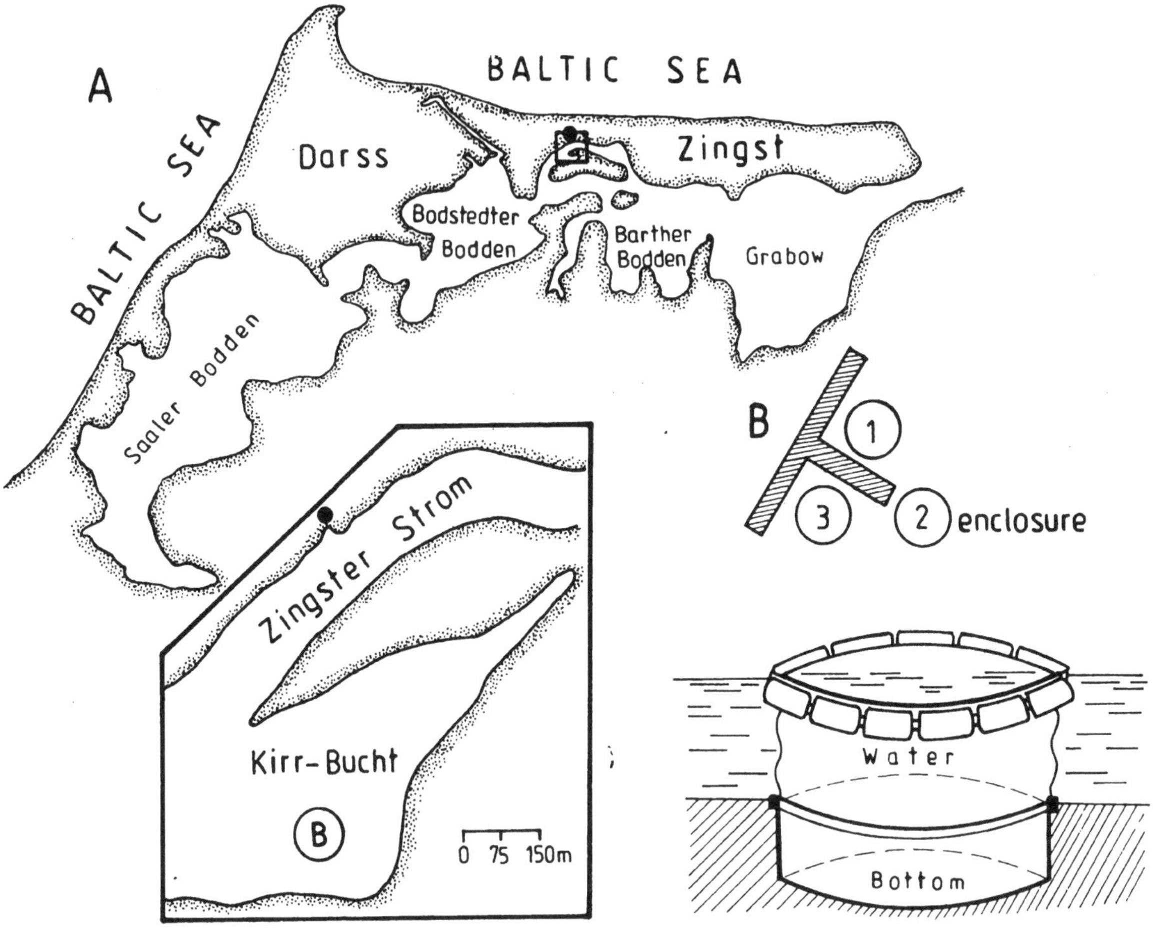

Fig. 1. Investigation area and mesocosm design.
A. Chain of boddens south of the Darss-Zingst peninsula.
B. Laboratory station Zingst, Kirr Bucht; mesocosm and enclosure design

Results

The ecosystem we studied is brackish and eutrophic. The food web is predominantly microbial in the pelagic zone ("microbial loop"). Large invertebrates (meiofauna, macrofauna) are important consumers only in the benthic zone. Figs. 2 and 3 show the measured pool sizes, input and output rates and the calculated fluxes between system components.

The detritus and phytoplankton pools are largest. The bacteria pool is significantly smaller. The sizes of these pools remained approximately constant during both periods. However, the turn-over rates of the detritus and bacteria pools were much higher during the second period than during the first. The pools comprising the other heterotrophic organisms varied considerably, but in all cases they were larger during the second period than during the first.

The crustaceans have virtually no influence on the turnover of matter. Small heterotrophic organisms are responsible for most of the carbon flux. This accelerates recycling within the ecosystem. The carbon flux to the sediment during the first period differs qualitatively from that during the second: sedimentation of phytoplankton predominated during the first, but the detritus pool, phytoplankton and heterotrophs metabolism contribute almost equally to this flux during the second.

The fluxes of matter between the various ecosystem components and the ways in which they can be influenced must be known if the processes taking place in the ecosystem are to be understood. We therefore analyzed the structural and functional variations in the different groups of organisms in order to elaborate a corresponding web.

Phytoplankton

Phytoplankton development was definitely nitrogen-limited when the experiment began. Loading with

Table 2. Phytoplankton species and percentage biomasses during the selected periods. (N + P)-loaded enclosure.

Period	% Biomass
1st—2nd week	
Crucigenia quadrata var. *secta*	46.9
Lyngbya contorta	17.6
Gomphosphaeria pusilla	14.8
Scenedemus quadricauda	6.5
Oocystis spec.	2.5
others	11.7
5th—6th week	
Scenedesmus quadricauda	16.0
Oscillatoria limnetica	12.9
Chroococcus limneticus	12.9
Synedra spec.	11.9
Crucigenia quadrata var. *secta*	11.3
others	34.9

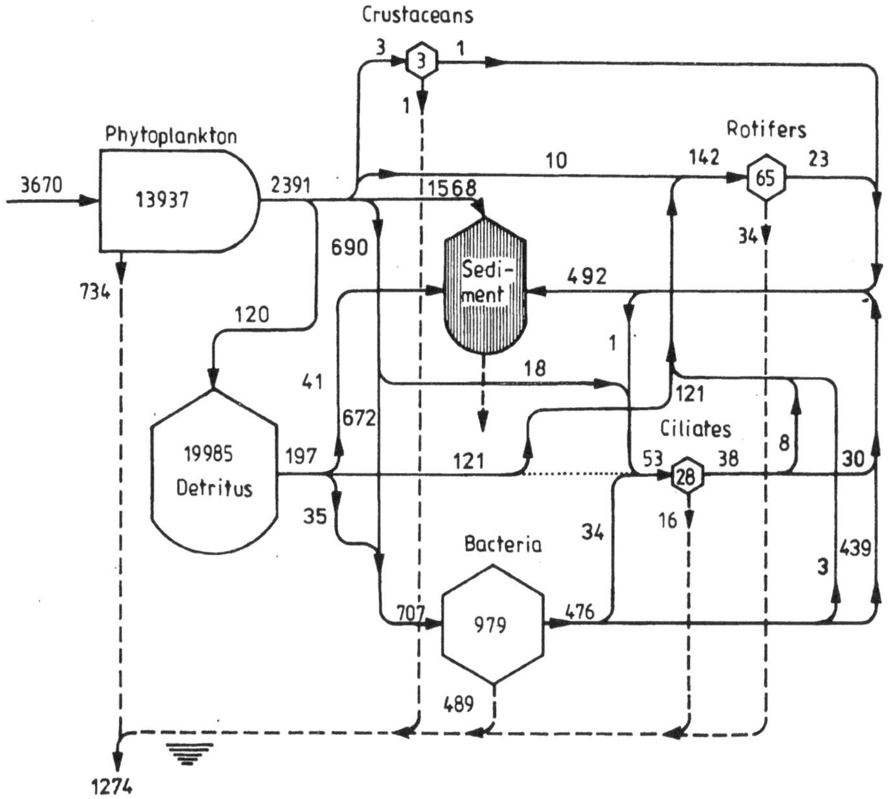

Fig. 2. Carbon pools and fluxes in the (N + P)-loaded enclosure, 1st-2nd week of the FLAK-experiment 1986.
Numbers in geometric symbols: carbon contents in mg · m^{-3} in the phytoplankton, detritus, crustacean, bacteria, ciliate and rotifer pools. Solid arrows: carbon fluxes in mg · m^{-3} · d^{-1} between the different pools. Broken arrows: respiration losses in mg · m^{-3} · d^{-1} carbon and total loss. Estimated values: pools, input and output; calculated: fluxes between the pools (see text). The sediment is considered as a sink only. All values are means for the 1st period

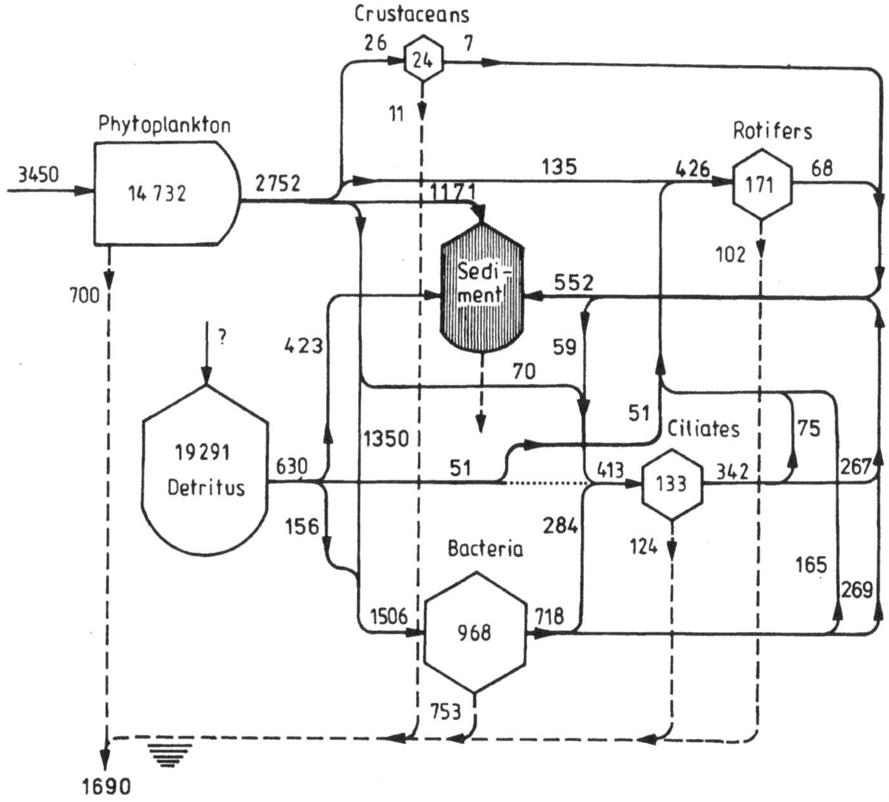

Fig. 3. Carbon pools and fluxes in the (N + P)-loaded enclosure, 5th—6th week of the FLAK-experiment 1986. (cf. Fig. 2 for further details)

(N + P) encouraged primary production and led to a larger, stable phytoplankton pool. Comparison of the two periods (Table 2) shows that the carbon pools and fluxes remain almost constant although the community structures change completely. During the 1st and 2nd week the green algae, particularly *Crucigenia*, predominate. On account of their high growth rate they outcompete the other algae. This and their relatively high sedimentation rate increases the flux of algae into the sediment.

During the second period the phytoplankton consisted of about equal biomasses of at least five algal species. Their co-existence indicates that growth conditions were about equally suitable for them all on account of their different growth and/or loss rates.

Bacteria

The structure of the bacterial populations was not analyzed. The different turnovers despite equal bacterial biomasses raises the question of the regulating mechanisms involved. There is evidently no direct link between either input rates or bacterial secondary production and phytoplankton biomass or productivity. This is contrary to the findings reported by FUHRMAN et al. (1980) for marine ecosystems and by HAGSTRÖM & LARSSEN (1984) for Baltic biotopes. The range of phytoplankton species and/or the detritus are obviously more important in the eutrophic brackish waters we studied.

Ciliates

Ciliates are always an important component of the ecosystem we studied (ARNDT 1986). The number of ciliate species it contains and the rapid changes in the composition of species are considerable (ARNDT et al. 1988). To avoid the complexity this would normally entail, we have split the ciliates into the three most important functional groups (Table 3). This also facilitated quantification. The bacterivorous and detritivorous group always had the highest feeding rate. The substantial decline in the algivorous group during the 5th—6th week was accompanied by growth of the omnivorous group.

Rotifers

A distinct change in dominance took place among the rotifers during our investigations (Table 3): *Filinia longiseta* was replaced as the dominant species by smaller rotifers. The other species played only a minor role. *Filinia longiseta* is a typical detritivour, and its loss of dominance was accompanied by a shift towards increased consumption of the bacteria by the smaller rotifers.

Crustaceans

Several crustacean species are found in the ecosystem although their total biomass is small. The copepods predominated initially, but cladocerans became more abundant later (Table 3). This change had no real impact on the ecosystem, however, because the potential turnover of the latter group is very small.

The above information regarding dominance and feeding rates was used to calculate the carbon fluxes between

Table 3. Functional groups of ciliates, species of rotifers and crustaceans and their percentage biomasses and feeding rates in the (N + P)-loaded enclosure during selected periods. FLAK-experiment 1986

	1st—2nd week		5th—6th week	
	Biomass (%)	Feeding rates (%)	Biomass (%)	Feeding rates (%)
Ciliates				
Bacterivores	33.8	64 6	66.5	68.8
Algivores	58.8	34.2	4.0	16.9
Omnivores	7.4	1.2	29.5	14.3
Rotifers				
Small rotifers (100 µm; *Trichocerca*, *Synchaeta* etc.)	1.5	1.7	79.9	82.0
Filinia longiseta	85.3	85.0	2.4	2.2
Brachionus urceolaris	5.3	5.3	8.0	7.2
B. quadridentatus	5.4	5.4	0.4	0.4
B. angularis, *Keratella tecta* and others	2.5	2.6	9.3	8.2
Crustaceans				
Copepods	100.0	100.0	68.2	58.7
Cladocera	0.0	0.0	31.8	41.3

the different compartments of the community. The sediment was considered only as a sink (Figs. 2 and 3). The respiration rate of the sediment as measured by the O_2 method was only about one third that of the pelagic zone. The ciliate population of the sediment increased 10-fold from the 1st—2nd week to the 5th—6th week. It consisted at all times mainly of bacterivours and detritivours. The meiozoobenthos, particularly the oligochaetes and harpacticoids, declined during the same interval. These changes were caused by increased organic inputs owing to the greater productivity of pelagic phytoplankton and accelerated the turnover of matter in the sediment.

During the 1st—2nd week only about a third of the phytoplankton production was used by pelagic heterotrophs, which were mainly bacteria. The remainder sedimented to the bottom. For ciliates the main food source is bacterioplankton, whereas the rotifers feed mainly on the detritus pool. With turnover times of $1.9 \cdot d^{-1}$ (ciliates) and $2.2 \cdot d^{-1}$ (rotifers) these two groups have considerable biomass turnover rates.

The carbon flux through the bacteria compartment was twice as high during the 5th—6th week. The larger amounts of dissolved organic matter needed for such an increase are probably released partly from the detritus and partly a result of increased biomass turnover by the ciliates and rotifers. The greater phytoplankton diversity probably also enhances bacterial metabolism by increasing the range of utilizable dissolved organic matter available. The increase in the proportion of omnivorous ciliates increases the density of the carbon flux web in the ciliate compartment. Bacteria and small phytoplankters constitute the main food of the rotifers during the second period of our study. The turnover time had increased for both the rotifers ($2.5 \cdot d^{-1}$) and ciliates ($3.1 \cdot d^{-1}$).

Discussion

Despite the virtual absence of large invertebrates as consumers, in the shallow pelagic zone 34% (1st—2nd week) and 55% (5th—6th week) of the net phytoplankton production were consumed during the experiments. This emphasizes the great importance of the "microbial loop" (AZAM et al. 1983) in this ecosystem. Compared with the 80% reported by OVERBECK (1983) and SIMON & TILZER (1987) for the top 5 m and 20 m of the pelagic zone in lakes, these turnover rates are amazingly high for a water depth of only 75 cm.

In the ecosystem we studied the decisive heterotrophic component is the bacterioplankton. According to JOST & BALLIN (1980) it consisted mainly of free bacteria and accounted for 38% and 45% of the total pelagic respiration in the first and second period of investigation respectively. Secondary bacterial production during these period totaled 20% and 26% respectively of the net primary production and resulted from a bacterial population of about 10^7 individuals per ml. This "summer plateau" (JOST & BALLIN 1980) was extremely stable and could not be influenced appreciably even by experimental manipulation. Bacterial growth therefore seems to be closely linked to corresponding loss processes. This view is supported by the increased bacterioplankton turnover during the 5th—6th week, which is accompanied by an increase in grazing pressure by ciliates and rotifers from 4% in the 1st—2nd week to 46%.

Dissolved organic carbon is the most important carbon source for the bacterioplankton. Excreted organic carbon can account for between 5% and 95% of that fixed by phytoplankton (cf. JENSEN 1985). However, since this can only be partly utilized, a considerable proportion of the carbon they need, at least during the 5th—6th week, must enter the water through the lysis of phytoplankton or be liberated as a result of the grazing activity of ciliates and rotifers. It seems likely that some also enters the water from the sediment, but this was not checked experimentally. In this case aerobic decomposition in the top few millimetres of the sediment would doubtless represent an important source.

The ciliates play an important, but imperfectly understood (cf. PORTER et al. 1985) role in the ecosystem described here. Present in about 50 species (ARNDT et al. 1988), they are obviously capable of forming complex substructures. The ciliate community includes all of the functional groups identified by PRATT & CAIRNS (1985). The groups shown in Table 3 were important during our studies. Owing to its considerable structural and functional variability, the ciliate community represents an important regulatory component of the ecosystem, enabling it to respond quickly to perturbations and thus ensuring its stability. Since population sizes are regulated "internally" by omnivorous species, they are less dependent upon regulation by other grazers. On the other hand, this only makes the overall system more difficult to understand. Microcosm experiments (CARON et al. 1985; ANDERSEN et al. 1986) have shown that ciliates play an important role in the recycling of nutrients. Bacteria, in contrast, compete with algae for nutrients (CURRIE et al. 1986).

The rotifers have some regulatory effect on the ciliates. Since crustaceans play no significant role, the rotifers can be regarded as the final link of the food web in the pelagic zone. Interacting with the bacteria and ciliates, they govern the relationships between the heterotrophic and photoautotrophic microorganisms. Our experiments show that control by biomanipulation from above through the top grazers is impossible in this ecosystem. The most promising approach for control appears to be through phytoplankton, although regulation through heterotrophic organisms may be possible in the case of ciliates. However, this will not be known definitely until our knowledge of these interesting organisms has increased.

Conclusions

1. The "microbial loop" is of decisive importance in the brackish eutrophic pelagic system we studied in the chain of boddens south of the Darss-Zingst peninsula.
2. Although the carbon pools and fluxes in the phytoplankton compartment were identical during both periods of investigation, the carbon pools, carbon fluxes and population structures of the heterotrophic compartments varied.
3. The variations were not governed by nutrient availability for phytoplankton since neither phosphorus nor nitrogen was a limiting factor. Temperature and light are the obvious factors causing seasonal changes.
4. The stability of the ecosystem is maintained quite generally through changes in fluxes, changes in biomass playing a relatively minor role.
5. The bacterial compartment is characterized by stable numbers of individuals and carbon pool, but flux rates in this compartment can vary considerably.
6. Considerable variability in carbon pool sizes and flux rates are characteristics of the ciliates and rotifers. Crustaceans are of minor importance. The ciliates evidently play an important regulatory role.
7. Structure and function are closely related. The functional redundancy is anchored largely in the species composition of the biotic components.

References

ANDERSEN, O. K., GOLDMAN, J. C., CARON, D. A., & DENNETT, M. R. (1986): Nutrient cycling in a microflagellate food chain: III. Phosphorus dynamics. Mar. Ecol. Prog. Ser. **31**, 47—55.

ARNDT, H. (1986): The relative importance of protozoans in a Baltic estuarine zooplankton community. Symposia Biologica Hungaria **33**, 257—261.

— SCHIEWER, U., JOST, G., WASMUND, N., WALTER, TH., HEERKLOSS, R., ARLT, G., & ARNDT, E. A. (1988): The importance of pelagic and benthic microfauna in a shallow water community of the Darss-Zingst estuary (Southern Baltic) during mesocosm experiments. *Limnologica* (Berlin) **20** (1), 101—106.

AZAM, F., FENCHEL, T., FIELD, J. G., GRAY, J. S., MEYER-REIL, L. A., & THINGSTAD, F. (1983): The ecological role of water column microbes in the sea. Mar. Ecol. Prog. Ser. **10**, 257—263.

CARON, D. A., GOLDMAN, J. C., ANDERSEN, O. K., & DENNETT, M. R. (1985): Nutrient cycling in a microflagellate food chain: II. Population dynamics and carbon cycling. Mar. Ecol. Prog. Ser. **24**. 243—254.

CURRIE, D. J., BENTZEN, E., & KALFF, J. (1986): Does algal-bacterial phosphorus partitioning vary among lakes? A comparative study of orthophosphate uptake and alkaline phosphatase activity in freshwater. Can. J. Fish. Aqu. Sci. **43**, 311—318

FUHRMAN, J. A., & AZAM, F. (1980): Bacterioplankton secondary production estimates for coastal waters of British Columbia, Antarctica and California. Appl. Environ. Microbiol. **39**, 1085—1095.

— — (1982): Thymidine incorporation as a measure of heterotrophic bacterioplankton production in marine surface waters: evaluation and field results. Mar. Biol. **66**, 109—120.

— AMMERMAN, J. W., & AZAM, F. (1980): Bacterioplankton in the coastal euphotic zone: distribution, activity and possible relationships with phytoplankton. Mar. Biol. **60**, 201—207.

GEORGI, F., SPITTLER, P., SCHNESE, W., & SCHLUNGBAUM, G. (1980): Untersuchungen zur Quantität und Qualität des Sestons in der Darss-Zingster Boddenkette. Wiss. Z. W.-Pieck-Univ. Rostock, Math.-nat. R. **29**, 27—31.

HAGSTRÖM, A., & LARSSON, U. (1984): Diel seasonal variation in growth rates of pelagic bacteria. In: J. E. HOBBIE & P. J. LEB. WILLIAMS (eds.), Heterotrophic activity in the sea, pp. 249—262. New York.

JENSEN, L. M. (1985): Characterization of native bacteria and their utilization of algal extracellular products by a mixed-substrate kinetic model. Oikos **45**, 311—322.

JOST, G., & BALLIN, G. (1980): Bestimmung der Bakterienzahl und Abschätzung der Bakterienbiomasse in der Darss-Zingster Boddenkette. Wiss. Z. W.-Pieck-Univ. Rostock, Math.-nat. R. **29**, 87—89.

— (1984): Seasonal changes in bacteriological parameters at a station in the chain of shallow waters (boddens) south of the Darss-Zingst peninsula (South Baltic). *Limnologica* (Berlin) **15**, 597—603.

MENZEL, D. W., & STEELE, J. H. (1978): The application of plastic enclosures to the study of pelagic marine biota. Rapp. P.-r. Reun. Cons. int. Explor. Mer **173**, 7—12.

MULLIN, M. M. (1982): How can enclosing seawater liberate biological oceanographers? In: G. D. GRICE & M. R. REEVE (eds.), Marine Mesocosms, pp. 398—410. New York, Heidelberg, Berlin.

OVERBECK, J. (1985): Stoffwechselkopplung zwischen Phytoplankton und heterotrophen Gewässerbakterien — ein Schlüssel zum Verständnis von Stoffkreisläufen im Gewässer. In: Rhein.-Westfäl. AdW, Vorträge N 337.

PORTER, K. G., SHERR, E. B., SHERR, B. F., PACE, M., & SANDERS, R. W. (1985): Protozoa in planktonic food webs. J. Protozool. **32**, 409—415.

PRATT, J. R., & CAIRNS, J. (1985): Functional groups in the protozoa: role in differing ecosystems. J. Protozool. **32**, 415—423.

RIEMANN, B. (1978): Absorption coefficients for chlorophyll a and b in methanol and a comment on interference of chlorophyll b in determinations of chlorophyll a. Vatten **3**, 187—194.

SCHIEWER, U., & JOST, G. (1986): Ecosystem research during the 1981/85 period in the Darss-Zingst estuary. Wiss. Z. W.-Pieck-Univ. Rostock, Naturwiss. R. **35**, 28—36.

SCHINDLER, D. W., SCHMIDT, R. V., & REID, R. A. (1972): Acidification and bubbling as an alternative to filtratation in determining primary productivity by the ^{14}C-method. J. Fish. Res. Bd. Can. **29**, 1627—1631.

SCHNESE, W., & HEERKLOSS, R. (1978): Nutritional-biological studies on the zooplankton of the chain of boddens south of the Darss-Zingst peninsula: determination of feeding and assimilation rates by means of ^{14}C under field conditions. Kieler Meeresforsch., Suppl. **4**, 267—274.

SIMON, M., & TILZER, M. M. (1987): Bacterial response to seasonal changes in primary production and phytoplankton biomass in Lake Constance. J. Plankton Res. **9**, 535—552.

TAKAHASHI, M., THOMAS, W. A., SEIBERT, D. L. R., BEERS, J., KOELLER, P., & PARSONS, T. R. (1975): The replication of biological events in enclosed water columns. Arch. Hydrobiol. **76**, 5—23.

UTERMÖHL, H. (1958): Zur Vervollkommnung der quantitativen Phytoplanktonmethodik. Verh, Int. Ver. Limnol. **9**, 1—38.

Address of the authors:

Prof. Dr. U. SCHIEWER
Sektion Biologie
Wilhelm-Pieck-Universität Rostock
Freiligrathstr. 7/8
Rostock
DDR-2500

Limnologica (Berlin) **20** (1), 101—106, April 1990

AdW der DDR, Institut für Geographie und Geoökologie, Bereich Hydrologie, Berlin, DDR; Wilhelm-Pieck-Universität Rostock, Sektion Biologie, Rostock, DDR

The Importance of Pelagic and Benthic Microfauna in a Shallow-Water Community of the Darss-Zingst Estuary, Southern Baltic, During Mesocosm Experiments

By Hartmut Arndt, Ulrich Schiewer, Günther Jost, Norbert Wasmund, Thomas Walter, Reinhard Heerkloss, Günther Arlt & Ernst Albert Arndt

With 4 Figures

Abstract

During enclosure experiments (3.1 m^2) in the shallow coastal waters south of the Darss-Zingst in June/July 1985 and 1986, respectively, the impact of nutrient loading (P, N) on ecosystem structure and functioning was studied. As a part of these studies the relative importance of protozoans (esp. ciliates) and micrometazoans (esp. small rotifers) in the pelagic as well as in the benthic habitat was investigated by estimation of their biomass and calculation of their activity using specific metabolic rates from literature. The results showed that zooplankton which passes common used plankton nets (56 µm mesh size) was the most active part of the whole animal community during the investigation period. In benthic habitat (muddy sand) microzoans were of minor importance compared to meiozoobenthos. Although the investigation site was very shallow (0.5—0.7 m depth) there was a clear difference in species composition of pelagic and benthic fauna. Apart from the dominance of different species in different enclosures nutrient loading had no considerable influence on total animal biomass. This points to the stability of the ecosystem. Pelagic ciliates were the metabolic most active faunal component of the investigated eutrophic waters. Their dynamics were characterized by short-term succession of species (in a few days) and by the cooccurrence of different functional groups like algivores, bacterivores, and omnivorous predators. Our results underline the demand for more protozoological studies to explain ecosystem functioning in estuaries.

Introduction

Though the knowledge on microfauna in pelagic and benthic estuarine habitats has increased much during the last decades, there are only a very few studies, where dynamics of micro-, meso-, and macrofauna have been registered simultaneously in the plankton and the benthos community as well. Results of studies dealing with benthic habitats have shown that biomass of microfauna is mostly of minor importance compared to meio- and macrofauna, but that microfauna production is often in the same range of that of meio- and macrofauna depending on the quality of substrate (e.g. Fenchel 1967; Jansson et al. 1984; Scharf & Schnese 1984). Data on pelagic estuarine microfauna are rather rare. The microfaunal component is mostly characterized by strong seasonal and short term changes ranging from overall importance to negligible importance in relation to total zooplankton biomass (e.g. Burkill 1982; Hagström & Larsson 1985; Arndt 1986).

In the shallow coastal waters south of the Darss-Zingst peninsula (southern Baltic) we carried out a series of mesocosm experiments in order to determine the influence of additional nutrient loading on functioning and stability of eutrophic estuarine ecosystems. Results regarding dynamics of nutrients, phytoplankton, zooplankton, and zoobenthos, respectively, have already been published (e.g. Schiewer & Baader 1982; Schiewer et al. 1986; Schlungbaum et al. 1983; Börner & Kell 1982; Wasmund 1988; Scharf et al. 1984; Schiewer et al. 1989). In this paper we present results on the relative importance of microfauna within the pelagic and benthic animal community.

Material and Methods

From the beginning of June till the end of July in both years, 1985 and 1986, respectively, three enclosures were exposed in the shallow waters of the Kirr-Bight (Darss-Zingst estuary, 3—7‰ S; a description of the aquatorium is given by Schiewer et al. 1985). Enclosures consisted of polyethylene foile with a polysterol filled ring at the upper part (cf. Fig. 1). The lower part is deepened into the sediment (muddy sand: medium corn size 0.2—0.5 mm; organic content of dry matter 1—2%; Nausch unpubl.). Each enclosure covers an area of about 3.1 m^2 with a mean water level of 0.55 m in 1985 and 0.72 m in 1986, respectively. In the course of the 8-week experimental period following nutrient supplies were given: 1985; enclosure A 10.6 gN · m^{-3}, B no additions, C 12.6 gN · m^{-3} and 1.7 gP · m^{-3}; 1986: enclosure A no additions, B 5.5 gN · m^{-3} & 2.6 gP · m^{-3}, C 23.0 gN · m^{-3} & 2.9 gP · m^{-3}. Dynamics of abiotic and biotic parameters were measured during the whole investigation period in all three enclosures and the surrounding water (Kirr). Most abiotic parameters were investigated nearly every day (e.g. temperature, turbidity, pH, dissolved nutrients), whereas biotic parameters were collected one time but mostly two times per week (e.g. chlorophyll content, biomass and production of phytoplankton, bacteria, protozoans, mesozooplankton, and meiofauna, respectively). In Fig. 2 and Fig. 3 only weekly mean values are shown. Some parameters were measured only occasionally like biomass and activity of microphytobenthos, macrophytes, and macrofauna, respectively. For other details of experiments see Schiewer et al. (1988, 1989).

Following methods were applied for the analysis of the shallow-water fauna:

1) Benthic macrofauna was sampled by means of an Ekman-grab (0.025 m^2). Five parallel samples were taken from all compartments at the beginning and at the end of experiments. All animals which were retained on a 0.5 mm sieve were counted after formalin fixation (4%) under a stereomicroscope. For each sample animal wet weights, dry weights (after 24 h at 60 °C), and ashfree dry weights (24 h at 500 °C) were determined.

2) Benthic meiofauna was sampled by means of a core sampler according to Arlt (1984) (core area 3 cm^3). The upper two centimeters were taken from three parallel samples and pooled together. 1985 formalin fixed samples (4%) were analyzed in a Bogorov-chamber under a stereomicroscope. 1986 subsamples were investigated by means of the living droplet method. Generally at least 20 length measurements were carried out for each species and biovolumes were estimated by approximations of body shapes to simple geometrical forms.

3) Mesozooplankton was collected by sieving 5 one-litre samples per enclosure through a plankton net (56 µm mesh size). Zooplankters were fixed with formalin (4%) and subsamples were analyzed in a

microscopic chamber (ARNDT 1985). Biovolumes were estimated as was done with meiofauna.

4) Microfauna (ciliates, rhizopods, small rotifers) was investigated by analysis of living samples. Benthic samples were taken like those for meiofauna analysis. A volume of 182 ml filtered biotope water was immediately added to three core samples ($3 * 3\ cm^2$; $18\ cm^3$). Subsampling was carried out by means of an automatic calibrated pipette during stirring of the sample. Plankton mixed samples consisting of 5 single samples were subsampled in the same way without dilutions. Abundant species were counted in a 0.35 ml microscopic chamber; scarcer species were counted by the living droplet method (cf. DALE & BURKILL 1982). Care was taken during manipulation of samples to reduce the time for stirring and counting of samples. Biovolumes were estimated like those for meiofauna. Length measurements were done to the nearest 5 μm interval.

5) Flagellates were analyzed only semiquantitatively by counting subsamples from benthic and pelagic microfauna in blood cell chambers without concentrations. Individuals were differentiated for pigmented and nonpigmented forms under a light microscope. Individuals were counted for size classes. By this procedure the amount of flagellates is probably underestimated.

Results and Discussion

1. Ecosystem structure

A view on the biotic structure within the enclosures is shown for the situation during the fourth week of July 1986 in the untreated control A for an example (Fig. 1). Since there are only a very few data for a nearly complete analysis of ecosystem structure (including microfauna) in literature we present estimations of all ecosystem components though our estimates are sometimes very rough and are typical only for the actual situation. As clearly to be seen phytoplankton and phytobenthos reach nearly the same production and even under the high turbidity (0.4 m Secchi depth) the microphytobenthos contributes significantly to total primary production. In June in both years phytoplankton consisted mainly of chlorophyceans and blue-greens. In the first phase chlorophyceans like *Scenedesmus* and *Crucigenia* and later coccal blue-greens like *Gomphosphaeria* and *Microcystis* and the filamentous *Lyngbya contorta* dominated phytoplankton biomass. High nutrient loading leads sometimes to specific blooms (cf. WASMUND 1989). Growth of macrophytes was of no importance in 1985, but was of high importance in 1986 when community was dominated by *Potamogeton pectinatus* and *Chara* spp. Microphytobenthos consisted mostly of sedimented but still active phytoplankters.

Regarding their metabolic activity bacteria were the most important heterotrophic component within the pelagic as well as within the benthic community. This points to the high level of eutrophication in Kirr-Bight. Total bacteria population could reach turnover times up to $4\ d^{-1}$.

Meiofauna was the most important animal component regarding its biomass but production compared to zooplankton was low. Meiofauna was dominated in both years by ostracods (mainly *Cyprideis litoralis*), oligochaetes, nematodes and harpacticoid copepods (mainly *Microarthridion littorale* and *Halectinosoma curticorne*). In addition turbellarians, hydrozoans (*Protohydra* sp.), cladocerans (mainly *Chydorus sphaericus* and *Alona rectangula*), bdelloids, and monogonont rotifers (mainly *Euchlanis*, *Colurella*, and *Lecane*).

Macrofauna was dominated by chironomid larvae. Of reduced importance were large oligochaetes, nemertines (*Prostoma obscura*), polychaetes (mainly *Hediste diversicolor* and *Spio* sp.), and crustaceans (mainly Gammaridae and *Sphaeroma*).

Though of minor importance regarding its biomass microfauna contributes to a significant part to benthic

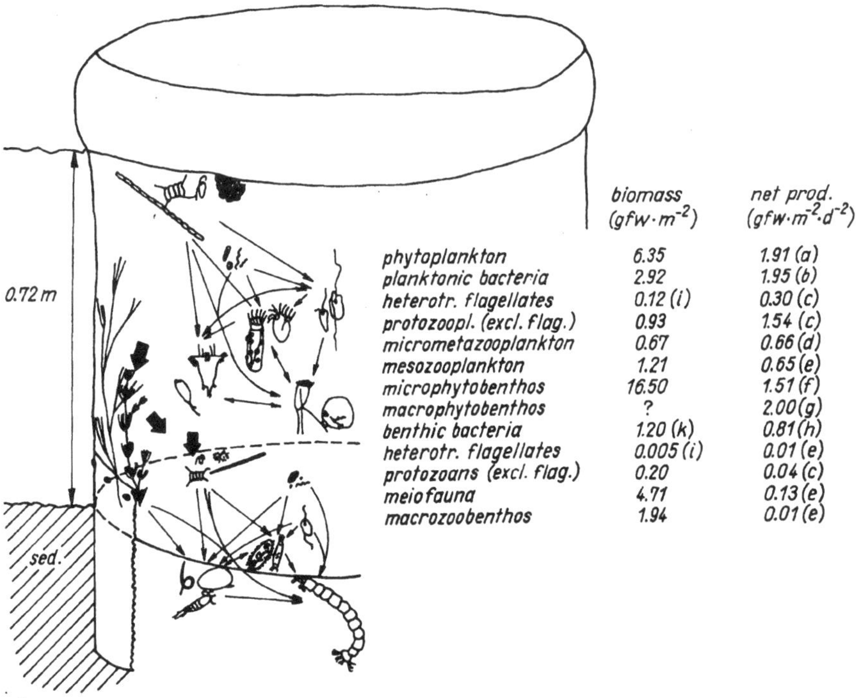

Fig. 1. Structure, biomass, and estimated production in the pelagic and benthic zone of brackish shallow-water enclosure A (without nutrient additions, fourth week of July 1986; small arrows stand for trophic interactions, large arrows point to the importance of sedimentation; a: ^{14}C-method; b: ^{3}H-thymidine incorporation method; c: assumed P/B-value from literature; d: egg-ratio method; e: assumed P/B-value derived from in situ growth rates; f and g: O_2-method, in situ incubation; h: specific production assumed to be the same as for planktonic bacteria; i: life counting in blood cell chamber; k: assumed specific biovolume)

animal production. Heterotrophic flagellates were very abundant but have only low biomasses. There were strong short term changes in abundances of flagellates. Ciliates were the most important component of microfauna, dominated by amphileptids, hymenostomes, and oxytrichids. Testaceans were generally of minor importance, but naked amoebae could reach very high abundances for short periods corresponding to high sedimentation rates from plankton.

Within the plankton community mesozooplankton dominated animal biomass. Rotifers were the most important group (mainly *Filinia longiseta*, *Keratella cochlearis*, *Brachionus quadridentatus*, *B. angularis*, *B. plicatilis*). Sometimes cladocerans (mainly *Chydorus sphaericus* and *Alona rectangula*) and copepods (*Eurytemora affinis* and *Acartia tonsa*) could significantly contribute to zooplankton biomass. Pelagic macrofauna (mysids, fishes) were excluded during experiments. Their metabolic activity would have been low relative to the other components, but their contribution to ecosystem functioning could have been of importance (cf. ARNDT et al. 1984; ARNDT 1986, in press).

We use the term micrometazooplankton for metazoans which pass common used zooplankton nets (e.g. 56 µm mesh size). These are generally very delicate rotifers (during experiments e.g. *Synchaeta kitina*, *S. cecilia*, *Trichocerca* cf. *dixon-nutalli*) difficult to register and determine in fixed samples. We investigated the delicate forms in living samples (droplet method). These small rotifers can be very productive (abundances > 10,000 ind. $\cdot$ l^{-1}) for short periods during summer in rivers, lakes, and estuaries (ARNDT, unpubl.). According to our opinion these delicate forms have been overlooked in many recent studies. Possible reasons are their small size, their delicate bodies, their generally short term occurrence in plankton, and their difficult taxonomy, respectively. During early summer these microzooplankters can compose the major part of zooplankton biomass in Kirr-Bight.

Protozooplankton often excluded from traditional plankton studies up to now was the most important component of the whole animal community regarding its metabolic activity during the investigation period. Numbers of heterotrophic flagellates like those of benthic flagellates showed strong short term changes in abundances. A more detailed analysis is necessary to have a realistic view on their importance in the shallow coastal waters south of the Darss-Zingst estuary. There are only a few reports in literature regarding the importance of heterotrophic flagellates in estuarine waters, but those which are available point to their high significance for ecosystem functioning (e.g. ANDERSON & FENCHEL 1985). The biomass of planktonic ciliates was dominated by gymnostomes (e.g. *Mesodinium*, *Monodinium*) and oligotrichs (e.g. *Strombidium*, *Strobilidium*, *Halteria*, *Tintinnidum*) (see Fig. 4).

Besides the pelagic and benthic habitat sometimes microflora and microfauna on macrophytes was an additional important compartment for ecosystem functioning. When sedimentation rates were high leaves of macrophytes were covered with a thin layer of sedimented planktonic algae. Within one day this layer could be very densely populated by naked amoebae and soon after it benthic ciliates and rotifers show high abundances (ARNDT, unpubl.). This epiphytic community has only a short life span till the next wind induced turbulence but can contribute to a fast degradation of sedimented phytoplankton especially due to the activity of naked amoebae. When sedimentation rates were high there was also found an enormous increase of naked amoebae populations (but other species than on macrophyte

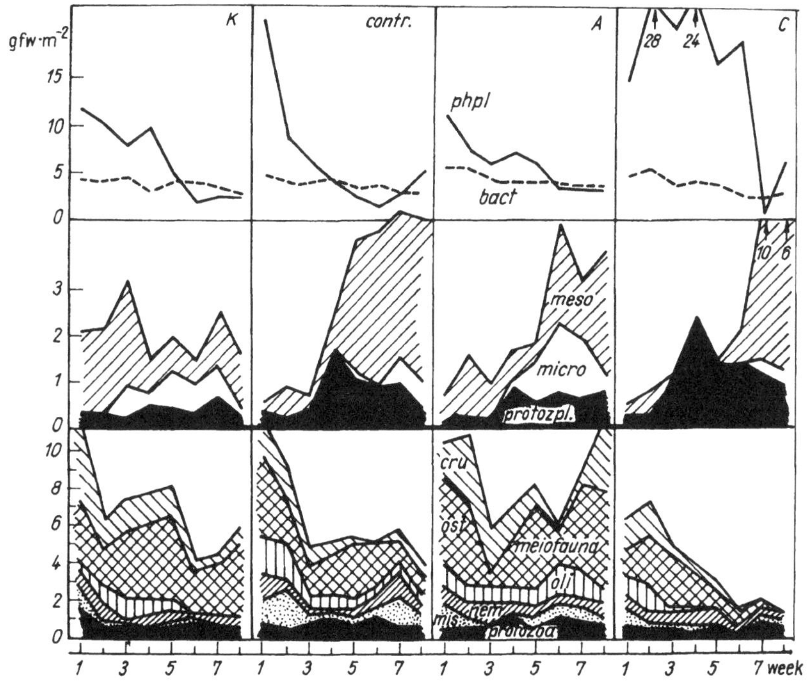

Fig. 2. Weekly fluctuations of animal biomass in the pelagic and benthic zone of four compartments in Kirr-Bight, Darss-Zingst estuary, in June/July 1985 (K: surrounding water; contr.: untreated control; A: stepwise addition of 10.6 gN $\cdot$ m^{-3}; C: stepwise addition of 12.3 gN $\cdot$ m^{-3} and 1.7 gP $\cdot$ m^{-3}; upper panel: biomass of phyto- and bacterioplankton; middle panel: biomass of zooplankton incl. protozoopl., microzoopl. and mesozoopl.; lower panel: biomass of protozoans and microfauna incl. cru — crustaceans, ost — ostracodes, oli-oligochaetes, nem — nematodes, and mis — miscellanous)

leaves) in the upper thin layer of sediment (esp. 8th week in 1986, about 2,080 naked amoebae per cm^2). Of course, there was also sessile fauna (e.g. peritrichs, hydrozoans) and flora (esp. pennate diatoms) but their contribution to ecosystem functioning during the investigation period seemed not to be of importance.

2. Influence of nutrient loading

As indicated in Fig. 2 and Fig. 3 dynamics of ecosystem components behave similarly in Kirr-Bight, control enclosure, and the enclosure with lower nutrient supply, respectively. The development of some special species in distinct enclosures (cf. WASMUND 1988) indicates that this similarity was not the result of exchange processes. Most similarities were found between surrounding water and the enclosure with low nutrient additions (enclosure A in 1985, B in 1986). This may be due to the continuous nutrient supply of the surrounding water by horizontal water exchange with Zingster Strom.

The general decline of phytoplankton biomass after the second and third week was most probably the combined effect of abiotic factors (e.g. nutrient depletion, irradiance, sedimentation, temperature). This decline was followed by increasing bacterial production (^{3}H-thymidine incorporation method, JOST unpubl.) which was nearly fully consumed by bacterivorous microzooplankton (see next chapter). From then microzooplankton was the metabolic most active component. In the last weeks in 1985 rotifers reached especially high abundances and were able to control phytoplankton production which had been already reduced by sedimentation processes.

The high degradation rates within the plankton community explains the reduced effect of phytoplankton production on benthic fauna. Exceptions were the short term pulses of naked amobae and heterotrophic flagellates (see above) as well as the increase in epibenthic cladocerans in enclosures C. The decreasing number of meiofauna and ciliates from late spring to summer is a typical phenomenon of the brackish waters south of Darss-Zingst (cf. ARLT 1984; SCHARF & SCHNESE 1984).

Only very high nutrient additions in enclosure C led to exceptionally high rates of primary production especially in 1985 where irradiance was higher than in 1986. In the first weeks in 1985 a dense bloom of *Scenedesmus* was established which collapsed in the sixth and seventh week by a sudden increase in temperature (cf. SCHIEWER et al. 1988). Thereafter high grazing pressure by rotifers prevented intensive phytoplankton growth. In 1986 a bloom of *Synedra* sp. was built up during the last weeks after a change in the relation of added nutrients (cf. WASMUND 1988). This very uncommon *Synedra* peak seems not to be a suitable food source for zooplankton. Detritus consuming cladocerans (*Chydorus*, *Alona*) made probably use of the sedimented algae.

In all enclosures the increase in the amount of sedimented material was not as high as would be expected from primary production (SCHIEWER & WASMUND, unpubl.). This was most probably due to the high activity of bacteria and microfauna. The low rate of increase in sedimented material in the shallow inner coastal waters (SCHLUNGBAUM & ARLT, pers comm.) was formerly mostly explained by the transport of sedimented material via water currents to the outer part of the estuary. In this respect our results underline the importance of fast degradation in the area itself supported by oxygenation and resuspension of presediment due to wind induced turbulences.

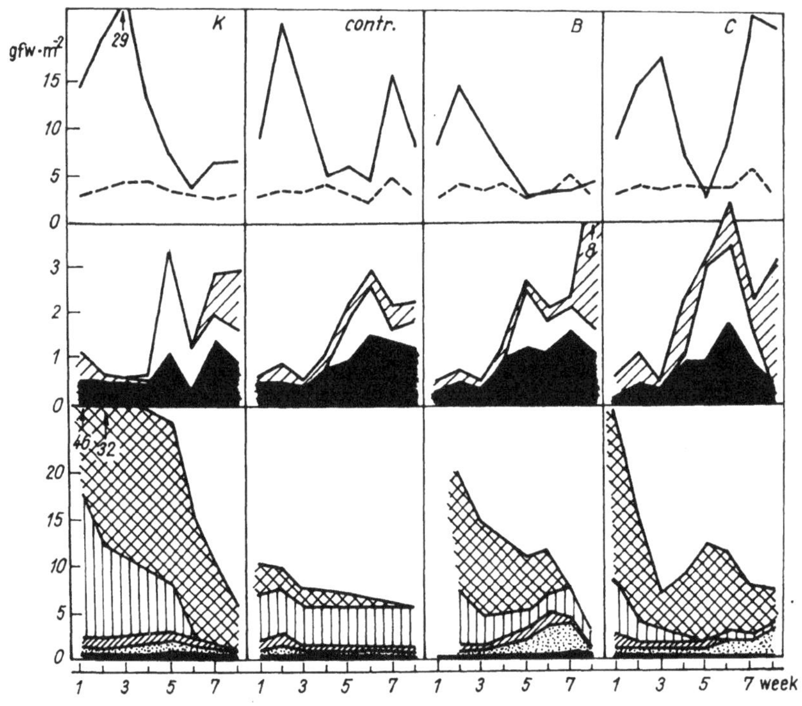

Fig. 3. Weekly fluctuations of animal biomass in the pelagic and benthic zone of four compartments in Kirr-Bight in June/July 1986 (K: surrounding water; contr.: untreated control; B: stepwise addition of 5.5 gN · m^{-3} and 2.6 gP · m^{-3}; C: stepwise addition of 23.0 gN · m^{-3} and 2.9 gP · m^{-3}; for description of graphs see Fig. 2)

3. Species diversity and ecosystem elasticity

Regarding species richness and diversity among the animal community protozoans (excl. flagellates which were not investigated into detail) were far the most diverse group. More than 50 benthic species and more than 40 pelagic species were occasionally abundant during the 8-week investigation period (ARNDT, unpubl.). Probably one major cause of the elasticity of many eutrophic ecosystems (cf. NILSSEN 1978) is microfauna diversity, a component which has been mostly excluded from ecosystem analysis up to now.

On the other hand, it seems that the dominance of euryphagous species in this highly eutrophic waters replaces the function of species diversity for achieving a certain level of ecosystem elasticity. Derived from our own observations (ARNDT & BURCKHARDT, unpubl.) each zooplankton group serves as a food source for the other (cf. small arrows in Fig. 1). Our knowledge on benthic interactions is not as good as for the pelagic habitat but derived from literature interactions should also be very intensive. There are also some interactions between benthic and pelagic fauna in the investigation area supported by migration patterns (cf. ARNDT et al. 1984) as well as due to resuspension especially of benthic bacteria. Furthermore, bioturbation especially by the activity of macrofauna should be of certain importance.

4. Temporal changes of planktonic ciliates

Since planktonic ciliates reach the highest production of secondary producers (excl. heterotrophic flagellates) during the investigation period their dynamics are

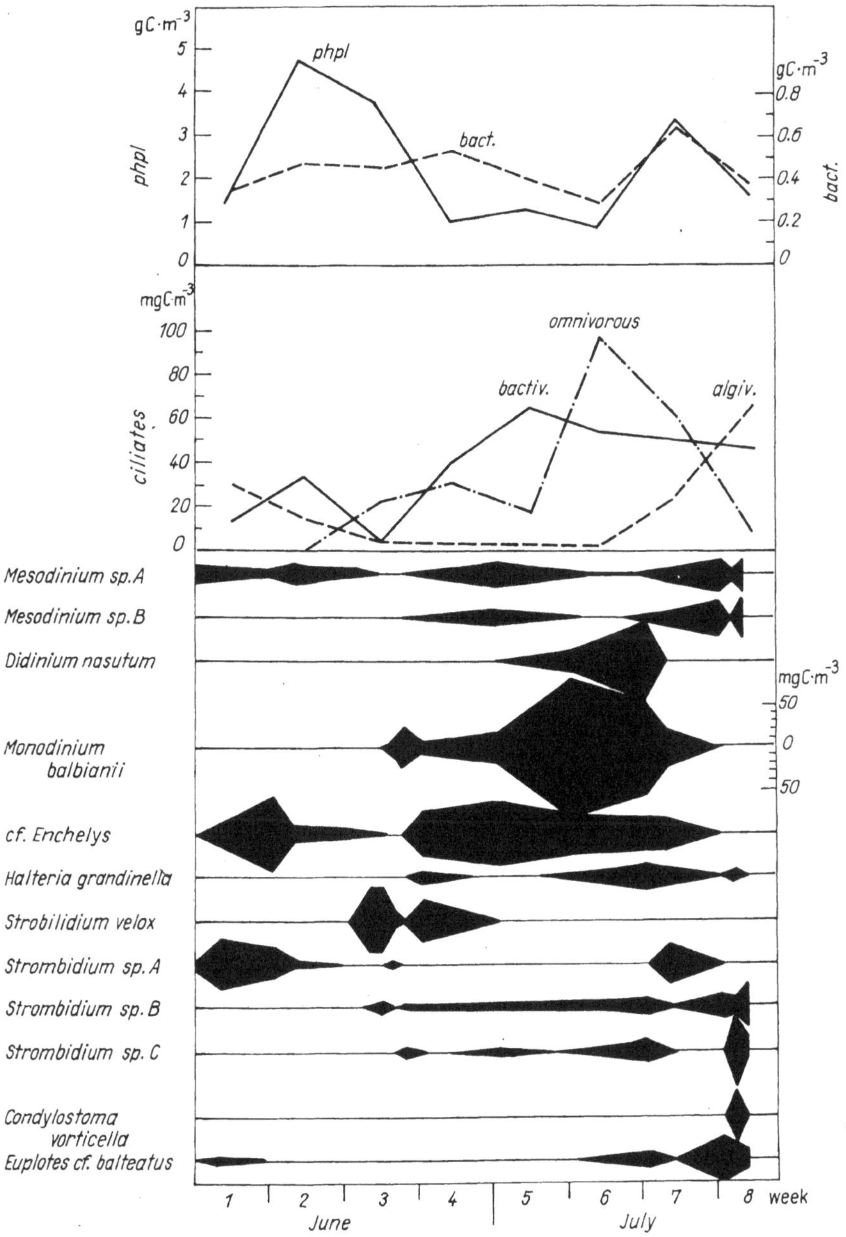

Fig. 4. Weekly fluctuations in biotic components in the pelagic zone of enclosure A (untreated control) in 1986 (upper panel: biomass of phytoplankton and planktonic bacteria; middle panel: biomass of planktonic ciliates separated for functional groups according to PRATT & CAIRNS (1985); lower panel: biomass of single ciliate species)

shown in more detail. For example, Fig. 4 presents the results of the control enclosure (1986). As already mentioned, the diversity of ciliates was much higher than total metazooplankton diversity. In Fig. 4 (lower part) the successions only of the dominant species are shown. The most interesting fact is the dominance of different functional groups. During degradation of phytoplankton and registered accompanying high production of planktonic bacteria numbers of bacterivorous ciliates increased remarkably. Calculated feeding rates of bacterivorous ciliates and heterotrophic flagellates were in the same range of the bacterial production (cf. ARNDT et al. 1988). Thus the production of bacteria should be under the control of microzooplankton. This perhaps may explain the relatively constant bacterial biomass.

From the third till the 7th week there were found high biomasses of omnivorous ciliates (*Strobilidium velox*, *Didinium nasutum*, *Monodinium balbianii*, *Condylostoma vorticella*) which are most probably able to control abundances of the small bacterivorous and algivorous species.

Thus the relatively stable ciliate biomass seems to be caused by the succession of several different species within short time intervals as well as the fact that the complexity of possible trophic relationships is already present within the Ciliophora.

The results point to the high importance of the matter flux via bacteria and microzooplankton in this eutrophic shallow waters and underline the necessity of more protozoological work for future ecosystem analysis in estuarine waters.

References

ANDERSON, P., & FENCHEL, T. (1985): Bacterivory by microheterotrophic flagellates in seawater samples. Limnol. Oceanogr. **30**, 198—202.

ARLT, G. (1984): Taxonomische und ökologische Untersuchungen am Meiozoobenthos von Brackgewässern. Promotion B, Wilhelm-Pieck-Universität, Rostock.

ARNDT, H. (1985): Eine Zählkammer für die mikroskopische Auswertung von Zooplanktonproben. WZ Rostock **34**, Meeresbiol. Beitr., 30—31.

— (1986): The relative importance of protozoans in a Baltic estuarine zooplankton community. Symp. Biol. Hung. **33**, 257—261.

— (in press): Zooplankton production and its consumption by planktivores in a Baltic inlet. Proc. 21st European Marine Biology Symposium, Gdansk, Sept. 1986.

— DEBUS, L., HEERKLOSS, R., & SCHNESE, W. (1984): Diurnal changes in the matter flux of a shallow-water ecosystem in a Baltic inlet. Ophelia, Suppl. **3**, 1—9.

— JOST, H., & WASMUND, N. (in press): Dynamics of pelagic ciliates in eutrophic estuarine waters: Importance of functional groups among ciliates and responses to bacterial and phytoplankton production. Arch. Hydrobiol., Beih. Ergebn. Limnol.

BÖRNER, R., & KELL, V. (1982): Einfluß von Nährstoffanreicherungen auf die Biomasse, Artensequenz und Primärproduktion des Phytoplanktons während einer Komplexanalyse im Zingster Strom (Juni 1981). WZ Rostock **31**, Meeresbiol. Beitr., 53—56.

BURKILL, P. H. (1982): Ciliates and other microplankton components of a nearshore food-web: standing stocks and production processes. Ann. Inst. Oceanogr., Paris **58** (S), 335—350.

DALE, T., & BURKILL, P. H. (1982): "Live counting" — a quick and simple technique for enumerating pelagic ciliates. Ibid., 267—276.

FENCHEL, T. (1967): The ecology of marine microbenthos. I. The quantitative importance of ciliates as compared with metazoans in various types of sediments. Ophelia **4**, 121—137.

HAGSTRÖM, A., & LARSSON, U. (1984): Diel and seasonal variation in growth rates of pelagic bacteria. In: HOBBIE, J., & LEB. WILLIAMS (eds.), Heterotrophic Activity in the Sea. New York.

JANSSON, B.-O., WILMOT, W., & WULFF, F. (1984): Coupling the sub-systems — the Baltic Sea as a case study. In: FASHAM, M. J. R. (ed.), Flows of Energy and Materials in Marine Ecosystems, pp. 549—595. New York.

NILSSEN, J. P. (1978): Eutrophication, minute algae and inefficient grazers. Mem. Ist. Ital. Idrobiol. **36**, 121—138.

PRATT, J. R., & CAIRNS, J., JR. (1985): Functional groups in the Protozoa: Roles in differing ecosystems. J. Protozool. **32**, 415—423.

SCHARF, E.-M., & SCHNESE, W. (1984): Abundance and reproduction potential of benthic ciliates in a β-mesosaprobic aquatorium. *Limnologica* (Berlin) **15**, 429—437.

— — & SCHLUNGBAUM, G. (1984): Short term response of benthic and pelagic ciliates to artificial eutrophication. Ophelia, Suppl. **3**, 195—200.

SCHIEWER, U., & BAADER, G. (1982): Pelagic NH_4-rhythm independent of the sediment in eutrophic shallow waters. Int. Rev. ges. Hydrobiol. **67**, 845—850.

— BÖRNER, R., & WASMUND, N. (1988): Deterministic and stochastic influence of nutrients on phytoplankton function and structure in coastal waters. Kieler Meeresforsch., Sonderh. **6**, 173—183.

— JOST, G., & HEERKLOSS, R. (1983): Vergleich des C-Umsatzes in regelmäßig und unregelmäßig durchmischten Bereichen der Darss-Zingster Boddenkette. WZ Rostock **32**, Meeresbiol. Beitr., 35—38.

— ARNDT, H., HEERKLOSS, R., JOST, G., & WASMUND, N. (1989); Carbon flux dynamics in a shallow eutrophic estuary. *Limnologica* (Berlin) **20** (1), 95—100.

— — BAADER, G., BALLIN, G., BÖRNER, R., EVERT, F.-K., GEORGI, F., HEERKLOSS, R., JOST, G., KELL, V., KRÜGER, B., & WALTER, T. (1986): The bounds and potential effects of NH_4^+ (loading) on the pelagic system of a Baltic estuary. *Limnologica* (Berlin) **17**, 7—28.

SCHLUNGBAUM, G., NAUSCH, G., ARLT, G., & SCHARF, E.-M. (1983): Erste Ergebnisse über die Belastbarkeit von flachen eutrophen Küstengewässerökosystemen. WZ Rostock **32**, Meeresbiol. Beitr., 20—28.

WASMUND, N. (1989): Characteristics of phytoplankton in brackish waters of different trophic levels. *Limnologica* (Berlin) **20** (1), 47—51.

Addresses of the authors:

Dr. HARTMUT ARNDT
AdW der DDR
Institut für Geographie und Geoökologie
Bereich Hydrologie, Hy 2
Müggelseedamm 260
Berlin
DDR-1162

Prof. Dr. ULRICH SCHIEWER
Dr. GÜNTHER JOST
Dr. NORBERT WASMUND
Dr. THOMAS WALTER
Dr. sc. REINHARD HEERKLOSS
Dr. sc. GÜNTHER ARLT
Prof. Dr. ERNST ALBERT ARNDT
Wilhelm-Pieck-Universität Rostock
Sektion Biologie
Freiligrathstr. 7/8
Rostock
DDR-2500

Limnologica (Berlin) **20** (1), 107–111, April 1990

Wilhelm-Pieck-Universität Rostock, Sektion Biologie, Rostock, DDR

Mischkulturen und die Analyse biotischer Wechselwirkungen

Mixed Cultures: Their Use for Analysing Interspecific Relationships

By B. Krüger, S. Mikkat & S. Fulda

Mit 3 Abbildungen und einer Tabelle

Key words: Mixed cultures; Turbidostat; Chemostat; Biotic interactions.

Abstract

Analysis of biotic interactions are becoming increasingly important for the clarification of complex causal relationships in aquatic ecosystems.

Here results are presented of a joint cultivation of a heterotrophic bacterial isolate or a green alga (*Scenedesmus quadricauda*), respectively, in each of both together with a species of cyanobacteria (*Synechocystis* spec. or *Microcystis firma*), in chemostat and turbidostat cultures.

In the chemostat incubated with mixed cultures of *Synechocystis* spec. and the heterotrophic bacterium, the bacterial biovolume was low ($< 1\%$) with respect to the total biovolume because only 2% of the carbon having been fixed by *Synechocystis* was exudated by way of organic compounds. The addition of sucrose caused the cyanobacterial biomass to decline and the bacterial biomass to increase at low dilution rates. In this situation simultaneous utilization of sucrose and cyanobacterial exudates by the bacteria was important to their biomass increase.

In the turbidostat incubated with mixed cultures of C-autotrophic organisms, nutrient limitations are insignificant in the competitive relationships. A mathematical model derived from separate turbidostat cultures of each of the two species, *M. firma* and *S. quadricauda*, does not describe fully all details of the light and temperature dependence of biomass formation of the two species growing jointly in one turbidostat culture. Promotion of growth of the cyanobacterium caused by exudates of the green alga induces slight shifts in the ranges of both coexistence and dominance.

The chemical nature of compounds responsible for the described biotic influences is unknown as yet.

Einleitung

Je intensiver und komplexer Ökologen aquatische Ökosysteme und deren Untereinheiten beschreiben, analysieren und modellieren, um so häufiger treffen sie auf Schwierigkeiten, bestimmte Phänomene, wie z. B. Artensukzessionen, ausschließlich mit abiotischen Faktorenkombinationen zu erklären. Ursachen können biotische Wechselwirkungen sein, deren Nachweis unter natürlichen Bedingungen aufgrund der Komplexität der wirkenden Umweltfaktoren praktisch unmöglich ist. Es wird daher die Forderung nach synökologischen Untersuchungen im Labor erhoben.

Im folgenden Beitrag sollen anhand zweier Beispiele einige Probleme angedeutet werden, die mit derartigen Laboruntersuchungen und ihrer Interpretation verbunden sind.

Abkürzungen

M. f. — *Microcystis firma*; *S. q.* — *Scenedesmus quadricauda*; D — Verdünnungsrate im Chemostaten; EOC — extrazellulärer organischer Kohlenstoff; I_r — Strahlungsflußdichte; NM — Nährmedium; T — Temperatur; P — Irrtumswahrscheinlichkeit; α — Stimulationsfaktor (α_M für *M. f.*., α_B für Bakterienstamm M); μ — relative Wachstumsrate (μ_M für *M. f.*, μ_s für *S. q.*); ν — Verdrängungsrate (ν_M, ν_S).

Mischkultivierung von Bakterien und Cyanobakterien im Chemostaten

Vom Phytoplankton wird in Abhängigkeit von der Populationsstruktur und den abiotischen Umweltfaktoren ein Teil der Photosyntheseprodukte als extrazellulärer organischer Kohlenstoff (EOC) abgegeben (Fogg 1983). Zum Teil wird dieser EOC über die „mikrobielle Schleife“ (Azam et al. 1983), deren Komponenten heterotrophe Bakterien, bakterivore Protozoen sowie Prädatoren dieser Protozoen sind, in die Weidenahrungskette zurückgeführt. Bei der Verwertung von EOC oder von anderen gelösten organischen Substanzen können Bakterien mit dem Phytoplankton um anorganische Nährstoffe konkurrieren (Currie & Kalff 1984; Bratbak & Thingstad 1985). Die Wechselwirkungen zwischen Phyto- und Bakterioplankton beschränken sich jedoch nicht nur auf trophische Beziehungen wie Kommensalismus und Konkurrenz, sondern können Ausdruck einer Vielzahl stimulierender oder inhibierender Effekte sein (Cole 1982).

Um die trophische Kopplung zwischen Cyanobakterien und heterotrophen Bakterien in einer Mischkultur zu untersuchen, isolierten wir einen gram-negativen Bakterienstamm (Stamm M), dessen Existenz durch Exsudate des chroococcalen Cyanobakterienstammes SAG 92.79 *Synechocystis* spec. ermöglicht wird (Mikkat 1986). Als Kultivierungsverfahren wurde der Chemostat gewählt, in dem sich Organismen mit ihrer Biomassekonzentration entsprechend der vorgegebenen Verdünnungsrate auf einen limitierenden Nährstoff einstellen. Zwei Arten können im Chemostaten koexistieren, wenn ihr Wachstum durch unterschiedliche Faktoren limitiert wird oder wenn die Existenz der einen von exsudierten Stoffwechselprodukten der anderen Art abhängig ist.

Bei den Versuchsansätzen gingen wir von der Feststellung aus, nach der Mischpopulationsanalysen die Kenntnis des Wachstumsverhaltens jeder Art in Einzelkultur voraussetzen (Hobson 1969).

Je eine Einzelkultur von *Synechocystis* und Stamm M sowie eine Mischkultur beider wurden in einem modifizierten Nährmedium (NM) nach Allen & Arnon (1955) kultiviert, wobei das Wachstum der Cyanobakterien nitratlimitiert war und die Bakterien in der Mischkultur nur auf der Grundlage der Exsudate der Cyanobakterien wachsen konnten. Die Einzelkultur von Stamm M ohne organische Kohlenstoffquelle hatte nur eine Kontrollfunktion und endete erwartungsgemäß mit der Auswaschung der Bakterien.

Eine zweite Serie Chemostatkulturen, deren NM Saccharose als zusätzliche organische Kohlenstoffquelle für die heterotrophen Bakterien enthielt, wurde parallel zu den oben genannten Kulturen untersucht.

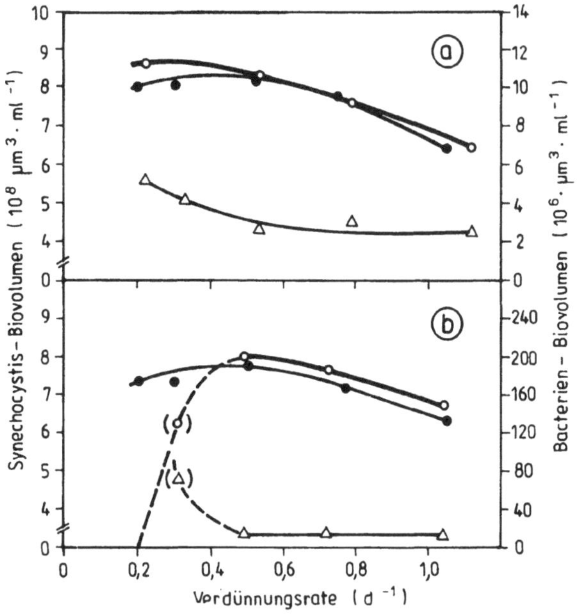

Abb. 1. Biovolumen von *Synechocystis* spec. und Stamm M in Abhängigkeit von der Verdünnungsrate im Steady-State der Chemostatkulturen
● *Synechocystis* spec. in Einzelkultur
○ *Synechocystis* spec. in Mischkultur mit Stamm M
△ Stamm M in Mischkultur mit *Synechocystis* spec.
Die Kultivierung erfolgte in nitratlimitiertem NM mit 2 mmol · l^{-1} · KNO_3 (a) bzw. in dem gleichen Medium mit Zusatz von 1,46 mmol · l^{-1} Saccharose (b). Die Mischkultur in (b) erreichte bei einer Verdünnungsrate von 0,31 d^{-1} kein Steady-State (eingeklammerte Symbole und unterbrochen gezeichneter Teil der Kurven)

Die axenische *Synechocystis*-Kultur blieb in ihrem Wachstum durch Saccharose unbeeinflußt (Abb. 1a und 1b). Die Einzelkultur von Stamm M wurde auch im Medium mit Saccharose bei allen untersuchten Verdünnungsraten ($D = 0{,}2^{-1}$ d bis 1,0 d^{-1}) ausgewaschen, obwohl dieser Bakterienstamm Saccharose als einzige Kohlenstoff- und Energiequelle verwerten kann. Die in diskontinuierlicher Kultur ermittelte relative Wachstumsrate betrug jedoch nur 0,1 d^{-1} bis 0,2 d^{-1}.

Im Gegensatz zu den Einzelkulturen kam es in den Mischkulturen mit *Synechocystis* nicht zur Auswaschung des Stammes M. Stellten die Exsudate von *Synechocystis* die einzige organische Kohlenstoffquelle für die Bakterien dar, lag der Anteil des Bakterienbiovolumens am Gesamtbiovolumen bei allen Verdünnungsraten unter 1% (Abb. 1a). Die Ursache ist in der geringen Exsudation durch *Synechocystis* spec. zu suchen. Wie die mit der ^{14}C-Methode gewonnenen Ergebnisse zeigten, werden durchschnittlich nur 2% der Photosyntheseprodukte als EOC ins Medium abgegeben. Die Bakterien hatten in der Mischkultur ohne Saccharose keinen Einfluß auf das Biovolumen von *Synechocystis* (Abb. 1a).

Stand den Bakterien in der Mischkultur Saccharose als zusätzliche organische Kohlenstoffquelle zur Verfügung, war bei Verdünnungsraten > 0,5 d^{-1} ihr Biovolumenanteil nur geringfügig erhöht (durchschnittlich 1,5% des Gesamtbiovolumens; Abb. 1b). Erst bei Verdünnungsraten < 0,5 d^{-1} wurde die Saccharose von Stamm M zum Biomassezuwachs genutzt. Die Zunahme des Biovolumens der Bakterien in der Mischkultur führte zur fortschreitenden Verdrängung von *Synechocystis* (Abb. 1b). Nachdem auf diese Weise das Biovolumen der Cyanobakterien auf 1/3 des Ausgangswertes dezimiert wurde, nahm nach Erreichen eines Maximums auch das Biovolumen von Stamm M ab. Als Ursache betrachten

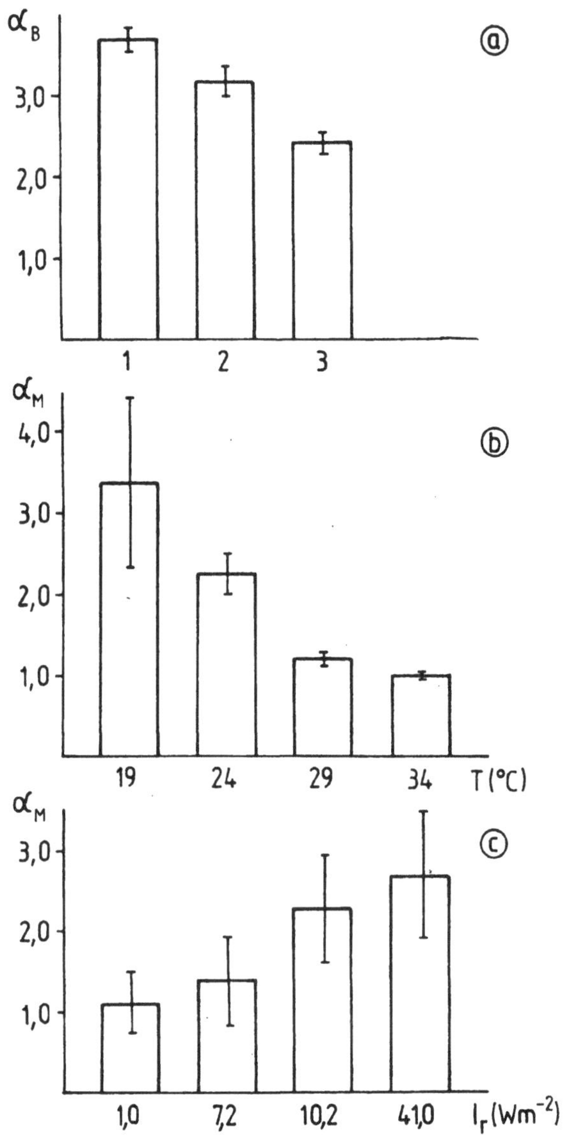

Abb. 2. Stimulationsfaktoren α_B für den Bakterienstamm M und für *M. firma* in Batch-Kultur nach Filtratzugaben. Mittelwerte dreier Parallelversuche, errechnet nach Formel (1), mit Konfidenzintervallen, P = 5%.

a — Einfluß von Filtrat der *Synechocystis*-Einzelkultur (1), Filtrat der Mischkultur von *Synechocystis* und Stamm M (2) und von 0,1 g · l^{-1} Hefeextrakt (3) auf α_B. Die Filtrate stammen aus Chemostatkulturen mit D = 0,2 d^{-1}. Alle Ansätze enthielten 20 mmol · l^{-1} KNO_3 und 5,84 mmol · l^{-1} Saccharose (MIKKAT 1986).

b — Einfluß der Kultivierungstemperatur auf α_M für *S. q.*-Kulturfiltrat (I_r = 41 W · m^{-2}). *M. f.* und *S. q.* wurden jeweils unter identischen Bedingungen kultiviert. Das Filtrat wurde mit NM im Verhältnis 1 : 1 gemischt.

c — Einfluß der Strahlungsflußdichte auf α_M für *S. q.*-Kulturfiltrat (T = 19 °C). Die *S. q.*-Filtratspenderkultur wurde bei I_r = 35 W · m^{-2} und T = 29 °C kultiviert. Das Filtrat wurde mit NM im Verhältnis 1 : 1 gemischt wir die mit der Verringerung der Cyanobakterienkonzentration einhergehende Reduzierung der EOC Konzentration. Letztendlich wurden sowohl *Synechocystis* als auch Stamm M aus der Kultur ausgewaschen.

Das Wachstumsverhalten dieser Mischkultur demonstriert in eindrucksvoller Weise den positiven Effekt der Exsudate von *Synechocystis* spec. auf das Wachstum von Stamm M, insbesondere bei der Verwertung von Saccharose.

Diese Wachstumsförderung konnte mit Hilfe sogenannter Batch-Filtratversuche bestätigt werden. Das Wesen der Methode besteht darin, eine zu testende Akzeptorkultur mit zellfreiem Filtrat einer Spenderkultur zu exponieren und mit Kontrollen ohne Filtratzusatz zu vergleichen. Die Quantifizierung erfolgte in Anlehnung an KRÜGER & LIBBERT (1988) mit Stimulationsfaktoren (α), die Differenzen im Wachstum zwischen Kontroll- und Filtratkulturen widerspiegeln.

$$\alpha = \frac{\mu\,(\text{Filtratansatz})}{\mu\,(\text{Kontrollansatz})} \qquad (1)$$

Nehmen Stimulationsfaktoren den Wert 1 an, so verhalten sich die Arten zueinander neutral, für $\alpha < 1$ liegt eine Inhibition und für $\alpha > 1$ eine Stimulation durch die Filtratspenderkultur vor.

Unter dem Einfluß des *Synechocystis*-Kulturfiltrates stieg die relative Wachstumsrate von Stamm M im Vergleich zum Wachstum im NM mit Saccharose um mehr als 250% ($\alpha_B = 3{,}7$) und wurde somit stärker stimuliert als durch die Zugabe von Hefeextrakt ($\alpha_B = 2{,}4$; Abb. 2a). Welche der von *Synechocystis* exsudierten organischen Substanzen diese Stimulierung bedingen, bedarf weiterer Untersuchungen.

Offen bleibt auch die Frage, inwiefern die Verdrängung der Cyanobakterien aus der Chemostatkultur (s. oben) das Ergebnis einer unterlegenen Konkurrenz um limitierenden Nitratstickstoff war oder durch die Exsudation spezifischer Wachstumsinhibitoren durch Stamm M hervorgerufen wurde.

Mischkultivierung von Cyanobakterien und Chlorophyceen im Turbidostaten

Im Gegensatz zum Chemostaten gestattet der Turbidostat, die mit spezifischen, extrazellulären Wirkstoffen interferierenden Nährstofflimitationen als Konkurrenzfaktor in Mischpopulationen auszuschließen. Dieses kontinuierliche Kultivierungssystem arbeitet bei Nährstoffsättigung. Regelgröße ist die trübungsäquivalente Biomassekonzentration, die bei intermittierender Bestimmung der optischen Dichte je nach Überschreitung eines fixierbaren Sollwertes durch äquivalente Verdünnung mit frischem Nährmedium konstant gehalten wird. Dieses Steuerungsprinzip bedingt, daß eine Koexistenz zweier oder mehrerer Arten nur bei Identität ihrer relativen Wachstumsraten möglich ist (KRÜGER & LIBBERT 1988). Ansonsten erfolgt unweigerlich die Auswaschung der wachstumsmäßig unterlegenen Komponente.

Unsere Aufgabe bestand darin, die Eignung dieses Systems zur Mischkultivierung und Analyse biotischer Wechselwirkungen am Beispiel zweier häufig vorkommender Phytoplanktongruppen, der Cyanobakterien (*Microcystis firma*) und Chlorophyceen (*Scenedesmus quadricauda*) in Abhängigkeit vom Licht und der Temperatur zu testen.

Ausgangspunkt war auch hier die detaillierte Untersuchung des Steady-State-Wachstums der Turbidostateinzelkulturen. Auf der Grundlage der μ-Mittelwerte für 12 Temperatur-Licht-Kombinationen wurden nach nichtlinearer Regression mathematische Modelle für die Beschreibung der $\mu(I_r, T)$-Abhängigkeit entwickelt, deren Parameter von einer guten Anpassung an die Versuchsergebnisse zeugen (KRÜGER et al. 1988):

$$\mu_M = 0{,}125\,\{\exp[4{,}1 \cdot 10^{-3}(37{,}5 - T)^2] + 10\,I_r^{-1}\}^{-1} \qquad (2)$$

$$\mu_S = 0{,}115\,\{\exp[7{,}5 \cdot 10^{-3}(30{,}5 - T)^2] + 28{,}5\,I_r^{-1}\}^{-1} \qquad (3)$$

Zur Veranschaulichung der Koexistenz- und Dominanzbereiche in einer Mischkultur definierten wir die Verdrängungsrate v, die als Differenz der relativen Wachstumsraten zweier Arten A und B berechnet wird:

$$v_A = \mu_A - \mu_B (= -v_B) \qquad (4)$$

Bei $v_A > 0$ verdrängt A den Partner B; bei $v_A < 0$ wird A durch B verdrängt und bei kontinuierlicher Koexistenz ist $v_A = v_B = 0$. Somit resultieren nach Gleichung (4) als Voraussagen für die Mischkultur folgende Verdrängungsraten mit der Einheit d^{-1} (KRÜGER & LIBBERT 1988):

$$v_M (= -v_S) = 3\{\exp[4{,}1 \cdot 10^{-3}(37{,}5 - T)^2] + 10 I_r^{-1}\}^{-1} - 2{,}76\,\{\exp[7{,}5 \cdot 10^{-3}(30{,}5 - T)^2] + 28{,}5 I_r^{-1}\}^{-1}. \qquad (5)$$

Die Gültigkeit dieser Gleichung ist an das Fehlen biotischer Wechselwirkungen gebunden.

Wir prüften die Übereinstimmung zwischen dem Modell und der tatsächlichen Biomasseentwicklung in Mischkulturen unter variablen Licht-Temperatur-Bedingungen. Drei Beispiele sind in Tab. 1 zusammengefaßt.

Tab. 1. Verdrängungsraten v_M und Stimulationsfaktoren α_M in Turbidostatmischkulturen unter verschiedenen Licht-Temperatur-Bedingungen.
Die Berechnungen basieren auf den Mittelwerten aller Messungen der Steady-State-Einzelkulturen bzw. während der gesamten Mischkulturen (MK) (aus KRÜGER & LIBBERT 1988).

MK	I ($W \cdot m^{-2}$)	T (°C)	Verdrängungsraten v_M (d^{-1}) errechnet nach Gleichung 5	gemessen in Mischkultur	Stimulationsfaktor α_M errechnet nach Gleichung 6
1	14	29	0,55	0,58	0,94
2	18	24	0,19	0,48	1,02
3	19,5	19	−0,01	0,07	1,21

Bei hohen Temperaturen (29 °C; Mischkultur 1) unterscheidet sich der theoretisch berechnete v_M-Wert kaum von dem im Versuch ermittelten. Die Grünalge wurde relativ schnell aus dem System verdrängt. Mit der Verringerung der Temperatur erhöht sich die Wahrscheinlichkeit einer Koexistenz (v nähert sich Null an), es vergrößert sich aber auch die Differenz zwischen theoretisch vorhergesagtem und praktisch bestimmtem Wert (Tab. 1). Für Mischkultur 3 ($T = 19$ °C) wurde theoretisch eine Auswaschung von *M. firma* erwartet ($v_M = -0{,}01\ d^{-1}$), praktisch dominierte aber auch hier das Cyanobakterium ($v_M = 0{,}07\ d^{-1}$). Nach KRÜGER & LIBBERT (1988) berechneten wir für beide Komponenten die Stimulationsfaktoren:

$$\alpha = \frac{\mu\,(\text{Mischkultur})}{\mu\,(\text{Einzelkultur})} \qquad (6)$$

Die Werte für *M. firma* beweisen, daß die Diskrepanz zwischen dem theoretisch berechneten und tatsächlich gemessenen ν_M durch eine Wachstumsförderung der Cyanobakterien in der Mischkultur verursacht wurde. Sie war bei niedrigen Temperaturen am stärksten ausgeprägt ($\alpha_M = 1{,}21$ für Mischkultur 3; Tab. 1). Über wachstumsstimulierende Wirkungen von *S. quadricauda* auf andere Phytoplankter berichten auch BEDNARZ & CIERNIAK (1983) und BURKIEWICZ (1987). *S. quadricauda* wurde demgegenüber von *M. firma* nicht beeinflußt (KRÜGER 1984).

Untersuchungen mit der oben beschriebenen Batch-Filtratmethode bestätigen diese Ergebnisse. Die Förderung des Wachstums von *M. firma* durch das Grünalgenfiltrat stieg durch Temperaturerniedrigung von 34 °C auf 19 °C um das 3fache (Abb. 2b). Demgegenüber scheint zwischen dem Lichtangebot und α_M eine direkte Proportionalität zu bestehen (Abb. 2c). Wechselnde Umweltbedingungen führen somit zu einer wechselnden Sensibilität der Cyanobakterien gegenüber den Grünalgenexsudaten, bewirken aber offensichtlich auch Unterschiede in Qualität und Quantität der Exsudate selbst [α_M (19 °C; 41 W · m^{-2}) = 3,4 für *S. q.*-Filtrat nach Kultivierung bei 19 °C und 41 W · m^{-2}, α_M (19 °C; 41 W · m^{-2}) = 2,7 für *S. q.*-Filtrat nach Kultivierung bei 29 °C und 35 W · m^{-2}; Abb. 2b, c]. *M. firma* wird als thermophiler und stark lichtempfindlicher Cyanobakterienstamm (KRÜGER et al. 1988) insbesondere unter für sie ungünstigen Bedingungen, wie geringe Temperatur und hohes Lichtangebot, im Wachstum durch die Grünalgenexsudate stimuliert. Die stoffliche Ursache dieser Förderung konnte bisher nicht eindeutig geklärt werden. Es werden der Einfluß spezifischer, stoffwechselaktiver Stimulatoren als auslösende Faktoren und im weiteren Kultivierungsverlauf pH-stabilisierende Eigenschaften des Grünalgenfiltrates diskutiert (KRÜGER 1984).

Auf der Grundlage der mathematischen Modelle zur Licht-Temperatur-Abhängigkeit der relativen Wachstumsraten von *M. firma* und *S. quadricauda* und deren Verhalten in Turbidostatmischkulturen konnte unter Berücksichtigung der Ergebnisse zum Filtrateinfluß ein Schema zur Konkurrenzbeziehung beider Arten entwickelt werden, welches das Verhältnis von Cyanobakterien und Chlorophyceen unter eutrophen bzw. hypertrophen Bedingungen widerspiegelt (Abb. 3). Die Bedeutung der von uns nachgewiesenen biotischen Interferenzen erscheint im Vergleich zu jener der abiotischen Faktoren gering. Ihr Einfluß beschränkt sich auf eine leichte Verschiebung der Isoklinen und ist somit hauptsächlich für das Koexistenzverhalten im Turbidostaten relevant.

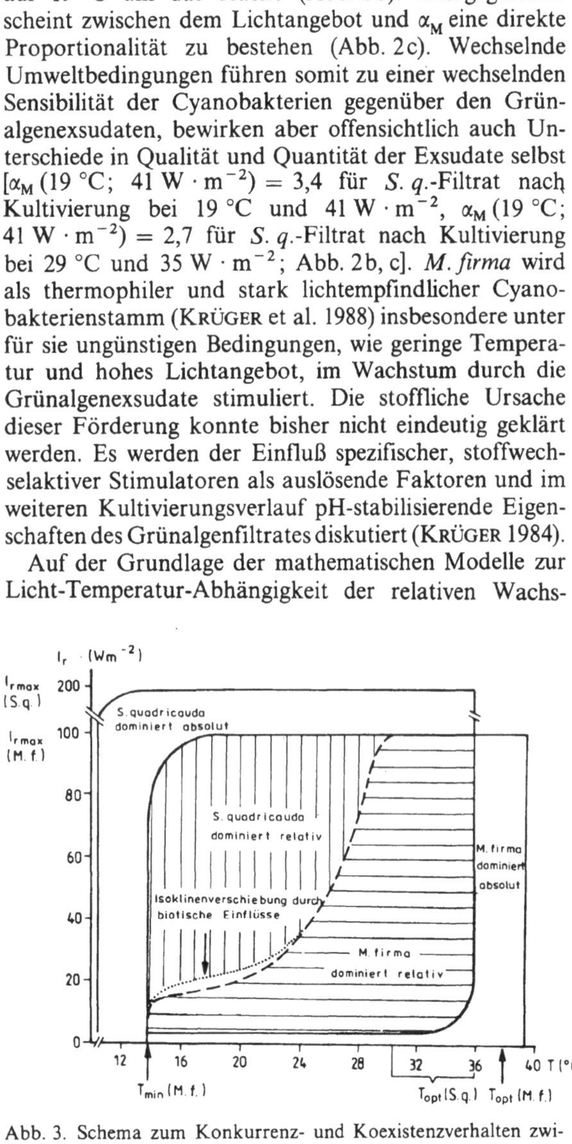

Abb. 3. Schema zum Konkurrenz- und Koexistenzverhalten zwischen *M. firma* und *S. quadricauda* in Abhängigkeit vom Licht und der Temperatur unter Berücksichtigung biotischer Interferenzen (nach KRÜGER 1984).
- waagerecht schraffierte Fläche = I_r — T-Kombinationen für die $\mu_M > \mu_S$
- senkrecht schraffierte Fläche = I_r — T-Kombinationen, für die $\mu_M < \mu_S$
- freie Flächen = absolute Dominanz einer Art, da die andere unter diesen Bedingungen nicht existieren kann.
- Isokline = I_r — T-Kombination, für die $\mu_M = \mu_S$ (----) bzw. Isoklinenverschiebung durch biotische Einflüsse (......)

Schlußfolgerungen

Chemostat und Turbidostat erlauben die kontinuierliche Mischkultivierung mehrerer Arten. Dabei ist der Chemostat besonders für die Untersuchung trophischer Kopplungen geeignet. In einer Chemostat-Mischkultur von *Synechocystis* spec. und einem heterotrophen Bakterienstamm, in der das Wachstum ersterer durch die anorganische N-Quelle und das Wachstum letzterer durch die organische C-Quelle, die von den Cyanobakterien in Form extrazellulärer organischer Substanzen abgegeben wurde, limitiert war, lag der Biomasseanteil der heterotrophen Bakterien unter 1%. Die Zugabe einer allochthonen C-Quelle (Saccharose) bewirkte bei niedrigen Verdünnungsraten eine mit der Zunahme der Bakterienbiomasse verbundene Verdrängung der Cyanobakterien. Dabei war für das Biomassewachstum der heterotrophen Bakterien die gleichzeitige Verwertung von Saccharose und EOC von entscheidender Bedeutung.

Im Turbidostaten ist die Analyse des Einflusses abiotischer Faktoren auf das Konkurrenzverhalten zweier Arten bei Nährstoffsättigung möglich. Biotische Wechselwirkungen, z. B. auf der Grundlage spezifischer Wirkstoffe, werden nicht durch Konkurrenz um limitierende Nährstoffe überlagert. Für ihre Identifikation sind vergleichende Analysen der Mischkulturen mit den jeweiligen Einzelkulturen unerläßlich. Ein aus den Einzelkulturturbidostaten abgeleitetes mathematisches Modell beschreibt das licht- und temperaturabhängige Konkurrenzverhalten von *M. firma* und *S. quadricauda* nicht in allen Details. Eine Wachstumsstimulierung der Cyanobakterien durch die Grünalgenexsudate ruft leichte Verschiebungen in den Koexistenz- und Dominanzbereichen hervor.

Untersuchungen zum gegenseitigen Einfluß der Kulturfiltrate in Batch-Kultur stellen eine sinnvolle Ergänzung zur Mischkultivierung in kontinuierlichen Systemen dar.

Die beschriebenen Ergebnisse der Laborexperimente können und wollen keinen Anspruch erheben auf unmittelbare Übertragbarkeit auf natürliche Bedingungen. Sie tragen Modellcharakter und versuchen einen Ausschnitt aus dem vielfältigen Wechselwirkungsgefüge im natürlichen Gewässer zu erfassen.

Literatur

ALLEN, M. B., & ARNON, D. I. (1955): Studies on nitrogen-fixing algae. II. The sodium requirement of *Anabaena cylindrica*. Physiol. Plant. **8**, 653—660.

AZAM, F., FENCHEL, T., FIELD, J. G., GRAY, J. S., MEYER-REIL, L. A., & THINGSTAD, T. F. (1983): The ecological role of water-column microbes in the sea. Mar. Ecol. Prog. Ser. **10**, 257—263.

Bednarz, T., & Cierniak, J. (1983): Extracellular excretions of algae as a factor regulating the growth of algae cultures. Oceanologia (PRL) **17**, 19—20.

Bratbak, G., & Thingstad, T. F. (1985): Phytoplankton-bacteria interactions: an apparent paradox? Analysis of a model system with competition and commensalism. Mar. Ecol. Prog. Ser. **25**, 23—30.

Burkiewicz, K. (1987): The influence of gibberellins and cytokinins on the growth of some unicellular baltic algae. Bot. Mar. **30**, 63—69.

Cole, J. J. (1982): Interactions between bacteria and algae in aquatic ecosystems. Ann. Rev. Ecol. Syst. **13**, 291—314.

Currie, D. J., & Kalff, J. (1984): Can bacteria outcompete phytoplankton for phosphorus? A chemostat test. Microb. Ecol. **10**, 205—216.

Fogg, G. E. (1983): The ecological significance of extracellular products of phytoplankton photosynthesis. Bot. Mar. **26**, 3—14.

Hobson, P. N. (1969): Growth of mixed cultures and their biological control. Symp. Soc. Gen. Microbiol. **19**, 43—64.

Krüger, B. (1984): Zur Wachstumsphysiologie diskontinuierlicher und kontinuierlicher Cyanobacterien — Chlorophyceen — Mischpopulationen. Diss. A, W.-Pieck-Univ. Rostock.

— & Libbert, E. (1988): Wachstum von Cyanobacterien und Chlorophyceen bei Einzel- und Mischkultivierung. II. Vergleich des Wachstums der Komponenten in Einzel- und Mischkultur. Biochem. Physiol. Pflanzen **183**, 317—327.

— Schiewer, U., & Libbert, E. (1988): Wachstum von Cyanobacterien und Chlorophyceen bei Einzel- und Mischkultivierung. I. Modellierung des Wachstums der einzelnen Komponenten unter kombinierter Licht-Temperatur-Einwirkung. Biochem. Physiol. Pflanzen **183**, 307—316.

Mikkat, S. (1986): Wachstum von Bacterien in Mischkultur mit *Synechocystis* spec. Dipl.-Arbeit, W.-Pieck-Univ. Rostock.

Anschrift der Verfasser:

Dr. B. Krüger
S. Mikkat
S. Fulda
Wilhelm-Pieck-Universität
Sektion Biologie
Doberaner Str. 143
Rostock
DDR-2500

Limnologica (Berlin) **20** (1), 113—117, April 1990

Wilhelm-Pieck-Universität Rostock, Sektion Biologie, Rostock, DDR

NH_4 rhythm of Cyanobacteria in Eutrophic Shallow Waters and in Laboratory Cultures

By S. ABARZUA & U. SCHIEWER

With 5 Figures

Key words: Cyanobacteria; *Microcystis firma*; NH_4uptake; NH_4assimilation; NH_4release.

Abstract

Process of pelagic NH_4 rhythm discovered by SCHIEWER & BAADER (1982) in eutrophic shallow waters independent of the sediment is inducible in batch cultures of *Microcystis firma*. Optimal physiological conditions for the inducibility of NH_4rhythm were found. The principal mechanism of NH_4rhythm was elaborated.

Introduction

Eutrophic shallow waters are often characterized by seasonal and diurnal changes of nutrients. This is sometimes also true for the brackish waters in the Darss-Zingst boddens (South Baltic) (SCHIEWER et al. 1984). An analysis of the influence of different ammonium concentrations on experimental enclosures of water samples from this eutrophic shallow brackish estuary (bodden) led to the discovery of a sediment independent NH_4 rhythm under conditions of an excess of NH_4 entry and of growth limitation: NH_4 is fixed during exposure to light and released in the dark (SCHIEWER & BAADER 1982). This rhythm seems to be related to cyanobacteria because they were dominant in the biocenosis of these water samples. 65% of dry matter of these samples consisted of cyanobacteria, and only 35% of other algae, pelagic bacteria and zooplankton. Probably ammonium rhythm of cyanobacteria plays an important ecological role in eutrophic and hypertrophic waters.

The aim of the present investigation was to explain this phenomenon discovered in field experiments and for this to find out (1) if this rhythm is inducible in laboratory cultures of cyanobacteria, (2) under which physiological conditions this rhythm occurs and (3) to find out via which transport systems and biochemical pathways the uptake, assimilation, and excretion of ammonium are realized.

Material and Methods

Organism

The organism used was the chroococcale cyanobacterium *Microcystis firma* (BREB. et LENORM.) SCHMIDLE, strain GROMOV/LEN. 398 from the culture collection GROMOV kept at the Institute of Biology of the State University "A. A. Shdanow" Leningrad.

Culture conditions

Maintenance culture

The suspensions of cyanobacteria of *Microcystis firma* were cultivated in cylindrical vessels illuminated from two sides under sterile conditions. Continuous light was used (30 W · m^{-2}). The inorganic

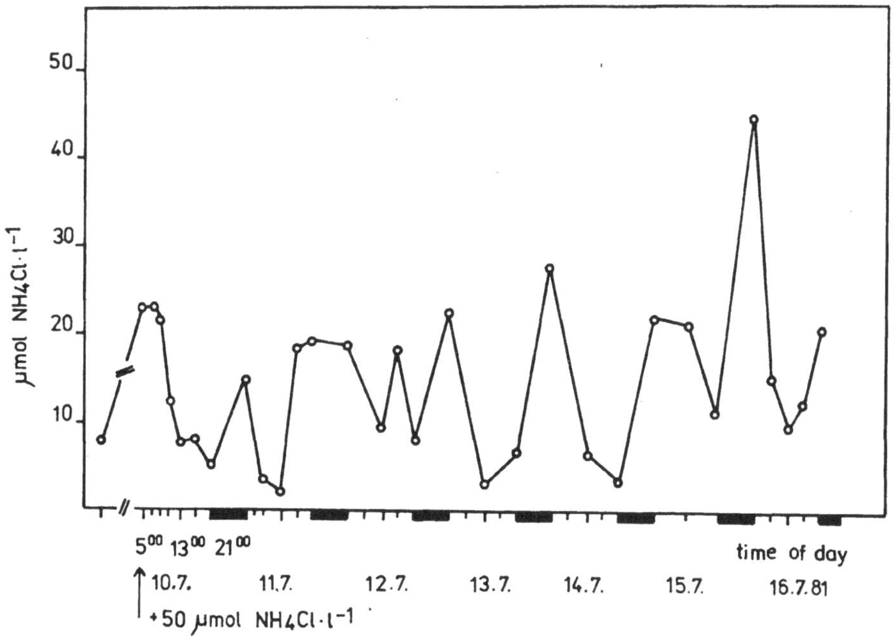

Fig. 1. Pelagic rhythm of ammonia (SCHIEWER et al. 1984)

nutrient solution was that described by ALLEN & ARNON (1955). The culture vessels were submersed in a water bath at a constant temperature of 29 °C. Inoculation density was 0.35 (E_{750}). Carbon dioxide concentration in the aerating gas mixture was maintained at 5% (v/v).

Culture for standard experiments

Cells of maintenance cultures with additional N-starvation for 24 h were harvested (E_{750} = 0.9—1.0) for standard experiments to study NH_4 rhythm. The cultures for standard experiments have the following modifications in comparison to maintenance cultures: 1/8 KNO_3 to reach a further N-starvation and thus a more significant NH_4uptake, NH_4Cl addition (+ 20 mg $NH_4Cl \cdot l^{-1}$ = 374 µmol $NH_4Cl \cdot l^{-1}$), light intensity: 10—15 W · m^{-2}, alternation of light/dark period (12/12 h), no aeration so that, by an absence of CO_2 supply, growth limitation was reached.

Content of ammonium

Ammonium in suspension supernatants was estimated by the phenol: alkaline hypochlorite method (SOLORZANO 1969). In all experiments the removal of ammonium from the medium was measured.

Cell number

The number of cells was estimated by the electronic particle counter "Laborscale" (Medicor Budapest).

Contents of pigments

The content of chlorophyll a and phycocyanin was estimated by SIGALAT & DE KOUCHOVSKY (1975).

Reference magnitudes

The cyanobacterial cell density (cell number, chlorophyll a, phycocyanin) is nearly constant during all presented experiments because of growth limitation so that NH_4concentration is presented per volume (l).

Enzyme activities

Enzyme extraction

50 ml centrifuged cells of *Microcystis firma* were homogenized with 0.5 ml of a 0.1 M Tris acetate buffer, pH 7.6, 0.1% β-mercaptoethanol and 2 ml glass beads (diameter 0.1—0.2 mm, VEB Trisola Steinach, G. D. R.) in a vibrogen homogenizer (Bühler, Tübingen, F. R. G.) for 2 min at 0—4 °C. After washing out with 0.1 M 3.5 ml Tris acetate buffer, pH 7.6, the homogenate was centrifuged at 20,000 g for 15 min at 0—4 °C. The supernatant was used as the enzyme and protein source.

Glutamine synthetase (GS) activity

The Mn^{2+}-dependent γ-glutamyl transferase activity (EC 6.3.1.2.) was assayed as described by SHAPIRO & STADTMAN (1970).

Glutamate synthase (GOGAT) activity (EC 1.4.7.1.)

GOGAT activity was determined by measuring the ferredoxin-dependent formation of L-glutamate from L-glutamine and 2-oxoglutarate as described by NICKLISCH et al. (1976).

Glutamate dehydrogenase (GDH) activity (EC 1.4.1.4.)

Reductive amination was measured by following the oxidation of NADPH at 340 nm described by SHATILOV & SUND (1983).

Protein assay

A LOWRY's Folin-phenol method was employed in protein estimation (LOWRY et al. 1951).

Results

Fig. 2 demonstrates that NH_4 rhythm is inducible in batch cultures of cyanobacteria under the following physiological conditions: NO_3^- prestarvation, growth limitation (e.g. CO_2 deficiency), light/dark alternation and optimum of pH, temperature and light intensity at 7—8; 29 °C and 10.5 W · m^{-2}, respectively. Ammonium is taken up at the beginning of light period and released in the middle or at the end of light period (Fig. 2). A decreasing tendency of ammonium concentration in the course of 7 days is demonstrated what refers to a storage of ammonium. This correlates with the discovered result of an increase of dry matter (Fig. 2).

The discovery of inducibility of NH_4 rhythm and optimal conditions for its appearance in cultures of cyanobacteria was followed by investigations about transport systems and biochemical pathways of ammonium uptake, -assimilation and -excretion.

Activities of the following enzymes were tested for the study of ammonium assimilation: glutamine synthetase (GS), glutamate synthase (GOGAT) and glutamate dehydrogenase (GDH). NH_4 rhythm is reversely correlated with an rhythm of GS activity (Fig. 3). The same result was found for GOGAT (results not shown) what demonstrates the cooperation of these 2 enzymes in ammonium assimilation. The alternative GDH pathway in ammonium assimilation, which is acting by utilization of high ammonium concentrations (> 1 mmol $NH_4Cl \cdot l^{-1}$) and in eukaryotic cells in the most cases, is of minor importance; the activity is very low and does not show any rhythm (Fig. 3). Our result that fixation of ammonium involves mainly the GS/GOGAT pathway is proved by experiments with special inhibitors of these 2 enzymes. L-Methionine-sulfoximine (MSX), inhibitor of GS activity, and azaserine, inhibitor of GOGAT activity, block the assimilation of ammonium and produce a significant excretion of ammonium (Figs. 4, 5). To study processes of ammonium excretion long-term tests with different inhibitors (DCMU, KCN, CCCP) were performed. By elimination of special reactions of metabolism we wanted to estimate the importance of this/these reactions for the formation of NH_4rhythm. It is seen that DCMU has no effect on the uptake and assimilation of ammonium, but inhibits the release of ammonium (Fig. 4). DCMU blocks the electron transport from the PS II to the PS I, e.g. inhibits formation of reduced equivalents like NADPH or reduced ferredoxin. That means that reduced equivalents are absolute necessary for the processes of NH_4 excretion. KCN, inhibitor of cytochromoxidase, supresses not only the assimilation but also the excretion of ammonium, an evidence for the energy requirement of both processes. Additional experiments with CCCP (uncoupler of electron transport and oxidative phosphorylation) confirm the energy requirement for assimilation and excretion of NH_4 (results not shown).

Short-term experiments with MSX, azaserine, KCN and CCCP to study processes of ammonium uptake demonstrate that inhibitors have not an inhibitory effect at the beginning of experiment but only after 15 min (Fig. 5). We suggest that the first, rapid phase of ammonium uptake (0—15 min) is independent of meta-

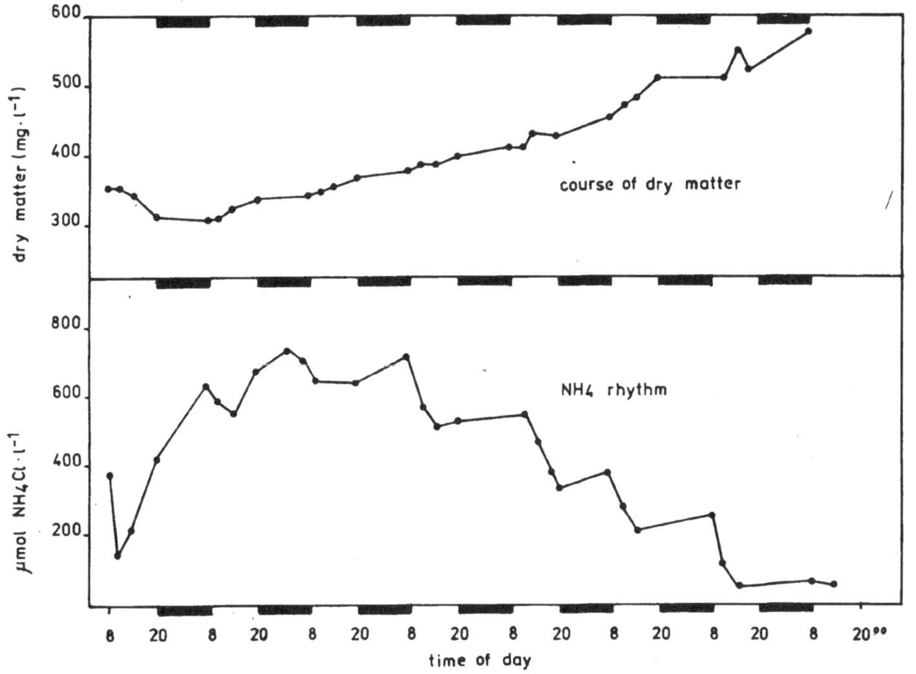

Fig. 2. Ammonium rhythm and adequate course of dry matter in batch cultures of *Microcystis firma*

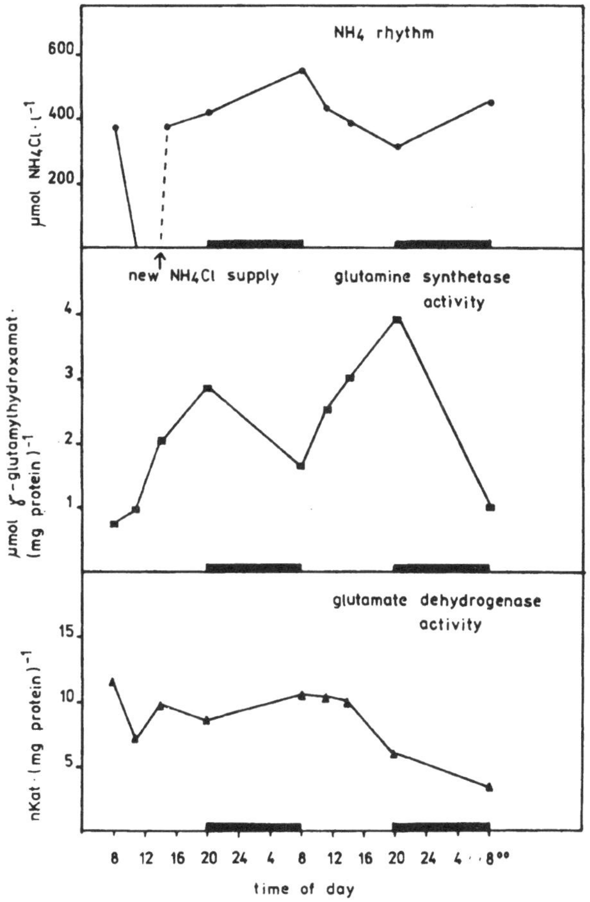

Fig. 3. NH_4 rhythm and adequate activities of GS and GDH in batch cultures of *Microcystis firma*

bolism of ammonium and can be attributed to uptake via a (NH_4) transport system, probably to passive diffusion. Only after this phase the inhibitors have a blocking effect, that means assimilation of ammonium occurs at this time under normal conditions.

Discussion

The present investigation proves that the process of NH_4 rhythm discovered by SCHIEWER & BAADER (1982) in situ can be induced in batch cultures of *Microcystis firma*. Optimal physiological conditions for the inducibility of NH_4 rhythm were found.

The results prove that the major route of primary ammonium assimilation in the non-nitrogen-fixing cyanobacterium *Microcystis firma* is the GS/GOGAT pathway (MIFLIN & LEA 1976; ROWELL et al. 1979). Experiments with MSX, inhibitor of GS activity (RONZIO et al. 1969), and azaserine, inhibitor of GOGAT activity (MORENO-VIVIAN et al. 1983), demonstrated that the assimilation of ammonium is suppressed by the addition of these inhibitors and a significant excretion of NH_4 occurs. Similar results were found in MSX-treated cells of *Anacystis nidulans* (BOUSSIBA & GIBSON 1985) and azaserine-treated cells of *Rhodopseudomonas capsulata* E1F1 (MORENO-VIVIAN 1983).

A possible reason for the formation of NH_4 rhythm is the rhythmical change of GS and GOGAT activity during light/dark alternation. A light regulation of GS is described by a lot of authors. TISCHNER & HÜTTERMANN (1980) described a sevenfold enhancement of GS-activity and a twofold increase of GOGAT activity after 1 hour illumination in cells of *Chlorella sorokiniana*. ROWELL et al. (1979) found a reversible inactivation of glutamine synthetase in *Anacystis nidulans* and *Anabaena cylindrica* in dark.

Process of NH_4 rhythm can be divided in 3 phases: the first, very rapid uptake phase, the second, slower phase, attributed to ammonium assimilation and the third phase of ammonium excretion. KERBY et al. (1986) demonstrated a biphasic pattern of $^{14}CH_3NH_3$ uptake in the cyanobacteria *Anabaena variabilis* and *Nostoc* CAN. We suggest for our experiments that ammonium is taken up passively in the first, quickly running phase following the electrical gradient through the plasmalemma, because treatment of *Microcystis firma* with MSX, azaserine, KCN, and CCCP had no inhibitory

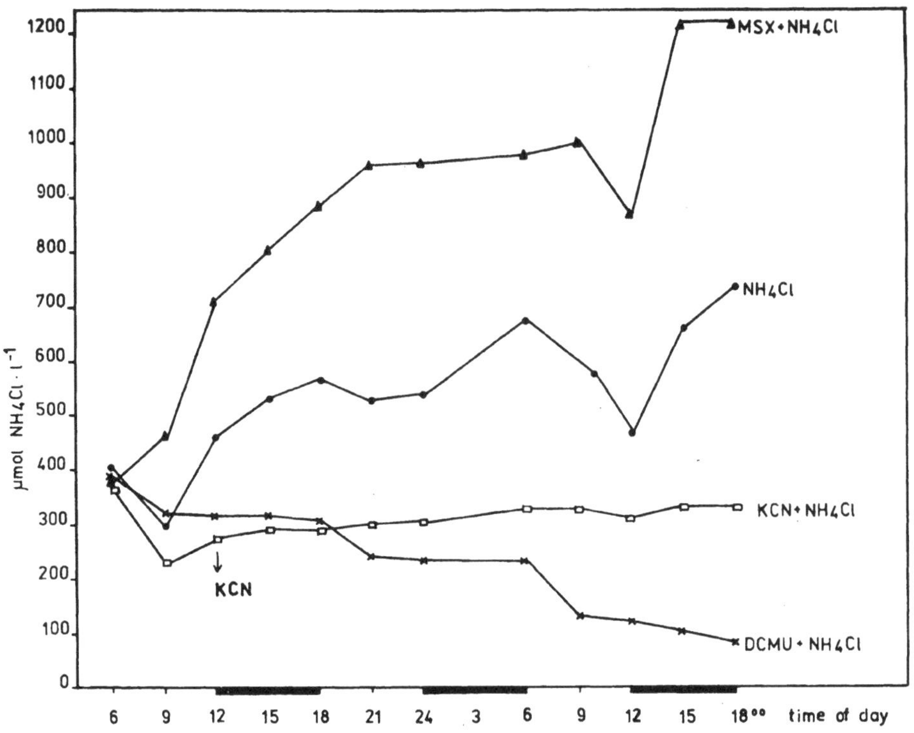

Fig. 4. Long term effects of MSX, KCN and DCMU on ammonium uptake, -assimilation and -release in cultures of *Microcystis firma*. Light/dark alternation: 6 h/6 h. DCMU- and KCN concentration: 100 µmol $\cdot$ l^{-1}, MSX concentration: 500 µmol $\cdot$ l^{-1}

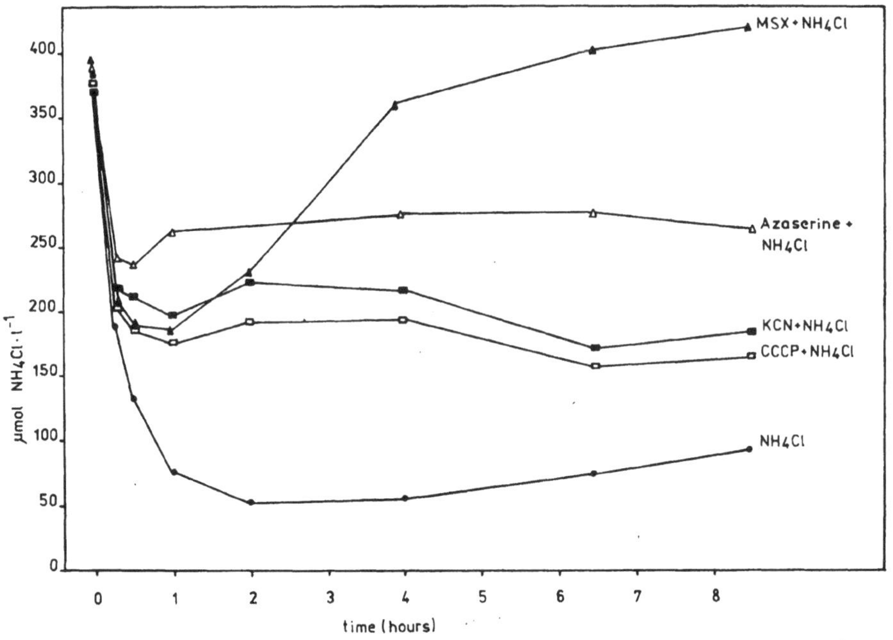

Fig. 5. Short-term effects of MSX, azaserine, KCN, and CCCP on NH_4uptake, -assimilation and -release in cultures of *Microcystis firma*.
KCN- and CCCP concentration: 100 µmol $\cdot$ l^{-1}.
MSX concentration: 500 µmol $\cdot$ l^{-1}.
Azaserine concentration: 200 µmol $\cdot$ l^{-1}

effect on the first, rapid phase of ammonium uptake and, consequently, energy (ATP) is not required. ULLRICH et al. (1984) postulated a passive uptake of ammonium in a uniport mechanism in *Lemna gibba*. The second, slower phase of ammonium assimilation was blocked by MSX, azaserine, KCN, and CCCP, an evidence for energy requirement of this process via the GS/GOGAT pathway. The third phase of NH_4 rhythm, the phase of ammonium excretion is inhibited by DCMU, KCN, and CCCP, an evidence for requirement of energy and reduced ferredoxin for the process of ammonium release. It is suggested that because of the non-efficiency of reduced ferredoxin in DCMU-treated cells of *Microcystis firma* nitrate assimilation occurring in spite of the presence of NH_4 is inhibited under conditions used (ABARZUA 1985) and thus release of NH_4 could not take place consequently. The same effect was found by DEMIDOV et al. (1986). By elimination of nitrate assimilation (utilization of KCl containing medium instead of KNO_3 containing one) only very small processes of ammonium excretion are observed which are to be referred to NH_4release in the photorespiratory nitrogen cycle, degradations of proteins and amino acids.

Thus, it was proved that NH_4 rhythm discovered in field experiments (SCHIEWER & BAADER 1982) is inducible at laboratory in the cyanobacterium *Microcystis firma* under optimal physiological conditions. The principal mechanism of NH_4 rhythm was elaborated.

Summary

The inducibility of NH_4rhythm discovered by SCHIEWER & BAADER (1982) in eutrophic shallow waters independent of the sediment was proved in batch cultures of cyanobacteria. Optimal physiological conditions for this inducibility were found. Process of NH_4rhythm can be divided in 3 phases: the first, very rapid uptake phase; the second, slower phase, attributed to ammonium assimilation and the third phase of ammonium excretion. Experiments were performed with special inhibitors (MSX, KCN, CCCP, DCMU, azaserine) to study these 3 phases. It is suggested that ammonium is taken up passively in the first quickly running phase following the electrical gradient through the plasmalemma; energy is not required. The second, slower phase is attributed to metabolism of ammonium via the GS/GOGAT pathway. Energy requirement (ATP) for assimilation processes of ammonium is proved. The processes of ammonium excretion require the availability of reduced equivalents and energy (ATP). Possible sources of ammonium release such as nitrate assimilation, photorespiratory nitrogen cycle and degradation of proteins and amino acids are discussed.

Acknowledgement

The authors are very grateful to Mrs. C. BAUER and Mrs. E. FISCHER for help with the technical work.

References

ABARZUA, S. (1985): NH_4-Rhythmik bei der photoautotrophen chroococcalen Blaualge *Microcystis firma*. Proceedings der 17. wissenschaftlichen Koordinierungstagung des RGW zum Thema 1. 18. 3 in Třeboň, 6.—12. 10. 1985.

ALLEN, M. B., & ARNON, D. I. (1955): Studies on nitrogen-fixing bluegreen algae II. The sodium requirement of *Anabaena cylindrica*. Physiol. Plant. **8**, 653—660.

BOUSSIBA, S., & GIBSON, J. (1985): The role of glutamine synthetase activity in ammonium and methylammonium transport in *Anacystis nidulans* R-2. FEBS LETT. **180**, 13—16.

DEMIDOV, E. D., PAVLOVA, E. A., & SMOLOV, A. P. (1986): Light-dependent nitrate reduction by *Chlorella* cells. Physiol. rast. **33**, 913—921 (russ.).

KERBY, N. W., ROWELL, P., & STEWART, W. D. P. (1986): The uptake and metabolism of methylamine by N_2-fixing cyanobacteria. Arch. Microbiol. **143**, 353—358.

LOWRY, O. H., ROSEBROUGH, N. J., FARR, A. L., & RANDALL, P. J. (1951): Protein measurement with the Folin phenol reagent. J. Biol. Chem. **193**, 265—276.

MIFLIN, B. J., & LEA, P. J. (1976): The pathway of nitrogen assimilation in plants. Phytochemistry **15**, 873—885.

MORENO-VIVIAN, C., CEJUDO, F. J., CARDENAS, J., & CASTILLO, F. (1983): Ammonia assimilation pathways in *Rhodopseudomonas capsulata* E1F1. Arch. Microbiol. **136**, 147—151.

NICKLISCH, A., GESKE, W., & KOHL, J.-G. (1976): Über die Bedeutung von Glutamatsynthase und Glutamatdehydrogenase für die N-Assimilation im Weizenprimärblatt. Biochem. Physiol. Pflanzen **170**, 85—90.

RONZIO, R. A., ROWE, W. B., & MEISTER, A. (1969): Studies on the mechanism of inhibition of glutamine synthetase by methionine sulfoximine. Biochemistry **8**, 1066—1075.

ROWELL, P., SAMPAIO, M. J. A. M., LADHA, J. K., & STEWART, W. D. P. (1979): Alteration of cyanobacterial glutamine synthetase activity in vivo in response to light and NH_4^+. Arch. Microbiol. **120**, 195—200.

SCHIEWER, U., & BAADER, G. (1982): Pelagic NH_4rhythm independent of the sediment in eutrophic shallow waters. Int. Rev. ges. Hydrobiol. **67**, 845—850.

— BÖRNER, R., KRÜGER, B., & EVERT, F.-K. (1984): Estuarine phytoplankton response to nutrient changes in controlled ecosystem enclosures. Ophelia, Suppl. **3**, 201—212.

SHAPIRO, B. M., & STADTMAN, E. R. (1970): Glutamine synthetase (*Escherichia coli*). Methods in Enzymology **17 A**, 910—922.

SHATILOV, V. R., & SUND, H. (1983): Glutamate dehydrogenase of the unicellular green alga *Scenedesmus acutus*. Planta **157**, 367—370.

SIGALAT, C., DE KOUCHOVSKY, Y. (1975): Fractionement et caractérisation de l'appareil et caracterisation de l'algue bleue unicellulaire *Anacystis nidulans*. I. Obtention de fractions membranaires par "lyse osmotique" et analyse pigmentaire. Physiol. Veg. **13**, 243—258.

SOLORZANO, L. (1969): Determination of ammonia in natural water by the phenol-hypochlorite method. Limnol. Oceanogr. **14**, 799—801.

TISCHNER, R., & HÜTTERMANN, A. (1980): Regulation of glutamine synthetase by light and during nitrogen deficiency in synchronous *Chlorella sorokiniana*. Plant Physiol. **66**, 805—808.

ULLRICH, W. R., LARSSON, M., LARSSON, C. M., LESCH, S., & NOVACKY, A. (1984): Ammonium uptake in *Lemna gibba* G1, related membrane potential changes, and inhibition of anion uptake. Physiol. Plant. **61**, 369—376.

Address of the authors:
Dr. S. ABARZUA
Prof. U. SCHIEWER
Wilhelm-Pieck-Universität Rostock
Sektion Biologie
Doberaner Str. 143
Rostock
DDR-2500

Limnologica (Berlin) **20** (1), 119—125, April 1990

Institute of Biology of the Latvian SSR Academy of Sciences, Salaspils, USSR

Protozoans as a Component of Biological Monitoring of the Baltic Sea

By Elmira Boikova

With 5 Figures and 3 Tables

Key words: Ciliated protozoa; Cultivation; Laboratory bioassays; Natural communities; Heavy metals

Abstract

The impact of heavy metals (Zn, Pb) on ciliates was demonstrated by standardized laboratory bioassays. On the basis of 3 ciliate species as test organisms it was shown that a unique biological organization of protozoans allowed to use a wide-spectrum of test criteria. Heterogeneity of toxic resistance of clones is considered as a mechanism of population adaptation to the toxicity.

The impact of lead and zinc on ciliate communities in isolated ecosystems in situ illustrates that natural populations of ciliates could be used for express control of marine environment. With the help of indices of similarity degree and dominance it was stated that changes in community structure occurred.

According to laboratory bioassays and ecotoxicological experiments in situ, protozoans are recommended as biomonitors to investigate the impact of heavy metals.

1. Introduction

For successful realization of the Baltic Monitoring Program (BMP) the state of the biological part of the ecosystem, the pollutant levels in the environment and possible effects of contaminants on the biota should be investigated simultaneously.

Incorporation of microheterotrophs into the classical paradigm of the planktonic food web (Williams 1981; Burkill 1982) shows that more attention should be paid to the effects of contaminants on protozoans.

For the second stage of the BMP protozooplankton is indicated as a tentative parameter of biota, however, the studies in the ecology of protozoans have recently been developed (Smetaček 1981; Boikova 1984a, b; Arndt 1986; Kivi 1986; Leppänen et al. 1986).

Information regarding pollution influences on the marine protozoans in comparison with other components of biota both of laboratory bioassays and ecotoxicological investigations in situ are still very scarse (Curds 1981; Stoecker et al. 1986).

Investigations of the concentrations of heavy metals in recent sediments of the accumulation areas of the Baltic Sea show that the levels of zinc and lead have increased threefold above the background levels (Melvassalo 1981) and as it was mentioned, without detailed studies of the effects of heavy metals on biota, it is impossible to evaluate their significance in the ecosystem.

The present study has the following aims: to develop methods of ciliate cultivation obtained from natural populations, to investigate the impact of heavy metals zinc and lead in laboratory bioassays by the applications of a spectrum of test criteria and to investigate the impact of metals on the structure of ciliate communities by an ecotoxicological experiment in situ.

2. Materials and Methods

2.1. Cultivation of ciliates

On the basis of ecological investigations of plankton and benthos ciliates in the Gulf of Riga the species composition and the dominant species were estimated. In this region salinity is low (3—6‰) and the formation of ciliate fauna is closely connected to this environmental factor. Analysis showed that the group of dominant species was formed by marine, brackish water and freshwater species. 3 species of mass cultures — *Paramecium putrinum*, *Stylonychia mytilus* and *Euplotes harpae* — were selected from natural populations. The first two species occur in fresh as well as in brackish waters, *Euplotes harpae* belongs to the marine species.

For the first step of cultivation the mass cultures of ciliates from natural populations were slowly adapted for growing in synthetic sea water (Sterba 1956) with a salinity 4‰, with the aim to obtain monocultures for each test species. After ciliate adaptation to standardized laboratory conditions the monocultures were free from flagellates. As a food source of organic matter sterilized hay infusion as well as microorganisms *Bacillus subtilis* and *Saccharomyces cereviseae* were added. From the stock cultures 10 to 30 clones were derived for every test species.

Mono- and clone cultures were subcultured in fresh media at the stationary phase once in 7—10 days. Log phase cultures were used for experiments. Under sterile conditions cultures were maintained for 3 years.

The heavy metals zinc and lead were introduced into the culture medium in the form of acetate. The levels of experimental concentrations were established in preliminary experiments, and they ranged within 0.05—10.0 mg/l.

2.2. Laboratory bioassays

The resistance of monocultures and clone cultures to the effect of lead and zinc was measured in short-term (48—120 h) and long-term experiments (to 5 months). In the short-term experiments it was done by counting the density of individuals in cultures treated with metals and in the control. In long-term experiments survival was observed.

The division rate of ciliates was measured in special counting chambers on clone cultures (the clone material provided genetic homogeneity and a relatively small variability) where the medium and daughter cells were renewed every day (Curds & Vandyke 1966).

The influence of heavy metals on phagocytosis was measured by the rate of food vacuole formation in *Euplotes harpae*. In these experiments the physiological state of protozoa is of great importance. Before the feeding experiments ciliates were starved for 8 hours. Suspension of coal was used as a model food for 5 minutes. After that ciliates were fixed in 4% formalin. The amount of food vacuoles in the treated ciliates was counted and compared with the corresponding controls.

The impact of heavy metals on laboratory populations of ciliates was also investigated with the application of conjugation phenomenon as a test criterion. In preliminary experiments 30 complementary types of mating of clone cultures were selected. The experiments were carried out at the temperature 18° ± 2 °C. Standard error and additional statistic calculations were made for each experiment series.

2.3. Ecotoxicological experiments in situ

Ecotoxicological experiments in controlled ecosystems in situ were carried out in June and July, 1981, in a shallow area in the Gulf of Riga (3 m deep).

In spherical 1.3 m^3 plastic (polyvinylchloride film) bags the natural plankton communities were completely isolated. Isolation of plankton in every bag takes about 10 minutes by the help of special devices. Once a day the bags were shaken.

To test the effect of lead and zinc by a single addition of metals, bacterioplankton, phytoplankton, mesozooplankton and protozooplankton, the pigment composition as well as the environmental parameters (temperature, salinity, pH, oxygen, biogenous elements, metal levels) were monitored for 3 weeks. There were also samples taken from control bags and the sea (Seisuma et al. 1986).

In the case of protozooplankton, species composition and density of organisms were used as criteria for estimating the changes in natural ciliate communities. Quantitative and qualitative observations of ciliates were carried out on nonpreserved material (Boikova 1984 a, b). For the analysis of the species composition of ciliates, the dominance index (Joganzen & Faizova 1978) and the index of similarity degree (Sörensen 1948) was calculated.

3. The impact of heavy metals on ciliates in laboratory bioassays

3.1. The resistance of mono- and clone cultures

In the first series of experiments the impact of heavy metals on the resistance of mono- and clone cultures of 3 species of ciliates was chosen as a test-criterion.

Results of short-term experiments (96 h) by using a sublethal dose of zinc and lead (Fig. 1) showed high susceptibility to the treatment. The relative intensity of growth of ciliates (except *Paramecium putrinum*) decreased after 24 h by 50% and more in comparison with the control. However, after a few days at the presence of 10 mg/l zinc the ciliates *Euplotes harpae* and *Stylonychia mytilus* reached the same level of growth as the control. For *Stylonychia mytilus* a significant stimulating effect in culture growth was obtained after 72 h. Appart from zinc, an inhibiting effect of lead on the growth intensity of 3 species was maintained.

Later on the monocultures of 3 ciliate species were cultivated under a chronic impact of 10.0, 1.0, and 0.1 mg/l zinc and lead. Results of the prolonged effect of toxicants showed that at the presence of 10.0 mg/l lead ciliates *Paramecium putrinum* and *Euplotes harpae* could not survive for more than 2 months. However, regular adding of heavy metals for 5 months to the subculturing of monocultures didn't show any impact on the survival of *Stylonychia mytilus* at the level of 10.0 mg/l Pb and of 10.0 mg/l Zn of 3 test species. With a regular introduction of toxicants it was easy to maintain the monocultures of ciliates under a prolonged impact of Zn and Pb at 1.0 and 0.1 mg/l levels for 5 months.

In order to explain the high susceptibility of monocultures of ciliates immediately after their treatment and the relatively high resistance of these organisms under a chronic impact, a structural analysis of population was done. As it is well known the ciliate populations in nature survive by agamic (mitotic) division and under laboratory conditions that allow to obtain clone cul-

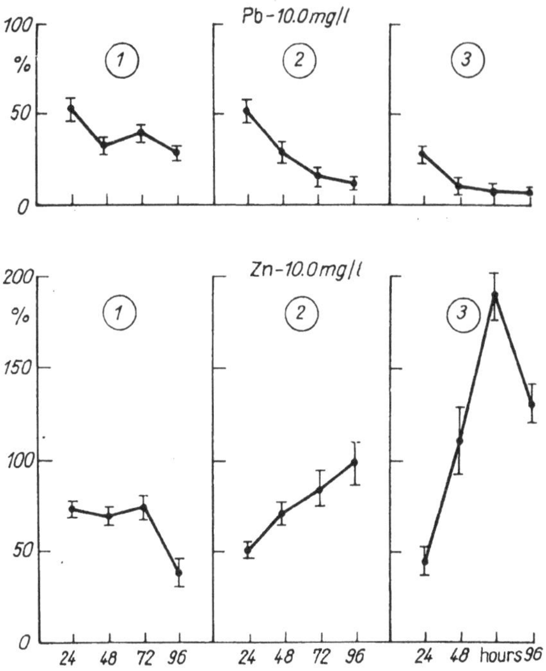

Fig. 1. The relative intensity of growth (% to control) of monocultures under toxic impact; 1 — *Paramecium putrinum*, 2 — *Euplotes harpae*, 3 — *Stylonychia mytilus*

tures. From a natural population 10—30 clone cultures for each test species were obtained. These clone cultures were used as discrete units of population in bioassays.

The resistance of clone cultures of ciliates *Paramecium putrinum* was investigated by the concentration of lead and zinc 10.0 mg/l at two different temperature levels. Fig. 2 shows that under an increased metal level the

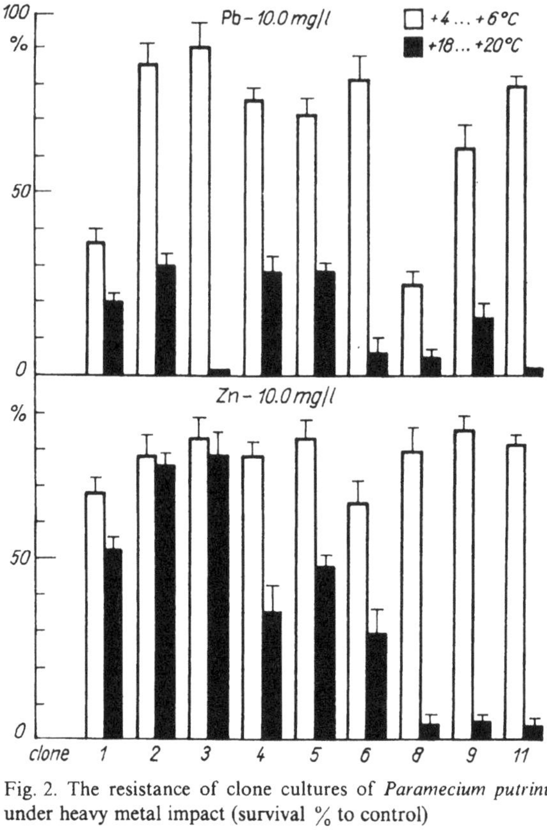

Fig. 2. The resistance of clone cultures of *Paramecium putrinum* under heavy metal impact (survival % to control)

resistance of separate clones after 120 h differed significantly. Three clones didn't survive, in others the relative growth rate was from 30 to 78% in comparison with the control. A similar effect was obtained also in the case of lead.

Heterogeneity of clone cultures of ciliates *Paramecium putrinum* was maintained also at the temperature +4 °C, although the toxic effect was lower. Therefore it can be supposed that the difference in the resistance of clone cultures to the toxic effect is inherited.

The response of clone cultures of the ciliates *Stylonychia mytilus* and *Euplotes harpae* was similar to that of *Paramecium putrinum*, although each species had also an individual response character (Boikova 1982). Obviously, the formation of clones of various degrees of toxic resistance provided the survival of the whole population under a toxic effect.

The results illustrated that 3 groups of clones were formed under a toxic effect: of high resistance, medium resistance and low resistance.

3.2. The division rate

The division rate of ciliates was used as a test-criterion for estimating the impact of lead on 2 clones characterized by high and low resistance.

The division rate of ciliates *Stylonychia mytilus* in the control was relatively constant for the whole period of experiment (24 days). The same clones treated with lead 10.0 mg/l showed a highly different character of cell division. Ciliates of low resistance clone culture stopped agamic reproduction on the 6th day, but ciliates of high

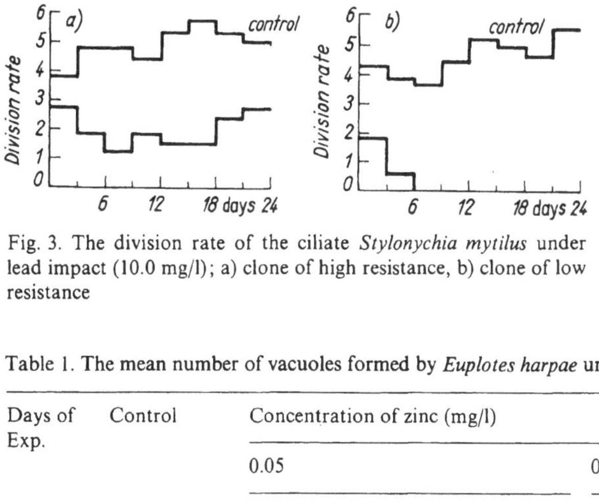

Fig. 3. The division rate of the ciliate *Stylonychia mytilus* under lead impact (10.0 mg/l); a) clone of high resistance, b) clone of low resistance

resistance clone maintained a relatively high division rate (Fig. 3).

Similar results were obtained by investigating the division rate of two other test species — *Paramecium putrinum* and *Euplotes harpae*. The results demonstrated that this test-criterion was more susceptible than the relative growth of cultures.

3.3. Phagocytosis

The intensity of phagocytosis depending on environmental factors also appears to be a useful criterion for estimating the impact of heavy metals. The formation intensity of food vacuoles was studied in ciliates of the clone *Euplotes harpae*, having a medium resistance for 15 days in experiments with added metals (Table 1).

According to the results a slow decrease of phagocytosis intensity was observed at the end of the experiment up to the level of 0.05 mg/l Zn and Pb. At 0.1 mg/l Zn and Pb the ciliates showed some stimulating effects in vacuole formation, especially in the case of lead, but on the 15th day the phagocytosis in control series and under the impact of metals was practically identical. With one order increase of the toxic level the formation of food vacuoles falls drastically or stops completely.

These results also showed that phagocytosis as a test criterion was characterized by fairly high susceptibility.

3.4. Conjugation phenomenon

The unique biological character of ciliate reproduction permit the use of the conjugation phenomenon as a test criterion to estimate the impact of metals. When ciliates of complementary clones were mixed in the control, at first the reaction of agglutination was observed and then the conjugation phenomenon which lasted for 6 to 24 hours. For every mating type 50—100 ciliates were mixed.

As it is shown in Table 2, after 5 days treatment with heavy metals at 0.1 mg/l the conjugation process was slightly disturbed and it was more prolonged in comparison with the control (48 h). More drastic influence was observed at the concentrations of 1.0 and 10.0 mg/l.

Table 1. The mean number of vacuoles formed by *Euplotes harpae* under heavy metal impact (df30 $t_{0.05}$ = 1.96)

Days of Exp.	Control	Concentration of zinc (mg/l)								
		0.05			0.10			10.0		
	X ± Sx	X ± Sx	$P_{0.05}$	%	X ± Sx	$P_{0.05}$	%	X ± Sx	$P_{0.05}$	%
3	10.3 ± 0.4	10.4 ± 0.5	—	102	11.0 ± 0.5	—	107	3.2 ± 0.1	+	33
5	10.4 ± 0.4	—	—	—	13.4 ± 0.7	+	129	2.1 ± 0.1	+	20
10	10.7 ± 0.5	12.0 ± 0.4	+	112	12.8 ± 0.5	+	119	3.8 ± 0.2	+	35
15	14.4 ± 0.5	12.2 ± 0.4	+	85	14.2 ± 0.5	—	99	4.5 ± 0.2	+	31

Days of Exp.	Control	Concentration of lead (mg/l)								
		0.05			0.10			10.0		
		X ± Sx	$P_{0.05}$	%	X ± Sx	$P_{0.05}$	%	X ± Sx	$P_{0.05}$	%
3	10.3 ± 0.4	8.9 ± 0.5	+	88	14.8 ± 0.5	+	144	5.3 ± 0.2	+	53
5	10.4 ± 0.4	—	—	—	12.6 ± 0.5	+	121	2.6 ± 0.1	+	25
10	10.7 ± 0.5	12.3 ± 0.4	+	114	15.0 ± 0.5	+	140	0.9 ± 0.0	+	8
15	14.4 ± 0.5	11.0 ± 0.3	+	77	13.4 ± 0.6	—	93	5.4 ± 0.3	+	37

Table 2. The impact of heavy metals on the conjugation phenomenon of the ciliate *Euplotes harpae*

Zinc mg/l	% to control		Lead mg/l	% to control	
	5 days	35 days		5 days	35 days
0.1	90	100	0.1	93	100
1.0	60	100	1.0	87	72
10.0	60′	40′	10.0	20′	0

60′, 40′, 20′ — conjugates appeared by an order lower in comparison to control.

At the presence of 10.0 mg/l zinc in 60% of complementary clones the phenomenon of conjugation was observed, but only 2—5 pairs of conjugates were formed (in control 50—100 pairs) for the corresponding mating types.

After a prolonged exposure (35 days) of ciliates of complementary clones to toxic effects and after repeated experiments, some adaptive reactions were noted, especially in the case of zinc.

Under a toxic impact of heavy metals the exchange of genetic informations among the complementary clones was disturbed: with the increase of metal level the amount of complementary ciliate pairs decreased, the phenomenon of conjugation became weaker and the conjugation period increased as compared with the control.

It should be also mentioned that the agamic reproduction was observed in spite of the interrupted conjugation by 10.0 mg/l of zinc and lead. It clearly demonstrates that the test criterion of conjugation is much more susceptible than the division rate of ciliates.

4. The effect of heavy metals on the ciliate community in ecotoxicological experiments in situ

4.1. Change in the structure of ciliate community according to the index of similarity degree

Altogether 30 ciliate species were identified during the experiment. The analysis was done on the basis of comparing the number of ciliate species in the sea, in the control bags and in bags with added metals. The index of similarity degree (S) can range from 1 (if species composition in 2 compared biocoenoses is identical) to 0 (no common species observed).

The results obtained at the comparison of sea samples and the samples of 8 control series bags illustrated that the similarity degree from the first to the fourth day was high ($S = 0.8—1.0$). In the following weeks a clear tendency was observed of decreasing the similarity between the control bags and the sea, and at the 19th day it ranged from 0.2—1.0. Thus by longer isolation of the marine environment the "bag effect" occurred.

In the experimental series of lead and zinc (Table 3) the control (S_0) was distinguished by high similarity with the sea from the first to the 14th day. Adding of metals at various concentrations and proportions considerably affected the species composition of ciliate communities at the 4th day. At the levels of Pb/Zn 0.1/1.0, 1.0/0.1 mg/l and zinc 1.0 mg/l ciliates in the samples were absent. On the 8th day, at some concentrations a partial recovery of ciliate species composition occurs in comparison with the control, however, further on with a decrease in the

Table 3. The index of similarity degree of species composition of ciliates by adding heavy metals in the enclosures

S	Pb/Zn	Day				
	(mg/l)	1	4	8	14	19
S_0		1.0	0.7	1.0	0.7	0.5
S	0.1/0	1.0	0.8	0.8	0.9	0.7
S	1.0/0	0.8	1.0	0.8	0.7	0.5
S	0.1/0.1	0.8	0.8	0.8	0.9	0.8
S	0.1/1.0	1.0	—	0.7	0.8	0.6
S	1.0/0.1	1.0	—	0.7	0.6	0.6
S	1.0/1.0	0.8	0.5	—	0.4	0.3
S	0/0.1	0.7	0.5	1.0	0.8	0.4
S	0/1.0	0.7	—	0.7	—	0.6

S_0: the similarity degree of species composition between the sea and control enclosures

similarity index both between the control bag and the sea, and the control and the enclosures with metals.

The toxic effect of heavy metals was illustrated only by the analysis of species composition.

4.2. Changes in the structure of the ciliate community according to the index of dominance

The dominant ciliate species were determined by the dominance index (I_d) that characterizes the role of species in an ecosystem according to their occurrence and density. If I_d ranges from 1 to 100 the species are dominating, below 1.0 — species are subdominating and below 0.1 — their role is insignificant.

According to biosamples of the sea, the dominant group consisted in fact of the ciliate *Strombidium* sp., and this picture is very typical for summer (Boikova 1984a, b). As it is shown in Fig. 4 the I_d ranges from 88—100 during the whole observation.

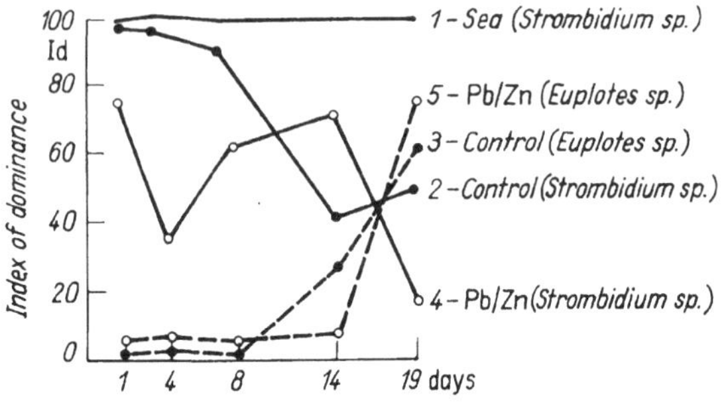

Fig. 4. Development of dominant species of ciliates in isolated ecosystems and in the sea

Another picture was observed in the control bag and the enclosures with added metals where 2 groups of dominant species were obtained: *Strombidium* sp. and *Euplotes* sp.

After adding heavy metals, the ciliate community reacted vigorously during the first 2—3 hours by lowering their density, and that was not observed with any other component of biota. The amount of the ciliate *Strombidium* sp. greatly decreases until the 4th day. Further on in all the enclosures with added metals an increase in density of the ciliates *Strombidium* sp. occurred by 1—2 orders up to the 14th day. At that time another group of ciliates — *Euplotes* sp. — appears to be dominating. A similar replacement of leading species occurred also in the control. Under a toxic effect the replacement of *Strombidium* sp. is slower.

Therefore, the reaction of the ciliate community from the 1st to the 14th day can be considered mainly as a response to toxicity, later it is covered by the "effect of the bag".

It is possible that the change of dominant species is connected with some changes in the plankton community metabolism after isolation.

4.3. Change in the relative density of the ciliate community

As it can be seen from the given data (Fig. 5), during the first day of experiment a sharp response of protozoan community was maintained to the added metals. Results of separate metal effects reveal that the toxicity of zinc is higher. Thus, e.g., at a zinc concentration of 1.0 mg/l no ciliates were observed in the samples within the 1—14 day, but at a concentration an order below, the toxic effect was less expressed.

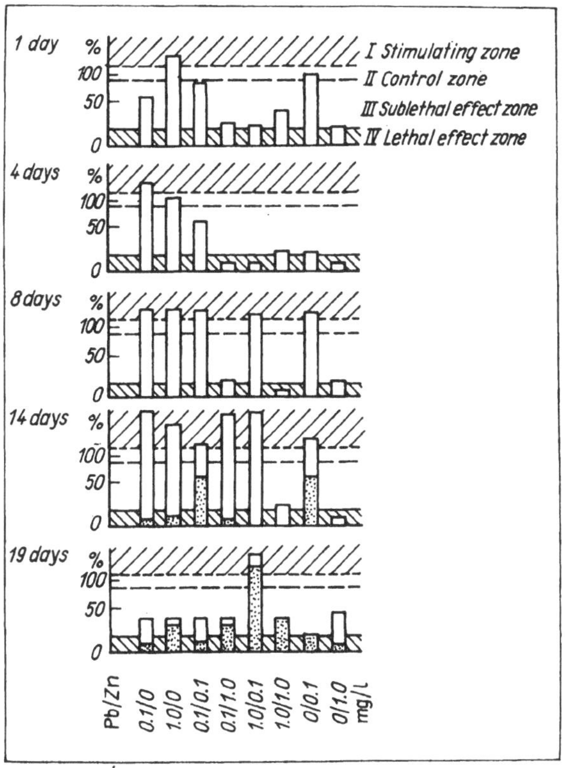

Fig. 5. Changes in the relative density (% to control) of ciliate communities by the adding heavy metals Pb and Zn
▭ *Strombidium* sp.; ▨ *Euplotes* sp.

The differences in metal toxicity may probably be related to the changes in toxicant concentrations in the isolated environments. Thus, the level of lead lowered drastically during the experiment in comparison with the initial concentration. For example, on the 14th day the level of lead 0.1 mg/l lowered by 50%, and at the level 1.0 mg/l its fall was only 12% of the initial one. An entirely different picture was observed in enclosures with zinc added to them. The levels of zinc remained relatively constant during the observation period.

The results of joint effect of lead and zinc on ciliates at the ratio 1:1, with the concentration of 0.1 mg/l, are similar to those of the separate effect of Zn at the concentration of 0.1 mg/l. Joint effect of metals on ciliates at the concentration of 1.0 mg/l is also similar to that of zinc separately. A changed ratio of metals causes essential changes in the character of response of the ciliate communities.

The response of the ciliate *Strombidium* sp. in ecotoxicological experiments in situ was similar to that obtained with monocultures, it can be explained by the physiological polymorphism of protozoans.

In the course of the experiment the levels of oxygen, ammonium nitrogen and pH changed due to the decay of dead plankton organisms.

Ciliates were the only plankton group the development of which was promoted by both concentrations of lead within the 4th and the 14th day as to the control. Obviously, protozoans were better adapted to lead impact since they could use the products of zoo-, phyto- and bacterioplankton decomposition for their development. In experimental bags containing zinc bacterioplankton was the most resistant group. The impact of combinations of heavy metals on the whole plankton varied from antagonistic with lowered toxicity to synergic that was growing in the course of the experiment (SEISUMA et al. 1986).

5. Discussion

The experiments on mono- and clone cultures show that formation of clones of various degrees of toxic resistance is of an importance in the species adaptation to the changing environmental conditions. Adaptive reactions, especially concerning zinc, have been stated under a prolonged influence of it (BOIKOVA 1982).

Heterogeneity of toxic resistance of clones underlines the physiological polymorphism of protozoan populations which is considered to be a mechanism of population adaptation to the toxicity. The inherited heterogeneity to toxic resistance of different clones should be considered when working with organisms of mitotic reproduction (ciliates, flagellates, unicellular algae), because the clone response to the stress can be extremely different, as it was demonstrated by the division rate.

Bioassays on phagocytosis clearly show that under a toxic stress the ability of ciliates to participate in self-purification processes in the ecosystem can be hindered.

The first results of application of the conjugation phenomenon as a test-criterion allowing to state the importance of toxicity for genetic renewal of population show that ciliates can be recommended as model organisms in studying the genetic consequences of contaminants.

Laboratory investigations of the impact of heavy metals show that the unique biological organization of protozoans allowing a wide spectrum of susceptible test-criteria, prove their advantage to stand above the other test organisms of marine environment. Most of these test criteria can be recommended also for in situ experiments.

Ecotoxicological experiments in controlled ecosystems in situ occupy a special place in the biomonitoring representing plankton community response to stress conditions in nature. It can be suggested that during the first hours of the reduction the population density is closely connected with survival, in the first days metals affected the reproduction of ciliates, but later a biological regulation of metal toxicity appears. By the opinion of

SIGEL (1986) in the metal-loaded container a higher level of amino acids and a shift in their composition was found. This suggests the possibility of regulation in metal uptake. Under the toxic impact the interactions between different trophic levels also seem to be changed.

The data of laboratory bioassays and in situ experiments with ciliates showed a difference in zinc and lead toxicity that can be explained by their different biological significance. As it is well known, zinc belongs to essential and lead to nonessential trace metals. The theory of BRANY (1984) suggests that essential metals (Zn, Cu, Mn . . .) can be regulated by living organisms, at least to some extent, but for non-essential metals (Cd, Pb . . .) the internal level of them depends on the external one. Experimental and in situ data of toxicity and regulation of some trace metals in various estuarine and coastal organisms (annelids, molluscs, crustaceans, fish) are in accordance with this theory (AMMIARD et al. 1985).

There is a large scale of environmental factors affecting the toxicity of heavy metals. The main physico-chemical factors are temperature, dissolved oxygen, pH, specific traits of metal, redox conditions, duration of exposition. Metal toxicity is strongly influenced also by a number of biotic factors, such as species-specific characteristics, accumulation, extracellular products, cellular activity, age and size of organisms, nutrient level, successive stage of community (SUGIMA et al. 1982; FOWLER 1985; LORCH & WEBER 1985; SIGEL 1986). This raises a number of consequences concerning metabolism, growth, reproduction, community structure. In this connection the standardized laboratory bioassays of ciliates are very useful tools for estimating the purely toxic effect of heavy metals based on material from natural populations. There are great feasibilities of field experiments to study ecological implications of pollution by heavy metals, but there are also some limits: these experiments, working with complex communities, do not allow to decide whether or not the organisms at higher trophic levels are affected directly or indirectly (via the organisms upon which they feed) (GÄCHTER 1983).

As it was demonstrated in a review paper by Prof. CAIRNS (1981), we are not using presently available practical biological monitoring methods on the scale that is needed for adequate environmental quality control. With the implementation of the existing methods new ones can be developed.

The results obtained from laboratory bioassays and ecotoxicological experiments in situ illustrate that protozoans can be recommended for future biomonitoring needs in the Baltic Sea.

6. Summary

On the basis of laboratory bioassays with lead and zinc on the resistance to mono- and clone cultures of 3 ciliate species, on the division rate, phagocytosis, conjugation phenomenon as well as the impact of ciliate communities in situ, a valuable information was obtained promoting the understanding of the effects of trace metal perturbation on food webs containing ciliates.

Laboratory bioassays reveal the purely toxic effect on cell-organism and population levels. Natural populations show an integrated community response under a toxic stress in situ where the impact of heavy metals with time can be covered by a large scale of biological interactions. However, ciliates demonstrated a rapid reaction to the impact of metals which allows them to be recommended for an express control of marine environment.

Investigations with protozoans are of a considerable practical importance for the assessment of water quality and represent a general interest in the effects of environmental conditions of ciliate communities. In this respect protozoans can be recommended as biomonitors and should be included in the programs of standard investigations of the Baltic Sea.

7. Acknowledgements

I am grateful to Prof. Dr. K. SUKHANOVA, Institute of Cytology, Leningrad, for her scientific advice concerning the cultivation and laboratory bioassays of ciliates. Kind thanks to Dr. Z. SEISUMA, Institute of Biology, for the possibility to take part in ecotoxicological investigations in situ and for providing data.

References

AMMIARD, J. C., AMMIARD-TRIQET, C., & METAYER, C. (1985): Experimental study of bioaccumulation, toxicity and regulation of some trace metals in various estuarine and coastal organisms. Symp. Biol. Hungarica **29**, 313—323.

ARNDT, H. (1986): The relative importance of protozoans in a Baltic estuary zooplankton community. Symp. Biol. Hungarica **33**, 257—261.

BOIKOVA, E. (1982): On adaptive reactions of unicellulars to the effect of toxicity. In: Experimental'naja vodnaja toksikologia, p. 19—32. Riga.

— (1984a): Benthos ciliates in the coastal area of the Gulf of Riga, p. 109—119. Riga.

— (1984b): Ecological character of protozoans (Ciliata, Flagellata) in the Baltic Sea. Ophelia, Suppl. **3**, 23—32.

BRYAN, G. W. (1984): Pollution due to heavy metals and their compounds. In: Marine ecology, Vol. 5, Part 3, p. 1289—1431.

BURKILL, P. H. (1982): Ciliates and other microplankton components of a nearshore food-web: standing stocks and production processes. Ann. Inst. Oceanogr. Paris **58**, 335—350.

CAIRNS, J. JR. (1981): Biological monitoring. Part VI-Future needs. Water Research **15**, Great Britain, 941—952.

CURDS, C. R. (1982): Pelagic protists and pollution — review of the past decade. Ann. Inst. Oceanogr. Paris **58**, 117—136.

— & VANDYKE, J. M. (1966): The feeding habits and growth rates of some fresh-water ciliates found in activated-sludge plants. J. Appl. Ecol. **3**, 127—137.

FOWLER, S. W. (1985): Heavy metal and radionuclide transfer and transport by marine organisms. Symp. Biol. Hungarica **29**, 191—206.

GÄCHTER, R., & URECH, J. (1983): Feasybilities and limits of field experiments to study ecological implications of heavy metal pollution. In: Trace element speciation in surface waters and its ecological implications, p. 137—158.

JOGANZEN, B., & FAIZOVA, L. (1978): Ob opredelenii pokazatelej vstrechaemosti, obiliya, biomassy i ikh sootnoshenie u nekotorykh gidrobiontov. In: Elementy vodnykh ekosistem, p. 215—225. Moskva.

KIVI, K. (1986): Annual succession of pelagic protozoans and rotifers in the Tvärminne Storfjärden, SW coast of Finland. Ophelia, Suppl. **4**, 101—110.

LEPPÄNEN, J. M., & BRUNN, J. E. (1986): The role of pelagic ciliates including the autotrophic *Mesodinium rubrum* during the spring bloom of 1982 in the open northern Baltic proper. Ophelia, Suppl. **4**, 147—158.

LORCH, D., & WEBER, A. (1985): Accumulation, toxicity and localization of lead in Cryptogams: Experimental results. Symp. Biol. Hungarica **29**, 51—60.

MELVASSALO, T. (1981): Baltic Sea environment proceedings No. 5B. Assessment of the effects of pollution on the natural resources of the Baltic Sea. Helsinki Commission, p. 425.

SEISUMA, Z., KULIKOVA, I., MARCINKEVICHA, S., RUDZROGA, A., BOIKOVA, E., & LEGZDINA, M. (1986): Impact of heavy metals on plankton in experimental ecosystems in situ. Zinatne (Riga), p. 252.

SIGEL, H. (1986): Metal ions in biological systems. In: Concepts on metal ion toxicity, Vol. 20, p. 66—100. New York, Basel.

SMETAČEK, V. (1981): The annual cycle of protozooplankton in the Kiel Bight. Mar. Biol. **63** (1), 1—11.

SORENSEN, T. (1948): A method of establishing groups of equal amplitude in plant society based on similarity of species content. K. Danske Vidensk. Selsk. **5**, 1—34.

STERBA, G. (1956): Aquarienkunde, Bd. II, S. 375. Leipzig, Jena.

STOECKER, D. K., SUNDA, W. G., & DAVIS, L. H. (1986): Effects of copper and zinc on two planktonic ciliates. Mar. Biol. **92**, 21—29.

SUGIMA, K., GOTO, M., & KURIHARA, Y. (1982): Effect of Cu^+ stress on an aquatic microcosm: A holistic study. Environm. Res. (Tokyo) **27**, 307—315.

WILLIAMS, P. J. (1981): Incorporation of microheterotrophic processes into the classical paradigm of the plankton food web. Kieler Meeresforsch., Sonderh. **5**, 1—28.

Address of the author:

ELMIRA BOIKOVA
Institute of Biology
of the Latvian SSR Academy of Sciences
Salaspils-229021
3 Miera str.
USSR

Limnologica (Berlin) **20** (1), 127—129, April 1990

Tartu State University, Estonian SSR, USSR; Institute of Applied Geophysics, Baltic Branch, Tallinn, USSR

Some Aspects of Microbiology of the Gulf of Finland

By KAI KÜNNIS & ASTRID SAAVA

With 2 Figures and one Table

Key words: Total bacterial number; Heterotrophic bacteria; Vertical distribution; Short-term fluctuation; Gulf of Finland.

Abstract

The vertical distribution and short-term fluctuations of both the total bacterial number and heterotrophic bacteria were studied at the entrance to the Gulf of Finland over two periods in summer 1986. The highest values for the studied parameters were recorded in the upper layer, with maximum at the depths of 5—10 m. In the deeper layers the bacterial numbers decreased significantly. The degree of short-term fluctuations in total bacterial numbers was roughly twice smaller compared to the fluctuations in the numbers of heterotrophic bacteria.

Introduction

Micro-organisms, especially bacteria, play an important role in the accumulation, transformation and decomposition of organic matter and, consequently, in water purification. They also form a major link in the production processes.

Most of the microbiological data obtained in the Baltic Sea area are based on point measurements. Information on the possible ranges of variation during short periods is available in only a few cases (VÄÄTÄNEN 1976; TSYBAN et al. 1981; RHEINHEIMER 1984).

The aim of the present investigation is to characterize both the vertical distribution and short term fluctuations in bacterial number at the entrance to the Gulf of Finland (59°41,6'N, 23°27,5'E, depth — 85 m).

Material and Methods

The water samples for the total bacterial number and for the plate counts of heterotrophic bacteria were collected over the periods 11.—18. 06. 1986 and 22.—28. 08. 1986 twice a day (at mid day and at night). Sampling depths were 0.5, 5, 10, 20, 30, 60 and 83 m. In August additional samples were collected in the thermocline layer at 15 m depth.

The total number of bacteria was estimated by direct brightfield microscopy. To concentrate the bacteria, membrane filters of 0.23 µm pore size (Synpor filters No. 8) were used. The cells retained on the filters were stained with a mixture of erytrocine (5%) and phenol (5%) in distilled water (RODINA 1965). Altogether 300 to 1200 bacteria were counted per filter.

The spread-plate technique and two different plating media, the peptone-yeast extract ZOBELL medium 2216 E (OPPENHEIMER & ZOBELL 1952) and the Soviet standard nutrient agar, were used for the enumeration of colony-generating units of heterotrophic bacteria. The plates were inoculated immediately after sampling and incubated first at room temperature, then in a thermostat at 20 °C. The first counting of colonies was performed after 3—4 days and repeated after 2 weeks (3 days are the standard incubation period for the heterotrophic bacteria on Soviet standard agar medium).

There were no significant changes in the hydrometeorological conditions during the study periods. Especially during the first study period in June the weathers were stable. However, some differences existed between the two investigation periods in the vertical distribution of temperature. In June the temperature ranged from 1.13 to 13.81 °C and in August from 1.76 to 16.15 °C.

In August the hydrographic conditions could be characterized by a pronounced thermocline at about 15 to 25 m depth. The salinity difference between the surface and the bottom layer was approximately 3‰ during the periods under consideration (6.4—9.9‰ in June and 6.7—10.0‰ in August).

Results and Discussion

The total bacterial counts and the counts of heterotrophic bacteria serve as trophic level indicators.

The total number of bacteria fluctuated within 0.1 to $1.83 \times 10^6 \cdot ml^{-1}$ range during the investigation period in June and within 0.06—2.07 $\times 10^6 \cdot ml^{-1}$ in August. The direct bacteria counts were the highest in the surface water from 0 to 14 m. In the cold water below the thermocline, bacterial numbers decreased 4 to 6 times.

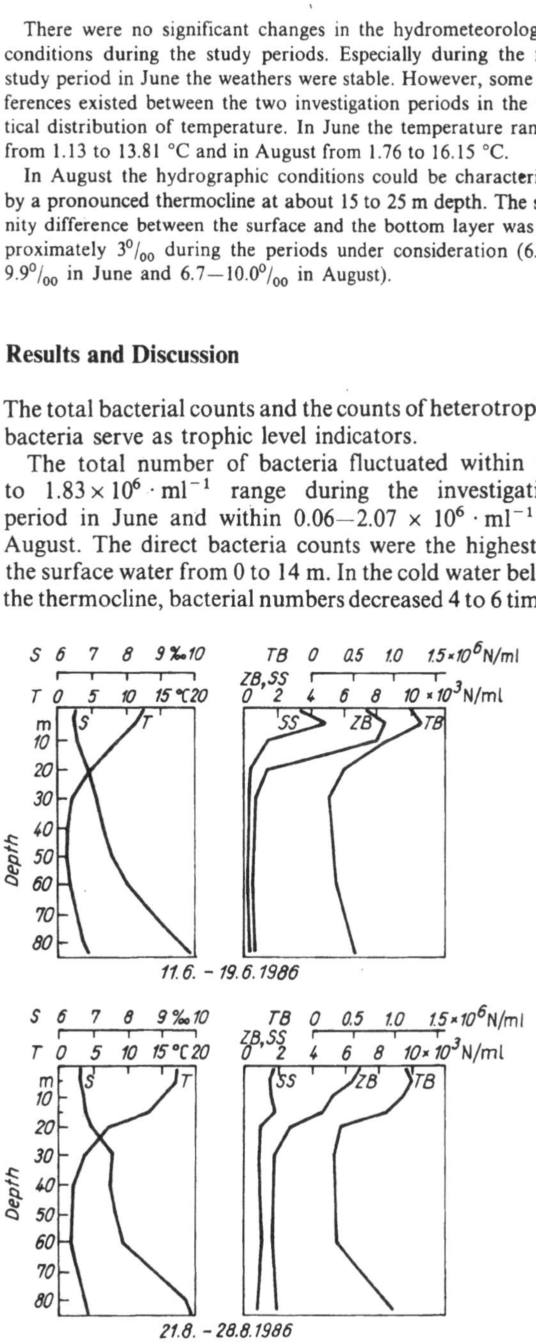

Fig. 1. Vertical distribution of total bacterial numbers (TB); ZOBELL 2216E-medium (ZB) and Soviet standard medium (SS) plate counts, temperature (T) and salinity (S) for two investigation periods

In the bottom water with a slight increase in temperature, bacterial numbers increased somewhat, and a change in the composition of the population was observed (Fig. 1).

The relatively higher total bacterial counts in the deeper water in August might have been due to the increase in the amounts of dead organic material during the vegetation period. The mean values ($\bar{x}$) of the total bacterial counts and the standard deviations (SD) for three water masses were found to be:

surface water	June	$\bar{x} = 1.14 \times 10^6 \cdot ml^{-1}$; SD = 0.31
	August	$\bar{x} = 1.09 \times 10^6 \cdot ml^{-1}$; SD = 0.37
intermediate water	June	$\bar{x} = 0.29 \times 10^6 \cdot ml^{-1}$; SD = 0.15
	August	$\bar{x} = 0.30 \times 10^6 \cdot ml^{-1}$; SD = 0.18
bottom water	June	$\bar{x} = 0.51 \times 10^6 \cdot ml^{-1}$; SD = 0.30
	August	$\bar{x} = 0.97 \times 10^6 \cdot ml^{-1}$; SD = 0.38

From the abundance of heterotrophic bacteria one can judge the presence and even quantity of easily acceptable organic matter.

Heterogeneity of vertical and temporal distribution of heterotrophic bacteria could also be clearly seen. The numbers of heterotrophic bacteria (ZoBELL-medium plate counts) varied between 10—2400 · ml^{-1} during the investigation period in June and 75—1200 · ml^{-1} in August. The highest concentrations and fluctuations in heterotrophic bacteria numbers were noted at 0.5—20 m depth (in June: $\bar{x} = 796 \cdot ml^{-1}$, SD = 675 and in August: $\bar{x} = 565 \cdot ml^{-1}$, SD = 226). The maximum plate counts were often obtained at 0.5 m and 5 m depths. In August the second increase was sometimes observed in the thermocline layer. In the layers below the thermocline their amounts decreased significantly (in June: $\bar{x} = 43 \cdot ml^{-1}$, SD = 28 and in August: $\bar{x} = 179 \cdot ml^{-1}$, SD = 42). In June the decrease was 16—28 times and in August 1—3 times. No increase in the quantity of heterotrophic bacteria was noted in the bottom layer (see Fig. 1). The increase in total bacterial number in deep water is assumed to be due to chemoautotrophic and/or anaerobic bacteria. TSYBAN et al. (1981, 1987) found an increased concentration of bacterioplankton near the bottom, where the oxygen content was minimum and hydrogen sulfide was present.

A comparison with results obtained by other scientists shows the amount of bacteria at the entrance to the Gulf of Finland to be roughly at the same level as the values for the Central and Northern Baltic and the open areas of the Gulf of Finland (VÄÄTÄNEN 1981; PANOV 1983; TSYBAN et al. 1987).

The temporal variability in the heterotrophic bacterial number was larger in June (49—116%) than in August (18—40%). The fluctuation was greatest in the surface water layers (Fig. 2). During both investigation periods

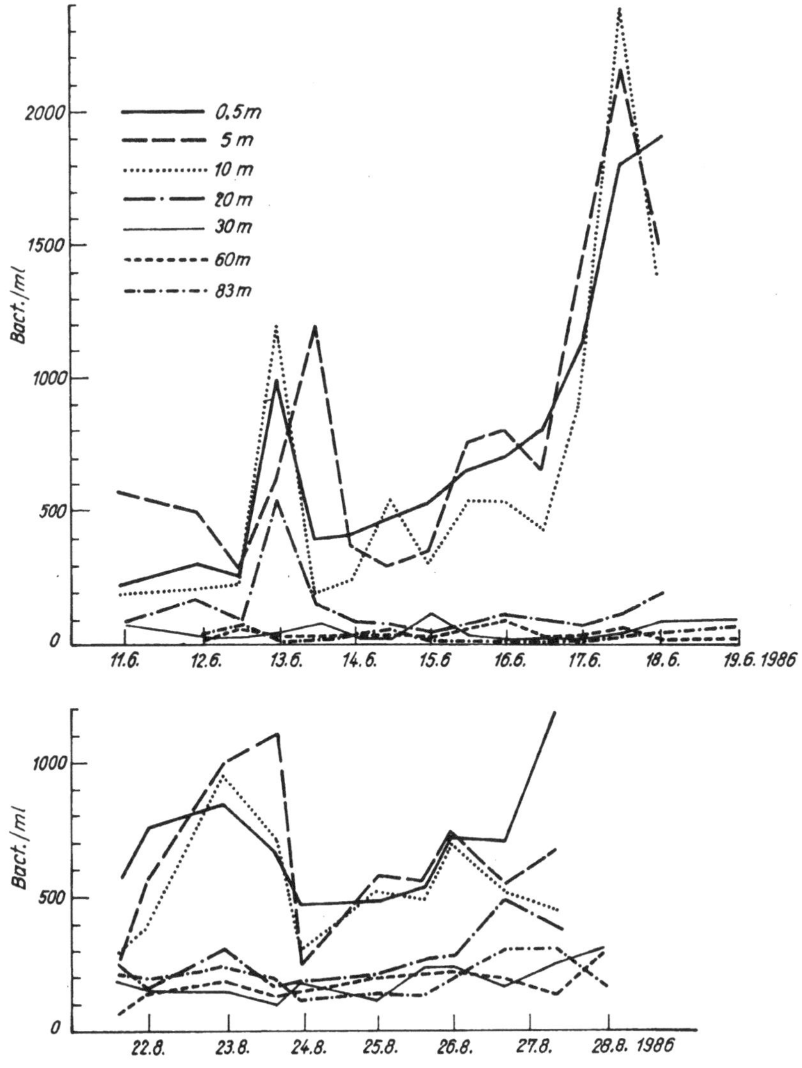

Fig. 2. Short-term fluctuations in heterotrophic bacterial numbers at different depths at the entrance to the Gulf of Finland

the increase in the numbers of heterotrophic bacteria in the surface layers occurred with the interval of 3—4 days. It's interesting to note, that the same interval has been obtained by GOCKE (1975) for saprophytic bacteria and ZIMMERMANN (1977) for the number of a certain bacterial population, which could be identified under the microscope by its characteristic fluorescence and morphology in Kiel Bight.

The temporal distribution of microbiological quantities is doubtless a very complex matter, since the biological quantities are influenced not only by the physical parameters, but they are also governed by their own lows.

The heterotrophic bacteria density constituted only 0.01—0.07 per cent of the total bacterial number.

When comparing the heterotrophic bacterial plate counts for ZOBELL medium 2216 E to Soviet standard nutrient agar medium, the first was noted to yield 1.1 to 1.8 (mean 1.4) times higher counts after 3-day incubation period and 1.2—4.2 (mean 2.3) times higher counts after 14-day incubation. Table 1 presents the mean plate counts of heterotrophic bacteria and ratios between ZOBELL medium 2216 E and Soviet standard agar medium plate counts.

Table 1. The comparison of the ZOBELL medium 2216 E to the Soviet standard nutrient agar medium plate counts

	11.—18. 06. 1986			22.—28. 08. 1986		
Depth (m)	Plate counts (bact./ml)		A/B ratio	Plate counts (bact./ml)		AB ratio
	ZOBELL medium 2216 E (A)	Soviet standard medium (B)		ZOBELL medium 2216 E (A)	Soviet standard medium (B)	
0.5	510	430	1.2	720	170	4.2
5	700	470	1.5	640	160	4.0
10	210	140	1.5	560	170	3.3
15	—	—	—	420	180	2.3
20	100	40	2.5	260	100	2.6
30	60	20	3.0	190	90	2.1
60	35	15	2.3	180	110	1.6
83	30	25	1.2	180	105	1.7

The higher organic matter concentration in Soviet standard agar medium was found to be inhibitory for autochthonous marine bacteria.

Conclusions

During the investigation periods the distinct changes in the bacterial numbers were observed, especially in the surface water layers; at the same time the fluctuations in water temperature and salinity were slight.

The highest bacterial counts were found in the surface water, with maximum values above the thermocline. The bacterial number strongly decreased in the cold water below the thermocline.

The organic matter content, the content of phytoplankton and the water temperature may be the main factors controlling the abundance and distribution of bacteria in the sea water.

The comparison of ZOBELL medium 2216 E with Soviet standard nutrient agar medium plate counts revealed some significant differences, which may obviously be caused by the fact that different types of heterotrophic bacteria have different nutrient requirements.

References

GOCKE, K. (1975): Studies on short-term variations of heterotrophic activity in the Kiel Fjord. Mar. Biol. **33**, 49—55.

OPPENHEIMER, C. H., & ZOBELL, C. E. (1952): The growth and viability of sixty-three species of marine bacteria as influenced by hydrostatic pressure. J. Mar. Res. **11**, 10—18.

PANOV, G. V. (1983): Dynåmics of microbiological processes in the open waters of the Baltic Sea in 1976—1982. In: Symposium on Ecological Investigations of the Baltic Sea Environment. Riga, USSR, 16—19 March 1983, pp. 350—370.

RHEINHEIMER, G. (1984): Bacterial ecology of the North and Baltic Sea. Bot. Mar. **27**, 277—299.

RODINA, A. G. (1965): Metody vodnoi mikrobiologii: Prakticheskoje rukovodstvo. Moskva-Leningrad.

TSYBAN, A. V., PANOV, G. V., DAKSH, L. V., & YURKOVSKAIJA, V. A. (1981): Bakterialnoje naselenije otkrytyh vod Baltiiskogo morja. V: Issledovanije ekosistemy Baltiiskogo morja. Vypusk 1, pp. 41—60. Leningrad.

— SAAVA, A., KUDRYAVTSEV, V., PANOV, G., RHEINHEIMER, G., & GOCKE, K. (1987): Microbiology. In: First Periodic Assessment of the State of the Marine Environment of the Baltic Sea Area, 1980—1985. Background Document. Baltic Sea Environment Proceedings No **17B**, 322—351.

VÄÄTÄNEN, P. (1976): Microbial studies in coastal waters of the Northern Baltic Sea. I: Distribution and abundance of bacteria and yeasts in the Tvärminne area. W. and A. de Nottbeck Found. Sci. Rep. **1**, 1—58.

— (1981): Microbial ecology of brackish waters off the Southern coast of Finland. W. and A. de Nottbeck Found. Sci. Rep. **3**, 1—22.

ZIMMERMANN, R. (1977): Estimation of bacterial number and biomass by epiflourescence microscopy and scanning electron microscopy. In: RHEINHEIMER, G. (ed.), Microbial Ecology of a Brackish Water Environment (Ecol. Stud. **25**), pp. 103—120. Berlin.

Addresses of the authors:

Prof. D. M. ASTRID SAAVA
Tartu State University
Ulikooli Str. 18
SU — 202400 Tartu
Estonian SSR
USSR

KAI KUNNIS
Institute of Applied Geophysics
Baltic Branch
Järvevana tee 5
SU-200001 Tallinn
Estonian SSR
USSR

Limnologica (Berlin) **20** (1), 131—134, April 1990

Department of Ecology, University of Education, Słupsk, Poland

Bacteriological Studies of Lake Łebsko

By ZBIGNIEW MUDRYK

With 4 Figures

Key words: Lake Łebsko; Bacteria; Physiological properties; Nutritional requirements.

Abstract

Microbiological research was carried out in the Lake Łebsko which is situated in Slowinski National Park. In bacteriological studies the following capabilities of bacteria were considered: physiological properties and nutritional requirements. Most numerous among the physiological groups of bacteria were proteolytic, lipolytic and amylolytic organisms. The smallest groups of them consist of bacteria reducing sulphates and producing H_2S from organic compounds. In Lake Łebsko predominated bacteria which required for the optimal growth a mixture of amino acids. Bacteria requiring simple nutritional compounds, vitamin and soil extract were not numerous among isolated strains.

Introduction

Heterotrophic bacteria are the most numerous group of microorganisms inhabiting aquatic ecosystems. Their main function in natural environments is the decomposition of organic matter (INNIS & YOUNG 1977; RIEMANN 1980). The development of heterotrophic bacteria depends primarily upon available food. The main source of food and energy for these bacteria in water basins are likely breakdown products of dead plant residues, animal bodies and organic substances excreted by living organisms (CHAN & MC MANUS 1969; COLER & GUNNER 1969).

Material and Methods

Microbiological studies were carried out in the Lake Łebsko (Fig. 1). This lake is situated in Slowinski National Park, it is the greatest Polish lake-estuarium. The surface of the lake is above 7000 ha. On the other hand Lake Łebsko is very shallow because its average depth is 1.6 m. Large River Łeba flows into the Lake Łebsko. At the same time Lake Łebsko is connected with Baltic Sea through channel 2 km long. Near this channel there is the small town Łeba, with lively tourist traffic during the summer time and also some fishing for recreation. During storms and northely winds from the open Baltic Sea water flows into the lake. The flows are usually largest in summer months with maximum in August.

Water samples for bacteriological studies were taken from six stations (Fig. 1) which represent different environmental conditions in Lake Łebsko. Stations 0, 1, 2, 3 were situated in the freshwater zone, stations 4 and 5 were located in the zone of mixing freshwater with sea water and station 6 was located in the brackish water zone. Samples were taken during the time from April to October 1986 every month. Water samples were taken in sterile glas bottles and placed in an ice-container and transported as soon as possible to the laboratory. Time between sampling and plating usually did not exceed 6—8 hours. In bacteriological studies the following characteristics of bacteria were considered: number of heterotrophic bacteria, physiological properties and nutritional requirements.

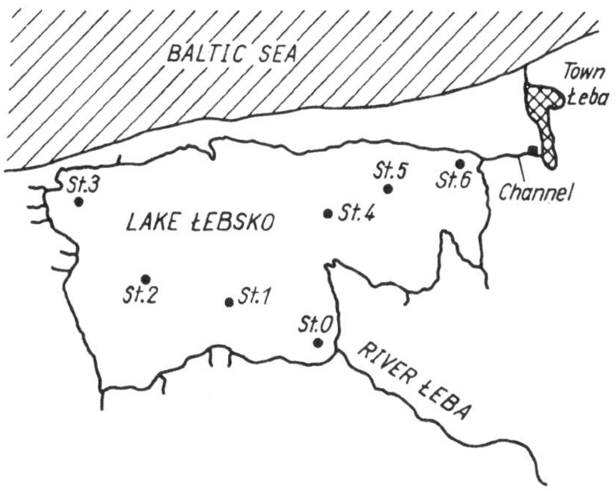

Fig. 1. Map of the Lake Łebsko with location of sampling sites

Number of heterotrophic bacteria in surface water in Lake Łebsko was determined by the spread plate method using the iron-pepton agar according to FERRER et al. (1963). Samples of water were diluted with sterile water from the Lake Łebsko. After an incubation period of 10 days at 26 °C the bacteria were counted. All tests were made in 5 replications.

Physiological properties of bacteria isolated from water were also investigated. The following physiological properties of the isolated bacteria were studied: ammonification, production of H_2S with organic compounds, starch hydrolysis, gelatine and lipid hydrolysis, formation of acid from glucose and reduction of sulphates. For the physiological tests different media were employed (DONDERSKI 1983). Inoculated media were incubated for 6 days at 26 °C. Only the media for indicating the sulphate reduction were incubated for 10 days.

Nutritional requirements of the bacteria were studied according to the technique of LOCHHEAD & CHASE (1943). Microorganisms were divided in 5 groups according to the intensity of growth in the following media:

medium B — basal medium containing mineral salts + glucose;
medium C — basal medium + alanine;
medium A — basal medium + amino acids;
medium G — basal medium + B group vitamin;
medium S — basal medium + soil extract.

All media were inoculated with one loop of bacterial suspension obtained from a 5 days old culture in liquid iron-pepton medium. The cultures were grown at 20 °C during 6 days. The growth response of each isolate was determined by assigning a 4 point turbidity value. A difference of not less than 2 points was considered significant.

Results

The results of the examination of the amount of heterotrophic bacteria in water are presented in Fig. 2. This figure shows the occurrence of bacteria in water at all sampling points.

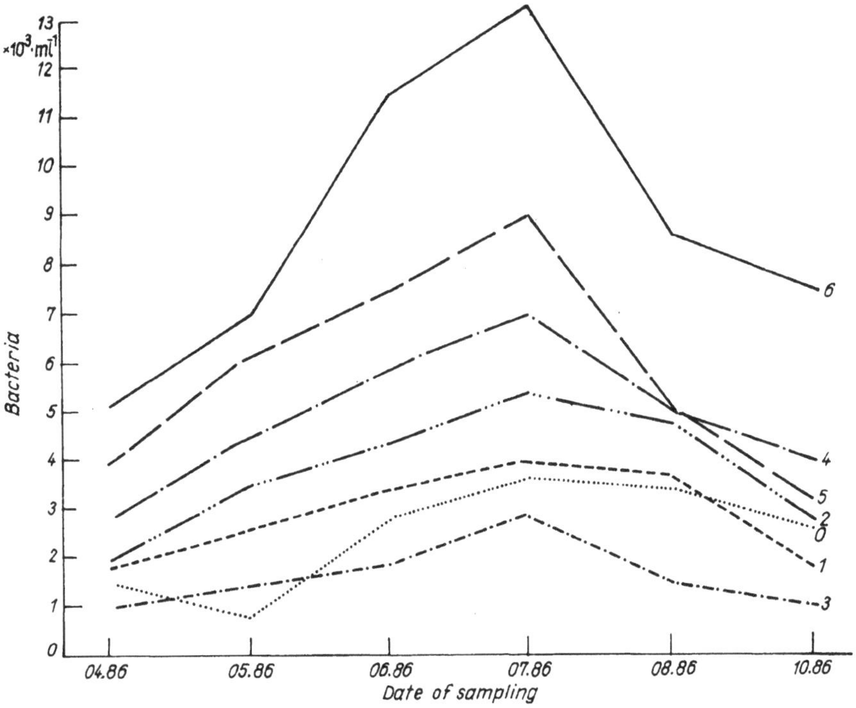

Fig. 2. Number of heterotrophic bacteria in surface water of Lake Łebsko (1—6: sampling sites)

The number of bacteria varies from one thousand to thirteen thousand. In the Lake Łebsko the maximum numbers of planktonic bacteria were noted in the summer time (June and July). In the spring and autumn months the number of bacteria decreased. As seen from this figure the highest number of bacteria in the water appeared in the zone of the channel connecting the lake with the Baltic Sea (stations 5 and 6). Lowest number of bacteria was noted on station 3 situated in the northern part of the lake.

The physiological groups of bacteria isolated from water are presented in Fig. 3. In general most numerous among the physiological groups of bacteria isolated in spring, summer and autumn at all stations were lipolytic organisms. The ability of lypolysis was found in above 90% of the total number of bacteria. Also the gelatine hydrolysis bacteria were more numerous than other physiological groups in Lake Łebsko. The most intensive development of amylolytic bacteria was observed in spring. On the other hand, proteolytic bacteria were more numerous in summer and autumn. Glucose acidifying and ammonifying organisms showed also high values among the studied planktonic bacteria. The smallest physiological groups were producing H_2S from organic compounds and sulphate reducing bacteria.

Fig. 4 shows the results of investigation of the nutritional requirements of bacteria isolated from water of the Lake Łebsko. It appears from this figure that in spring the largest amounts of bacteria required alanine for their best growth. The second largest group were

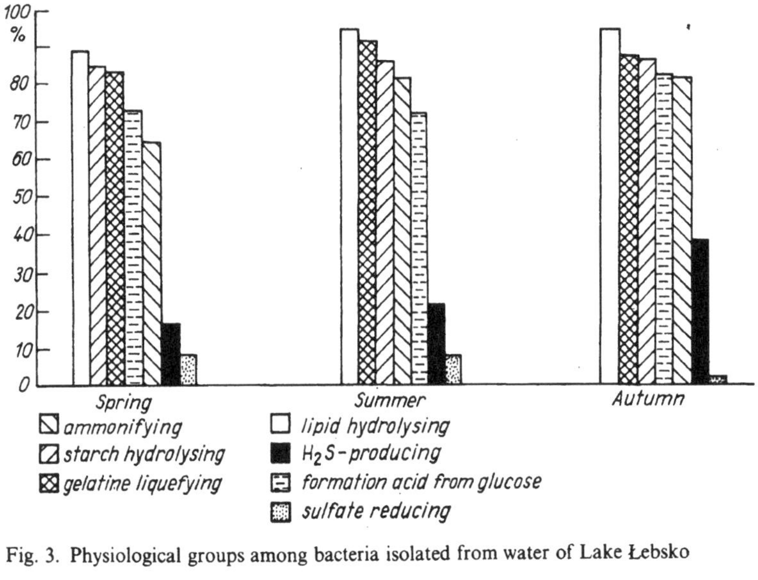

Fig. 3. Physiological groups among bacteria isolated from water of Lake Łebsko

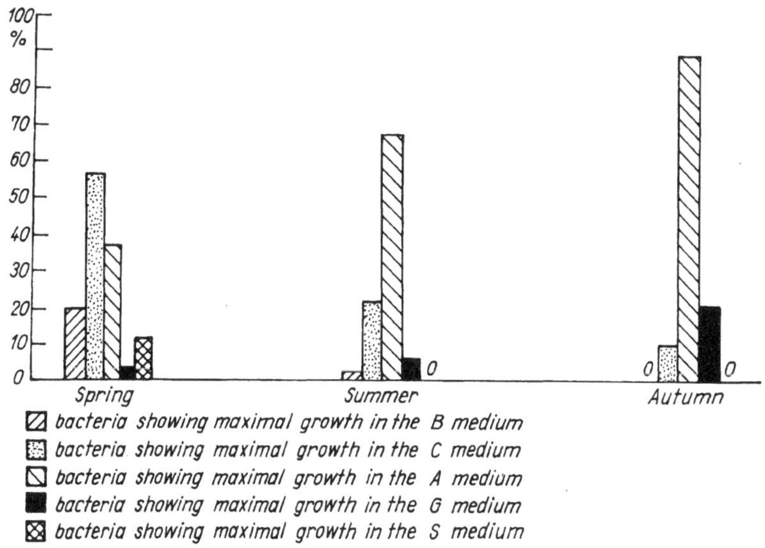

Fig. 4. Nutritional groups among bacteria isolated from water of Lake Łebsko

organisms requiring amino acids. Bacteria having most simple nutritional requiring were relatively numerous among the isolated strains.

In the summer time the number of bacteria requiring alanine decreased to 20% whereas bacteria requiring amino acids increased to 68%. The number of B and G groups of bacteria were low. Organisms requiring soil extract were entirely absent.

In autumn the predominating group of bacteria in Lake Łebsko were organisms requiring the presence of amino acids for their optimal development. Alanine was necessary only for 10% of bacteria. Relatively numerous were bacteria requiring vitamins for their optimal growth. In autumn the prototrophic and soil extract requiring bacteria were not found.

Discussion

The results obtained in our work show that maximum numbers of planktonic bacteria in the Lake Łebsko were noted in the summer time. The occurrence of large numbers of bacteria in the surface water during summer might have been caused by development of the phytoplankton and coastal plant excretions which may be an additional source of food for bacteria (OLAH 1969; RAI & HILL 1981). The larger numbers of bacteria at the sampling points 5 and 6 might have been resulted by pollution coming from sewages of the nearby town Łeba.

In general lipolytic, proteolytic and amylolytic bacteria were most numerous among the physiological groups. The occurrence of large quantities of those bacteria in the other lakes was also observed earlier by DONDERSKI (1971), JONES (1971), and MUDRYK & DONDERSKI (1986). MURCHALENO & BROWN (1970) found that the physiological types of heterotrophic bacteria in the Long Island Sound varied seasonally, but proteolytic, lipolytic and amylolytic bacteria appeared to be independent of the season. Numerous bacteria which are capable of decomposing proteins, lipids and starch are the evidence that complex organic substances are being attacked by microoorganisms in the Lake Łebsko. The number of bacteria glucose acidifying, ammonifing and producing H_2S from organic compounds was much lower. Similar results were obtained by SIEBURTH (1967), STRZELCZYK et al. (1976) and DONDERSKI (1983). The smallest group among them consists of bacteria reducing sulphates. These results are in agreement with those published by other authors (NIEWOLAK 1968; GELDERICH et al. 1980).

Nutritional requirements of water bacteria have been studied seldom. Most bacteria isolated from water of the Lake Łebsko in summer and autumn required amino acids for their best growth. Similarly FONDEN (1969) found that most bacteria isolated from water of the Lake Hjämaren required amino acids for their growth. Also STRZELCZYK et al. (1967) and DONDERSKI (1972) found in water of the Lake Jeziorak in summer and autumn that the highest number of bacteria required amino acids for their growth. The occurrence of these bacteria in summer and autum is undoubtedly connected with the development of phyto- and zooplankton and higher plants in water basin. These organisms have the ability to exude such organic compounds as amino acids, sugars and vitamins (TAHA & REFAI 1962) which in effect stimulate the development of bacteria.

References

CHAN, E. C., & MCMANUS, E. A. (1969): Distribution, characterization and nutrition of marine microorganisms from algae *Polysyphonia lanosa* and *Ascophyllum nodosum*. Can. J. Microbiol. **5**, 409—419.

COLER, E. A., & GUNNER, H. B. (1969): The rhizosphere of an aquatic *Lemna minor*. Can. J. Microbiol. **15**, 964—966.

DONDERSKI, W. (1971): Incidence of physiological types among bacteria isolated from water and mud of the lake Jeziorak. AUNC Toruń. Limnol. Papers **6**, 15—37.

— (1972): Nutritional requirements of bacterial isolates from water and mud of the lake Jeziorak. AUNC Toruń. Limnol. Papers **7**, 13—37.

— (1983): Tlenowe bakterie heterotroficzne jezior o różnej trofii (Heterotrophic aerobic bacteria in lakes of different trophy). UMK Toruń.

FERRER, E. B., STAPERT, F. M., & SOKOLSKI, W. T. (1963): A medium for improved recovery of bacteria from water. Can. J. Microbiol. **9**, 420—422.

FONDEN, R. (1965): Nutritional classification of some limnic bacterial population. Oikos **20**, 373—383.

GELDERICH, E. E., NASH, H. D., SPANO, D. F., & REASMER, D. J. (1980): Bacterial dynamics in a water supply reservoir: A case study. Rès. and Technol., 31—40.

INNIS, W. E., & YOUNG, M. (1977): Metabolic activity in cores of lake bottom sediments. Water Res. **11**, 75—77.

JONES, J. G. (1971): Studies on freshwater bacteria factors which influence the population and its activity. J. Ecol. **59**, 593—613.

LOCHHEAD, A. G., & CHASE, F. E. (1943): Qualitative studies of soil microorganisms. V. Nutritional requirements of the predominant bacterial flora. Soil Sci. **55**, 185—195.

MURCHALENO, R. A., & BROWN, C. (1970): Heterotrophic bacteria in Long Island Sound. Marine Biol. **1**, 1—6.

MUDRYK, Z., & DONDERSKI, W. (1986): Distribution and physiological properties of heterotrophic bacteria in the estuary lake Gardno. AUNC Toruń. Limnol. Papers **17** (in press).

NIEWOLAK, S. (1968): Seasonal changes of the heterotrophic microflora of the Iława lakes bottom deposits. Pol. Arch. Hydrobiol. **3**, 211—224.

OLAH, J. (1969): A quantitative study of the saprophytic and total bacterioplankton in the open water and the littoral zone of the lake Balaton in 1968. Ann. Inst. Biol. Tihany **30**, 197—212.

RAI, H., & HILL, G. (1981): Observations on heterotrophic activity in Lago Jannari: a Ria/Varzea lake of Central Amazon. Int. Ver. Limnol. **21**, 715—720.

RIEMANN, B. (1980): Diurnal variations in the uptake of glucose by attached and free-living microheterotrophs in lake water (Lake Mosso Denmark). Develop. Hydrobiol. **3**, 153—160.

SIEBURTH, J. M. (1967): Seasonal selection of estuarine bacteria by water temperature. J. Exper. Mar. Biol. Ecol. **1**, 98—127.

STRZELCZYK, E., RZEMIENIEWSKI, F., & WIERZBICKI, K. (1967): Comparison of different media for enumeration of bacteria in water. AUNC Toruń. Limnol. Papers **3**, 45—56.

— DONDERSKI, W., & STOPIŃSKI, M. (1976): Studies on physiological properties of heterotrophic bacteria isolated from water and mud of three lakes. AUNC Toruń. Limnol. Papers **9**, 27—39.

TAHA, E. E., & REFAI, A. E. (1962): Physiological and biochemical studies on nitrogen fixing blue-green algae. I. On the nature of cellular and extracellular nitrogenous substances formed by *Nostoc commune*. Arch. Mikrobiol. **41**, 307—314.

Address of the author:

Dr. ZBIGNIEW MUDRYK
Department of Ecology
University of Education
Arciszewskiego 22 b
76-200 Słupsk
Poland

Limnologica (Berlin) **20** (1), 135—139, April 1990

AdW der DDR, Zentralinstitut für Mikrobiologie und experimentelle Therapie, Jena, DDR

Methanogenesis in Different Parts of a Brackish Water Ecosystem

By JÜRGEN HEYER, URSULA BERGER & REINHARD SUCKOW

With 2 Figures and 9 Tables

Summary

This study deals with the quantification of methanogenesis and its relation to sulfate reduction in sediments of the shallow water area between the islands of Rügen and Hiddensee in the southern Baltic.

Methane production rates differ greatly between the sediments of the open water area (23—84 µmol $CH_4/m^2 \cdot h$) and those in a creek and basins (787—4618 µmol $CH_4/m^2 \cdot h$) because of the different availabilities of organic matter. The substrate limitation of methanogenesis was demonstrated by the increase in methane production in sediment slurries after addition of methane precursors and complex organic substances (algal matter). In the presence of high sulfate concentrations, methanogenesis is low because the methanogenic bacteria are outcompeted by sulfate reducing bacteria for the limited substrates acetate and H_2/CO_2. The competition was eliminated by inhibition of sulfate reducing bacteria, by low sulfate concentrations and by an excess of organic matter. Methanogenesis was calculated to account for 1.5—67% of mineralization.

The ecological importance of anaerobic decomposition is growing in the coastal area of Baltic Sea, particularly in backwaters and estuaries, because of the increasing eutrophication. The examination of these microbiological processes, their calculation and the assessment of their consequences are tasks of high urgency.

The anaerobic decomposition of organic matter occurs in aquatic ecosystems where oxygen as the terminal electron acceptor is absent, i.e. mainly in sediments. Here, complex organic substances are gradually decomposed into low molecular compounds by different groups of anaerobic microorganisms. It is characteristic of this anaerobic food chain that the organisms of the microbial community are linked by many metabolic interactions. The most important terminal steps are denitrification, dissimilatory sulfate reduction and methanogenesis. In these processes the carbon is either completely oxidized to carbon dioxide or reduced to methane.

In marine ecosystems, sulfate reduction is the predominant process, because seawater contains high sulfate concentrations. However the enrichment of methane in the anoxic waters of fjords, basins and in the Black Sea, like the high methane concentrations in marine sediments, show that methanogenesis can be ecologically significant under certain conditions. This can also be expected in brackish water ecosystems and estuaries with lower sulfate concentrations and a high sedimentation rate of particulate organic matter. Methane is produced by the activity of a special group of strictly anaerobic bacteria, the methanogenic bacteria, which use the anaerobic decomposition products of polymeric organic substances (hydrogen and carbon dioxide, acetate, formate, methylamines, methanol or carbon monoxide) as substrates for the energy metabolism.

This study deals with methane formation in a brackish water ecosystem with the aim of calculating the relative importance of methanogenesis in the anaerobic decomposition of organic matter and to elucidate the interactions between methanogenic and sulfate reducing bacteria.

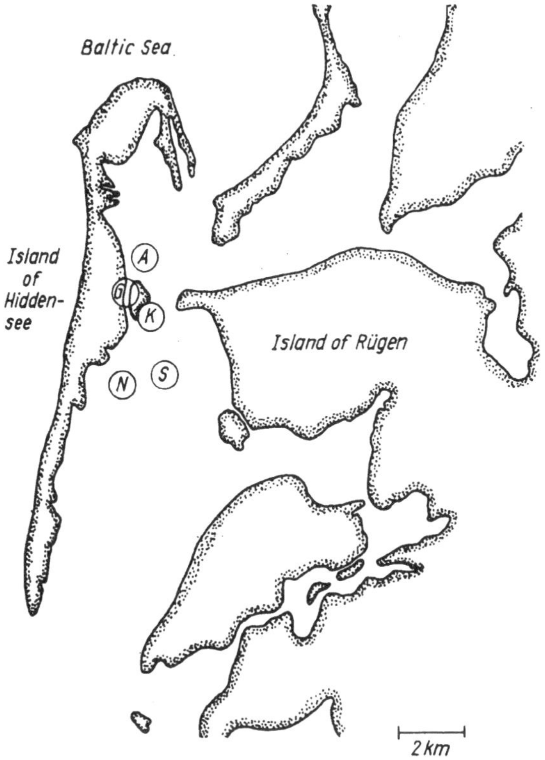

Fig. 1. Study area with the position of sampling points

Our investigations were performed in the surroundings of our microbiological coastal station situated in the shallow water area between the islands of Rügen and Hiddensee in the southern Baltic.

We selected 5 sampling stations with different conditions (Fig. 1): two stations represent the open water area with strong water movement (A, S); one station is in a creek (K) and two stations are small basins (G, N). Some characteristics of the sediments are summarized in the Table 1.

The first indication of the methane production in the sediments consists in bubble formation. The ascending gas bubbles contain 65—96% of methane.

We measured the rates of methane production in sediment samples by monitoring the increase in the methane concentrations in the head spaces of closed serum bottles with sediment slurries incubated under in situ conditions.

Table 1. Characteristics of the sediments at different sampling stations

Sampling station	Water depth (m)	Water content (%)	Volume of 100 g sediment (ml)	Volatile solids (% of dry sediment)	Methane concentration of interstitial water (μmol/l)
Open water area with strong water movement					
A	0.4	28	57	0.9	18
S	3	49	70	3.9	n. d.
Creek					
K	0.4	45	68	3.4	108
Basins					
G	0.6	78	89	12.6	178
N	3	85	92	20.2	613

Table 2. Methane production rates in brackish water sediments (Calculations from in vitro experiments; sediment layer: 0—10 cm)

Sampling station	Methane production rates (μmol CH_4/kg dry sed. · h)	(μmol CH_4/m^2 · h)	Volatile solids (% of dry sediment)
Open water area			
A	0.18— 0.35	23—42	0.9
S	1.1	84	3.9
Creek			
K	11.1— 51.5	1042—4618	3.4
Basins			
G	44.6— 72.6	787—1671	12.6
N	48.7—193.4	853—3553	20.2

The methane formation rates calculated from these in vitro experiments (Table 2) show great differences of three orders of magnitude at different sites in the same shallow water ecosystem. The highest rates were measured in the creek and basins, where large amounts of organic matter have accumulated, and the lowest in the open water area with strong water movement.

What is the cause of these differences?

The prerequisites for methanogenesis are
- anoxic conditions with a negative redox potential;
- exclusion of oxygen, sulfate and nitrate as electron acceptors;
- presence of precursor substrates.

According to in situ measurements the redox potential is negative in all sediments at depths greater than 2—4 cm. The studies of REVSBECH et al. (1980) showed, that oxygen can penetrate into coastal marine sediments only to a depth of few millimeters.

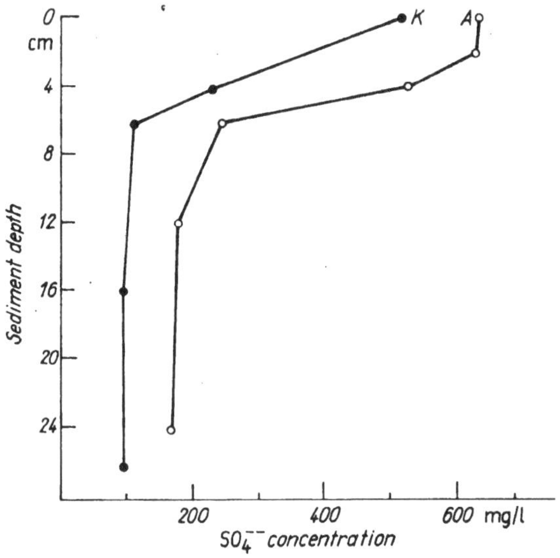

Fig. 2. Sulfate concentrations at different sediment depth

By measuring sulfate profiles using a system of dialysis chambers we could confirm that sulfate concentrations decrease with increasing sediment depth. Fig. 2 shows the differences between the sediments from the open area (A) and those from the creek (K) and basins.

The most important prerequisite for methanogenesis in sediments is the presence of organic matter which is decomposible to methane precursor substrates by the anaerobic microorganisms. In fact, sediments with the highest methane production rates have a high volatile solid content, but the correlation is not strong.

Table 3. Methane production in sediment slurries supplemented with organic substances after different incubation times (μmol CH_4/flask)

Sediment sample	Incubation time (days)	Control without additions	Acetate (1 mmol/flask)	Methanol (1 mmol/flask)	Algal matter (1 g/flask)
A	2	0.10	0.13	0.32	0.079
	10	2.0	2.4	564	9.8
S	2	0.043	0.72	0.64	1.83
	10	2.8	21.5	633	81.6
K	2	48.5	112	95.3	241
	10	502	931	719	1198
G	2	2.0	74.6	42.3	113
	10	100	848	616	1204
N	2	50.1	268	202	334
	10	537	1461	941	2036

Methanogenesis in these sediments is limited by the amount of organic matter. After the addition of methane precursors such as acetate, methanol or complex organic substances (algal matter), the activity increased (Table 3). During the first two days, the methanogenic activities of the supplemented slurries were two to 60 times higher than those of the controls. After 10 days incubation the differences in the stimulation of methanogenesis showed that methanol is the preferred substrate for methane formation in all sediments. 75—100% of the added methanol was reduced to methane. In the sediments with the highest activity, acetate was transformed into methane in the same quantity. Moreover, methanogenesis in the sediments was also stimulated by the addition of lactate, hydrogen and algal matter.

The methane values after two days incubation showed the potential methanogenic activity of the sediments with an excess of organic substrates. After ten days incubation the microorganisms had adapted to the high substrate supply, for example by multiplication.

Table 4. Methanogenic activity in different sediment layers (sample station K)

Sediment layer	Methane formation rate (μmol $CH_4/m^2 \cdot h$)	Volatile solids (% of dry sediment)
0— 5 cm	1102	3.4
5—10 cm	147.2	1.8
10—15 cm	13.5	2.9
15—20 cm	1.0	4.2

Table 5. Influence of sulfate on methane production rates in sediments (μmol CH_4/kg dry sediment · h)

Sediment sample	Sediment slurries with seawater (6—7 mmol SO_4^{--}/l)	Sediment slurries with seawater + 10 mmol molybdate/l	Sediment slurries with sulfate-reduced seawater (2—3 mmol SO_4^{--}/l)
A	0.040	0.26 (6.5×)	0.13 (3.3×)
S	0.031	0.75 (25×)	0.034
K	0.47	71.3 (152×)	35.5 (76×)
G	4.5	134 (30×)	50.0 (11×)
N	5.6	461* (82×)	111 (20×)

In brackets: Increase in methane production rate relative to control

The distribution of methanogenic activity in different sediment layers was investigated in the creek (Table 4). The highest methane formation rate was measured in the uppermost layer, where freshly deposited organic matter is present. This sample also shows that the nature of the organic matter is more important for the methanogenic activity than the total quantity.

The second significant factor influencing methane formation in this habitat is the sulfate concentration. In the presence of high sulfate concentrations (6—7 mmol SO_4^{2-}/l) the methanogenesis was very low (Table 5). After addition of 10 mmol molybdate/l, a specific inhibitor of sulfate reduction, to the sediment slurries with seawater, methane production rates increased dramatically. The methane formation rates were to 6.5—152 times the levels measured in the seawater slurries without molybdate. A corresponding effect was obtained in slurries with reduced sulfate concentrations.

These results can be explained in the following way: In the presence of high sulfate concentrations, the methanogenic bacteria are outcompeted by sulfate reducing bacteria for the limited substrates (Ward & Winfrey 1985). The main products of anaerobic fermentation are acetate and hydrogen, which can be used by both methanogenic and sulfate reducing bacteria. Because the use of these substrates for dissimilatory sulfate reduction is thermodynamically more efficient than for methanogenesis and because sulfate reducing bacteria have a greater affinity for acetate and hydrogen in high sulfate environments, the sulfate reduction is the predominant terminal process. The preferred utilization of acetate and hydrogen was confirmed by the measurement of hydrogen sulfide levels in sediment slurries supplemented with substrates (Table 6). The addition of acetate, hydrogen and complex organic matter induced an increase in hydrogen sulfide production.

Table 6. Effect of substrate additions on sulfate reduction in sediment slurries (10 d)

	μmol H_2S/flask (% utilization of added substrate)	
	Sediment S	Sediment G
Control	161	326
Acetate 1 mmol/flask	460 (30)	523 (20)
Methanol 1 mmol/flask	171 (1.4)	376 (6.7)
Hydrogen 0.36 mmol/flask	442 (79)	n.d.
Algal matter 1 g/flask	496 (2)	574 (1.5)

After inhibition of sulfate reducing bacteria with molybdate the competitive substrates acetate and hydrogen are metabolized by methanogenic bacteria. Under low sulfate conditions, sulfate reduction is limited and the methanogens are more competitive. After depletion of the sulfate, the total quantity of competitive substrates is used for methanogenesis, and the metabolic function of the sulfate reducing bacteria changes to that of hydrogen producing actogens.

The competitive interactions between sulfate reducing and methanogenic bacteria are only restricted to the main products of the anaerobic decomposition, i.e. to acetate and hydrogen. Besides, some substrates such as methanol and methylamines, cannot be used by sulfate reducing bacteria but only by methanogenic bacteria. According to Oremland & Polcin (1982) these noncompetitive substrates are present in marine and estuarine environments and are metabolized only to methane in high sulfate regions of these habitats. We assume, that the low methane production rates in high sulfate slurries in our experiments resulted from the use of such noncompetitive substrates by methanogenic bacteria. The preferred utilization of added methanol as a substrate for methane formation in all sediments supports this view.

The competitive inhibition of methanogenic bacteria by sulfate reducing bacteria in high sulfate environments is partially removed by an excess of competitive substrates or organic matter. The methanogenic activity of the sediment slurries was very low at seawater sulfate concentrations but the methane production rates increased in the supplemented slurries (Table 7).

The significance of methanogenesis for anaerobic decomposition in brackish water sediments follows from the calculation of the mineralization rates for organic substances (Table 8). According to NEDWELL (1982) the production of one micromole methane or hydrogen sulfide is equivalent to the decomposition of 0.024 mg carbon of organic matter.

The role of methanogenesis in anaerobic mineralization is slight in poor sediments with a low methanogenic activity. However, in the sediments of the creek and basins with a high methanogenic activity the methanogenesis is more important and can exceed the values for sulfate reduction.

Table 7. Activities of methanogenic bacteria in sediment slurries after addition of organic substances (µmol CH_4/kg dry sediment · h; 2 days incubation)

Sediment sample	Seawater control	+Acetate 1 mmol/ 100 ml	+Methanol 1 mmol/100 ml	+Algal matter 1 g/100 ml
A	0.04	0.04	0.10	0.03
S	0.03	0.4	0.5	1.2
K	0.5	82.3	69.8	176
G	4.5	181	103	275
N	5.6	595	449	743

Table 8. Mineralization rates of organic substances by methanogenesis and sulfate reduction in brackish water sediments

	A	S	K	G	N
Methanogenesis					
µmol CH_4/kg dry sed. · h	0.35	1.14	51.5	72.6	193
mg C mineralization/kg dry sed. · h	0.008	0.027	1.24	1.74	4.64
Sulfate reduction					
µmol H_2S/kg dry sed. · h	21.8	18.2	25.5	166	215
mg C mineralization/kg dry sed. · h	0.52	0.44	0.61	3.98	5.15
Total mineralization					
mg C/kg dry sed. · h	0.53	0.46	1.85	5.72	9.79
mg C/m² · h	64.2	33.8	166	93.5	180
Portion of methanogenesis (%)	1.5	5.8	67	30	47

The results of in vitro experiments lead to the following conclusions regarding the distribution of methanogenesis and dissimilatory sulfate reduction in this brackish water ecosystem.

Methanogenesis occurs at low rates in the uppermost sediment layer with high sulfate concentrations by the utilization of the non-competitive substrates such as methanol and methylamines. The main products of the anaerobic decomposition, acetate and hydrogen, are metabolized through sulfate reduction, and the methanogenic bacteria are outcompeted for these substrates in this layer by sulfate reducing bacteria. Sulfate reduction is the cause of the decrease in sulfate concentrations in the deeper sediment layers. Therefore, sulfate reduction is limited by the availability of sulfate. In these low sulfate regions of the sediment, the methanogens are more competitive and, after the total depletion of sulfate, methanogenesis is the sole terminal process in anaerobic decomposition. The scale over which stratification is observed is a function of how rapidly the sulfate is depleted in the sediment. This in turn is a function of the sulfate concentration of overlying water and of the availability of organic matter in the sediments.

In the creek and basins of our ecosystem
- the water movement is restricted,
- sedimentation increases, because these regions act as traps for the horizontally transported particulate organic material,
- the content of organic substances in the sediment is high,
- microbial activity is high, so oxygen is depleted by aerobic decomposition of organic material,
- rapid sulfate consumption provides a steep sulfate gradient,
- hydrogen sulfide inhibits bioturbation and macroinfaunal irrigation,
- sulfate limitation of sulfate reduction eliminates the competitive inhibition of methanogenic bacteria, and
- methane production in the upper layers of the sediment is high.

In the open water area on the other hand,
- there is a strong water movement,
- sedimentation is disturbed by movement,
- the content of organic substances in the sediment is low,
- oxygen and sulfate are consumed slowly,
- the penetration of sulfate and oxygen into the sediment is supported by bioturbation and macroinfaunal irrigation, and
- methane production is low and is restricted to the deeper layers of sediment.

Table 9. Methane production rates in sediments of marine and brackish water environments (µmol CH_4/m² · h)

Environments	Methane production rates	Authors
Coastal area of the southern Baltic		this study
open area	23— 84	
creek and deepenings	787—4618	
Baltic Sea		LEIN et al. (1981)
Golf of Riga	129	
Gotland depression	2.19	
Black Sea (1000 m water depth)	0.4—0.5	HEYER et al. (1989)
Santa Barbara Basin	7.5	BARNES & GOLDBERG (1976)
Cariaco Trench	18.2	REEBURGH (1976)
Chesapeake Bay		LIPSCHULTZ (1982)
salt marsh	51— 224	
Sapelo Island salt marsh		KING & WIEBE (1978)
short Spartina	27—1164	
Cape Lookout Bight	560—2000	CRILL & MARTENS (1983)

The comparison of the methane production rates in our ecosystem with those of other marine environments (Table 9) confirms that the methanogenesis is important for anaerobic decomposition in brackish water and estuarine habitats because of the lower sulfate concentrations than in marine systems and the large amounts of available organic matter.

References

BARNES, R. O., & GOLDBERG, E. D. (1976): Methane production and consumption in anoxic marine sediments. Geology **4**, 297—300.

CRILL, P. M., & MARTENS, C. S. (1983): Spatial and temporal fluctuations of methane production in anoxic coastal marine sediments. Limnol. Oceanogr. **28**, 1117—1130.

HEYER, J., BERGER, U., MALASHENKO, YU. R., & BUDKOVA, E. N. (1989): Methanogenesis and Sulfate-reduction in the Black Sea Sediments. Abstracts of the 9th International Symposium on Environmental Biogeochemistry, p. 109. Moscow.

KING, G. M., & WIEBE, W. J. (1978): Methane release from soils of a Georgia salt marsh. Geochim. Cosmochim. Acta **42**, 343—348.

LEIN, A. YU., NAMSARAEV, B. B., TROTSYUK, V. YA., & IVANOV, M. V. (1981): Bacterial methanogenesis in holocene sediments of the Baltic Sea. Geomicrobiol. J. **2**, 299—315.

LIPSCHULTZ, F. (1981): Methane release from a brackish intertidal salt marsh embayment of Chesapeake Bay, Maryland. Estuaries **4**, 143—145.

NEDWELL, D. B. (1982): The cycling of sulphur in marine and freshwater sediments. In: NEDWELL, D. B., & BROWN, C. M. (eds.), Sediment Microbiology, pp. 73—106. London.

OREMLAND, R. S., & POLCIN, S. (1982): Methanogenesis and sulfate reduction: competitive and noncompetitive substrates in estuarine sediments. Appl. Environm. Microbiol. **44**, 1270—1276.

REEBURGH, W. S. (1976): Methane consumption in Cariaco Trench waters and sediments. Earth and Planetary Science Letters **28**, 337—344.

REVSBECH, N. P., SORENSEN, J., & BLACKBURN, H. (1980): Distribution of oxygen in marine sediments measured with microelectrodes. Limnol. Oceanogr. **25**, 403—411.

WARD, D. M., & WINFREY, M. R. (1985): Interactions between methanogenic and sulfate-reducing bacteria in sediments. Advances in Aquatic Microbiology **3**, 141—179.

Addresses of the authors:

Dr. JÜRGEN HEYER
Dr. URSULA BERGER
Akademie der Wissenschaften der DDR
Zentralinstitut für Mikrobiologie und experimentelle Therapie
Beutenbergstraße 11
Jena
DDR-6900

Dr. REINHARD SUCKOW
Akademie der Wissenschaften der DDR
Zentralinstitut für Mikrobiologie und experimentelle Therapie
Außenstelle Fährinsel/Hiddensee
Fährinsel
Vitte/Hiddensee
DDR-2359

Limnologica (Berlin), 141—144, April 1990

AdW der DDR, Zentralinstitut für Mikrobiologie und experimentelle Therapie, Jena, DDR

Distribution and Activity of Methanotrophic Bacteria in a Brackish Water Ecosystem

By URSULA BERGER & JÜRGEN HEYER

With one Figure and 11 Tables

Key words: Brackish water ecosystem; Methane cycling; Methanotrophy; Methane oxidation.

Summary

In this study of the methane cycle in the shore region of the Baltic Sea, first investigations concerned the ecology of methanotrophic bacteria in the shallow brackish water ecosystem between the islands of Rügen and Hiddensee.

Methanotrophic bacteria were found widely distributed in the water and sediment of the habitat, with the highest numbers in organic-rich sediments ($1.7 \cdot 10^3$—$1.2 \cdot 10^5$ cells/g). The cupper sediment layer (0—1 cm) was found to be the main zone of methanotrophic activity. The oxidation rates were controlled by the methane concentration according to MICHAELIS-MENTEN-kinetics.

The measured potential rates permitted the calculation of in situ activities for the actual methane concentrations in the environment (4.8—546 μmol/m² · h). The resulting oxygen demand was estimated. Between 2 and 68% of the produced methane were recycled by oxidation.

Methane cycling in the shore region of the Baltic Sea involves methane oxidation as a significant process. The aim of our investigations is to obtain fundamental knowledge about the distribution and activity of methanotrophic bacteria in a brackish water habitat. The data can help to calculate the relative importance of methane oxidation in the carbon cycle and to establish a sensitive microbiological monitoring system to reveal changes in ecological conditions.

In aquatic ecosystems, methane is formed by microorganisms in the anaerobic layers of the sediment. On its way to the atmosphere by diffusion and ebullition, the methane enters aerobic regions of the habitat where aerobic methane oxidation can take place. The ability to use methane as sole source of carbon and energy is mainly restricted to a group of specialists, the methanotrophic bacteria. These are aerobic, heterotrophic Eubacteria of the genera *Methylomonas*, *Methylococcus*, *Methylobacter*, *Methylosinus* and *Methylocystis*. Methanotrophic bacteria are widely distributed in nature, but, because of their dependence on the substrate methane, only where methanogenesis takes place.

The ecological importance of methane oxidation is summarized in the following: Methane oxidation

— functions as an important barrier to the methane flux from the sediments to the atmosphere and returns the methane carbon to the carbon cycle,
— can greatly affect the oxygen budget of the habitat by its high consumption of oxygen,
— has importance for the nitrogen cycle in certain habitats through the ability of methanotrophic bacteria to fix molecular nitrogen and oxidize ammonia, and
— is assumed to be implicated in the mineralization of complex organic substances or xenobiotics in nature by co-oxidation.

The studies reported here were carried out at our coastal station on a small island (Fährinsel) in the southern Baltic. A sampling grid reflecting the ecological diversity of the habitat was established between the islands Hiddensee and Rügen. A map of the sampling stations is given in HEYER & BERGER (1989, this volume).

Methane-assimilating bacteria have been shown to be widely distributed in this brackish water environments. By an enrichment technique we were able to prove their presence in water and sediments at all sampling sites with water depths from 0.5 to 3 metres. We isolated many strains in pure cultures. The most common groups were *Methylocystis*, *Methylosinus* and *Methylomonas*, but *Methylobacter* was also found. In a few samples we found organisms growing at 55 °C. All these thermophilic isolates belong to *Methylobacter*. We did not succeed in enriching halophilic methanotrophs, but some halotolerant *Methylobacter* strains isolated can grow with up to 4% NaCl in the medium.

For quantitative investigations the most probable numbers of the methanotrophic bacteria in water and sediment were estimated by selective enrichment cultivation (Table 1). The cell numbers varied greatly among the sampling points. The water contains significantly less cells (2—3 orders of magnitude) than the sediments. The numbers vary in consequence of water movement. The highest numbers were found in the organic-rich sediments of the trench and creek. In the poor, sandy sediment near the open sea, the cell density was markedly lower. Cell densities depend on various ecological factors, but mainly on the availability of carbon and energy sources, i.e. on the methane concentration.

Table 1. Cell numbers of methanotrophic bacteria (geometric means with 95% confidence limits)

Sampling station	Water (cells/l)	Sediment (cells/g)
A open area	$8.3 \cdot 10^2$ ($4.0 \cdot 10^2$—$1.1 \cdot 10^3$)	$5.2 \cdot 10^2$ ($1.1 \cdot 10^2$—$9.7 \cdot 10^2$)
G trench	$2.3 \cdot 10^4$ ($6.0 \cdot 10^3$—$5.5 \cdot 10^4$)	$4.4 \cdot 10^4$ ($3.2 \cdot 10^3$—$1.2 \cdot 10^5$)
K creek	$4.6 \cdot 10^4$ ($6.3 \cdot 10^3$—$6.0 \cdot 10^5$)	$1.4 \cdot 10^4$ ($1.7 \cdot 10^3$—$4.0 \cdot 10^4$)
N basin	$1.3 \cdot 10^3$ ($3.0 \cdot 10^3$—$3.3 \cdot 10^3$)	$3.9 \cdot 10^3$ ($2.5 \cdot 10^3$—$7.0 \cdot 10^3$)
A open area, after a storm	$1.2 \cdot 10^5$	

Measurement of the methane concentration by gas chromatography yielded information about the distribution of methane in this ecosystem (Table 2). In open water, methane concentrations are always very low. Methane can be enriched in the deep water above basins and in bights during periods of stagnation in summer or under the ice cover in winter.

Table 2. Methane concentrations in the water (μmol/l)

Sampling station	Open water	Stagnant water	
		in summer	under ice cover
A	0.02—0.8		5.2
G	0.1 —1.0	1.1— 1.4	2.3— 3.4
K	0.04—1.1	2.2—27.0	0.5— 1.8
N	0.3 —1.0	1.8— 2.7	7.6—81.8

In the interstitial water of the sediments, methane concentrations are considerably higher (Table 3). They show great differences and can even rise to saturation levels. Gas bubbles are often seen above trenches and in bights.

Table 3. Methane concentrations in the sediment water (μmol/l)

A	open area	0.1— 26.0
G	trench	2.6—527.3
K	creek	0.1—363.0
N	basin	160.7—965.0

In the sediments, a correlation has been found between methane concentrations and cell numbers, and, on principle, between methane concentration and decomposible organic matter as the basis for methane production also (Table 4).

Table 4. Relations between numbers of methanotrophic bacteria in sediments and sediment characteristics

Sampling station	Number of cells per g	Methane concentration in sediment water (μmol/l)	Water content (%)	Organic matter (% volatile solids of dry sediment)
A open area	110—970	0.1—26	25—29	0.8—1
G trench	3200—124000	3—527	64—89	7—18
K creek	1700—40000	0.1—363	35—56	2—7
N basin	2500—7000	161—965	81—89	16—25

There are no correlations between the numbers of methanotrophic bacteria and the concentrations of inorganic nitrogen, phosphate and oxygen in the water of this habitat.

The majority of organisms are able to tolerate the natural variations in the salinity of the water in this environment, which ranges from 0.6 to 1.2%. The pH value is tolerated over a wide range from 6 to 9, including the natural pH range 6.5—8.8.

Quantitative studies in the sediment profile (Table 5) showed nearly constant cell numbers in the upper 3 cm of sediment and considerable numbers even down to 10 or 15 cm depth. It can be assumed that oxygen and

Table 5. Numbers of methanotrophic bacteria in different layers of sediments (geometric means with 95% confidence limits)

Sediment depth (cm)	Station A (cells/g)	Station K (cells/g)
0—1	351 (193— 638)	8970 (4930—16330)
1—2	608 (306—1211)	10360 (5690—18860)
2—3	412 (226— 750)	11960 (6570—21800)
3—4	147 (81— 268)	3660 (2010— 6660)
4—6	160 (88— 290)	3910 (1960— 7780)
6—8	98 (54— 178)	3460 (1900— 6300)
8—10	178 (98— 324)	2110 (1160— 3840)

bacteria are carried into the sediment by bioturbation and macroinfaunal irrigation.

In order to estimate the importance of methanotrophic bacteria in this ecosystem, it was necessary to measure their activity. The rates of methane consumption were calculated by measuring the decrease in methane concentration in samples of water and sediment incubated for 24 h in closed serum bottles under in situ temperature and light conditions. By adding methane, it was possible to calculate potential methane oxidation rates as a function of methane concentration.

In water with low cell densities and low natural methane concentrations in situ, methane consumption was rarely detectable (Table 6). The rates given in the table are maximal values measured under conditions of stagnation.

Table 6. Methane oxidation potential rates in samples with addition of methane

Water samples	Methane consumption		
	(μmol/l · h)	(μmol/m² · h; water column)	
A open area	not measurable	—	
G trench	0 ... 0.01	5 (0.60 m)	
K creek	0 ... 0.02	9 (0.40 m)	
N basin	0 ... 0.01	33 (3.00 m)	
Sediment samples	**Methane consumption (μmol/m² · h)**		
	in summer 17—24 °C	in autumn 10 °C	in winter 0 °C
A	3.9— 62.7	—	—
G	6.8—166.9	42.7	14.7
K	7.3—164.7	30.5	5.2
N	35.7—267.6	—	47.1

The rates measured in the sediment were greatly higher than those in the water. The differences in activity at the different sampling points correlate with the cell densities and rates of methane production. In this way we found the lowest rates of methane consumption at station A, where the sandy sediment is poor in organics and cell numbers are low.

The activity of the methanotrophic bacteria from deeper layers of sediment was also considerable (Table 7). The presence of great numbers of metabolically active methanotrophic bacteria in anoxic layers cannot be explained ecologically at the moment.

Table 7. Investigations on the sediment profile at station G

	Sediment layers		
	0—5 cm	5—10 cm	10—15 cm
Water content (%)	80	67	46
Organic matter (% of dry weight)	14	10	4
Methane content (μmol/l sediment water)	277	143	22
Number of methanotrophs (cells/g)	$8.4 \cdot 10^4$	$1.0 \cdot 10^5$	$7.8 \cdot 10^4$
Rate of methane consumption (methane conc. 7.0 μmol/l) (μmol/kg · h)	15.5	7.1	6.7

The influence of several ecological factors on methanotrophic activity was analyzed.

The presence of oxygen is a prerequisite for aerobic methane oxidation. Normally, oxygen is available in the whole water body down to the sediment because of the water movement. Additionally, oxygen is produced in the upper sediment layers by the photosynthesis of benthic algae and can penetrate into the sediment through faunal activity or plant roots. From the results of our redox measurements, we consider the top 1 cm of the sediment to be the active zone of aerobic methane oxidation.

The measurement of methane consumption rates in sediments at different pH values shows that activity is nearly constant within the pH range of 5.5—8.5. Significant inhibition was observed only at pH values lower than 5 and higher than 9. The pH tolerance of methane oxidation is greater than the natural fluctuations in the pH value of the ecosystem.

The dependence of methane oxidation on temperature is shown in Table 8. The relatively high rates at 10 °C show that this process is ecologically effective also at low temperatures. Methanotrophic activity could even be measured at 1 °C in winter.

Table 8. Methane consumption rates at different temperatures (starting methane concentration 19—20 μmol/l)

Temperature of incubation	Consumption of methane (μmol methane per kg dry sediment and hour)
10 °C	57.59
20 °C	65.43
30 °C	93.99

The methane concentration is a significant factor influencing methane oxidation. The relationship between methane concentration and oxidation rate is governed by the substrate saturation kinetics (Fig. 1) and permits the calculation of the maximal rates (v_{max}) and the half-saturation constants (K_M). The in situ activity can be calculated from the in situ methane concentration of the environment.

Table 9 presents values calculated for the different parts of the habitat.

Comparison with the rates of methane production (Table 10) shows the amounts of methane that are oxidized and transformed into CO_2 and biomass. These percentages vary from 2 to 68%. If we suppose that in most cases only part of the methane produced is released from the sediment, methane oxidation probably disposes of an even greater proportion of the escaping methane.

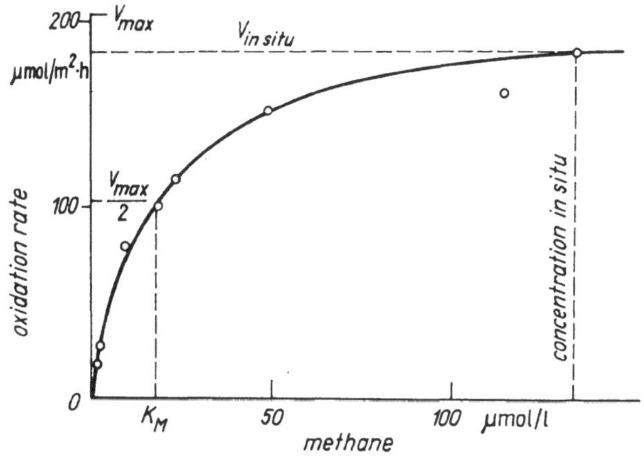

Fig. 1. Dependence of the methane concentration rates on the substrate concentration (MICHAELIS-MENTEN-kinetics). Sample: Sediment station G

Table 9. Calculated methanotrophic activities in the sediments of different parts of the brackish habitat (results of 2 experiments)

Sampling station	v_{max} (μmol/m² · h)	K_M (μmol/l)	Methane concentration in situ (μmol/l)	$v_{in\ situ}$ (μmol/m² · h)
A	8.0	1.5	10	7.5
open area	5.3	1.1	26	4.8
G	213	27	122	177
trench	309	31	420	288
K	87	10	42	70
creek	104	10	108	95
N	187	14	161	171
basin	605	146	1346	546

Table 10. Calculated importance of methanotrophic activity in relation to methane production and oxygen consumption in the habitat

Sampling station	Methane consumption (total activity: water and sed.) (μmol/m² · h)	Methane production (μmol/m² · h)	Proportion oxidation: production of methane (%)	Oxygen demand for methane oxidation (mg/m² · d)
A open area	5— 8	23— 42	11—33	7— 12
G trench	177—293	787—1671	11—37	272—451
K creek	70—104	1042—4618	2—10	108—160
N basin	171—579	853—3553	5—68	263—889

Apart from its importance for the recycling of methane carbon, the main impact of methane oxidation is on the consumption of oxygen from the environment. Table 10 shows the calculated amounts of oxygen consumed for methane oxidation. In view of the high oxygen production by algae and the oxygen import from the atmosphere into the shallow water, methane oxidation cannot significantly affect the oxygen budget of the water, except under the ice cover in winter. But in the sediment, methane oxidation is one of the most important oxygen consuming processes.

Our results show the significance of methane oxidation in the brackish water environment. Cell numbers and activities of methanotrophic bacteria vary greatly in space in the habitat we studied. Methanotrophic activity in this brackish water environment is greatest in the upper layer of sediment, where relatively high methane and oxygen concentrations meet.

Table 11. Summary of published rates of methane oxidation in sediments of aquatic ecosystems

	Methane oxidation	Author
Limnic systems		
Lake Washington	0.27—2.5 μmol/l · h	LIDSTROM & SOMERS (1984)
Pond of Netzbach	2.5 μmol/kg · h	ZAISS (1981)
Methansee	100—150 μmol/l · h	WELCH et al. (1980)
Lake Stechlin	48—119 μmol/l · h	HEYER & BABENZIEN (1985)
Lungernsee	637—911 μmol/l · h	BOSSARD & GÄCHTER (1981)
Lake Dolgoye	1—5 μmol/l · h	ABRAMOCHKINA et al. (1987)
Netzbach	23 μmol/kg · h	ZAISS (1981)
River sediments	42—111 μmol/kg · h	TAM et al. (1981)
River Saale	20—240 μmol/l · h	BERGER & HEYER (1989)
Marine systems		
Bering Sea	0.00002—0.01 μmol/l · h	GRIFFITH et al. (1982)
Baltic Sea, coastal regions	0.5—55 μmol/l · h	this study

Comparison of the methane consumption rates found by us with values for other sediments (Table 11) shows the high methanotrophic activity of this habitat.

Our studies represent the first in which the methanotrophic bacteria of a brackish water habitat have been investigated quantitatively and isolated in pure culture. We have measured methane oxidation in a Baltic environment for the first time and have determined the in situ rates in the sediment.

References

ABRAMOCHKINA, F. N., BEZRUKOVA, L. V., KOSHELEV, A. V., GALCHENKO, V. F., & IVANOV, M. V. (1987): Microbial methane oxidation in a freshwater reservoir. Mikrobiologiya **56**, 464—471 (in Russ.).

BERGER, U., & HEYER, J. (1989): Untersuchungen zum Methankreislauf in der Saale. J. Basic Microbiol. **29** (4), 195—213.

BOSSARD, P., & GÄCHTER, R. (1981): Methan- und Sauerstoffhaushalt im mesotrophen Lungernsee. Schweiz. Z. Hydrol. **43**, 219—252.

GRIFFITH, R. P., CALDWELL, B. A., CLINE, J. D., BROICH, W. A., & MORITA, R. Y. (1982): Field observations of methane concentrations and oxidation rates in the southeastern Bering Sea. Appl. Environm. Microbiol. **44**, 435—446.

HEYER, J., & BABENZIEN, H.-D. (1985): Untersuchungen des Methankreislaufes in einem oligotrophen See (Stechlinsee). *Limnologica* Berlin) **16** (2), 267—276.

LIDSTROM, M. E., & SOMERS, L. (1984): Seasonal study of methane oxidation in Lake Washington. Appl. Environm. Microbiol. **47**, 1255—1260.

TAM, T.-Y., MAYFIELD, C. I., & INNISS, W. E. (1981): Nitrogen fixation and methane metabolism in a stream sediment — water system amended with leaf material. Can. J. Microbiol. **27**, 511—516.

WELCH, H. E., RUDD, J. W. M., SCHINDLER, D. W. (1980): Methane addition to an arctic Lake in winter. Limnol. Oceanogr. **25**, 100—113.

ZAISS, U. (1981): Natural ebullition of mine gas and its microbial oxidation in the Netzbach brook, Saarland. Verh. Int. Verein. Limnol. **21**, 1381—1385.

Address of the authors:

Dr. URSULA BERGER
Dr. JÜRGEN HEYER
AdW der DDR
Zentralinstitut für Mikrobiologie und experimentelle Therapie
Arbeitsgruppe Mikrobenökologie
Beutenbergstr. 11
Jena
DDR-6900

Limnologica (Berlin), 145—148, April 1990

Ernst-Moritz-Arndt-Universität Greifswald, Sektion Biologie, Biologische Station Hiddensee, Kloster/Hiddensee, DDR

Denitrification in Sediments of North Rugian Estuaries, Preliminary Communication

By S. DAHLKE

With 3 Figures and one Table

Key words: Denitrification; Acetylen block technique; Preconcentration; Annual periodicity; Limitation

Summary

The acetylene block technique was used to assay denitrification in undisturbed cores of sediments obtained from various parts of a backwater chain. A method is described that includes preconcentration of the produced N_2O for estimating in situ denitrification using a gas chromatograph with lower sensitivity than being in general use. First measurements carried out through nearly one year are reported. It was found an annual periodicity of denitrification. A limitation of the process by nitrate was demonstrated and a hypothetical pattern of limitations is discussed. Following a first coarse calculation, the loss of nitrogen from the area is nearly 5 g $N \cdot m^{-2} \cdot a^{-1}$, that means about the half of the anthropogenic input.

Introduction

Denitrification is the reduction of nitrogen oxides, usually nitrite or nitrate, to gaseous products such as nitrous oxide and molecular nitrogen (FOCHT 1982). The process is respected as the major way of biogenic release of gaseous nitrogen from terrestrial and aquatic ecosystems and, therefore, should counteract permanently increasing eutrophication.

During the recent years it was begun to investigate this in the North Rugian estuaries not well known process. The purpose was to collect several ecological data which make it possible to quantify denitrification at various parts of the waters during a year. Moreover, the data could be used to indicate some interrelationships.

Materials and Methods

Study site

The sampling area is described elsewhere by other authors (GESSNER 1937; CORRENS & JAEGER 1979, 1982). Briefly, it is a backwater chain in the north of the island of Rügen. The geographical situation and the stations are shown at Fig. 1. This and similar systems are characterized by a little average water depth and a lower salinity according to a greater distance from the inlet to the Baltic Sea. In an opposite tendency to this the level of trophy increases. According to SCHLUNGBAUM (1979) the system can be named an estuary with regard to the salinity pattern following the definition given by CAMERON & PRITCHARD (1963).

Sampling

Sediment cores were taken aboard a boat under utilization by polycarbonate tubes (50 by 5.4 cm, ID) coverd with a simple check valve changeable for rubber stoppers for transporting the samples. Before sampling the tubes were fixed at a shaft (up to 5 m long) using pipe clamps.

Experiments with cores

The tubes containing nearly undisturbed (i.e., not visibly disrupted) cores were brought to the laboratory and the following experiments were done within not more than 24 h. A plunger (material: plastic with a gasket ring made of silicone rubber) was inserted at the bottom of the tube in such a way that the nearly 6 cm upper core segment remains in the tube and the lower part of the core is cut off and discarded. The overlaying water was removed by a membrane pump. The remained core was pushed upwards until a distance of 2 cm to a rubber stopper pressed into the top end of the tube. Through two channels placed in the stopper the residual air was nearly entirely replaced by the overlaying water derived from the tubes and saturated with acetylene before to inhibit nitrous oxide reduction. The relative extreme saturation was used because it is known that acetylene fails to block N_2O reduction at low concentrations of inhibitor and nitrate (KASPAR 1982) and/or the presence of sulfid (KNOWLES 1982). The samples were incubated then in an upright position for 2 h at nearly in situ temperature. Following the incubation a slurry was made by addition of hydrogen peroxide solution (0.1% in water) in an 1:1 ratio (v/v). Nearly 50 ml air were added to the sample to achieve a "head space". The plunger was drew down during this manipulation so that the pressure inside the tube stayed nearly atmospheric. Then the sample was vigorously shaken.

Because the interruption of denitrification in this manner is not completely the following step was done immediately.

Preconcentration of nitrous oxide

The produced nitrous oxide is usually quantified by gas chromatography. The sensitivity of detection is sufficient for investigations in natural environments if the used device is equiped with a high temperature electron capture detector or a some more sensitive quadrupol mass spectrometer (FOCHT 1982). Without this a preconcentration of nitrous oxide often seems to be indispensable. That was done as follows: Sparging the slurry with cleaned argon the nitrous oxide and some other gases were adsorbed at molecular sieve 5A (Fig. 2). The molecular sieve (Wolfen Zeosorb 5A, spheroidal form) was activated by heating to 450 °C for at least 6 h before each use and then allowed to cool in the adsorption tubes (standard fittings of the gas chromatograph, made of chromium-plated brass, 21.5 ml volume). Similar methods were used by COHEN (1977) and RYDEN et al. (1979). The pattern of method applied for recovery of nitrous oxide was taken from RYDEN et al. (1979) with some modifications. Using a syringe the adsorbed gases were displaced from the molecular sieve in an evacuated serum bottle of known volume, usually nearly 150 ml, by addition of 50 ml destilled water. After a equilibration time of 18 h (over night) the pressure inside the bottle was returned to atmospheric by back filling with air through a supply line conceted to a simple manometer based on the principle of spirit level. To reach overpressure 5 ml of destilled water were supplied using a syringe immediately before gas samples were removed.

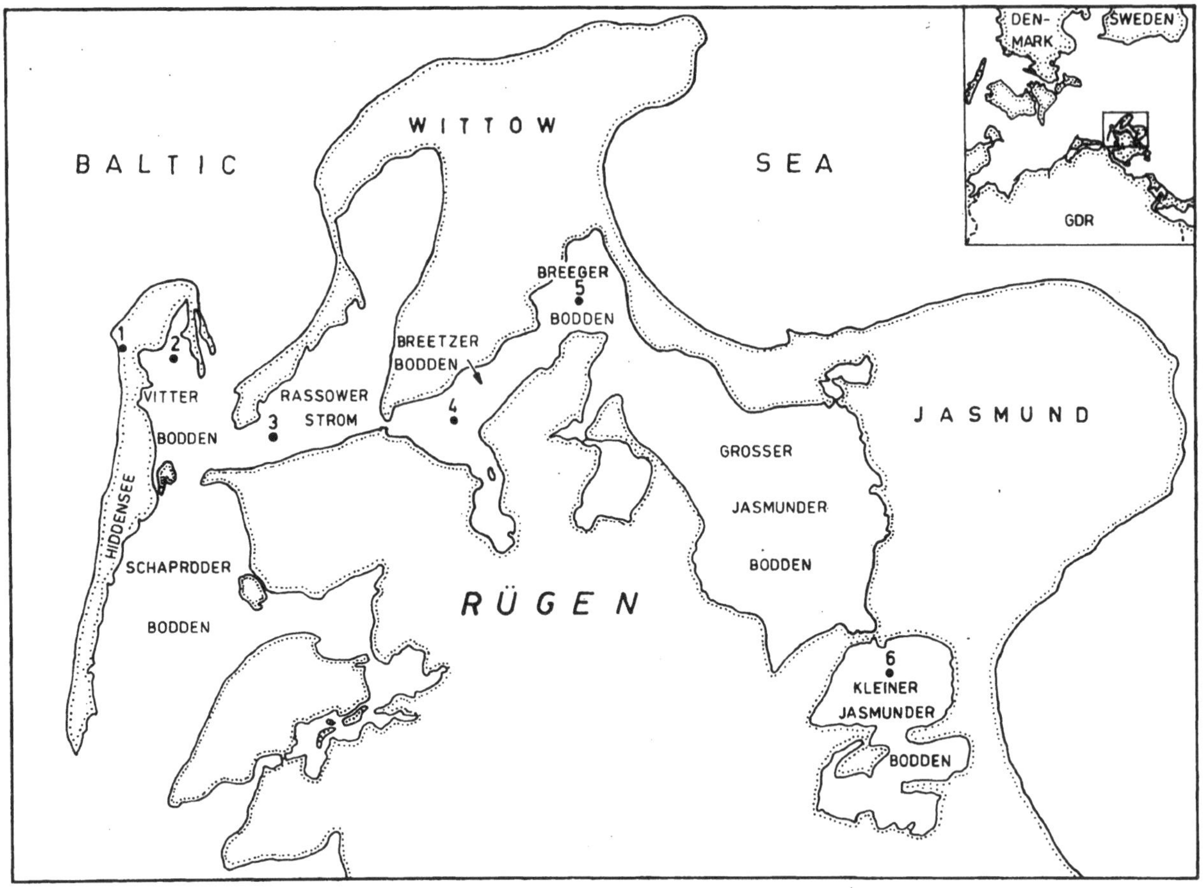

Fig. 1. Map of the North-Rugian estuary system. 1 = Location of the institute. 2—6 = Sampling stations

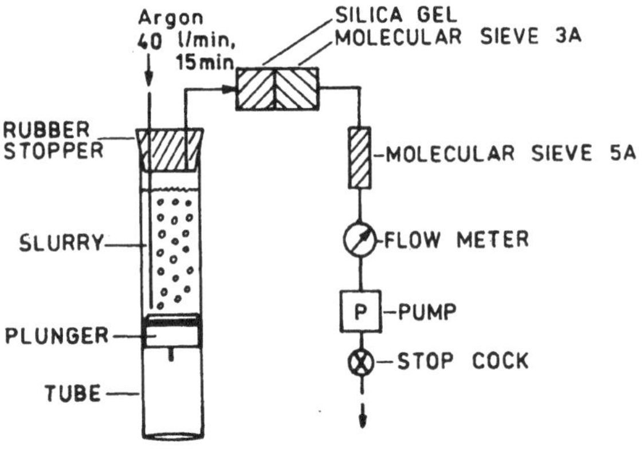

Fig. 2. Schematic representation of the sparging system

Gas chromatographic analysis

Gas samples (up to a volume of 2 ml) were obtained through the septum seal of the serum bottle with a gas-tight syringe and injected into a Chromatron GCHF 18.3 gas chromatograph (Chromatron, Berlin, GDR) with the following operating conditions: Nitrous oxide was separated from other gases in a stainless steel column (100 by 0.3 cm, ID) of molecular sieve 5A (Wolfen Zeosorb 5A, spheroidal form, crushed, 0.5/1.0 mm). The column and the ^3H-electron capture detector were heated at 200 °C. Technical argon cleaned by molecular sieve 5A filled filter groups (standard fittings of the chromatograph) was used as carrier gas at a flow-rate of 60 ml $\cdot$ min^{-1}.

Serial measurements were made in intervals of 12 min for avoidance of peak interferences from preceding injections.

Production of N_2O was determined as the difference between the samples and the controls treated with acetylene, too. Always some nitrous oxide was measureable because of adsorbtion by the molecular sieve during manipulation before use. Correction was made for dissolved N_2O using a solubility coefficient of $2.875 \cdot 10^{-2}$ mol $\cdot$ l^{-1} reported by WEISS & PRICE (1980). A loss of nitrous oxide was taken into consideration determinig a recovery rate of about 60% with a standard deviation of 6%. The molecular sieve may be repeadly used after drying and heating. The recovery rate was then occasionally checked by passing of standardized N_2O samples through the activated material filled in adsorption tubes because the recovery of N_2O tended to decrease. Standardized samples of the gas for the calibration procedures were taken from a self-made mixture of medical N_2O to 0.5% (v/v) in argon.

Results

The up today available data do not allow to distinguish the denitrification of the different stations. So these are not distinguished in Fig. 3. The highest denitrification rates were detected in April and in September, up to 200 μmol N $\cdot$ m^{-2} $\cdot$ h^{-1} and more. In April some nitrate still was measurable in the water. The temperature of the water raised from nearly 0 °C under ice cover in March to more than 10 °C in April during a short period of only 14 days. In September the nitrate level in the waters was very low (all data not shown). In summertime the measured denitrification rates were relatively low and tend to values below the detection limit at the beginning of autumn. Sometimes in this period the dispersion of the data was very high so that often no distinction to the controls was possible. To demonstrate a limitation of

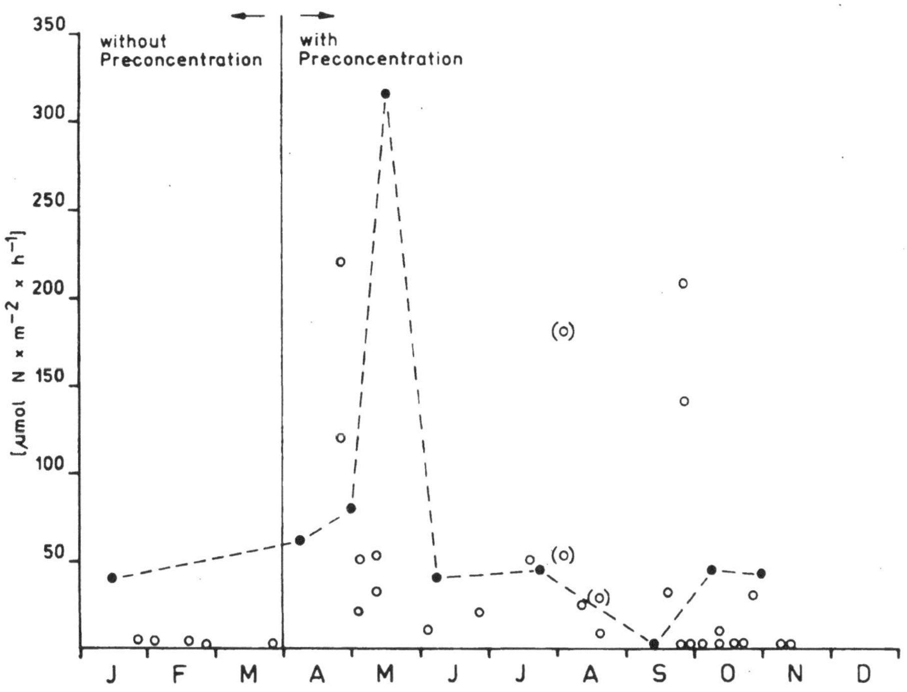

Fig. 3. Seasonal variation of measured denitrification rates obtained by the author (open circles; mean values of usually 5 parallel samples are shown) and published by SØRENSEN (1984) for the Kysing Fjord, Denmark (full circles).
The parenthesizes have the meaning of "not significant"

denitrification by nitrate in summertime some samples with nitrate-amended water from the study site were incubated. The detected increase of denitrification was greater than one order of magnitude. An example is shown in Table 1.

Registrated rates in wintertime are low (below 5 μmol $N \cdot m^{-2} \cdot h^{-1}$). The levels of nitrate in the water were much higher at this time than in summertime (not shown).

Table 1. Denitrification at in situ concentration and with nitrate-amended water (data of 10. August 1987, station "Vitter Bodden", in situ concentration of nitrate was below detection limit of a standard method)

	In situ samples	Nitrate-amended samples*)
Admixture ($\mu mol \cdot KNO_3 \cdot l^{-1}$)	—	+20
Denitrification**) ($\mu mol\ N \cdot m^{-2} \cdot h^{-1}$)	12.1 (± 5.92)	325.3 (± 29.62)

*) Nitrate-amended samples were preincubated for 4 h and then acetylene was added
**) Mean and standard deviation of 5 samples, respectively

Discussion

Possibly the denitrification rates detected in wintertime are underestimated because at this time no preconcentration of nitrous oxide was carried out but necessarily a longer incubation (18 h). Acetylene also inhibits other microbial reactions (for a review see KNOWLES 1982 and PAYNE 1984). Nitrification is there of special interest. Because of this drawback it is not recommendable to assume a linear relation between period of incubation and production of nitrous oxide especially if the nitrate-pool in the sediment is very low. Therefore, FOCHT (1982) and SÓRENSEN (1984) recommend incubation periods not longer than a few hours.

Apart from the findings in wintertime and some values detected in September there is a remarkable conformity with results obtained from danish waters by SØRENSEN (Fig. 3). This includes both the order of magnitude and the annual periodicity of the denitrification rates. A peak in spring and relatively low rates in winter and in summer as well as a minimum in autumn seem to be characteristically also for the waters investigated in this study.

Data of nitrification published by danish workers (HANSEN et al. 1981) may indicate a narrow connection between this process and denitrification, comparing the annual periodicity of the both processes. Particularly while nitrate limitation of denitrification occurs the amount of nitrate (and nitrite) produced by nitrification at the aerobic-anaerobic interface in the sediments should become a restricting factor (for more information see TIEDJE et al. 1982). To be shure some more systematic investigations are necessary. That means the estimate of nitrification and denitrification at the same station and time.

The effect of the main factors restricting denitrification in the investigated waters may be as follows: In winter low water temperatures confine denitrification. This state of affairs is quickly displaced by a nitrate limitation in spring. The reason is a fast growing consumption of nitrate not only by denitrification but also by other processes (assimilation, ammonification of nitrate) if the water temperature goes up. In summertime nitrification is out of order to compensate for the big consumption possibly because of limitation by oxygen and/or ammonium (HANSEN et al. 1981). In late autumn/early winter the restriction of denitrification by nitrate changes slowly in a kind like that by temperature. Additional work has to be done to verify this hypothesis.

The acetylene block technique of measuring denitrification is useful and high sensitive if modern devices for the detection of the nitrous oxide are used. But even in this case the method is not without drawbacks. Kaspar (1982) carefully estimated in a similar study like this the errors of his investigation and published a wide range between overestimation and underestimation.

Following a first coarse calculation with the up today available data the loss of nitrogen from the investigated waters is nearly $5 \text{ g N} \cdot m^{-2} \cdot a^{-1}$, that means about a half of the anthropogenic input.

Conclusions

A systematical sampling of in situ data is still necessary for real calculation of nitrogen release by denitrification from the ecosystem and also for a certainly interpretation of causal interrelationships. The use of devices with higher sensitivity of detection is the supposition for decrease in expense of manipulation and for increase of permeability of samples. The usefulness of the acetylene block technique for assay denitrification has some drawbacks, especially in environments where nitrogen is an limiting factor and, therefore, should be proved at several environmental conditions when possible. The reception of investigations about nitrification in the working programme is a desirable completion.

Zusammenfassung

Die Azetylen-Blockierungstechnik wurde genutzt, um Denitrifikationsraten an ungestörten Sedimentkernen aus verschiedenen Gebieten einer Boddenkette zu bestimmen. Die beschriebene Methode zur Erfassung der in situ-Denitrifikation im Sediment mittels eines Gaschromatographen niedrigerer Sensitivität als für diesen Zweck allgemein üblich, schließt eine Anreicherung des gebildeten N_2O ein.

Erste Ergebnisse von Messungen über einen Zeitraum von fast einem Jahr werden vorgestellt. Es wurde ein Jahresgang der Denitrifikation gefunden. Eine Limitierung des Prozesses durch Nitrat konnte gezeigt werden und ein hypothetisches Muster von die Denitrifikation begrenzenden Prozessen wird diskutiert.

Die jährliche Freisetzung von Stickstoff aus dem Untersuchungsgebiet beträgt einer ersten Grobkalkulation zufolge etwa $5 \text{ g N} \cdot m^{-2} \cdot a^{-1}$, was ungefähr die Hälfte des anthropogenen Eintrages ausmacht.

References

Cameron, W. M., & Pritchard, D. W. (1963): Estuaries. In: The Sea. Vol. 2, pp. 306—324. New York. quoted in Schlungbaum (1979).

Cohen, Y. (1977): Shipboard measurements of dissolved nitrous oxide in seawater by electron capture gas chromatography. Anal. Chem. **49**, 1238—1240.

Correns, M., & Jaeger, F. (1979): Beiträge zur Hydrographie der Nordrügenschen Bodden. I. Einführung in das Untersuchungsgebiet, Wasserstandsverhältnisse und Wasserhaushalt. Acta hydrophys. **24**, 149—177.

— (1982): Beiträge zur Hydrographie der Nordrügenschen Bodden. II. Strömungsverhältnisse, Salz- und Sauerstoffhaushalt. Acta hydrophys. **27**, 5—22.

Focht, D. D. (1982): Denitrification. In: R. G. Burns & J. H. Slater (eds.), Experimental Microbial Ecology, pp. 194—211. Oxford.

Gessner, F. (1937): Hydrographie und Hydrobiologie der Brackgewässer Rügens und des Darß. Kieler Meeresforsch. **2**, 1—80.

Hansen, J. I., Henriksen, K., & Blackburn, T. H. (1981): Seasonal distribution of nitrifying bacteria and rates of nitrification in coastal marine sediments. Microb. Ecol. **7**, 297—304.

Kaspar, H. F. (1982): Denitrification in marine sediment: measurement of capacity and estimate of in situ rates. Appl. Environ. Microbiol. **43**, 522—527.

Knowles, R. (1982): Denitrification. Microbiological Reviews **46**, 43—70.

Payne, W. J. (1984): Influence of acetylene on microbial and enzymatic assays. Journal of Microbiological Methods **2**, 117 to 133.

Ryden, J. C., Lund, L. J., & Focht, D. D. (1978): Direct in-field measurement of nitrous oxide flux from soils. Soil Sci. Soc. Am. J. **42**, 731—737.

Schlungbaum, G. (1979): Untersuchungen über die Sedimentqualität in den Gewässern der Darß-Zingster Boddenkette unter besonderer Berücksichtigung der Stoffaustauschprozesse zwischen Wasser und Sediment. Diss., Wilhelm-Pieck-Universität Rostock.

Sørensen, J. (1984): Seasonal variation and control of oxygen, nitrate, and sulfate respiration in coastal marine sediments. In: M. J. Klug & C. A. Reddy (eds.), Current Perspectives in Microbial Ecology, pp. 447—453. Washington, D.C.

Tiedje, J. M., Sextone, A. J., Myrold, D. D., & Robinson, J. A. (1982): Denitrification: ecological niches, competition and survival. Antonie van Leeuwenhoek **48**, 569—583.

Weiss, R. F., & Price, B. A. (1980): Nitrous oxide solubility in water and seawater. Marine Chemistry **8**, 347—359.

Address of the author:

Dipl.-Biol. Sven Dahlke
Ernst-Moritz-Arndt-Universität Greifswald
Sektion Biologie
Biologische Station Hiddensee
Kloster/Hiddensee
DDR-2346

Limnologica (Berlin) **20** (1), 149—152, April 1990

Wilhelm-Pieck-Universität Rostock, Sektion Biologie, Rostock, DDR; AdW der DDR, Institut für Meereskunde, Rostock-Warnemünde, DDR

Der Einfluß von Schadstoffen auf den Denitrifikationsprozeß in flachen Küstengewässern — Untersuchungen zur Mikrobiologie und Stoffbilanz

Influence of Pollution Factors on the Denitrification Process in Shallow Coastal Waters — Experiments to Microbiology and Nutrient Balance

Von G. KLINKENBERG, G. SCHLUNGBAUM & L. BRÜGMANN

Mit 4 Abbildungen und 2 Tabellen

Key words: Cadmium; Copper; Denitrification; Nitrogen Cycle; Sediment quality.

Abstract

Most diverse loadings occur in connection with the very intensive use of the surface waters. Besides eutrophication (inflow of nitrogen and phosphorous nutrients) an anthropogenic influx of heavy metals can take place. This paper is designed to give information on possible effects of the interrelation between denitrification processes and the metals copper and cadmium applicated as $CuSO_4 \cdot 5\,H_2O$ respectively as $CdSO_4 \cdot 8\,H_2O$, on the sediment-water-interface. The denitrification reaction, determined by means of nitrate reduction, is affected by copper in dissolved concentrations C_{Cu} 2 mg · l^{-1}. Cadmium has no influence in concentrations up to C_{Cd} 61 mg · l^{-1}. The quality of the sediment (organic matter, grain size) and the amount of the suspended material (detritus) exert an essential influence on the masking of the heavy metal ions and reduce consistently the biologically available amount.

1. Einleitung

In stickstoffbelasteten Gewässern stellt die Stickstoffelimination durch Denitrifikation einen wichtigen Prozeß dar. Bis zu 60% des Stickstoffeintrages können durch die vorwiegend in der Sediment-Wasser-Kontaktzone ablaufenden Denitrifikationsreaktionen in Stickoxide (NO, N_2O) und gasförmigen Stickstoff (N_2) umgewandelt werden und das Gewässer verlassen. Das Reaktionsprinzip beruht auf der Fähigkeit der fakultativ anaeroben Denitrifikanten unter Sauerstoffmangelbedingungen ihre Atmung auf Nitrat als Elektronenendakzeptor umzustellen, wenn ausreichend leicht abbaubare organische Substanz, die als H^+-Donator dient, vorhanden ist. Die Reaktionsgeschwindigkeit wird maßgeblich von den Größen

- Temperatur,
- Nitratkonzentration,
- organische Substanz und Reaktionsraum; diese beiden Größen können durch die Sedimentqualität mit den Parametern organischer Gehalt und Korngröße charakterisiert werden,

beeinflußt. Neben der zunehmenden Belastung der Küstengewässer durch Stickstoff, der hauptsächlich als Nitrat und Ammonium aus landwirtschaftlichen Abwässern durch die Zuflüsse eingetragen wird und zu Eutrophierungserscheinungen führt, kann Kontamination durch Schwermetalle eine zusätzliche Belastung darstellen. Ziel der vorliegenden Untersuchungen war es, die bisher wenig untersuchten Wechselwirkungen zwischen Eutrophierung und Kontamination näher zu betrachten.[1]) Dazu wurde der Einfluß von ausgewählten Schwermetallen auf den Denitrifikationsprozeß geprüft. Auf zwei verschiedenen Gebieten wurde dazu experimentell gearbeitet. Zum einen wurde die Wirkung der Schwermetalle Kupfer und Cadmium auf die Denitrifikationsleistung (als Maß diente die Nitratabnahme) untersucht und zum anderen die Wirkung dieser Metalle auf die denitrifizierende Bakterienflora.

[1]) vgl. Abb. 1

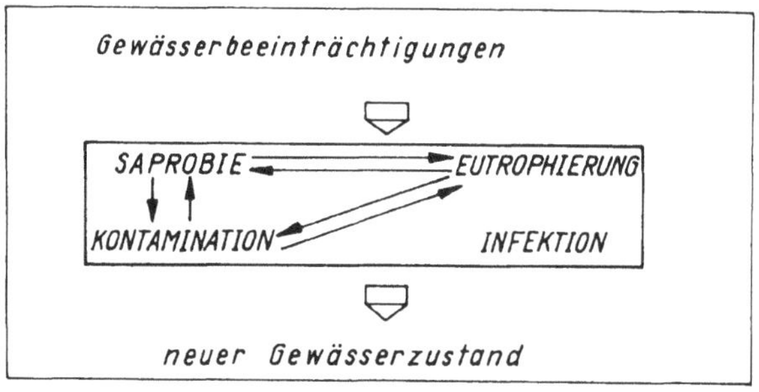

Abb. 1. Formen der Gewässerbeeinträchtigung
(⇄ Wechselwirkungen zwischen den Formen der Beeinträchtigung, die im Text erläutert werden)

2. Material und Methoden

2.1. Zu den mikrobiologischen Untersuchungen

Die Laborversuche wurden auf Nähragarplatten, die mit den Schwermetallen Kupfer bzw. Cadmium in den Konzentrationen 0, 0,01, 0,05, 0,1, 0,5, 1,0, 5,0, und 10 mmol · l^{-1} belastet waren, durchgeführt. Appliziert wurde Kupfer als $CuSO_4 \cdot 5\,H_2O$ und Cadmium als $CdSO_4 \cdot 8\,H_2O$ aus jeweils konzentrierten Stammlösungen. Je Konzentrationsstufe erfolgten 3 Wiederholungen. 0,1 ml einer Bakterienmischpopulation-Suspension wurden ausgespatelt. Die Inkubation erfolgte 10 Tage bei einer Temperatur von 28 °C im Brutschrank. Als Grenzwert galt die Konzentrationsstufe, auf der gerade noch ein Wachstum von Bakterienkolonien sichtbar war.

Charakteristik der Bakteriensuspension: 10^{-3} verdünnt;
Sedimententnahmeort: Zingster Strom/Kirrbucht;

Sedimentparameter:	Trockensubstanzgehalt	67,5%
	organischer Gehalt	0,55%
	mittlere Korngröße	0,222 mm

2.2. Zu den Experimenten zur Denitrifikationsleistung

Sämtliche hier vorgestellten Untersuchungen fanden in Flaschensystemen (s. Abb. 2) im Labor bei einer Temperatur von 28 °C unter Dunkelbedingungen statt. Die Sedimententnahme erfolgte mit einem Ekman-Bodengreifer. Eine gut durchmischte Probe (400 cm³) der 5 cm-Oberflächenschicht, die als aktivste Zone der Denitrifikation gilt (KASPAR 1983), kam zum Einsatz; überschichtet wurde mit 3000 ml Biotopwasser. Ein Vorteil des Flaschensystems besteht darin, daß eine im Biotop vorhandene Sedimentaufwirbelung simuliert werden kann. Um die potentielle Denitrifikationsleistung zu bestimmen, kam eine Nitratzugabe von

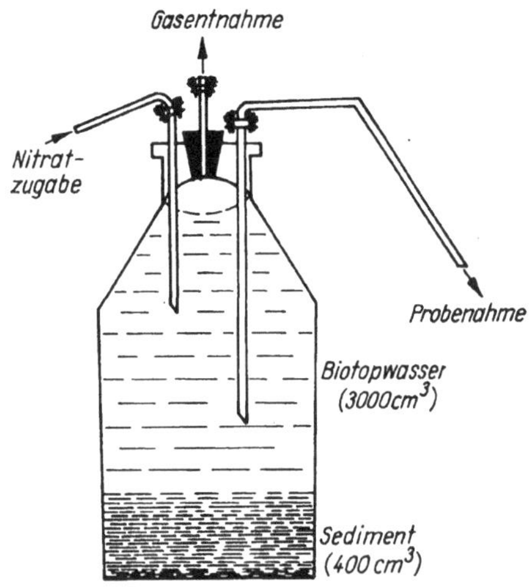

Abb. 2. Schema der verwendeten Versuchsapparatur

200 µmol · l^{-1} zur Anwendung. Als Maß für die Denitrifikationsleistung wurde die Nitratabnahme und N_2-Bildung erfaßt. Über den zu den verschiedenen Jahreszeiten aus dem Zingster Strom entnommenen Sedimenten (Charakteristik s. Tab. 1) wurden bei der täglichen Probenahme folgende Parameter ermittelt:

- pH-Wert
- Nitrat-, Nitrit-, und Ammoniumkonzentration in µmol · l^{-1}
- Schwermetallkonzentration — a) gelöst in mg · l^{-1} nach Filtration durch 0,4 µm am FLAAS (BRÜGMANN 1979)
  - b) aktiv in mg · l^{-1} Bestimmung an einer ionenselektiven Elektrode Me^{++}.

Kupfer wurde als $CuSO_4 \cdot 5\,H_2O$ aus einer konzentrierten Stammlösung bzw. als kristallines Salz appliziert und in Konzentrationen von 0,01, 0,02, 0,2, 0,4, 0,5 und 1,0 mmol · l^{-1} in Kurzzeitexperimenten (die maximale Kontaminationsdauer betrug 28 Tage) eingesetzt. Cadmium wurde als $CdSO_4 \cdot 8\,H_2O$ appliziert und kam in Konzentrationen von 0,05, 0,5 und 1,0 mmol · l , ebenfalls als Kurzzeitkontamination (maximal 28 Tage) zur Anwendung. Tab. 2 zeigt eine Übersicht der eingesetzten Schwermetallkonzentrationen.

Tab. 1. Sedimentcharakteristik
(TS = Trockensubstanz, Trocknung bei 105 °C; organischer Gehalt = Veraschung bei 550 °C als Glühverlust; mittlere Korngröße = m KG, Ermittlung durch Naßsiebung)

Entnahmezeit	TS (%)	organischer Gehalt (%)	m KG (mm)
7. 5.	55,9	2,3	0,215
17. 7.	60,0	1,7	0,220
11. 9.	64,8	1,5	0,345
8. 10.	68,4	1,6	0,321

Tab. 2. Schwermetallzugaben

Element	mmol · l^{-1}	mg · l^{-1}
Kupfer	0,01	0,635
	0,02	1,270
	0,20	12,7
	0,40	25,4
	0,50	31,75
	1,00	63,50
Cadmium	0,05	5,62
	0,50	56,2
	1,00	112,4

3. Ergebnisse und Diskussion

3.1. Zu den mikrobiologischen Untersuchungen

Der Grenzwert für das Wachstum von Bakterienmischpopulationen aus dem Sediment des Zingster Stromes (Kirrbucht) auf künstlich schwermetallkontaminierten Nährböden wurde für Kupfer zwischen 0,5 und 1,0 mmol · l^{-1} Cu ermittelt. Dieser Wert stimmt gut mit der Konzentration überein, die in Experimenten zur Denitrifikationsleistung über mineralischem Sediment die Nitratreduktion beeinflußte. Für Cadmium liegt der Grenzwert für das Wachstum von Bakterienkolonien zwischen 0,05 und 0,1 mmol · l^{-1} Cd. Dieser Wert stimmt nicht mit den in Versuchen zur Denitrifikationsleistung gefundenen Ergebnissen überein. Eine mögliche Ursache können die verschiedenen Komplexbildungspartner für Cadmium sein, die sich auf synthetischem Nährmedium von denen unter Biotopbedingungen unterscheiden.

3.2. Zu den Untersuchungen zum Einfluß auf die Denitrifikationsleistung

a) Zum Einfluß von Kupfer
Eine verminderte Denitrifikationsleistung, wir wählten zur Darstellung die Nitratabnahme und die Nitratzehrungsrate, konnte bei gelösten Cu-Restkonzentrationen von $C_{Cu} \geqq 2{,}0$ mg · l^{-1} im Wasserkörper beobachtet werden. Diese traten nach Applikation von 0,01 mmol · l^{-1} und 0,02 mmol · l^{-1} Cu nicht auf, sondern waren erst nach Cu-Zugaben von $\geqq$ 0,2 mmol · l^{-1} über den geprüften mineralischen Sedimenten zu beobachten; vgl. dazu Abb. 3. Es ist eine Anfangsverzögerung im Nitratabbau zu verzeichnen. Während in den Kontrollansätzen die Abnahme von Nitrat (NO_3^-) vom 0. bis zum

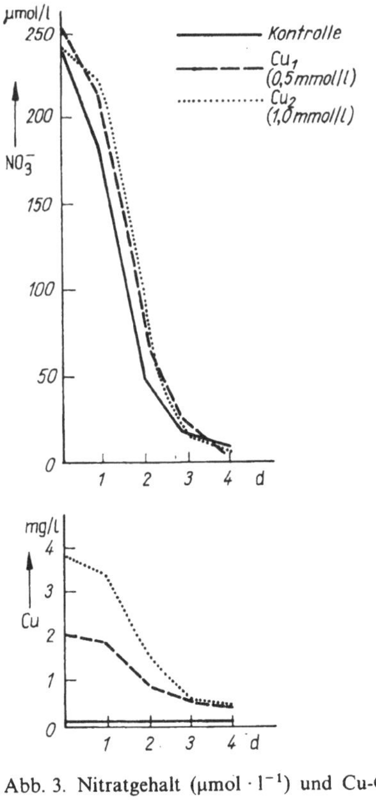

Abb. 3. Nitratgehalt (µmol · l^{-1}) und Cu-Gehalt (mg · l^{-1}) nach einmaliger Cu-Belastung

1. Tag 62 µmol · l^{-1} beträgt, ist bei Cu(1) (≙ Zugabe von 0,5 mmol · l^{-1} Cu) eine Abnahme von 48 µmol · l^{-1} NO_3^- und bei Cu(2) (≙ Zugabe von 1,0 mmol · l^{-1} Cu) eine Abnahme von 30 µmol · l^{-1} NO_3^- zu beobachten. Über den Gesamtzeitraum unterscheiden sich die berechneten Nitratzehrungsraten jedoch nicht:

Kontrolle	52,6 mg N · m^{-2} · d^{-1}
Cu(1)	58,5 mg N · m^{-2} · d^{-1}
Cu(2)	56,8 mg N · m^{-2} · d^{-1}.

Eine einmalige Cu-Belastung über mineralischem Sediment beeinflußt die Nitratzehrung nur in den ersten Versuchstagen, wenn die gelöste Restkonzentration noch ≧ 2,0 mg · l^{-1} Cu beträgt. Nach einer doppelten Cu-Belastung tritt die Beeinflussung der Nitratzehrung deutlicher hervor, geht aber mit erhöhten Cu-Restkonzentrationen von $C_{Cu} > 2{,}0$ mg · l^{-1} konform; vgl. Abb. 4. Bei der höheren Cu-Zugabe [Cu(2)] ist die hemmende Wirkung bis zum Versuchsende zu beobachten. Die Nitratzehrung ist gegenüber der Kontrollvariante um 30% vermindert. Über den Gesamtverlauf betrachtet betragen die Nitratzehrungsraten:

Kontrolle	28,5 mg N · m^{-2} · d^{-1}
Cu(1) (0,5 + 0,5)	29,0 mg N · m^{-2} · d^{-1}
Cu(2) (1,0 + 1,0)	19,0 mg N · m^{-2} · d^{-1}.

Die verfügbare Restkonzentration an Kupfer hat einen Einfluß auf den Nitratabbau. Die größten Anteile der zugefügten Mengen werden jedoch vom Sediment adsorbiert (90—99% der Zugabe, je nach Höhe und Sättigungsgrad des spezifischen Sediments für das ausgewählte Element; WIESE 1984).

b) Zum Einfluß von Cadmium

Bei den angewendeten Cd-Konzentrationen von 0,05, 0,5 und 1,0 mmol · l^{-1} Cd, die bis zu gelösten Restkonzentrationen von 61,0 mg · l^{-1} Cd führten, konnte die Nitratreduktion nicht beeinflußt werden. Eine Ursache kann in der elementspezifischen Wirkung von Cd auf die Bakterienpopulation gesehen werden und mit einer „Nichtaufnahme" durch die Denitrifikanten verbunden sein. Möglich ist auch eine Bindung an gelöste Sestonfraktionen, so daß der gelöst ermittelte Anteil biologisch nicht verfügbar ist.

4. Schlußfolgerungen aus den Ergebnissen

Die ökologische Wirkung der in das Gewässer gelangten Schwermetallverbindungen wird nicht durch deren Gesamtkonzentration, sondern vom biologisch wirksamen Anteil bestimmt. Wie sich dieser genau ermitteln läßt, ist nicht vollständig geklärt. Die gelöst vorliegende Restkonzentration (nach Filtration durch einen 0,4 µm Filter) kann nur einen Anhaltspunkt bieten. Auch die mit einer ionensensitiven Elektrode ermittelte aktive Restkonzentration muß mit Vorbehalt diskutiert werden. Der biologisch verfügbare Anteil ist von den physikalisch-chemischen Vorgängen im Biotop, wie z. B. Schwan-

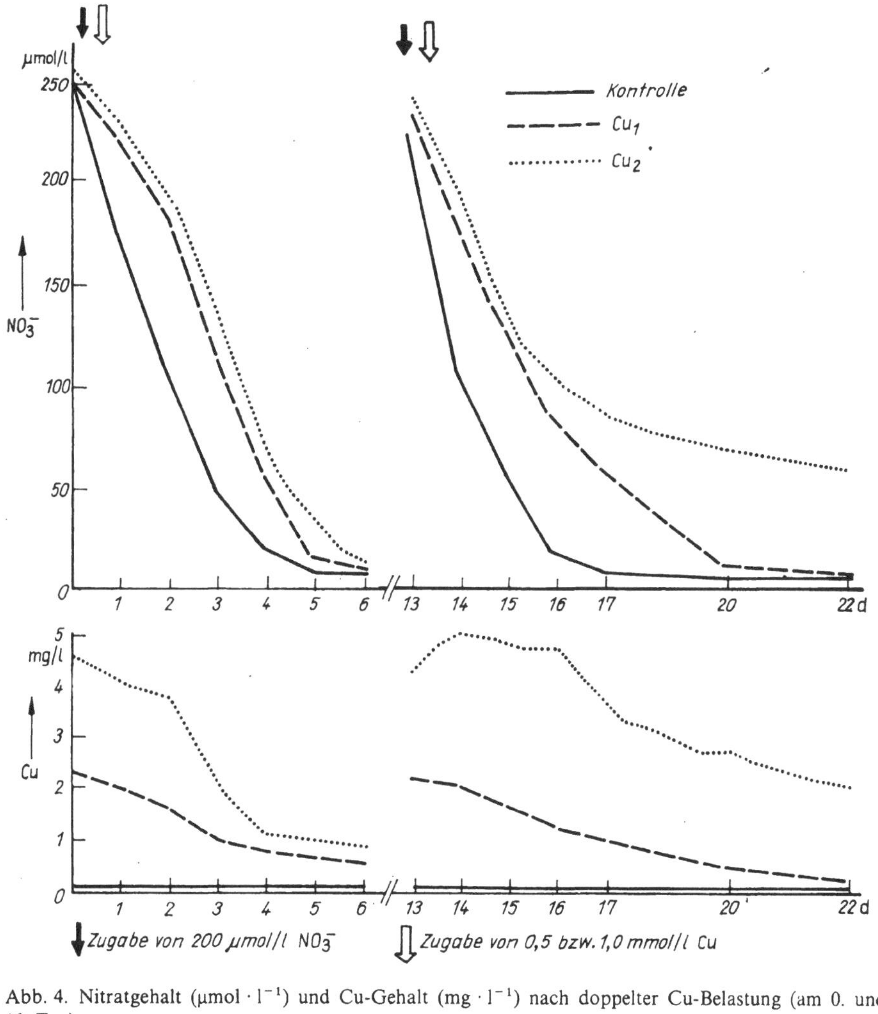

Abb. 4. Nitratgehalt (µmol · l^{-1}) und Cu-Gehalt (mg · l^{-1}) nach doppelter Cu-Belastung (am 0. und 13. Tag)

kungen von Temperatur, pH-Wert und Salinität, abhängig. In Flachgewässern spielen Sorptionsgleichgewichte zwischen den komplexbildenden Substanzen des Wasserkörpers bzw. des Sediments (erfaßbar z. B. als Sestongehalt bzw. organischer Gehalt) die Hauptrolle. Beeinflussende Faktoren wie Wind und die dadurch induzierte Wellenbewegung, die Sedimentaufwirbelungen hervorruft, sowie Strömungen, Turbulenzen und Bioturbation sind zu berücksichtigen und haben einen Einfluß auf die Sorptionsgleichgewichte, die sich zwischen Sediment und Schwermetall ausbilden. In jedem Fall sind elementspezifische Besonderheiten zu berücksichtigen.

5. Zusammenfassung

Stickstoffeliminationsprozesse, wie die Denitrifikation, können in den oft stoßartig belasteten Gewässern der Darss-Zingster-Boddenkette wesentlich zur Verbesserung der Gewässerbeschaffenheit beitragen. Das beweisen auch die in Laborexperimenten ermittelten Nitratzehrungsraten von bis zu 163.8 mg $\cdot$ N $\cdot$ m^{-2} $\cdot$ d^{-1}, wenn die entsprechenden Bedingungen (ausreichend Nitrat, leicht abbaubare organische Substanz, optimale Temperaturverhältnisse, 25—28 °C) vorhanden sind (Petzoldt 1986). Die im Sediment der Darss-Zingster-Boddenkette gegenwärtig vorhandenen Schwermetallkonzentrationen (vgl. Leipe 1986) üben keinen nachweisbaren Einfluß auf die Denitrifikationsleistung aus. Kupfer beeinflußt von gelösten Restkonzentrationen von $C_{Cu} \geqq 2{,}0$ mg $\cdot\, l^{-1}$ die Nitratzehrung, während Cadmium in den angewendeten Konzentrationen von 0,05, 0,5 und 1,0 mmol $\cdot\, l^{-1}$ über mineralischem Sediment keine sichtbaren Effekte auf die Nitratreduktion ausübt. Entscheidend für eine Wirkung auf den Prozeß der Denitrifikation ist der biologisch wirksame Schwermetallanteil. Dieser wird stark von der Sedimentqualität und den physikalischen Faktoren, wie Wind, Strömung und Turbulenz sowie der Bioturbation beeinflußt. Der Gehalt an komplexbildenden Substanzen, der im Gewässer einem Jahresgang unterworfen ist, muß ebenfalls berücksichtigt werden, wenn das Sorptionsgleichgewicht Sediment — Schwermetall diskutiert wird.

Literatur

Brügmann, L. (1979): Spurenmetalle in marinen Proben — Formen ihres Auftretens und ihre analytische Bestimmung. Diss. B, AdW der DDR, S. 284.

Kaspar, H. F. (1983): Denitrification, nitrate reduction to ammonia and inorganic nitrogen pools in intertidal sediments. Mar. Biol. **74**, 133—139.

Klinkenberg, G. (1987): Untersuchungen zum Einfluß von Schwermetallen auf die Denitrifikationsleistung von Gewässern. Diss. A, Fak. Biol. Chem. Agrarwissenschaften, Wilhelm-Pieck-Univ. Rostock.

Leipe, T. (1986): Beiträge zur Geochemie und Geoökologie rezenter Sedimente der Boddengewässer im Nordosten der DDR. Diss. A, Ernst-Moritz-Arndt-Univ. Greifswald.

Petzoldt, H. (1986): Stickstofftransformationen an der Sediment-Wasser-Kontaktzone eutropher Flachgewässer unter besonderer Berücksichtigung der Denitrifikation. Diss. A, Fak. Biol. Chem. Agrarwissenschaften, Wilhelm-Pieck-Univ. Rostock.

Wiese, G. (1984): Untersuchungen zur Nitratzehrungs/Denitrifikationsleistung eutropher Brackwasser-Sediment-Systeme unter dem Einfluß potentieller Störung durch Schwermetallkontamination. Dipl.-Arb., Wilhelm-Pieck-Univ. Rostock.

Anschrift der Verfasser:
Dr. G. Klinkenberg
Doz. Dr. G. Schlungbaum
Wilhelm-Pieck-Universität Rostock
Sektion Biologie
WB Experimentelle Ökologie
Freiligrathstr. 7/8
Rostock
DDR-2500

Prof. Dr. L. Brügmann
AdW der DDR
Institut für Meereskunde
Seestr. 15
Rostock-Warnemünde
DDR-2530

Limnologica (Berlin) **20** (1), 153—156, April 1990

Research Institute for Nature Management, Leersum, The Netherlands

Anthropogenic Influences and Management of Estuaries

By W. J. Wolff

Abstract

Estuaries have been considered to be fragile ecosystems, but nowadays all information present points to the conclusion that they are robust systems. Many stresses can be absorbed and after disturbance recovery occurs rapidly.

A number of anthropogenic influences are discussed, viz. exploitation of living resources, pollution, and embankment and reclamation. Their effects on estuarine ecosystems are described and examples are given from the Netherlands Wadden Sea and Delta area, and several other estuaries all over the world.

The management of estuaries should be based on the robust character of estuarine ecosystems and on the conclusion that these systems are self-maintaining. Hence, only the anthropogenic influences have to be managed. Management systems based on these conclusions are now in operation for the German 'Wattenmeer' and the Dutch Wadden Sea. These systems are described and their effectiveness is evaluated.

Are estuaries vulnerable ecosystem?

It has been stated in the past that estuaries are fragile and vulnerable ecosystems (e.g. Odum 1971). However, several authors (e.g. Copeland 1970; Boesch 1972, 1974; Jernelöv & Rosenberg 1976) have shown that this is not the case. Estuaries are able to absorb considerable stresses without difficulty and if perturbed, they rapidly return to their original state. Some examples will be given below.

Effects of stresses on estuarine ecosystems

One of the most important factors in the ecology of estuaries is the movement of water. Hummel et al. (1986a, 1986b) in a series of experiments studied the effects of changes in the pattern of tidal movement on estuarine tidal flat fauna. Under normal conditions this fauna experiences emergence and submergence twice a day. Hummel (op. cit.) changed this situation in three different ways, viz.

— permanent submergence;
— change of the frequency of the tidal rhythm;
— permanent emergence.

All changes were studied both for single animals in the laboratory and for the entire tidal-flat community in large-scale outdoor experiments. His results may be summarized as follows.

Permanent submergence has no immediate effect on the tidal flat ecosystem. Only after long periods (> 100 tides) changes become apparent, apparently because ecological equilibria have been changed.

Change of the frequency of the tidal rhythm has no noticeable impact, at least not for the particular changes studied.

Permanent emergence has an effect, which is only natural for a fauna consisting nearly entirely of aquatic organisms. What is noteworthy, however, is the tolerance of many members of this fauna for quite long periods of exposure. A few species die after two or three tidal periods of exposure, but many easily withstand 5—10 periods. Air temperature, however, is an important factor modifying the response.

Another example of a stress which can be absorbed by an estuarine ecosystem is organic pollution. Essink (1978, 1984) describes how the discharge of 40—55 t BOD · day^{-1} of organic waste into the Dutch Wadden Sea in the period September—December caused significant effects only in an area of a few hectares around the outfall. Moreover, after the period in which the waste was discharged the original situation was largely restored after eight months.

Many more examples how estuaries are able to absorb relatively large stresses can be mentioned, but is believed that these two instances are sufficient demonstration. However, it should be very clear that the capacity of estuaries to cope with stresses is not unlimited. Several examples are at hand (e.g. Rosenberg 1972; Pearson 1975) to show that too large stresses are able to destroy the estuarine ecosystem.

Reactions of estuaries and estuarine ecosystems to perturbations

The reaction of estuarine ecosystems and even entire estuaries to major perturbations has been documented many times.

Van den Heiligenberg (1987) and McLusky et al. (1983) studied the reaction of estuarine tidal-flat faunas to lugworm (*Arenicola marina*) digging. This activity results in a large-scale destruction of the benthic fauna. Recovery, however, is rapid and after about half a year most of the species had recovered because of immigration and settlement of larvae.

Similar observations were made by De Vlas (1982) who studied the cockle (*Cerastoderma edule*) fishery. This fishery results in the disappearance of the benthic fauna in the top 5 cm of the sediment. Also here recovery is rapid and after about one year most of the damage had been restored.

Large-scale examples are to be derived from the effects of very cold winters. Many authors (e.g. Kristensen 1957; Wolff 1973a) showed that after massive mortality of the benthic fauna of estuaries, large-scale settlement occurred in the next summer so that after 2—3 years the original situation was more or less restored.

Another example of recovery after a climatic catastrophe is given by BOESCH et al. (1976) who studied the recovery of the benthic fauna after the passage of a hurricane.

McNULTY & (1970) described how an estuarine fauna recovered quickly after pollution abatement in the Biscayne Bay, Florida, USA. A similar result was obtained by DEAN & HASKIN (1964) for the Raritan River Estuary, USA.

WOLFF (1973b, 1974) described how the Krammer-Volkerak estuary in the Netherlands was colonized by estuarine organisms after salinity had markedly increased due to civil engineering works.

The examples given so far are sufficient to demonstrate that estuarine ecosystems readily recover after a major perturbation. The recovery time of plankton communities is in the order of magnitude of weeks to months, of benthic communities months to years, and it is only for the long-living components of system, such as mangrove trees, birds and marine mammals, that recovery times may reach tens of years. For example, the tern populations of the southern North Sea have not yet recovered completely from a catastrophic mortality in 1965 (ROOTH 1989).

Two points have to stressed. In the first place recovery occurs only after the perturbing factor has disappeared. Chronic pollution or continuous overexploitation of estuarine systems will result in degraded systems as long as such a factor is operating. Secondly, recovery is not always 100% complete. Sometimes, one or a few species may be lacking from the ecosystem after all other species have returned and ecosystem functioning has been restored completely. An example of this phenomenon are the oyster beds in the German Wattenmeer. Overexploitation led to disappearance of the beds in the 1920-ies and since then they have not returned, nor is it likely that they will do so in future.

Effects of anthropogenic influences

Exploitation of living resources

Living resources are normally considered to be renewable resources, implying that exploitation without lasting damage is possible. Several forms of exploitation of living resources in estuaries will be shortly discussed below.

Seagrass (*Zostera marina*) **exploitation in the Netherlands.** Until the wasting disease in the 1930-ies seagrass vegetation in the western Wadden Sea, The Netherlands, was harvested for use in matresses, furniture etc. About 15,000 ha of seagrass vegetation was harvested annually, resulting in an annual yield of about 300—700 t of dry seagrass (pers. comm. Dr. N. DANKERS). This appears to have been a sustainable exploitation.

Saltmarsh grazing. Salt marshes along all European coasts are being grazed by livestock. Some marshes are grazed very intensively, e.g. in Nordfriesland, Federal Republic of Germany, and Morecambe Bay, England, but others are used for only limited number of cattle or sheep. Overexploitation is certainly possible, but normally is prevented by the owners of the marsh. Different grazing pressures result in very different types of vegetation.

Shellfish fishing and cultivation. Shellfish are harvested from estuaries all over the world. The US annual yield of soft-shelled clams (*Mya arenaria*) is about 35 million kg (ST. AMANT 1973). The Dutch cockle (*Ceratoderma edule*) fisheries land annually some 7 million kg (DE VLAS 1982). However, free fisheries, in which nobody feels responsible for the harvest of next year, tend to result in overexploitation, especially of the longer living species such as oysters. This explains the success of shellfish cultivation, since this is based on the principle that a fisherman gets the exclusive right of the use of a part of the seabed. If he invests in the harvest of next year, he is sure that he will be the only one who reaps the benefit. Thus the production of oysters (*Ostera edulis*) in the Oosterschelde estuary, The Netherlands, increased from about 0.5 million oysters per year obtained in a free fishery to 30 million per year after introduction of a culture (KORRINGA 1976). Nowadays, shell fish cultures are found all over the world. Major species are the Japanese oyster (*Crassostera gigas*), blue mussel (*Mytilus edulis*) and hard-shelled clam or quahog (*Mercenaria mercenaria*). The blue mussel production of Spain is nowadays about 120,000 t per year, whereas the Netherlands produce some 70—100,000 t per year. It is interesting to compare the shellfish production of the Dutch Wadden Sea, which is about 2 g dry organic matter per m^2 per year to the fishery yield of the adjacent North Sea which is about 1 g (dry organic matter) per m^2 per year.

The effects of these fisheries and cultures are relatively small. Damage done by the fisheries usually disappears within a few months up to maximally a few years. Cultivation has its natural upper limits, since too large populations of shellfish result in poor growth.

Other fisheries. Fisheries aiming for other species than shellfish are widespread in estuaries and of a very diverse nature. Many different species of fish and prawns are caught, and normally this exploitation affects only the species fished for. Overexploitation is widespread, but easily to remedy.

Pollution

It is convenient to distinguish various forms of pollutions, each of which has its own characteristic effects, viz.

— toxic micropollutants;
— organic waste;
— eutrophication;
— oil pollution;
— thermal pollution;
— other forms of pollution.

Toxic pollutants. The best known toxic pollutants are heavy metals and chlorinated hydrocarbons. Usually they are more or less specific in their action. Not the ecosystem as a whole is affected, but particular species, often the final links of food chains. Examples are the decline of brown pelicans (*Pelecanus occidentalis*) in the USA due to DDT and other pesticides, massive mortality of coastal birds in the Netherlands due to poisoning by the pesticides telodrin and dieldrin (KOEMAN 1971), the decline of harbour seals (*Phoca vitulina*) in the Netherlands due to PCBs (REIJNDERS 1980, 1986) and the mortality among Japanese fishermen and their families due to mercury poisoning ("Minamata disease") in 1960.

Organic waste. Large quantities of organic waste may exceed the capacity of an estuary to absorb this stress and the final result will be anaerobic conditions and disappearance of the original estuarine ecosystem. The pollution and the subsequent recovery of the Thames estuary is a well-known example. Another example is the present condition of Scheldt estuary in Belgium. About 20 km of this estuary are anoxic and lifeless.

Eutrophication. Extensive use of fertilizers, primary and secondary treatment of organic waste, atmospheric deposition and organic pollution all result in the addition of phosphorus and nitrogen compounds to the estuarine environment. This usually results in excessive blooms of phytoplankton and benthic algae. The mineralization of such blooms may lead to oxygen shortages and kills of fish and other estuarine organisms.

Oil pollution. Based on field observations all over the world GUNDLACH & HAYES (1978) proposed a scale to judge the damage and time needed for recovery of coastal habitats after oil pollution. On their scale estuarine habitats appear to be the ones which are more affected by oil pollution than any other habitat. The major reasons for this are that estuaries have no physical mechanisms to remove the oil once it has entered, and the fact that oil which has entered the sediment is not easily degraded microbiologically because of lack of oxygen. For these reasons oil can have far reaching and long lasting effects on estuarine ecosystems.

Thermal pollution. The use of estuarine waters for cooling purposes is widespread and has two major effects: increase of the temperature of the estuary and removal of part of the populations of planktonic and fish populations. Additionally, effects may occur because of measures to reduce the fouling of the water circuits.

Other pollution. Any matter discharged in amounts larger than natural and having a negative effect on living organisms can be considered as pollutant. Examples of this are potassium in the Weser estuary, plastic waste along all European shores and the enormous quantities of mud which reached San Francisco Bay in the period of hydraulic gold mining (1853—1884) (HEDGPETH 1979).

Embankment and reclamation

Through embankment and reclamation (parts of) estuarine ecosystems are destroyed and changed into completely different, often agricultural systems. In doing so we loose a very productive natural system. The magnitude of this loss has been discussed by several authors (e.g. DANKERS 1982; WOLFF & ZIJLSTRA 1982) and therefore will not be discussed here in detail. One example will be given, however.

The closure of the Zuiderzee estuary, The Netherlands, in 1932 resulted in the virtual extinction of a local race of herring (*Clupea harengus*) and the loss of a fishery which yielded about 3 million tons of herring per year (HAVINGA 1954).

The management of estuaries

From the previous paragraphs it has become clear that we have a wealth of information on the effect of human interventions on estuarine ecosystems. One conclusion to be drawn from these data is that estuarine ecosystems are not very vulnerable and that sustained exploitation is possible without lasting damage to the ecosystem.

With this information at hand it is now possible to develop management systems for estuaries. Prerequisite for this is a good knowledge of the estuarine system. Information should be available on its hydrological and geomorphological processes, its populations of plants and animals and especially on the ecological processes in the estuary.

Next a management objective has to be defined. In doing so it is important to realize that an estuary cannot fullfil all possible functions at the same time. Discharge of wastewater cannot be combined with shellfish cultivation, for example. In some way it has to be determined which functions will have priority and which not (WOLFF & ZIJLSTRA 1980).

A further step is the definition of the boundaries of the system to be managed. An estuary is an hydrological, sedimentological and ecological entity and this has to be reflected in the boundaries chosen. Normally this should result in the entire estuary being included in the management system.

For this area a management plan has to be developed and a management organisation to be set up. Two examples will be discussed below, viz. the Wadden Sea in the Netherlands and the Nationalpark Niedersächsisches Wattenmeer in the Federal Republic of Germany.

The Wadden Sea, The Netherlands

The Dutch government has published a central planning decision on the Wadden Sea in 1981. In this policy document the boundaries of the Wadden Sea area are defined in such a way that the entire aquatic part of the area is included as well as the salt marshes along the coasts. Not included are the inhabited parts of the islands.

The principal management objectives for the Wadden Sea are:

1. The protection, the conservation and, if necessary, the restoration of the area as a nature reserve;
2. The safety of the inhabitants of the adjacent areas against storm surges;
3. The protection of the characteristic open landscape.

Next to these principal objectives a series of subordinate objectives has been formulated, implying that, as long as no damage is done to the principal objectives, many human activities are possible in the area. This relates to traffic (ferries!) and transport, harbours, extraction of natural gas, winning of sand, discharge of waste water, tourism, fisheries, shellfish cultures, and military training.

A complicated management structure has been set up, consisting of:

1. A coordinating committee of the various state departments;
2. A coordinating committee of state, provinces and communities;
3. An advisory committee representing the various groups of people with a interest in the area, e.g. fishermen, tourists, conservationists etc., and comprising some scientists.

These committees have developed in the first place a management strategy for the Wadden Sea area. This management strategy defines the principal lines. In the second place many management plans have been developed for different areas of human activities, such as fisheries, oil pollution, tourism, land reclamation etc.

The actual management is carried out by several government services such as the State Forestry Service (responsible for nature conservation), the Department of Public Works, the Police etc. in a closely coordinated effort.

An important characteristic of the Wadden Sea reserve is that many human activities are possible in the area. Only some small areas where breeding colonies of birds and seals exist, are strictly protected. In all other areas human activities are possible, in as far they do not interfere with the principal management aims. Of course, some management problems still have not been solved, but the main lines are not subject to discussion.

The Nationalpark Niedersächsisches Wattenmeer

For the Nationalpark Niedersächsisches Wattenmeer only one management objective has been defined, viz. the protection of nature and landscape.

The national park is divided in three zones, viz. a strictly protected zone, a transition zone and a touristic zone. The strict protection in the first zone does not apply to agricultural activities (salt marshes), some forms of hunting, fisheries, various forms of recreational activities, and several forms of management activities.

For the management of the national park a mangement board exists. An advisory committee, in which local authorities, private interests and scientists are represented, advices the management board.

An interesting difference with the Dutch system is that the German park attempts to exclude as many human activities as possible, whereas the Dutch reserve allows as many activities as possible.

References

BOESCH, D. F. (1972): Species diversity of marine macrobenthos in the Virginia area. Chesapeake Sci. **13**, 206—211.

— (1974): Diversity, stability and response to human disturbance. In: A. J. CAVÉ (ed.), Proc. 1st Int. Congr. Ecol., The Hague, The Netherlands, Pudoc, Wageningen, 109—114.

— DIEZ, R. J., & VIRNSTEIN, R. W. (1976): Effects of tropical storm agues on soft-bottom macrobenthic communities of the Fames and York estuaries and the Lower Chesapeake Bay. Sci. **17**, 246—259.

COPELAND, B. J. (1970): Estuarine classification and response to disturbances. Trans. Am. Fish. Soc. **99**, 826—835.

DANKERS, N. (1982): Quantifying the loss of functions of a natural area as a means of impact assessment for reclamation projects. In: ILRI, Papers Int. Symp. Polders of the World, Lelystad, The Netherlands, Vol. **2**, 639—652.

DEAN, D., & HASKIN, H. H. (1964): Benthic repopulation of the Rariton river estuary following pollution abatment. Limnol. Oceanogr. **9**, 551—563.

ESSINK, K. (1978): The effects of pollution by organic waste on macrofauna in the eastern Dutch Wadden Sea. Neth. Inst. Sea Res., Publ. Ser. 1—1978, 135 p.

— (1984): The discharge of organic waste into the Wadden Sea: local effects. Neth. Inst. Sea Res., Publ. Ser. 10—1984, 164—178.

GUNDLACH, E. R., & HAYES, M. O. (1978): Vulnerability of coastal environments to oil spill impacts. Mar. Techn. Soc. J. **12**, 18—27.

HAVINGA, B. (1954): Vissen. In: L. F. DE BEAUFORT (ed.), Veranderingen in de flora en fauna van de Zuiderzee (thans) IJsselmeer na de afsluiting in 1932. Ned. Dierk. Ver., Den Helder, 253—267.

HEDGPETH, J. W. (1979): San Francisco Bay — the unsuspected estuary. In: T. J. CONOMOS (ed.), San Francisco Bay, the urbanized estuary. Pacific Div. Am. Ass. Advancement Sci., San Francisco, 9—30.

HEILIGENBERG, T. VAN DEN (1987): Effects of mechanical and manual harvesting of lugworms *Arenicola marina* L. on the benthic fauna of tidal flats in the Dutch Wadden Sea. Biol. Conservation **39**, 165—177.

HUMMEL, H., BRUMMELHUIS, E. B. M., & DE WOLF, L. (1986a): Effects on the benthic fauna of embanking an intertidal flat area (the Markiezaat, Eastern Scheldt estuary, The Netherlands). Neth. J. Sea Res. **20**, 397—406.

— MEIJBOOM, A., & DE WOLF, L. (1986b): The effects of extended periods of drainage and submersion on condition and mortality of benthic animals. J. Exp. Mar. Biol. Ecol. **103**, 251—266.

JERNELÖV, A., & ROSENBERG, R. (1976): Stress tolerance of ecosystems. Environ. Conserv. **3**, 43—46.

KOEMAN, J. H. (1971): Het voorkomen en de toxicologische betekenis van enkele chloorkoolwaterstoffen aan de nederlandse kust in de periode van 1965 tot 1970. Thesis, University of Utrecht, 136 p.

KORRINGA, P. (1976): Farming marine organisms low in the food chain (Developments in aquaculture and fisheries science). Amsterdam.

KRISTENSEN, T. (1957): Differences in density and growth in a cockle population in the Dutch Wadden Sea. Arch. néerl. Zoo. **12**, 351—453.

MCLUSKY, D. S., ANDERSON, F. E., & WOLFE-MURPHY, S. (1983): Distribution and population recovery of *Arenicola marina* and other benthic fauna after bart digging. Mar. Ecol. Prog. Ser. **11**, 173—179.

MCNULTY, J. K. (1970): Effects of abatment of dramatic sewage pollution on the benthos, volumes of zooplankton, and the fouling organisms of Biscayne Bay, Florida. University of Miami Press, Miami, 107 p.

ODUM, E. P. (1971): Fundamentals of ecology. Philadelphia.

PEARSON, T. H. (1975): The benthic ecology of Loch Linnke and Loch Eil, a sea loch system on the west coast of Scotland. IV. Changes in the benthic fauna attributable to organic enrichment. J. exp. Mar. Biol. Ecol. **20**, 1—41.

REIJNDERS, P. J. H. (1980): On the causes of the decrease in the harbour seal (*Phoca vitulina*) population in the Dutch Wadden Sea. Thesis, Agric. University Wageningen, Dept. Nature Conservation.

— (1986): Reproductive failure in common seals feeding on fish polluted coastal waters. Nature **324**, 456—457.

ROOTH, J. (1989): De Nederlandse broedpopulatie van de grote stern *Sterna sandvicensis* in 1961—1988. Limosa **62**, 121—124.

ROSENBERG, R. (1972): Benthic faunal recovery in a Swedish Fjord following the closure of a sulphite pulp mill. Oikos **23**, 92—108.

ST. AMANT, L. S. (1973): Shellfish and crustacean productivity in marshes and estuaries. In: H. CHABRECK (ed.), Proc. Coastal Marsh and Estuary Management Symp., Louisiana State Univ., Baton Rouge, 151—161.

VLAS, J. DE (1982): De effecten van de kokkelvisserij op de bodemfauna van Waddenzee en Oosterschelde. Rijkinstitut voor Natuurbeheer, Texel. RIN-rapport **82/19**, 99 p.

WOLFF, W. J. (1973a): The estuary as habitat. Zool. Verh. Leiden **126**, 242 p.

— (1973b): Changes in intertidal benthos communities after an increase in salinity. Thalassia Jugoslavica **7**: 429—434.

— (1974): Benthic diversity in the Rhine Meuse estuary. Hydrobiol. Bull. Amsterdam **8**: 242—252.

— & ZIJLSTRA, J. J. (1980): Management of the Wadden Sea. Helgol. Meeresunters. **33**: 596—613.

— & — (1982): Effects of reclamation of tidal flats and marshes in the Netherlands on fishes and fisheries. In: ILRI: Papers International Symposium Polders of the World, Lelystad, The Netherlands. ILRI, Wageningen, vol. 3: 57—67.

Address of the author:

Prof. Dr. W. J. WOLFF
Research Institute for Nature Management
P.O. Box 46
3956 ZR Leersum
The Netherlands

Limnologica (Berlin) **20** (1), 157—163, April 1990

Wilhelm-Pieck-Universität Rostock, Sektion Biologie, Rostock, DDR; Sektion Technische und Biomedizinische Kybernetik der Technischen Hochschule Ilmenau, Ilmenau, DDR; Bereich Forschung der Wasserwirtschaftsdirektion Magdeburg, Magdeburg, DDR

Ecosystem Management in Estuaries: Costs and Benefits

By U. VIETINGHOFF, H. PUTA, H. KLAPPER, M. STENDER, M. L. HUBERT & B. KUNTZE

With 13 Figures and one Table

Key words: Ecosystem management; Mathematical model; Scenario analysis; Optimization technique; Costs; Benefits; Estuar; Baltic Sea.

Abstract

The improvement of the water quality of eutrophied water bodies generally causes costs of millions of marks. Therefore, the choice and timing of an adaequate restoration method is a difficult task.

A short list of restoration methods for eutrophied estuarine waters is presented here; we differentiate between processes that cause costs, and processes that lead to benefits; all elementary processes of restoration are expressed in monetary terms.

We describe two computer-based methods to assist in the difficult decision process. The ecological processes are quantitatively described by mathematical models. In order to find the restoration method that shows best ecological results connected with a minimum of costs we have applied scenario analyses and optimization techniques (variable metric method).

1. Introduction

All countries of the world with highly developed industries and agriculture are faced with the **eutrophication** of their water bodies and its disadvantageous effects (VOLLENWEIDER 1979; WELCH 1980).

Therefore, for more than two decades a lot of effort has been put into the investigation of **rehabilitation techniques** and their introduction in practical water management (DUNST 1974; KLAPPER 1978 and others). In the shallow, eutrophied estuaries of the South Baltic Sea some of the well-established restoration techniques like aeration, selective discharge or nitrate dissimilation, belonging to deep waters, are not applicable. We have summarized in part 2 of this article the restoration techniques for eutrophied estuaries that are on the short list. We have checked costs and benefits of the different techniques (VIETINGHOFF 1982) and tried to express all values in monetary terms, at least as shadow prices or willingness to pay. Fortunately we have estimates of the costs (STERGER 1984) for a great number of processes that are more or less connected with water management in our Republic; an example is shown in part 3 of this article.

The choice and timing of appropriate actions for the restoration of the water quality leads to an optimization problem. In order to find an optimal solution, different ways are possible (WESTMAN 1985). We used two ways of solution. Firstly, with a complex mathematical model of the ecosystem of the Barther Bodden, a shallow estuarine water body of the South Baltic Sea (VIETINGHOFF 1982; WIEDEMANN 1985), we made scenario-analysis for the decision process (part 4). Secondly, we used a simplified ecosystem model and two numerical methods of dynamic optimization (LOOTSMA 1972; FÖLLINGER 1985; PUTA 1987; KUNTZE 1988, in prep.); first results of our team work are shown and the problems discussed in part 5.

2. Survey of rehabilitation techniques for estuaries — costs and benefits

The costs of restoration purposes are given very often in millions of marks. Therefore, with the choice of adaequate restoration techniques 3 points have to be considered:

- their effectiveness and comparably low negative effects on other components or processes,
- the possibility of their practical application, and
- their economical side: costs and benefits should be well balanced.

Table 1 shows a short list of restoration techniques for eutrophied estuarine waters. The peculiarity of this table lies in the distinction between steps for which one has to pay as well as steps that bring benefits.

In our country, we are in the fortunate position of having estimates of the monetary costs and benefits of the steps specified in the table; an example is given in the next part.

3. An example of the economical index figures

One of the powerful restoration techniques for estuarine (and shallow inland) waters is dredging; the nutrient-rich sediments are taken away. Of course, this causes costs for dredger performance and the transport of the sediments, the latter sometimes over longer distances. On the other hand, benefits result from improved water quality (decreased internal loading), improved shipping traffic, and (when toxic substances are absent) the provision of organic fertilizers for agriculture. The price list for dredging (compiled by SCHWEIGER in STERGER 1984) contains costs for dredging in reservoirs (5.70 M—22.60 M per m^3), large lakes (25 M—40 M per m^3), small lakes (3.50 M—10.00 M per m^3) and fish ponds (45 TM/ha) as well as estimates for benefits from sediments as fertilizers (300—500 m^3/ha, 30 to 50% more yield/ha, total benefits about 20—40 Marks per m^3 sediment). These estimates are a good basis to appraise the economical side of this restoration technique.

4. Searching for optimum management variants by scenario-analyses with a complex mathematical model

For the restoration of the chain of shallow water bodies "Boddenkette West" (VIETINGHOFF 1982; WIEDEMANN 1985) our water authority insists on control of nutrient import and dredging. By means of scenario-analyses with a complex mathematical model we checked the efficacy

Table 1. Restoration techniques for eutrophied estuaries

Management technique	Costs for	Benefits from
1. Control of nutrient import	restoration steps: * waste water treatm. plants in towns and villages * closure of water cycles in industry * closure of nutrient cycles in agriculture * changing land use practices * control of soil erosion	— increased water quality for industry and agriculture — rise in category of holiday camps — economy on fertilizer — recreation (sports fishery, swimming . .)
	economy: index figures for * starting and current costs for treatment plants * recycling of water and valuable materials in several different industries * treatment of liquid manure	— prices for industrial waters, depending on water class — list of prices for holiday-camps according to category — lists of prices for fertilizers — lists of prices for different spraying systems, timing, increase in yield depending on soil and fruit
2. Dredging	restoration steps: * dredging * loss of bottom fauna * building sediment traps	— use of sediments as organic fertilizer — increase in shipping traffic
	economy: index figures for * price lists of different dredging methods and transport distances	— sediments as organic fertilizers — micronutrients for soil — transport by ships cheaper than train/car
3. Financial measures against not planned discharge of waste water	restoration steps:	— payment for waste water discharge
	economy: index figures for	— list of prices for water discharge
4. Planned use of selfpurification capacity	economy: index figures for	3.50 M for 1 kg BSB_5
5. Shoreland protection around lakes and alongside rivers	restoration steps: * auxiliary ecosystems	— increase in water quality (s. point 1) — increase in recreation value
	economy: index figures for * price lists of young trees	— (s. point 1) — shadow prices (willingness to pay) for recreation value and socio-aesthetic value of a landscape
6. Biomanipulation	restoration steps: * rearing young fish	— increase in fish production — increase in water quality (s. p. 1)
	economy: index figures for * price list of young fish	— price list for commercial fish
7. Biotic harvesting	restoration steps: * harvesting of *Phragmites c.*	— supplementary food for cows — elimination of nutrients (s. p. 1)
	economy: index figures for * wages * ships and cars	— list of prices for agricultural products

of these techniques. Here the model results of nutrient import reduction are reported.

The model of the "Barther Bodden", part of the "Boddenkette West", consists of a system of 24 differential equations for the main components of the said ecosystem: diatoms, greens, bluegreens and their content of nitrogen and phosphorus (9 equations), inorganic nitrogen and organic nitrogen of the pelagial and the upper layer of the sediment, as well as the energy content of both organic nitrogen compartments (6 equations), easily and with difficulty destroyable detritus of the upper layer of the sediment as well as pelagic detritus and the nitrogen content of these detritus fractions (6 equations), pelagic bacteria as well as bacteria and ciliates of the upper sediment layer (3 equations).

In order to check the role of nutrient import reduction we diminished in the model the import of detritus and nitrogen compounds by 5% each year for a time of 21 years. The results of the scenario analysis are shown in Figs. 1—9. As shown in Fig. 1 some components show small effects (e.g. hardly destroyable detritus of the upper layer of the sediment). We can see dramatic decreases in the figures of DIN, DON and bacteria of the pelagial (Figs. 2—4). Median effects are shown in Figs. 5—9 for pelagic detritus, benthic easily destroyable detritus, diatoms, greens and bluegreens. The conclusion we have drawn from these figures is that nutrient import

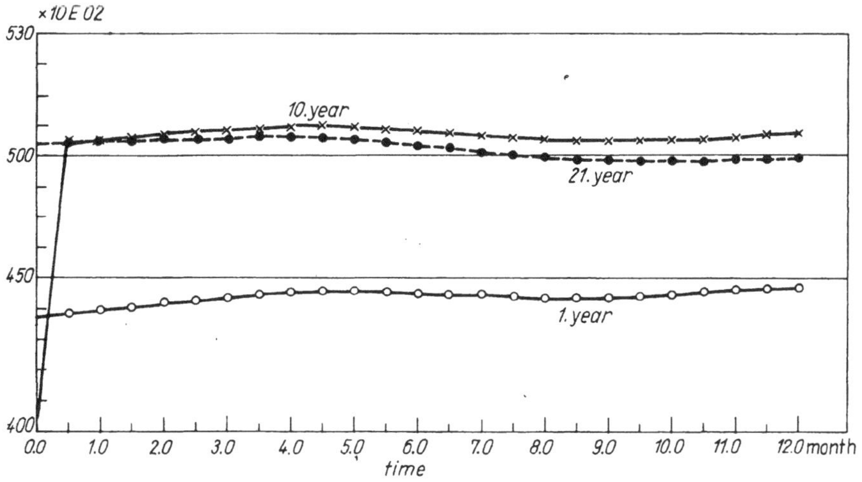

Fig. 1. Hardly destroyable benthic detritus [kJ · m^{-2}]

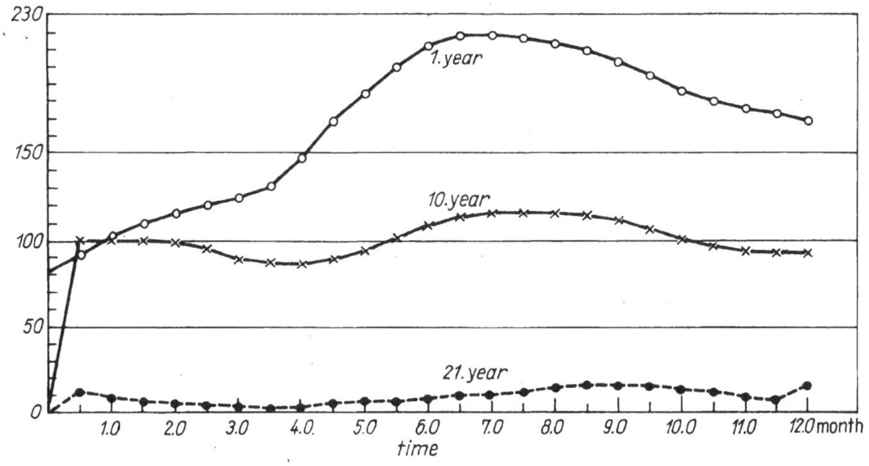

Fig. 2. Pelagic DON [mg-at · m^{-3}]

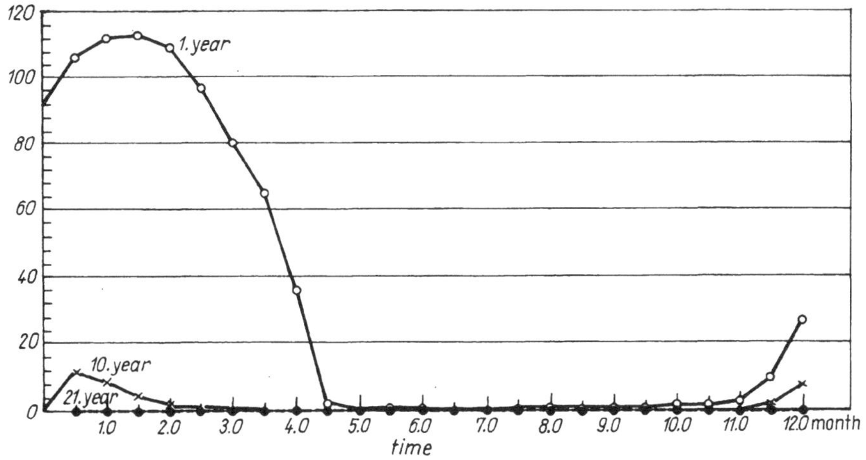

Fig. 3. Pelagic DIN [mg-at · m^{-3}]

Fig. 4. Biomass of pelagic bacteria [kJ · m^{-3}]

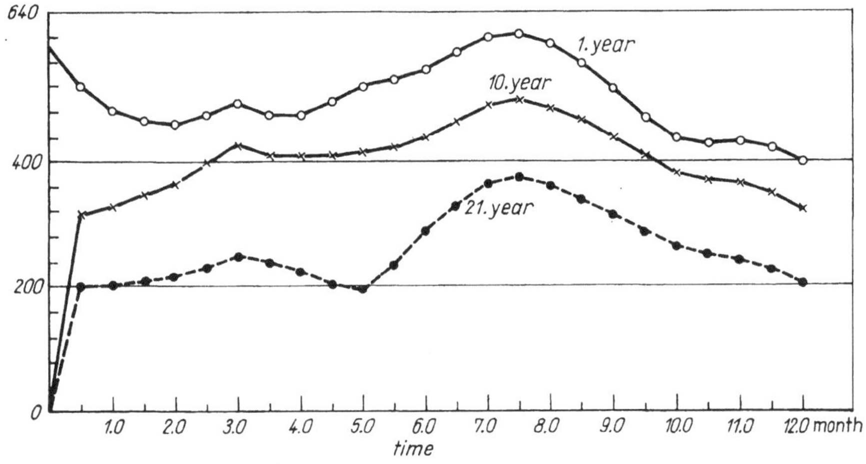

Fig. 5. Pelagic detritus [kJ · m^{-3}]

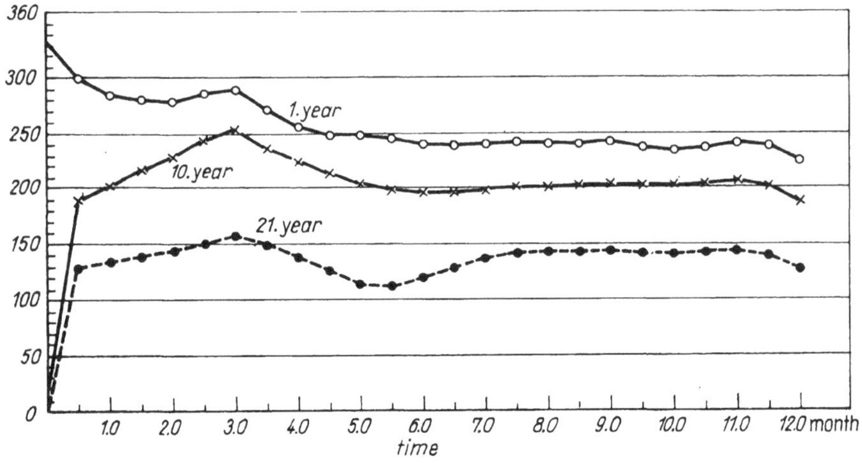

Fig. 6. Benthic easily destroyable detritus [kJ · m^{-2}]

reduction slowly restores the water quality; it should be supported by other methods, e.g. dredging. Without doubt, such scenario analyses can give good arguments in the decision process.

5. Computation of optimal mangement variants by numerical optimization techniques

The **optimization techniques,** the economical background and the mathematical model must be explained briefly. Of the optimization techniques reported in literature (LOOTSMA 1972; SINGH & TITLI 1978; FÖLLINGER 1985) we used the method of hierarchical optimization (HASSAN & SINGH 1981) and especially the variable metric method (gradient method, LOOTSMA 1972); details of these algorithms are given by PUTA (1987).

The **economical background** is depicted in Fig. 10: on the right is shown the **curve of costs** (in millions of mark) **for damages** during eutrophication (less recreation, less fishery); the curve starts on the transition from class 2 to class 3 and amounts to about 11.2 millions of mark

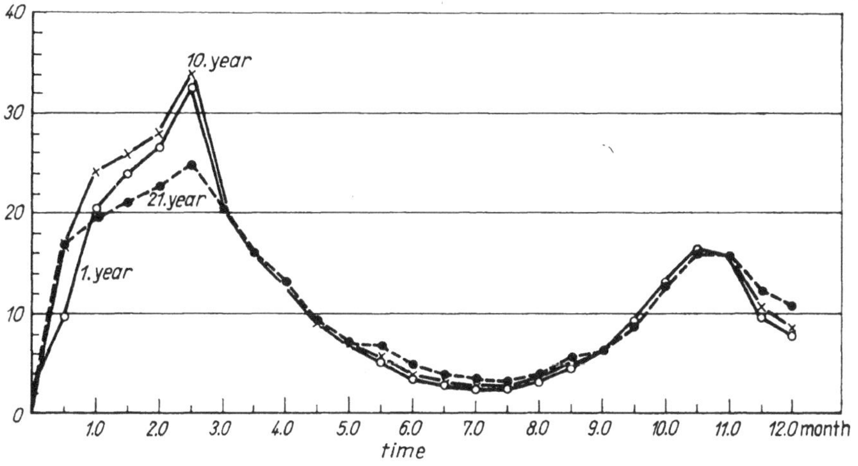

Fig. 7. Biomass of diatoms [kJ · m^{-3}]

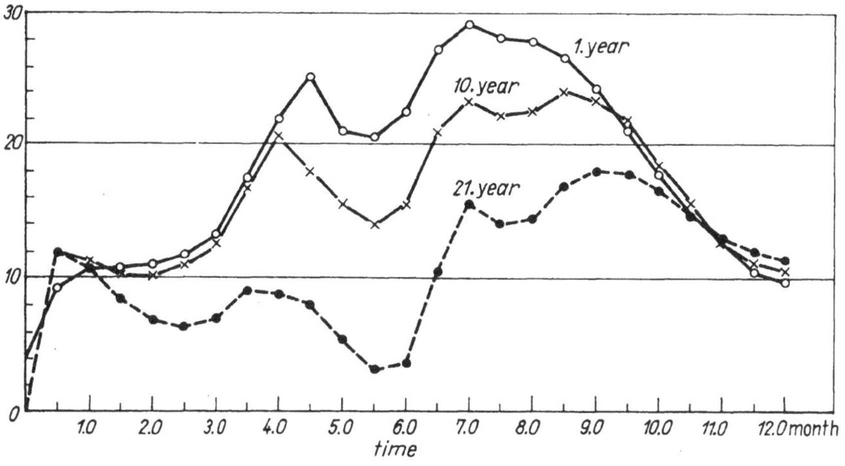

Fig. 8. Biomass of green algae [kJ · m^{-3}]

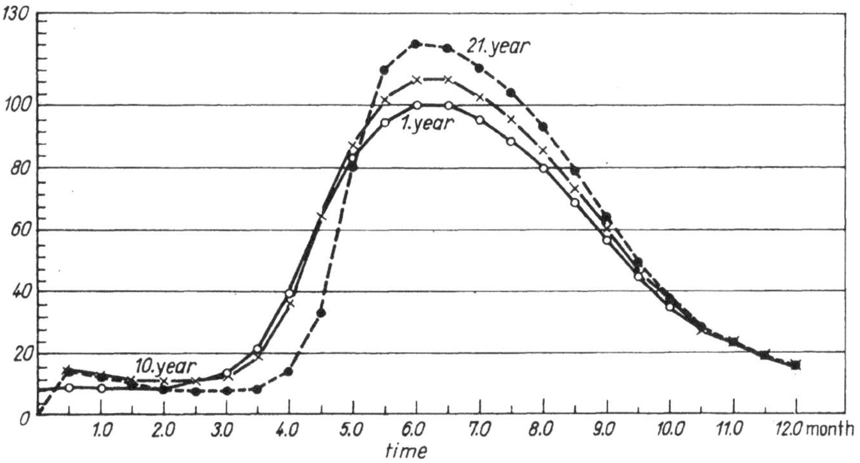

Fig. 9. Biomass of bluegreens [kJ · m^{-3}]

per year on the border of class 3 to class 4. These two points serve to construct a curve of the form $y = k \cdot x^2$ (y = monetary costs, x = parameter of water quality), in this case the curve of costs of damages. If the water is eutrophied and restoration carried out, the costs will raise according to the **curve of costs for restoration** (on the left); this curve starts on the border between class 5 and 4 (no restoration, no costs). For the first step of restoration the water authority plans to invest 18 million marks; experts assume that this investment will lead to a water quality on the border of class 4 to class 3.

Again, these two marked points serve to construct the curve of costs for restoration (on the left, Fig. 10). Of course, in practice, the costs build n-dimensional mountains, and we have to find out the deepest valley within these mountains of costs. This target function (minimum of total costs) is computed iteratively with the variable metric optimization method.

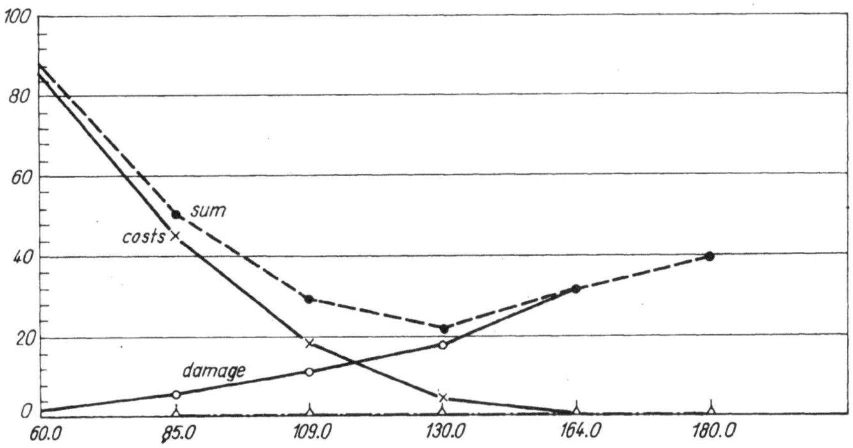

Fig. 10. Curve of costs for damage, curve of costs for restoration and curve of total costs (for further explanation see text). Abszissa: the 5 classes of water quality (parameter is biomass of bluegreens)

To get over the initial difficulties with introduction of the big optimization programmes we have used up to now only a very reduced mathematical model. This "minimodel" (KUNTZE 1989, in prep.) consists of 6 ordinary differential equations for the following 6 components of the aquatic ecosystem:
inorganic nitrogen,
organic nitrogen,
biomass of bluegreens,
nitrogen content of bluegreens
detritus, and
nitrogen content of detritus.

A test result, done with the minimodel and the variable metric method is shown in Figs. 11—13. As a (more theoretical) restoration technique the biomass of bluegreens is decreased. Fig. 11 shows the course of the damages

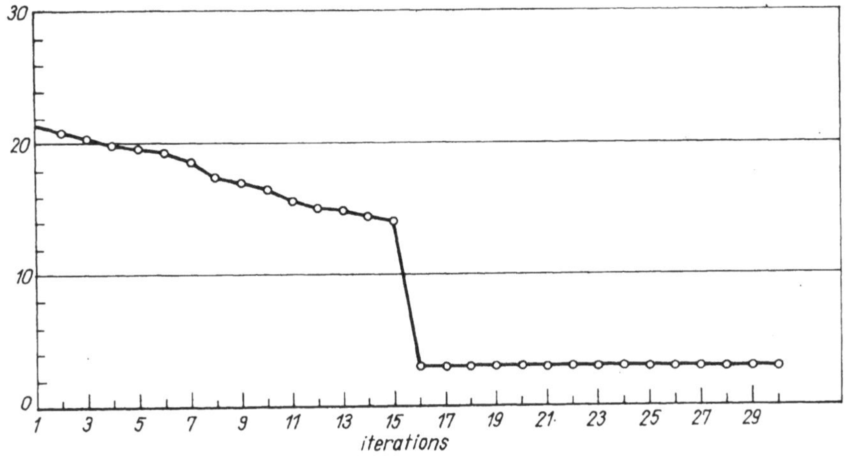

Fig. 11. Damage function [10^5 M · a^{-1}]

Fig. 12. Costs for restoration [10^5 M · a^{-1}]

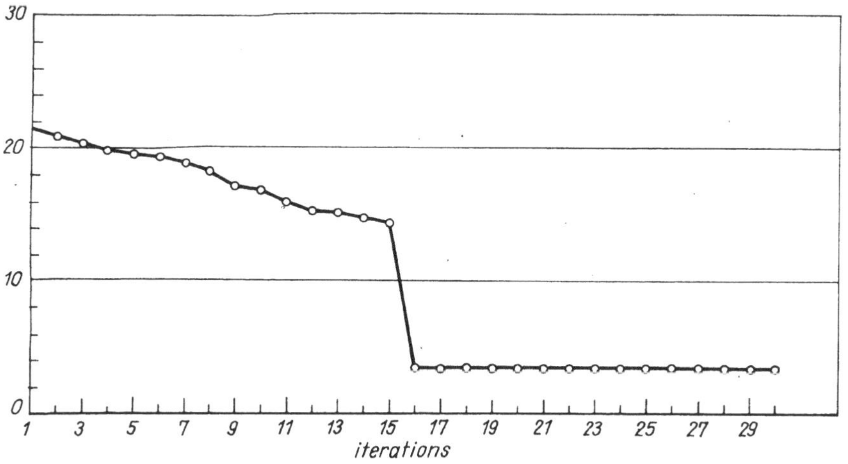

Fig. 13. Total sum of costs $[10^5 \text{ M} \cdot a^{-1}]$

function. One can see the favourable effects of the restoration; the costs of damages drop down to the optimum. Fig. 12 shows the development of the costs for restoration; starting from zero, the costs for the improvement of the water quality rise up to the 16th iteration, in which optimal (minimum total) costs are reached.

Fig. 13 shows the value of the target function, the total sum of costs. The costs drop down to the 16th iteration and are constant in the following iterations. The optimum of total costs in reached. The costs of restoration/are in this example about 1/10 of the costs for damates. To look for more favourable variants, we have to test other restoration techniques, or to change the weighting factors in the target function, or to introduce restrictions in the state variables.

6. Conclusions

The improvement of the water quality of eutrophied water bodies generally causes costs of millions of marks. The choice of an adaequate restoration method, its effectiveness, secondary effects, modes of practical application and cost/benefit relationships are of greatest importance, and the decision process a difficult task. In this paper we describe two computer based methods to assist this difficult decision process.

A short list of restoration techniques for eutrophied, estuarine waters is presented; we differentiate between processes that cause costs, and processes that lead to benefits. For a lot of such elementary processes associated with water management, the costs and benefits respectively, expressed in monetary terms, are known for the water management situation in the GDR (STERGER et al. 1984). In this way, all elementary processes of restoration are expressed in Marks of the GDR. The ecological processes are quantitatively described by mathematical models. The choice and timing of the restoration method that shows the best ecological results connected with a minimum of costs is carried out by scenario analyses or optimization techniques (variable metric method). Both methods are exemplified.

References

DUNST, R. C. (1974): Survey of lake rehabilitation techniques and experiences. Madison.

FÖLLINGER, O. (1985): Optimierung dynamischer Systeme. München.

HASSAN, M. F., & SINGH, M. G. (1981): Hierarchical successive approximation algorithms for non-linear systems. Large scale systems **2**: 65—95.

KLAPPER, H. (1978): Möglichkeiten und Effektivität der Sanierung von Talsperren und Seen. Promotion B, TU Dresden.

KUNTZE, B. (1989): Anwendung von Optimierungsverfahren auf das mathematische Modell des Barther Boddens zur Ermittlung optimaler Steuerstrategien. Diss., Universität Rostock.

LOOTSMA, F. (1972): Numerical methods for non-linear optimization. London.

PUTA, H. (1987): Lösung von dynamischen Steuerungsproblemen durch Ein- und Zweiebenen-Optimierungsmethoden und Anwendung auf Land- und Wasserwirtschaft. Promotion B, Ilmenau.

SINGH, M. G., & TITLI, A. (1978): Systems: Decomposition, optimisation and control. Oxford.

STERGER, O. (1984): Kennzahlen zur ökonomischen Bewertung wasserwirtschaftlicher Maßnahmen als Rationalisierungsmittel für die Arbeit der staatlichen Gewässeraufsicht. WWD Obere Elbe, Dresden.

VIETINGHOFF, U. (1982): Die mathematische Modellierung des Ökosystems Barther Bodden als Beitrag zur Ökosystemanalyse der Boddenkette West. Promotion B, WPU Rostock.

VOLLENWEIDER, R. A. (1979): Das Nährstoffbelastungskonzept als Grundlage für den externen Eingriff in den Eutrophierungsprozeß stehender Gewässer und Talsperren. Z. Wasser- u. Abwasserforschung **12** (3/4), 10—20.

WELCH, E. B. (1980): Ecological effects of waste water. Cambridge.

WESTMAN, W. E. (1985): Ecology, impact assessment, and environmental planning. New York.

WIEDEMANN, B. (1985): Simulation des dynamischen Ökosystemmodells Barther Bodden auf dem Digitalrechner unter besonderer Berücksichtigung des Stickstoffkreislaufs als Beitrag zur Ökosystemanalyse der Boddenkette West. Promotion A, WPU Rostock.

Addresses of the authors:

ULRICH VIETINGHOFF
MATTHIAS STENDER
MARIA-LUISE HUBERT
BERND KUNTZE
Sektion Biologie der Wilhelm-Pieck-Universität Rostock
Universitätsplatz 2
Rostock
DDR-2500

HORST PUTA
Sektion Technische und Biomedizinische Kybernetik der Technischen Hochschule Ilmenau
Am Ehrenberg
Ilmenau
DDR-6300

HELMUT KLAPPER
Bereich Forschung der Wasserwirtschaftsdirektion Magdeburg
Wallstr. 17
Magdeburg
DDR-3010

Limnologica (Berlin) **20** (1), 165—182, April 1990

Wasserwirtschaftsdirektion Küste, Stralsund, DDR

Untersuchungen über das Phytoplankton der Unterwarnow 1979/1980

Investigations on the Phytoplankton of the Unterwarnow in 1979/1980

Von INGEBORG SCHMIDT

Mit 3 Abbildungen und 8 Tabellen

Key words: Estuary; Phytoplanktonbiomass; Spatial and seasonal variations; Relations to salinity and nutritive load.

Abstract

During 1979 and 1980 190 taxa of phytoplanktonorganisms in the estuary Unterwarnow were stated, 100 of them are diatoms. The distribution of net-phytoplankton is closely related to salinity. In the bordersphere between limnetic and marine influences a species minimum — as characteristic for brackish waters — was found. Most of the dominating species with importance for primary production are nanoplanktonorganisms. In the observationperiod *Stephanodiscus hantzschii* was the predominating species of the upper part of the Unterwarnow, in the lower course also Flagellata s.l. temporary got importance. Spatial and seasonal variations of phytoplankton-biomass are determined by the populationdynamics of the dominating species, which are themselves influenced by factors as salinity, season, waterexchange and zooplanktondevelopment. Biomassmaxima occurred in the lower course later than in the upper course. Biomassmaxima were spatially related to the load sources. The trophic level in the estuary decreases in the direction of the current.

Inhalt

1. Einführung

Im Rahmen von Routineuntersuchungen der Wasserwirtschaftsdirektion Küste zur Kontrolle der Beschaffenheit der Unterwarnow wurden in den Jahren 1979 und 1980 zeitgleich mit der Probenahme für chemische Untersuchungen auch Planktonnetz- und Schöpfproben gezogen, deren Untersuchung erste Informationen über die Verteilung des Phytoplanktons im Verlauf des Ästuars liefern sollte. Der Darstellung der Ergebnisse sei eine kurze Charakteristik der Unterwarnow als Ökosystem vorangestellt. Sie basiert im wesentlichen auf Auswertungen von BACHOR (1988, 1989), dessen Arbeiten auch weitere Angaben zu den Untersuchungsstationen zu entnehmen sind (s. Abb. 1, S. 30 dieses Heftes).

Das Unterwarnow-Ästuar ist ein Gewässer mit starkem ökologischen Gefälle zwischen Warnowzufluß im Süden und Eintritt in die Ostsee im Norden. Auf Grund der hydrographischen Bedingungen nimmt der mittlere Salzgehalt von der Warnowmündung in Richtung Ostsee zu. Nach der Brackwassernomenklatur des Venedig-Systems (s. CASPERS 1959) ist die Unterwarnow in eine oligohaline und eine mesohaline Zone zu gliedern. Die Grenze zwischen beiden Zonen liegt im Mittel etwa im Bereich der Station 2, kann aber weit in nördliche oder auch in südliche Richtung verschoben sein. Für das gesamte Gewässer sind starke Salzgehaltsschwankungen charakteristisch. Ihre Amplitude bewegte sich im Zeitraum 1977—81 im Verlauf der Unterwarnow zwischen ca. 8 und 12,5 ‰. Das Maximum der Salzgehaltsschwankungen lag an der Station 3. Bei Einstrom von Ostseewasser kommt es zu starken vertikalen Salzgehaltsschichtungen. Entsprechend ihrer größeren Dichte schichten sich dabei Wasserkörper höheren Salzgehalts in Grundnähe ein. Die vertikalen Salzgehaltsdifferenzen nehmen im Verlauf der Unterwarnow ab. Für den Zeitraum 1977—81 wurden für die Stationen 1—3 mittlere vertikale Differenzen zwischen 4,4 ‰ (Stat. 3) und 5,7 ‰ (Stat. 2) berechnet, für die Stationen 4 und 5 Differenzen um 2,8 ‰. Im Breitling, einem von den Wasseraustauschprozessen zwischen Ostsee und Warnow weniger betroffenen Gewässerabschnitt im Mündungsbereich der Unterwarnow sind Salzgehaltsschwankungen und vertikale Salzgehaltsschichtung am wenigsten ausgeprägt.

Hauptbelastungsquellen der Unterwarnow sind Warnowzufluß oberhalb Stat. 1 und Kläranlagenablauf im Bereich der Stat. 3. Eingetragen werden sowohl Pflanzennährstoffe als auch den Sauerstoffhaushalt belastende organische Schwebstoffe. Der Lage der Belastungsquellen entsprechend, ist der Trophiegrad im oberen Teil der Unterwarnow (Stat. 1—3) hoch und nimmt dann, der Salzgehaltsänderung entgegengesetzt, in Richtung Ostsee ab. Das Ausmaß der Trophiegradänderung umfaßt bei vorsichtiger Interpretation der Ergebnisse von BACHOR (1989) mindestens eine Trophiestufe.

Wie die Charakteristik zeigt, ist die Unterwarnow ein Gewässer mit starker zeitlicher und räumlicher Variabilität der Faktoren, die auf die Zusammensetzung und Populationsdynamik des Phytoplanktons entscheidenden Einfluß haben, Ziel der Untersuchungen war daher neben der Erlangung von Informationen über Artbestand und Verteilung des Phytoplanktons in der Unterwarnow auch die Erfassung von Zusammenhängen zwischen Phytoplanktonparametern und diesen Einflußfaktoren.

2. Material und Methoden

Das Stationsnetz entspricht den Angaben von BACHOR (1989). Die Probenahme erfolgte durch Abfischen der Wassersäule (Vertikalhol) mit dem 25er Planktonnetz und durch Entnahme von Wasser-

Tab. 1. Phytoplankton der Unterwarnow 1979 und 1980. Gesamtartenliste mit Angaben zur Verbreitung und ökologischen Charakteristik (+ vorhanden; D Dominante)

	Jahr		Stationen												Halobität	Salztoleranz	Saprobie	Indikat.-gewicht	Sonstige Angaben
			1		2		3		4		5		6		(PANKOW et al. 1976)		(Ausgew. Methoden)		
	79	80	79	80	79	80	79	80	79	80	79	80	79	80					
CYANOPHYTA SCHUSNIG **1926:**																			
cf. *Achroonema*	+							+											
Anabaena cylindrica LEMM. 1896		+						+							oligohalob				
Anabaena sp.	+	+	+	+	+	+	+			+		+							
Aphanizomenon flos-aquae (L. 1753) RALFS ex BORNET et FLASH. 1888	+	+	+	+	+	+	+	+	+	+					oligo-	pleioeury- halin	β-meso- saprob	4	
Aphanothece sp.	+						+												
Chroococcus limneticus LEMM. 1898	+		+												oligo-	meso-	oligo-/ β-meso-	3	
Gomphosphaeria aponina KÜTZ. 1836	+		+												oligo-	meso-			
Gomphosphaeria pusilla (VAN GOOR) KOMAREK 1958	+	+				+	+				+				oligo-	meso-			
Lyngbya limnetica LEMM. 1898	+	+		+		+		+	+	+		+		+	oligo-	meso-			
Merismopedia sp.		+								+					oligo-				
Microcystis aeruginosa (KÜTZ. 1833) KÜTZ. 1845—49		+		+		+									oligo-	meio-/ meso-	β-meso-	3	
Microcystis reinboldii (RICHTER in REIN. 1889) FORTI in DE TONI 1907		+		+											oligo-	meio-/ meso-	β-meso-	3	
Microcystis viridis (A. BR.) LEMM.		+		+		+									oligo-				
Microcystis wesenbergii (KOMAREK 1958) STARMACH 1966		+		+		+		+							oligo-	meio-/ meso-			
Nodularia spumigena MERT. in JÜRGENS 1822 ex BORNET et FLASH. 1888	+	+	•						+			+		+	oligo-	pleio-			Kosmopolit
Oscillatoria lacustris (KLEBAHN 1895) GEITLER 1925, nicht zu trennen von *Oscillatoria agardhii* GAMONT 1892	+	D	+	+	+	+	+	+		+		+			oligo-	meio-			
Oscillatoria limosa C.A.AG. 1812 ex GAMONT 1892		+		+		+									oligo-	meso-	β-/α-meso-	2	Kosmopolit
Oscillatoria redekei VAN GOOR	+	+	+	+	+	+	+	+	+						oligo-		β-meso-/ oligo-	3	
CHRYSOPHYTA PASCHER **1914:**																			
Flagellata s. l.:																			
Dictyocha speculum EHRENBERG 1837		+										+		+	poly-	meso-			
Dinobryon divergens IMHOF 1890		+		+											oligo-	meio-	β-meso-	3	Kosmopolit
Dinobryon sertularia EHRENBERG 1838	+	+		+		+		+	+						oligo-	meio-	oligo-	4	
Ebria tripartita (SCHUMANN 1867) LEMM. 1900	+	D		+	+	+	+	+	+	+	+	+	+	+	poly-	meso-			heterotroph
Synura uvella EHRENBERG 1838		+		+		+									oligo-	meio-	β-meso-	3	
Diatomaphyceae:																			
Actinocyclus octonarius EHRENBERG 1838	+	D							+	+		+		+	poly-	meso-/ meio-			Kosmopolit
Actinoptychus senarius (EHRENBERG 1838) EHRENBERG 1843		+										+			poly-	meio-			
Cerataulina pelagica (CLEVE 1889) HENDEY 1937	+	+				+		+	+	+	+	+	+		poly-	meio-			Kosmopolit
Chaetoceros affinis LAUDER 1864	+	+			+						+	+			poly-	meio-			Kosmopolit
Chaetoceros compressus LAUDER 1894		+										+			poly-	meio-			Kosmopolit
Chaetoceros curvisetus CLEVE 1889		D						+		+		+		+	poly-	meio-			Kosmopolit
Chaetoceros danicus CLEVE 1889	+	+							+		+	+			meso-	pleio-			Kosmopolit
Chaetoceros decipiens CLEVE 1873	+	+							+	+	+	+	+	+	poly-	meio-			Kosmopolit
Chaetoceros densus CLEVE 1901		+										+			poly-	meio-			
Chaetoceros cf. *eibenii* GRUNOW in v. H. 1880—85	+													+	poly-	meso-/ meio-			Kosmopolit
Chaetoceros holsaticus SCHÜTT 1895		+										+		+	meso-				
Chaetoceros cf. *laciniosus* SCHÜTT 1895	+										+				poly-	meio-			
Chaetoceros mitra (BAILEY 1856) CLEVE 1896		+										+			poly-	meio-			
Chaetoceros cf. *simplex* OSTENFELD 1901		+												+	poly-	meso-			
Chaetoceros wighamii BRIGHTWELL 1856	+	+				+	+			+	+	+		+	meso-	(meso-?)			Kosmopolit kaltsteno- therm
Chaetoceros sp.	+	+							+		+	+		+					
Coscinodiscus concinnus W. SMITH 1856		+		+		+				+		+			poly-				
Coscinodiscus excentricus EHRENBERG 1840	+	+		+		+	+			+		+		+	poly-	meio-/ meso-			
Coscinodiscus granii GOUGH 1905	D	D	+	+	+	+	+	+	+	+	+	+	+	+	poly-	pleio-/ meso-			Kosmopolit
Coscinodiscus radiatus EHRENBERG 1839	+	+	+						+	+				+	poly-	pleio-			
Coscinodiscus rothii (EHRENBERG 1854) GRUNOW 1878	+	+	+	+		+	+	+	+						oligo-	pleio-			Kosmopolit
Coscinodiscus sp.	+						+		+		+								
Cyclotella meneghiniana KÜTZ. 1844	+		+		+		+								oligo-	pleio-	α-/β-meso-	3	Kosmopolit
Detonula confervacea (CLEVE 1896) GRAN 1900	D	D	+		+	+	+	+	+	+	+	+	+	+	poly-	meio-/ meso-			Kosmopolit

Tab. 1 (Fortsetzung)

	Jahr		Stationen 1		2		3		4		5		6		Halobität (Pankow et al. 1976)	Salztoleranz (Pankow et al. 1976)	Saprobie (Ausgew. Methoden)	Indikat.-gewicht (Ausgew. Methoden)	Sonstige Angaben
	79	80	79	80	79	80	79	80	79	80	79	80	79	80					
Ditylium brightwellii (T. West 1860) Grunow in v. H. 1880–85	+	D				+		+		+		+	+	+	poly-	meio-			
Guinardia flaccida (Castracane 1886) Perag. 1892		+										+		+	poly-	meso-			Kosmopolit
Lauderia borealis Gran 1900	+	+	+		+		+		+		+	+	+	+	poly-	meio-			Kosmopolit
Melosira ambigua (Grunow in v. H. 1880–85) Müller 1903	+	D	+	+	+	+	+	+	+	+	+	+	+	+	oligo-	meio-	oligo-/ β-meso-	3	
Melosira arctica (Ehrenberg 1854) Dickie in Pritchard 1861	+	+	+	+	+	+	+	+	+	+	+	+	+		poly-	pleio-			kaltsteno-therm
Melosira granulata (Ehrenberg 1843) Ralfs in Pritchard 1861	D	D	+	+	+	+	+	+	+	+	+	+			oligo-	meio-	β-meso-	4	Kosmopolit
Melosira italica (Ehrenberg) Kütz.	D	+	+	+	+	+	+	+	+		+				oligo-		oligo-/ β-meso-	3	
Melosira jürgensii C.A.Ag. 1824	+	+	+	+	+	+		+	+	+		+		+	meso-	meso-			
Melosira lineata (Dillw.) Ag.		+								+		+		+	Brackw.	(Cleve-Euler)			
Melosira moniliformis (O. F. Müller 1783) C.A.Ag. 1824	+								+		+				poly-	pleio-			
Melosira nummuloides (Dillw. 1809) C.A.Ag. 1824		+										+			meso-				
Melosira varians C.A.Ag. 1817	+	+	+	+	+	+									oligo-	meso-	β-meso-	2	Kosmopolit
Rhizosolenia alata Brightwell 1858	D	+						+	+		+	+		+	poly-	meio-/ meso-			Kosmopolit
Rhizosolenia fragilissima Berger 1903	+										+				poly-	meso-			Kosmopolit
Rhizosolenia hebetata f. *semispina* (Hensen 1887) Gran 1905	+	+			+							+	+		poly-	meio-			
Rhizosolenia pungens A. Cl. 1937	+								+						marin/Brackw. (Cleve-Euler)				
Rhizosolenia setigera Brightwell 1858	+	D				+		+	+	+	+	+		+	poly-	meio-			Kosmopolit
Sceletonema costatum (Grev. 1866) Cleve 1878	D	D				+	+	+	+	+	+	+	+	+	poly-	pleio-			Kosmopolit
Sceletonema subsalsum (Cleve-Euler 1912) Bethge 1928	D	+	+		+		+	+	+		+				β-meso-/Brackw.				
Stephanodiscus astraea (Ehrenberg) Grunow	+	+	+	+	+		+								leicht brack. Gew. (Cl.-E.)		β-meso-/ oligo-	3	
Stephanodiscus hantzschii Grunow in Cleve et Grunow 1880	D	+	+	+	+	+	+	+	+		+		+	+	oligo-	meio-	α-/β-meso-	3	Kosmopolit
Terpsinoë americana (Bail.) Ralfs		+										+			Brackw. (Cleve-Euler)				
Thalassiosira baltica (Grunow in Cleve et Grunow) Ostenfeld 1901	+						+		+						meso-				
Thalassiosira decipiens (Grunow in v. H. 1880–85) Jörg. 1905	+	+			+				+	+		+	+	+	poly-	meso-			
Thalassiosira nana Lohm. 1908	+								+		+				meso-				
Thalassiosira nordenskjöldii Cleve 1873	+								+						poly-	meio-			
Achnanthes taeniata Grunow in Cleve et Grunow 1880	+	D	+	+		+	+	+	+	+	+	+	+	+	poly-	pleio-			kaltsteno-therm
Amphiprora alata (Ehrenberg) Kütz. 1844	+												+		α-meso-				
Amphiprora paludosa W. Smith 1853	+				+		+		+		+		+		meso-	euryhalin			
Amphiprora sp.		+		+										+					
Amphora sp.		+				+		+		+									
Asterionella formosa Hass. 1850	+	D	+	+	+	+	+	+	+	+	+	+	+	+	oligo-	meio- (meso-)	oligo-/ β-meso-	3	
Bacillaria paradoxa Gmelin 1778	D	+	+	+	+	+	+	+							meso-	hol-	β-meso-	4	
Caloneis amphisbaena (Bory 1824) Cleve 1894	+								+						oligo-	meio-	β-/α-meso-	2	
Caloneis sp.	+						+												
Campylodiscus echeneis Ehrenberg 1840		+										+			α-meso-				
Cylindrotheca sp.	+								+										
Cymatopleura elliptica (Breb. 1849) W. Smith 1856		+		+											oligo-		β-meso-	3	
Cymatopleura solea (Breb. 1849) W. Smith 1856	+	+	+	+	+		+								oligo-	meso-	β-/α-meso-	2	
Diatoma elongatum (Lyngb. 1819) C. A. Ag. 1824	D	D	+	+	+	+	+	+	+	+	+	+	+	+	oligo-	pleio-	oligo-/ β-meso-	3	halophil
Diatoma vulgare Bory 1828		+								+					oligo-	meso-	β-meso-	2	
Diploneis didyma (Ehrenberg 1841) Cleve 1894	+	+								+	+			+	meso-	euryhalin			
Fragilaria capucina Desmaz. 1825	+						+								oligo-	meio-	β-meso-/ oligo-	3	
Fragilaria cf. *construens* (Ehrenberg 1841) Grunow 1862	D	+			+	+	+				+				oligo-	meio-/ meso-	β-meso-	3	
Fragilaria crotonensis Kitton 1883	+	D		+	+	+	+	+		+	+	+			oligo-	meio-	oligo-/ β-meso-	3	
Fragilaria sp.	+	+	+	+	+	+	+	+		+									

Tab. 1 (Fortsetzung)

	Jahr		Stationen												Halobität	Salztoleranz	Saprobie	Indikat.-gewicht	Sonstige Angaben
			1		2		3		4		5		6		(PANKOW et al. 1976)		(Ausgew. Methoden)		
	79	80	79	80	79	80	79	80	79	80	79	80	79	80					
Gyrosigma fasciola (EHRENBERG 1840) GRIFF. et HENFREY 1856	+						+				+				α-meso-				
Gyrosigma sp.	+	+	+	+				+				+							
Licmophora sp.		+								+		+							
Mastogloia sp.	+						+												
Navicula cuspidata KÜTZ. 1844		+		+											oligo-	indifferent	β-/α-meso-	3	
Navicula rhynchocephala KÜTZ. 1844	+								+		+				oligo-	meio-/ meso-	α-meso-	4	
Navicula salinarum GRUNOW in CLEVE et GRUNOW	+								+						meso-	euryhalin			
Navicula sp.	+	+	+	+	+		+		+		+		+						
Nitzschia acicularis W. SMITH 1853	+	+	+	+	+		+		+					+	oligo-	meso-/ pleio-	α-meso-	4	
Nitzschia acuminata (W. SMITH 1853) GRUN. in CLEVE et GRUNOW 1880	+										+				poly-	meio-			
Nitzschia apiculata (GREG.) GRUN. in CLEVE et GRUN. 1880		+												+	meso-	euryhalin	α-meso-	5	
Nitzschia seriata CLEVE 1883		+								+		+		+	poly-	meso-			
Nitzschia sigma (KÜTZ. 1844) W. SMITH 1853	+	+	+	+	+	+	+	+	+	+	+				meso-	hol- (?)			
Nitzschia sigmoidea (NITZSCH 1817) W. SMITH 1853	+	+	+	+			+		+		+				oligo-	meso-(?)	β-meso-	4	
Nitzschia sp. 1 — Aufwuchs auf Blaualgen	+		+		+														
Nitzschia sp. 2	+		+		+				+		+		+						
Pinnularia sp.	+		+		+														
Pleurosigma sp.	+								+										
Surirella biseriata BREB.		+		+											oligo-		β-meso-		
Surirella elegans		+		+											oligo-				
Surirella gemma (EHRENBERG 1840) KÜTZ. 1844	+	+		+		+			+		+			+	α-meso-				
Surirella ovalis var. *crumena* (BREB. ex KÜTZ. 1849) v. H. 1880—85	+	+							+	+					oligo-	meso-	β-meso-		
Surirella striatula TURPIN 1828	+	+	+	+	+	+	+	+			+		+		β-meso-				
Synedra acus KÜTZ. 1844	+	D	+	+	+	+	+	+		+	+	+	+	+	oligo-	meio-	β-meso-	3	
Synedra gaillonii (BORY 1822) EHRENBERG 1830		+		+											poly-	pleio-			
Synedra pulchella RALFS ex KÜTZ	+	+	+	+		+		+							meso-	euryhalin	β-meso-	4	
Synedra tabulata (C. A. AG. 1830—32) KÜTZ. 1844	+	+	+	+			+	+	+	+		+		+	meso-	euryhalin	α-meso-	4	
Synedra ulna (NITZSCH 1817) EHRENBERG 1838	+		+		+		+								oligo-	meso-	β-meso-	1	
Synedra sp.	+	+	+		+		+		+		+	+	+	+					
Thalassionema nitzschioides (GRUN. 1862) GRUNOW in v. H. 1880—85	+	D					+	+	+	+	+	+	+	+	poly-	meio-/meso			
PYRRHOPHYTA:																			
Cryptomonas sp.	+		+		+		+		+		+		+						
cf. *Rhodomonas minuta*	D		+		+		+		+		+		+						
Amphidinium crassum LOHMANN 1908	+								+		+		+		poly-	meio-			
Amphidinium sp.	+	+		+							+		+						
Ceratinum fusus (EHRENBERG 1834) DUJARDIN 1841	+	+					+		+		+	+	+	+	poly-	meso-			
Ceratium fusus (EHRENBERG 1834) CLEVE 1899		+										+			poly-	meio-			
Ceratium longipes var. *baltica* OSTENFELD 1903	+										+		+		poly-	meso-			
Ceratium tripos var. *subsalsum* (OSTENFELD) PAULSEN 1907	D	D				+	+		+	+	+	+	+	+	poly-	meso-			
Cochlodinium sp.	+										+								
Dinophysis acuminata CLAP. et LACH. 1858—59	+	+				+				+	+	+			poly-	pleio-			
Dinophysis norvegica CLAP. et LACHM. 1858—59	+	+							+	+	+	+			poly-	pleio-			
Gymnodinium sp.	+								+		+								
Peridinium depressum BAILEY 1985	+										+				poly-	meso-			
Peridinium oceanicum VANHÖFFEN 1897	+	+							+			+			poly-	meio-			
Peridinium pellucidum (BERGH 1881) SCHÜTT 1895	+	+		+		+		+		+	+	+		+	meso-	pleio-			
Peridinium sp.	+	+	+		+	+	+		+		+		+	+					
Prorocentrum micans EHRENBERG 1835	+	+							+	+		+		+	poly-	meso-			
CHLOROPHYTA:																			
Flagellata s. l.:																			
Carteria sp.	+								+		+		+						

Tab. 1 (Fortsetzung)

	Jahr		Stationen												Halobität	Salz-toleranz	Saprobie	Indikat.-gewicht	Sonstige Angaben
			1		2		3		4		5		6		(PANKOW et al. 1976)		(Ausgew. Methoden)		
	79	80	79	80	79	80	79	80	79	80	79	80	79	80					
Chlamydomonas sp.	+		+		+		+		+		+				oligo-(?)				
Eudorina sp.		+				+		+							oligo-		β-meso-	3	
Euglena acus EHRENBERG 1830	+	+		+			+								oligo-		β-meso-		
Euglena ehrenbergii KLEBS		+		+											oligo-		β-meso-		
Eutreptia sp.	+	+		+		+		+	+	+	+		+						
Gonium pectorale O. F. MÜLLER 1773		+		+											oligo-	meio-	α-meso-/ poly-	2	
Pandorina morum (O. F. MÜLLER) BORY 1824	+	+	+	+		+	+	+	+			+		+	oligo-	meio-	β-meso-	3	
Phacus pleuronectes (O. F. MÜLLER 1786) DUJARDIN 1841		+										+			oligo-	meio-/ meso-			
Phacus torta (LEMM.) SKVOR.		+		+											oligo-				
Chlorococcales:																			
Actinastrum hantzschii LAGERH. 1883	+	+	+	+			+								oligo-	meso-	β-meso-	4	
Ankistrodesmus setigerus (SCHROEDER) G. WEST	+						+		+						oligo-				
cf. *Ankyra*	+		+		+		+		+		+								
Coelastrum microporum NÄG. in KÜTZ. 1849	+	+	+	+	+	+	+	+							oligo-	meio-	β-meso-	4	
Coelastrum proboscideum BOHLIN		+		+											oligo-				
Crucigenia quadrata MORREN 1830	+		+		+		+								oligo-	meio-			
Dictyosphaerium ehrenbergianum NÄG. 1849		+		+											oligo-	meio-	β-meso-	5	
Dictyosphaerium pulchellum WOOD 1874	+	+	+	+		+		+				+			oligo-	meio-	β-meso-	3	
Elakatothrix gelatinosa WILLE	+				+										oligo-				
Kirchneriella sp.	+		+		+		+				+								
Lagerheimia genevensis CHODAT 1895	+	+	+	+			+				+				oligo-				
Lagerheimia wartislawiensis SCHROEDER		+		+											oligo-				
Micractinium pusillum FRESENIUS 1856—58	+	+	+	+								+			oligo-	meio-	β-meso-	4	
Monoraphidium contortum (THURET in BREB. 1856) KOMARKOVA-LEGN. 1969		+		+											oligo-	meio-			
Monoraphidium sp.	+		+		+		+		+		+								
Oocystis borgei SNOW 1903	+		+								+				oligo-	meio-			
Oocystis lacustris CHODAT	+		+		+		+								oligo-		β-meso-	3	
Oocystis solitaria WITTROCK in WITTROCK et NORDSTEDT 1879		+				+									oligo-	meio-			
Oocystis sp.	+	+	+			+		+											
Pediastrum boryanum (TURPIN 1828) MENEGH. 1840	D	D	+	+	+	+	+	+	+	+	+	+	+	+	oligo-	meso-	β-meso-	3	
Pediastrum duplex MEYEN 1829	D	D	+	+	+	+	+	+	+	+	+	+	+		oligo-	meio-	β-meso-	4	
Pediastrum integrum NÄG. 1849	+	+	+		+	+	+	+	+	+		+		+	oligo-	meio-			
Pediastrum tetras (EHRENBERG 1838) RALFS 1844	+	+	+	+	+		+								oligo-	meio-	β-meso-	3	
Scenedesmus acuminatus (LAGERH. 1883) CHODAT 1902	+	+	+	+	+	+	+	+	+	+					oligo-	meio-	β-meso-	4	
Scenedesmus acutus MEYEN 1829	+	+	+	+	+	+	+	+						+					
Scenedesmus arcuatus LEMM. 1899		+						+									β-meso-	4	
Scenedesmus carinatus (LEMM.) CHODAT 1913	+	+	+	+	+	+	+				+				oligo-	meio-			
Scenedesmus opoliensis P. RICHTER 1896	+	+	+	+	+	+	+	+	+		+				oligo-	meio-			
Scenedesmus quadricauda (TURPIN 1820) BREB. et GODAY 1835	+	+	+	+	+	+	+	+				+		+	oligo-	pleio-	β-meso-	3	
Scenedesmus-Arten n. d.	+		+		+		+		+		+								
Tetraëdron caudatum var. *incisum* LAGERH. 1883	+						+								Süßw./leicht brackig				
Tetraëdron minimum (A. BR. 1855) HANSG. 1888	+		+		+		+				+				Süßw./leicht brackig		β-meso-	3	
Tetraëdron trigonum (NÄG. 1849) HANSG. 1888	+		+		+										Süßw./leicht brackig				
Tetrastrum staurogeniaeforme (SCHROEDER 1897) LEMM. 1900	+		+		+		+		+						oligo-	meio-			
Sonstige Chlorophyta:																			
Binuclearia lauterbornii (SCHMIDLE 1903) PROSCHKINA-LEVRENKO 1966	+	D	+	+	+	+		+		+	+	+		+	oligo-	meso- (?)			
Closterium moniliferum (BORY) RALFS 1848		+		+											oligo-		β-meso-	3	
Closterium sp.	+	+		+		+	+	+	+										
Staurastrum gracile RALFS 1845 ex RALFS 1848		+		+											Süßw./leicht brackig		oligo-/ β-meso-		
Staurastrum paradoxum MEYEN	+	+	+	+		+									Süßw./leicht brackig				
Gesamtzahl der Taxa			110		93		99		89		108		69						

proben mit dem RUTTNER-Wasserschöpfer aus dem Oberflächenwasser. Insgesamt wurden 18 Netzprobenserien bearbeitet, die in monatlichen Abständen im Zeitraum März bis November 1979 und 1980 gezogen worden waren, sowie 6 Schöpfprobenserien von März bis Juli und November 1979.

Die Netzproben wurden in Sedimentierkammern ohne Aufsatz mit dem Umkehrmikroskop untersucht, um die Vergleichbarkeit der subjektiven Einschätzung der relativen Häufigkeiten der Arten zu verbessern. Die Schätzung der relativen Häufigkeit und die Bestimmung der Phytoplanktonbiomasse als Phytoplanktonvolumen erfolgten gemäß den Vorgaben der für die Laboratorien der Wasserwirtschaft verbindlichen „Ausgewählten Methoden der Wasseruntersuchung" (BREITIG & VON TÜMPLING 1982). Das für die Berechnung des Phytoplanktonvolumens erforderliche mittlere Artvolumen der verschiedenen Phytoplankter wurde in der Regel direkt bestimmt. Für die taxonomische Zuordnung der beobachteten Arten standen die Bestimmungswerke von PANKOW et al. (1976) und CLEVE-EULER (1968), die „Süßwasserflora von Mitteleuropa" (begründet von PASCHER), sowie der Teil „Peridiniales" aus dem Bestimmungswerk „Nordisches Plankton" (PAULSEN 1908) zur Verfügung. Unsichere Bestimmungen sind mit cf. gekennzeichnet.

3. Ergebnisse

3.1. Artenbestand und ökologische Charakteristik des Phytoplanktons

In den untersuchten Proben wurden insgesamt 190 Taxa nachgewiesen, von denen 149 bis zur Art bestimmt werden konnten (s. Tab. 1). Die Zahl der tatsächlich vorkommenden Arten ist wahrscheinlich noch größer, da die Nanoplankter mit den wenigen Schöpfproben sicher nicht vollständig erfaßt wurden. Als Ursache dieser Artenvielfalt ist der Zustrom von Wassermassen aus Oberwarnow und Ostsee anzusehen, die als Gewässer unterschiedlichen Salzgehalts von ganz verschiedenen Planktonpopulationen besiedelt sind. Hinsichtlich der Gesamtartenzahl sind zwischen den einzelnen Stationen nicht sehr große, aber deutliche Unterschiede erkennbar (Abb. 1). Das Artenminimum lag im Untersuchungszeitraum an der Station 6 im Breitling (69 Taxa), d. h. in dem vom Wasseraustausch am wenigsten betroffenen Gewässerabschnitt. Von den Stationen im Stromverlauf erbrachten Stat. 1 (Warnowzufluß, 110 Taxa) und 5 (Mündung in die Ostsee, 108 Taxa) relativ hohe Artenzahlen, die dazwischen liegenden Stat. 2—4 etwas niedrigere. Das Minimum lag an der Stat. 4 (89 Taxa), d. h. in einem stärker salzwasserbeeinflußten Bereich mit Salzgehaltsschwankungen von 3,5 bis 12,5$^0/_{00}$ im Oberflächenwasser und 8,7 bis 16$^0/_{00}$ in Grundnähe.

Im Artenspektrum dominieren die Kieselalgen mit 100 Taxa. Es folgen die Chlorophyta mit 48 Taxa, die Cyanophyta mit 19 Taxa und die Pyrrhophyta mit 17 Taxa. Bei den determinierten Cyanophyta und Chlorophyta handelt es sich ausschließlich um limnische, bei den Pyrrhophyta ausschließlich um Salzwasserarten. Nur die Chrysophyta rekrutieren sich aus Oligo- (26), Meso- (25) und Polyhalobien (34). Die polyhaloben Kieselalgen gehören überwiegend zu den Centricae. Insgesamt sind von den determinierten Arten 82 Oligohalobien, 25 Mesohalobien und 48 Polyhalobien. Für die Mehrzahl der Arten ist eine mehr oder weniger ausgeprägte Salztoleranz bekannt. Die Oligo- und Polyhalobien sind meist meio- oder mesoeuryhalin, seltener pleioeuryhalin. Die Salztoleranz der Mesohalobien liegt — sofern überhaupt eine Einstufung vorgenommen wurde — meistens in einem mittleren Bereich (Angaben in Tab. 1 nach SIMONSEN aus PANKOW et al. 1986, CLEVE-EULER 1968).

Etwa 50% der festgestellten Arten sind Saprobien. Entsprechend ihrer Einstufung in das Saprobiensystem (BREITIG & VON TÜMPLING 1982) indizieren sie fast ausschließlich mesosaprobe (zu 60% β-mesosaprobe) Verhältnisse (s. Tab. 1). Die Mehrzahl der Arten ist euryök und hat daher nur geringen Zeigerwert. Abgesehen davon, daß die Saprobie von Salzwasserarten bisher nicht bestimmt ist, kann daher die ökologische Analyse des Phytoplanktonartenspektrums im Hinblick auf die organische Belastung der Unterwarnow nur zu sehr allgemeinen Aussagen führen.

Für eine ganze Reihe von Arten geben PANKOW et al. (1976) kosmopolitische Verbreitung an. Mindestens 3 Arten sind kaltstenotherme Salzwasserformen.

Nach dem Ergebnis dieser Analyse überwiegen im Phytoplanktonartenspektrum der Unterwarnow Arten mit breiter ökologischer Amplitude. Auch diese Erscheinung ist als unmittelbare Folge der besonderen hydrographischen Bedingungen, insbesondere des starken Salzgehaltsgefälles und der breiten Amplitude der Salzgehaltsschwankungen, aber auch der Belastung anzusehen.

3.2. Dominanten des Phytoplanktons

Welche Arten in einer Planktonprobe als Dominanten erscheinen, wird maßgeblich von der Probenahme bestimmt. Bei der Darstellung der Dominanzverhältnisse wird daher im folgenden zwischen Netz- und Schöpfproben unterschieden. Die Liste der Dominanten des Netzplanktons enthält alle Arten, die im Untersuchungszeitraum wenigstens einmal mit einer geschätzten Häufigkeit $\geq$ 5 auftraten (BREITIG & VON TÜMPLING 1982), weiterhin einige in einer Vielzahl von Proben, aber in geringerer Individuendichte (Häufigkeitswert 3 = mehrfach) beobachtete Arten, sowie einige wenige, die in phytoplanktonarmen Perioden zwar nur in geringer Individuendichte nachgewiesen wurden, aber die Hauptmenge des gefangenen Phytoplanktons ausmachten (s. Tab. 2). Die Hauptmenge des im Untersuchungszeitraum eingebrachten Netzplanktons wird von relativ wenigen Arten gebildet, die überwiegend zu den Kieselalgen gehören. An der Station 1 handelt es sich vor allem um Oligohalobien, 2 Brackwasserarten (*Sceletonema subsalsum*, *Synedra tabulata*) und 2 pleioeuryhaline Polyhalobien (*Sceletonema costatum* und *Coscinodiscus granii*). Im weiteren Verlauf der Unterwarnow verlieren die Oligohalobien als Dominanten des Netzplanktons sehr schnell an Bedeutung, und es treten zunehmend Salzwasserarten an ihre Stelle. Schon an der Station 2 dokumentiert das vermehrte Auftreten von Mesohalobien wie *Bacillaria paradoxa* (holeuryhalin) und *Melosira jürgensii* (mesoeuryhalin), sowie von Polyhalobien wie *Detonula confervacea* und *Melosira arctica* verstärkte Salzwassereinflüsse. An der Station 3 indiziert *Achnanthes taeniata* das Fortschreiten dieser Tendenz, an der Station 4 sind es die marinen Plankter *Rhizosolenia alata*, *Thalassionema nitzschioides*, *Ditylum brightwellii* und *Ceratium tripos*. Die Dominanten der Stat. 5 schließlich gehören fast ausschließlich zum marinen Plankton. Nur die halophile, pleioeuryhaline Süßwasserart *Diatoma elongatum* zeigt den noch vorhandenen Süßwassereinfluß an. An der etwas abseits vom stärksten Wasseraustausch gelegenen Stat. 6 spielen neben den auch an der Stat. 5 beobachteten marinen Phytoplanktern nur wenige Vertreter der Oligohalobien wie *Diatoma*

Tab. 2. Räumliche Verteilung der aspektbestimmenden Netzplankter, Stat. 1—6

	1	2	3	4	5	6
Synedra tabulata	+					
Synura uvella	+					
Surirella striatula	+	+				
Melosira ambigua	+	+	+			
Synedra acus	+	+	+			+
Fragilaria crotonensis	+	+	+			
Stephanodiscus hantzschii	+	+	+			
Sceletonema subsalsum	+		+			
Binuclearia lauterbornii	+		+			
Asterionella formosa	+	+	+	+		
Pediastrum boryanum	+	+	+	+		
Pediastrum duplex	+	+	+	+		
Melosira granulata	+		+	+	+	
Diatoma elongatum	+	+	+	+	+	
Sceletonema costatum	+		+	+	+	+
Coscinodiscus granii	+			+	+	+
Melosira jürgensii		+				
Bacillaria paradoxa		+	+			
Fragilaria sp.		+	+	+		
Melosira italica		+	+	+		
Oscillatoria cf. *lacustris*		+	+	+	+	
Detonula confervacea		+	+	+	+	+
Melosira arctica		+		+		+
Scenedesmus-Arten			+			
Lyngbya limnetica			+			
Coscinodiscus radiatus				+		
Rhizosolenia alata				+	+	
Thalassionema nitzschioides				+	+	+
Ditylum brightwellii				+	+	+
Ceratium tripos				+	+	+
Lauderia borealis					+	
Rhizosolenia fragilissima					+	
Ceratium longipes					+	
Actinocyclus octonarius					+	
Nitzschia seriata					+	
Melosira lineata					+	
Chaetoceros-Arten					+	+
Ceratium fusus					+	+
Thalassiosira decipiens						+
Peridinium pellucidum						+

Tab. 3. Räumliche Verteilung der produktionsbiologisch wichtigen Phytoplankter in Schöpfproben, Stat. 1—6

	1	2	3	4	5	6
Melosira ambigua	+					
cf. *Ankyra*	+	+				
Cyclotella meneghiniana	+	+				
Melosira granulata	+	+				
Scenedesmus-Arten	+	+				
Diatoma elongatum	+	+	+			
Sceletonema subsalsum	+	+	+	+	+	
Chlamydomonas-Arten	+	+	+	+	+	
Cryptomonas-Arten	+	+	+	+	+	+
Monoraphidium sp.	+	+	+	+	+	+
Kirchneriella sp.	+	+	+	+	+	+
Stephanodiscus hantzschii	+	+	+	+	+	+
cf. *Rhodomonas minuta*	+	+	+	+	+	+
Detonula confervacea		+	+	+	+	+
Amphidinium sp.				+	+	+
Carteria sp.				+	+	+
Eutreptia sp.				+	+	+
Sceletonema costatum					+	+

elongatum, *Asterionella formosa* und *Synedra acus* kurzzeitig als Dominanten eine Rolle.

Die Schöpfproben zeigten völlig andere Dominanzverhältnisse (Tab. 3). Bei der Mehrzahl der Arten handelt es sich um Nanoplankter, die mit dem Planktonnetz überhaupt nicht oder nur unvollständig erfaßt werden. Die Kieselalgen waren im Gegensatz zum Netzplankton mit relativ wenigen Arten vertreten, und Salzwasserarten spielten nur eine untergeordnete Rolle. Die größten Biomassen entwickelte *Stephanodiscus hantzschii*, eine oligohalobe, sicher zu unrecht als nur meioeuryhalin eingestufte, α- bis β-mesosaprobe Kieselalge und typischer Bewohner stark belasteter bzw. eutrophierter Gewässc· Beobachtet wurden sowohl Einzelzellen als auch mehr oder weniger lange Zellketten, die z. T. der Beschreibung von *St. subtilis* (VON GOOR emend.) A. CL. in CLEVE-EULER (1968) nahe kamen. Die chlorokokkalen Grünalgen und Flagellaten, einschließlich nicht einzuordnender flagellatenähnlicher Zellen, waren nicht determinierbar und daher ökologisch auch nicht einzustufen. Ähnlich wie bei den Netzproben änderten sich die Dominanzverhältnisse im Verlauf der Unterwarnow. Im oberen Abschnitt waren im Untersuchungszeitraum vor allem Kieselalgen Träger der planktischen Primärproduktion, im unteren Abschnitt neben *Stephanodiscus hantzschii* insbesondere Flagellaten, flagellatenähnliche Zellen und winzige chlorokokkale Grünalgen unsicherer taxonomischer Zugehörigkeit.

3.3. Räumliche und zeitliche Variabilität des pflanzlichen Netzplanktons

Abb. 1 zeigt die Verbreitung der Arten unterschiedlicher Halobität und Salztoleranz in der Unterwarnow. Die Veränderung des Halobitätsspektrums im Verlauf des Gewässers ist eindrucksvoll. Die Oligohalobien nehmen in Stromrichtung ab, die Polyhalobien zu. Die Zahl der Mesohalobien verändert sich vergleichsweise nur wenig. Die Veränderung des Halobitätsspektrums entspricht dem Salzgehaltsgefälle, das durch Zunahme des mittleren Salzgehalts in Richtung Ostsee charakterisiert ist (s. o.). Die Verbreitung meio- und mesoeuryhaliner Arten wird stärker vom Salzgehalt beeinflußt als die Verbreitung der pleioeuryhalinen Formen.

Im Hinblick auf die Gesamtartenzahl zeichnet sich zwischen den beiden Extremen im Verlauf der Unterwarnow (Stat. 1 und 5) ein Artenminimum ab, das als Brackwasserminimum zu deuten ist. Die relative Artenarmut der Stationen 2—4 ist an der Station 3 am wenigsten ausgeprägt, d. h. im Bereich der größten Salzgehaltsschwankungen und einer Belastungsquelle. Daß die Unterschiede zu den Randstationen nicht größer ausfallen, liegt an der mit z. T. enormen Salzgehaltsschwankungen und z. T. erheblichen vertikalen Salzgehaltsgradienten verbundenen Gewässerdynamik (vgl. Tab. 5). Das tatsächliche Brackwasserminimum der Unterwarnow wird durch die Artenzahl der Stat. 6 im Breitling deutlicher widergespiegelt. Bemerkenswert ist hier vor allem der geringe Anteil der Oligohalobien am Gesamtphytoplankton, da sich der mittlere Salzgehalt nur wenig von dem der Stat. 5 unterscheidet. Er dürfte auf den geringeren Eintrag allochthoner Arten zurückzuführen sein.

Die Verteilung der mit dem Planktonnetz gefangenen Saprobien läßt, im Gegensatz zum Verhalten der Halobien gegenüber dem Salzgehaltsgefälle, kaum Beziehungen zu den Hauptbelastungsquellen (Warnowzufluß und Kläranlagenablauf) als Steuergrößen der Saprobität erkennen, sondern sie wird ebenfalls maßgeblich durch die Salzverhältnisse bestimmt (Tab. 4). Ihrer Salztoleranz

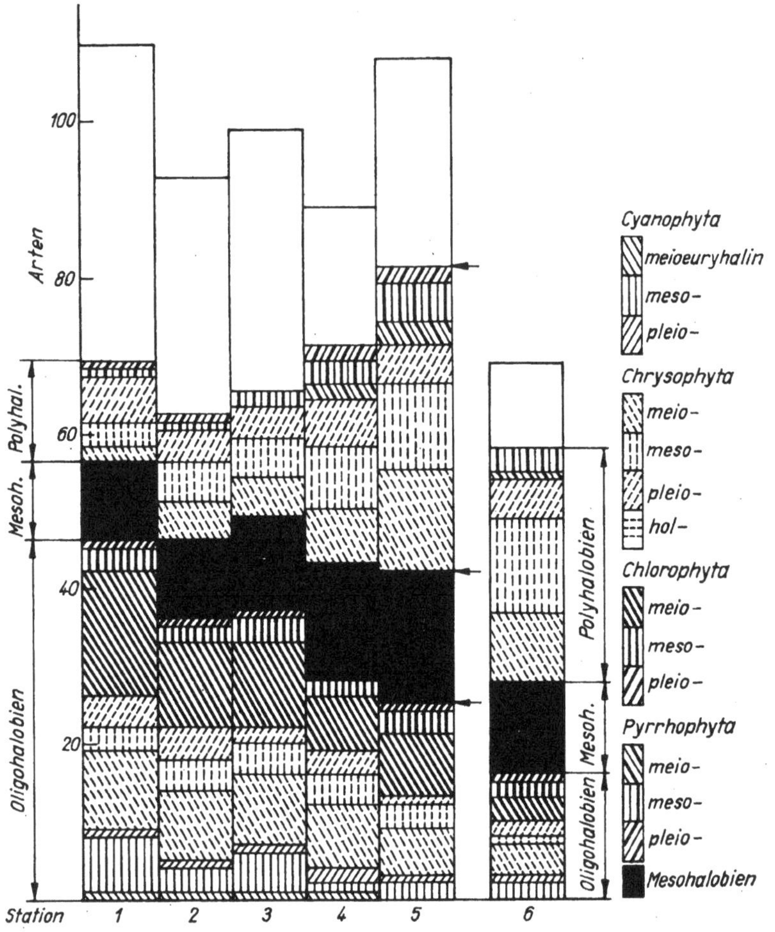

Abb. 1. Verbreitung der Arten unterschiedlicher Halobität und Salztoleranz in der Unterwarnow

Tab. 4. Räumliche Verteilung der Saprobien, Stat. 1—6

	1	2	3	4	5	6	Saprobie
Chroococcus limneticus	+						oligo-/β-meso-saprob
Microcystis reinboldii	+						β-meso-
Dinobryon divergens	+						β-meso-
Cymatopleura elliptica	+						β-meso-
Navicula cuspidata	+						β-/α-meso-
Surirella biseriata	+						β-meso-
Euglena ehrenbergii	+						β-meso-
Gonium pectorale	+						α-meso-/poly-
Dictyosphaerium ehrenbergianum	+						β-meso-
Closterium moniliferum	+						β-meso-
Staurastrum gracile	+						oligo-/β-meso-
Synura uvella	+	+					β-meso-
Melosira varians	+	+					β-meso-
Microcystis aeruginosa	+		+				β-meso-
Oscillatoria limosa	+		+				β-/α-meso-
Actinastrum hantzschii	+		+				β-meso-
Euglena acus	+		+				β-meso-
Cyclotella meneghiniana	+	+	+				α-/β-meso-
Stephanodiscus astraea	+	+	+				β-meso-/oligo-
Bacillaria paradoxa	+	+	+				β-meso-
Cymatopleura solea	+	+	+				β-/α-meso-
Synedra pulchella	+	+	+				β-meso-
Synedra ulna	+	+	+				β-meso-
Oocystis lacustris	+	+	+				β-meso-
Pediastrum tetras	+	+	+				β-meso-
Oscillatoria redekei	+	+	+	+			β-meso-/oligo-
Dinobryon sertularia	+	+	+	+			oligo-
Nitzschia acicularis	+	+	+	+		+	α-meso-

Tab. 4 (Fortsetzung)

	1	2	3	4	5	6	Saprobie
Coelastrum microporum	+	+	+	+			β-meso-
Aphanizomenon flos-aquae	+	+	+	+			β-meso-
Scenedesmus acuminatus	+	+	+	+			β-meso-
Melosira granulata	+	+	+	+	+		β-meso-
Melosira italica	+	+	+	+	+		oligo-/β-meso-
Fragilaria crotonensis	+	+	+	+	+		oligo-/β-meso-
Nitzschia sigmoidea	+		+	+	+		β-meso-
Dictyosphaerium pulchellum	+	+	+		+		β-meso-
Tetraedron minimum	+	+	+		+		β-meso-
Micractinium pusillum	+				+		β-meso-
Melosira ambigua	+	+	+	+	+	+	oligo-/β-meso-
Stephanodiscus hantzschii	+	+	+	+	+	+	α-/β-meso-
Asterionella formosa	+	+	+	+	+	+	oligo-/β-meso-
Diatoma elongatum	+	+	+	+	+	+	oligo-/β-meso-
Synedra acus	+	+	+	+	+	+	β-meso-
Pandorina morum	+	+	+	+	+	+	β-meso-
Pediastrum boryanum	+	+	+	+	+	+	β-meso-
Pediastrum duplex	+	+	+	+	+	+	β-meso-
Synedra tabulata	+		+	+	+	+	α-meso-
Scenedesmus quadricauda	+	+	+		+	+	β-meso-
Fragilaria construens		+	+		+		β-meso-
Eudorina sp.		+	+				β-meso-
Fragilaria capucina			+				β-meso-/oligo-
Scenedesmus arcuatus			+				β-meso-
Caloneis amphisbaena				+			β-/α-meso-
Diatoma vulgare				+		+	β-meso-
Surirella ovalis				+			β-meso-
Navicula rhynchocephala				+	+		α-meso-
Nitzschia apiculata						+	α-meso-

entsprechend, nimmt die Zahl der überwiegend oligohaloben Saprobien von Stat. 1—6 ab. Die gegenüber der Stat. 2 wenig erhöhte Saprobienzahl an der Stat. 3 kann als Hinweis auf den Kläranlagenablauf gedeutet werden. Die relativ große Zahl der schlechter als β-mesosaprob eingestuften Arten an den Stat. 1 und 3 könnte entsprechend gewertet werden. Die Unterschiede zu den übrigen Stationen sind jedoch minimal und durch den Salzwassereinfluß verwischt. Die globale Einschätzung des Gewässers nach der Saprobie der mit dem Planktonnetz gefangenen Phytoplankter ist daher durch Erfassung der Verbreitung in der Unterwarnow nicht zu konkretisieren. Aus der Dominanz β-mesosaprober Organismen auf einen guten Abbau der Belastung im Gewässer zu schließen, ist aus mehreren Gründen nicht zulässig. So sind z. B. für die Berechnung des Saprobitätsgrades auch Glieder anderer Lebensgemeinschaften als des Phytoplanktons heranzuziehen. Entscheidend ist jedoch, daß in einem rückgestauten Fließgewässer wie der Unterwarnow Parameter der Trophie die mit dem Abbau

Tab. 5. Räumliche und zeitliche Verteilung der dominierenden Netzplankter; Angaben als relative Häufigkeit (O Oberfläche, T Tiefe)

Station 1	1979									1980							
	III	IV	V	VI	VII	VIII	IX	X	XI	III	IV	V	VI	VII	VIII	IX	X
Salzgehalt ($^0/_{00}$) O:	0,1	1,6	5,1	0,3	2,2	0,9	2,2	7,2	2,9	9,0	4,7	8,5	3,2	1,6	6,8	0,1	5,9
T:	0,1	10	9,4	8,2	6,7	9,4	5,1	8,0	10	13	11	10	4,1	4,9	5,5	0,7	12
Melosira ambigua	3	3	2		3	n. e.										5	
Melosira granulata	3				5					2			9	5	3	5	
Sceletonema subsalsum	2				3				5—7								
Sceletonema costatum				kaum Phytoplankton		kaum Phytoplankton					2	3	5				kaum Phytoplankton
Asterionella formosa		2			2			3	3	3		2	3	2		3	
Diatoma elongatum		3	3					3			3—5	5	3			2	
Synedra tabulata		3	2														
Synedra acus											2	2—3	2	3	3	3	
Coscinodiscus granii					3			2		2				2	2		
Surirella striatula								3	3	2—3					2		
Binuclearia lauterbornii					3												
Pediastrum boryanum					3							2	2—3	3	2	2	
Stephanodiscus hantzschii			2									2—3	2			3	
Synura uvella													2—3				
Pediastrum duplex													3	2	2	2	
Fragilaria crotonensis													5	3			
Station 2 Salzgehalt ($^0/_{00}$) O:	0,8	1,7	5,0	5,7	2,2	1,1	7,4	11	11	9,0	8,6	7,3	3,5	2,3	4,7	0,6	10
T:	11	13	11	9,1	10	11	9,8	13	12	14	12	12	10	7,8	7,2	11	12
Detonula confervacea	7	3				n. e.				3							
Melosira ambigua	3				3					3		2				5	3
Melosira arctica	3	2										2					
Melosira granulata		2	kaum Phytoplankton	kaum Phytoplankton	3		5						3	3—5	7	5	3
Melosira italica							5						2				
Melosira jürgensii							3							2			
Bacillaria paradoxa							3—5										2
Fragilaria crotonensis							3				2	2	3	2			
Surirella striatula							3										
Fragilaria sp.							5	3		2	2						
Pediastrum boryanum							3						2—3	3	2	2	2
Stephanodiscus hantzschii								5		2	3	3					
Asterionella formosa								3		3		3			2	2	2
Oscillatoria cf. *lacustris*									3			2		3			
Diatoma elongatum								3	2	3	5	7					
Synedra acus											3	3	2		2	2	2
Pediastrum duplex							2						2—3	3	2	3	2
Station 3 Salzgehalt ($^0/_{00}$) O:	1,5	5,5	6,8	6,1	8,8	3,5	5,2	11	7,8	8,3	5,9	8,3	7,3	5,1	7,7	5,6	12
T:	13	11	10	8,5	12	11	9,9	15	12	13	13	11	11	11	11	10	12
Detonula confervacea	9	3—5				n. e.				5							
Melosira ambigua	3		3		3											5	3
Melosira granulata	2—3	3	2		3		3	2		3		2	3	5	5	3	2
Achnanthes taeniata	2	3								3							
Melosira italica		2		kaum Phytoplankton			5										
Sceletonema costatum		3						3	3	3		3					
Stephanodiscus hantzschii		3	3					2		2	2						
Diatoma elongatum		5	5					3	3	3	7	7					
Synedra-Arten		3	3								3	2	2	2		3	3
Pediastrum boryanum			2				3						2—3	3	2	2	2
Pediastrum duplex			2				3						2—3	3	3	2	2
Scenedesmus-Arten			3—5										2	3			
Bacillaria paradoxa							3—5	3	3—5								3
Fragilaria sp.							5	2									
Sceletonema subsalsum	2									3							
Asterionella formosa		2	2							5—7		2					
Fragilaria crotonensis			2				2					2	3		2		
Oscillatoria cf. *lacustris*									2			2		3	3		
Lyngbya limnetica															3		

Tab. 5 (Fortsetzung)

Station 4	1979									1980								
	III	IV	V	VI	VII	VIII	IX	X	XI	III	IV	V	VI	VII	VIII	IX	X	XI
Salzgehalt (‰) O:	3,8	5,6	6,9	8,9	11	5,7	9,6	12	12	9,4	9,8	9,7	8,1	8,2	8,5	9,7	11	n. e.
T:	11	14	11	9,2	11	10	11	14	12	15	15	12	12	11	9,9	13	11	
Detonula confervacea	9	3				n. e.				5								
Melosira arctica	2	2—3										2						
Sceletonema costatum	2							3	5	3		2					2	
Achnanthes taeniata		3		kaum Phyto- plankton						3								
Diatoma elongatum		5	5				2	2		3	5	7—9						
Pediastrum boryanum			2				3						2	2	2	2	2	
Pediastrum duplex			2				3						2	2	2	2	2	
Coscinodiscus radiatus				2	3													
Coscinodiscus granii					3								2	3	2	3	2	
Rhizosolenia alata					3										2	3	2	
Melosira italica							3								3			
Fragilaria sp.							5											
Asterionella formosa										5		2					2	
Binuclearia lauterbornii													2	3	2		2	
Oscillatoria lacustris													2	3	3			
Thalassionema nitzschioides														2		2	2	
Ditylum brightwellii															2	3	2	
Ceratium tripos															3			
Station 5 Salzgehalt (‰) O:	8,0	9,2	9,9	9,2	12	9,2	11	14	12	13	13	12	10	10	14	12	14	n. e.
T:	15	16	13	12	17	15	13	16	11	16	17	16	13	14	17	12	15	
Detonula confervacea	9	3				n. e.		n. e.		5	2							
Lauderia borealis	3	2																
Thalassionema nitzschioides	2	2		kaum Phytoplankton	2				3	2—3	2		kaum Phytoplankton	2	2	2		
Sceletonema costatum		2—3	2						3	5	2							
Achnanthes taeniata		3								2	3	7						
Diatoma elongatum		3	5							3								
Chaetoceros-Arten					5		2								5	5	5	2
Coscinodiscus granii					3		2		5					3	3	3		
Rhizosolenia alata					9											3		
Rhizosolenia fragilissima					3													
Ceratium longipes					3													
Ceratium tripos					3				5					5	2			7
Ceratium fusus					2				3									3
Melosira granulata										2	2	2		5	2			
Actinocyclus octonarius													3					
Oscillatoria cf. *lacustris*														3	3			
Ditylum brightwellii															3—5	5		
Nitzschia seriata																3	2	
Melosira lineata																	3	
Station 6 Salzgehalt (‰) O:	5,3	10	8,8	8,4	11	7,3	9,7	8,5	12	14	9,9	12	9,3	9,1	12	9,4	11	9,8
T:	10	10	9,3	8,8	12	10	9,8	14	12	14	13	12	9,9	9,1	12	9,4	12	9,8
Detonula confervacea	n. e.	2—3				n. e.				3								
Sceletonema costatum		2		kaum Phytoplankton					3			7	kaum Phytoplank- ton	kaum Phytoplank- ton	3	2	2	2
Diatoma elongatum		2—3	5								3	5—7						
Coscinodiscus granii					2			9	5				2	2	3	3	3	2
Thalassionema nitzschioides					2			3	3	2								3
Thalassiosira decipiens								3	3									
Ceratium tripos								7	3						2			5
Achnanthes taeniata										3	2	2						
Asterionella formosa			2							3								
Melosira arctica											2	3						
Peridinium pellucidum											2	2—3						
Chaetoceros-Arten												3						
Synedra acus												3						
Ditylum brightwellii															3	3	2	
Ceratium fusus								2										2—3

anthropogener Belastungen verbundenen Selbstreinigungsprozesse besser widerspiegeln als der Saprobitätsgrad.

Die jahreszeitliche Sukzession der aspektbestimmenden Netzplankter zeigt Tab. 5. Im Frühjahr dominierten in beiden Untersuchungsjahren Kaltwasserplankter der Ostsee wie *Detonula confervacea*, *Achnanthes taeniata* und *Sceletonema costatum*, die Brackwasserart *Sceletonema subsalsum*, sowie mehr oder weniger euryhaline oder halophile Süßwasserarten wie *Melosira ambigua* und *M. granulata*, *Asterionella formosa* und *Diatoma elongatum*. 1979 folgte auf die Frühjahrsentwicklung an allen Stationen ein ausgesprochen phytoplanktonarmes Intervall, 1980 nur an den Stationen 5 und 6, das zeitlich mit Zooplanktonmassenentwicklungen zusammenfällt, insbesondere des Brackwassercopepoden *Eurytemora affinis* (alle Stationen) und verschiedener *Synchaeta*-Arten (Stationen 4—6). Den sommerlichen Aspekt bestimmten oligohalobe Kieselalgen wie *Melosira granulata* und *M. italica* und *Fragilaria crotonensis*, oligohalobe Grünalgen (*Pediastrum*-Arten), weiterhin meso- und polyhalobe Kieselalgen (*Chaetoceros*-, *Rhizosolenia*-Arten, *Bacillaria para-*

doxa, Ditylum brightwellii) und *Ceratium tripos*. Einige der genannten Phytoplankter, insbesondere die Ceratien, erreichten erst im Spätherbst ihr Populationsmaximum. Gleiches gilt für *Coscinodiscus granii*. Von den Frühjahrsformen erschienen die *Sceletonema*-Arten und *Diatoma elongatum* am Ende der Vegetationsperiode erneut in kleinen Populationen. Die phytoplanktonarmen Netzproben im Herbst und Spätherbst waren in der Regel auch arm an Zooplankton.

Wie ein Vergleich der Artenlisten in Tab. 5 mit dem jeweils zugehörigen Salzgehalt zeigt, wird die Populationsdynamik an den 6 Untersuchungsstationen durch die Salzverhältnisse (Salzgehalt, Schichtung, Salzgehaltsschwankungen) maßgeblich beeinflußt. Salzgehaltsänderungen im Jahresverlauf können, aber müssen nicht, mit einem vollständigen Aspektwechsel verbunden sein. Signalisiert wird damit weniger ein Einfluß der Salzverhältnisse auf die Organismen als die Bewegung von Wasserkörpern unterschiedlichen Salzgehalts, die naturgemäß auch von verschiedenen Planktonpopulationen besiedelt sind. Auf Grund der angewendeten Probenahmetechnik (Vertikalhol) fanden sich in den untersuchten Netzproben nicht selten Bewohner verschiedener Salzgehaltsbereiche nebeneinander. Die Populationsstärke ließ in diesen Fällen in begrenztem Umfang auch Aussagen über die Stärke der übereinander liegenden Wasserkörper zu (s. Stat. 2, März 1979).

An der Station 1 ist der 1979 und 1980 vorherrschende Süßwassereinfluß zu allen Jahreszeiten an der Halobität der Dominanten klar erkennbar. Die stärkere Versalzung des Bereichs von Oktober 1979 bis Mai 1980 wird durch größere Populationen der Salzwasserarten *Sceletonema subsalsum, S. costatum* und der halophilen Kieselalge *Diatoma elongatum* indiziert, der erhöhte Warnowabfluß in den Sommermonaten durch Populationen verschiedener *Melosira*-Arten. Die z. T. erheblich salzigeren Wassermassen in Grundnähe werden nur durch einzelne Indikatorarten widergespiegelt. Sie können daher nicht sehr umfangreich gewesen sein (vgl. die Salzgehaltsangaben in Tab. 5).

An den Stationen 2—5 war vor allem der Frühjahrsaspekt stärker salzwasserbeeinflußt. Die Befunde der beiden Jahre weisen jedoch auf graduelle Unterschiede hin. Im März 1979 war zunächst die Ostseealge *Detonula confervacea* Prädominante im Netzplankton aller Stationen, wobei allerdings die relativ großen Mengen in den Netzproben der Stationen 2 und 3 einem wahrscheinlich umfangreichen salzreichen Tiefenwasser zuzuschreiben sind. Im März 1980 traten im Zusammenhang mit dem bereits erwähnten Salzwasserzustrom neben *Detonula* weitere Ostseealgen wie z. B. *Achnanthes taeniata* in der Unterwarnow in Erscheinung. Nicht zu den aktuellen Salzverhältnissen passen die relativ großen *Asterionella*-Populationen an den Stationen 3 und 4 im März 1980. Ein ursächlicher Zusammenhang zum Kläranlagenablauf ist denkbar, aber schwer zu erklären. Im April/Mai zeigte das Netzplankton in beiden Untersuchungsjahren im gesamten Bereich der Stationen 2—5 bei Salzgehalten zwischen ca. 5 und 10% im Oberflächenwasser einen gegenüber dem zeitigen Frühjahr stark veränderten Aspekt mit *Diatoma elongatum* als der dominierenden Art. Als Ursache des Aspektwechsels kommt vor allem das schnelle Ansteigen der Wassertemperatur in Frage, das ein Verschwinden der Kaltstenothermen bewirkt (s. Tab. 1). Im Sommerplankton der Stationen 2 und 3 dominierten vor allem Oligohalobien, die auch an der Station 4 noch einen wesentlichen Teil des pflanzlichen Netzplanktons ausmachten. Erst an der Station 5 traten Salzwasserplankter der Gattungen *Chaetoceros, Rhizosolenia* und *Ceratium* in größeren Mengen auf. Die Grenze zwischen Süßwasser- und Salzwasserindikatoren war vor allem in den Sommermonaten 1980 gegenüber dem zeitigen Frühjahr (März) in Stromrichtung verschoben. Die aktuellen Salzgehalte entsprechen im wesentlichen dieser Tendenz. Die tolerierte Grenzkonzentration ist für die oligohaloben Sommerplankter jedoch wahrscheinlich niedriger als für *Diatoma elongatum*, die Dominante der ausklingenden Frühjahrsentwicklung.

Ausgesprochene Herbst- oder Spätherbstformen des Süßwassers wurden nicht beobachtet. Bei sehr niedrigem Salzgehalt an den Stationen 1 und 2 waren im September 1980 noch die Sommerformen präsent. *Sceletonema subsalsum* trat ab Sept./Okt. bei einem Salzgehalt $\geqq 3^0/_{00}$ in größeren Populationen auf (Stat. 1, Nov. 1979), *Sceletonema costatum* bei einem Salzgehalt $\geqq 8^0/_{00}$ (Stat. 3, Okt./Nov. 1979), desgl. *Bacillaria paradoxa*. Vermehrtes Auftreten von *Coscinodiscus granii, Ceratium tripos* und anderen Ostseeplanktern war mit Salzgehalten um $12^0/_{00}$ und mehr verbunden (Stat. 5, 1979 und 1980).

An der Station 6 wies nicht nur das Artenspektrum, sondern auch die jahreszeitliche Sukzession der Netzplankter einige Besonderheiten auf. Im Frühjahr fehlten trotz des relativ hohen Salzgehalts nennenswerte Populationen von Ostseeformen wie *Achnanthes taeniata*. Das Sommerplankton war ausgesprochen arten- und individuenarm. Die im Juli 1979 an der Station 5 (Oberfl. $12{,}2^0/_{00}$, Grundn. $16{,}9^0/_{00}$) in großen Mengen beobachteten *Rhizosolenia*-Arten fehlten im Breitling vollständig (Oberfl. $10{,}5^0/_{00}$, Grundn. $11{,}5^0/_{00}$). *Coscinodiscus granii* und *Ceratium tripos* erreichten 1979 nach einem 4-monatigen netzplanktonarmen Intervall im Oktober hohe Populationsstärken, die 1980 bei vor allem in Grundnähe niedrigem Salzgehalt nicht wieder erreicht wurden. Träger der Sukzession waren eigentlich nur 4 Arten: im Frühjahr *Diatoma elongatum* und *Sceletonema costatum*, im Herbst *Coscinodiscus granii* und *Ceratium tripos*. Von diesen sind die beiden Kaltwasserplankter des Frühjahrs als echte Brackwasserbewohner anzusehen. Die 1979 hohen Konzentrationen der beiden Herbstplankter verdankten ihre Anwesenheit im Breitling wahrscheinlich einem Salzwasserzustrom, der durch einen Salzgehalt von $14{,}1^0/_{00}$ (Tiefenwasser) im Oktober angezeigt wird. Die Befunde belegen ein Artenminimum für das Netzplankton in einem Salzgehaltsbereich zwischen 8 und $12^0/_{00}$, das offensichtlich in den Sommermonaten besonders ausgeprägt ist.

Räumliche und zeitliche Variabilität des mit dem Planktonnetz erfaßbaren Phytoplanktons werden also im wesentlichen durch den Wechsel der Jahreszeiten und die mit Salzgehaltsänderungen verbundene Gewässerdynamik bestimmt. Ausschlaggebend für die zu einem bestimmten Zeitpunkt an einer Untersuchungsstation zu erwartenden Netzplanktonpopulationen sind Salzgehalt, Schichtung, aber auch aktueller Netzplanktongehalt benachbarter Gewässerabschnitte, sowie Halobität und Salztoleranz der Dominanten der betreffenden Jahreszeit. Die beobachtete Verbreitung der Arten unterschiedlicher Halobität und Salztoleranz in der Unterwarnow bestätigt im wesentlichen deren Einstufung in das Halobiensystem (s. Tab. 1). Abweichungen betreffen vor allem Oligohalobien (z. B. *Melosira ambigua, M. granulata, Stephanodiscus hantzschii*), deren Salztoleranz bisher wahrscheinlich unterschätzt wurde. Bemerkenswert ist das Verhalten von *Diatoma elongatum*. Es belegt die auch aus Beobachtungen in anderen Gewässern bekannte

Fähigkeit der Süßwasserart, die durch die Brackwasserscheu vieler Plankter entstandene ökologische Nische in Übergangsgewässern zwischen Süßwasser und Meer zu besetzen, bei Salzgehalten, die außerhalb des arteigenen Optimums liegen. Insgesamt erwies sich das pflanzliche Netzplankton als geeigneter Indikator für die aktuellen Salzverhältnisse an den Stationen der Unterwarnow und für den Einfluß benachbarter Gewässer.

3.4. Räumliche und zeitliche Verteilung des pflanzlichen Schöpfplanktons

Die Tab. 6, 7 und 8 und die Abb. 2 und 3 vermitteln ein Bild von der Populationsdynamik der produktionsbiologisch wichtigen Arten und deren Beziehungen zu verschiedenen Einflußfaktoren im Frühjahr und Sommer 1979. Die erhebliche räumliche und zeitliche Variabilität des Biovolumens der Dominanten, der Algengruppen bzw. des Gesamtphytoplanktons ist nicht zu übersehen. Sie ist das Ergebnis mehrerer, z. T. gleichsinnig, z. T. entgegengesetzt wirkender Einflußfaktoren und der ökologischen Valenz der einzelnen Arten und daher nicht leicht zu interpretieren.

Das zeitlich gestaffelte Auftreten der Dominanten (s. Tab. 6) dürfte in erster Linie auf jahreszeitlich gebundene Einflüsse (z. B. Licht- und Temperaturverhältnisse) zurückzuführen sein. Die beobachteten Kieselalgen sind auch in anderen Küstengewässern Frühjahrs-

Tab. 6. Biomasse der Schöpfplanktondominanten 1979, Stat. 1—6, Oberfläche (10^6 μm^3/l)

	14. 3.	10. 4.	23. 5.	27. 6.	26. 7.	9. 11.
Station 1						
Melosira ambigua	92					
Stephanodiscus hantzschii	6	15580	14170	14000	840	n. e.
Flagellata s. l.	20	885	8580		157	
Diatoma elongatum		128	780			
cf. *Ankyra*			480			
Scenedesmus-Arten			288	240	38	
Monoraphidium, Kirchneriella				1800	840	
Sceletonema subsalsum					1170	
Cyclotella meneghiniana					200	
Melosira granulata					180	
Station 2						
Detonula confervacea	10	100				n. e.
Flagellata s. l.	66	308	176		1635	
Stephanodiscus hantzschii		7140	1230	14000	67	
Diatoma elongatum		71	246			
Scenedesmus-Arten			60	240	54	
cf. *Ankyra*			120			
Monoraphidium, Kirchneriella				1800	250	
Melosira granulata					511	
Cyclotella meneghiniana					268	
Sceletonema subsalsum					1170	
Station 3						
Stephanodiscus hantzschii	60	7000	980	7840	67	74
Detonula confervacea	40	80				
Flagellata s. l.	256	460	601		5747	
Diatoma elongatum		73	204			
Monoraphidium, Kirchneriella				1800	1320	
Sceletonema subsalsum						72
Oscillatoria n. d.						53
Station 4						
Detonula confervacea	730	75				
Flagellata s. l.	184	147	2389	2445	7920	
Stephanodiscus hantzschii		4995	426	4480		
Monoraphidium, Kirchneriella				1800	7200	
Amphidinium sp.				240	32	
Sceletonema subsalsum						72
Station 5						
Sceletonema costatum	64					
Stephanodiscus hantzschii	81	2968	73			
Detonula confervacea		180				
Flagellata s. l.			1353	617	3048	
Monoraphidium, Kirchneriella				300	564	
Amphidinium sp.				132	36	
Sceletonema subsalsum						72
Station 6						
Stephanodiscus hantzschii	n. e.	122				
Detonula confervacea		195				
Flagellata s. l.			1025	284	6740	28
Monoraphidium, Kirchneriella				600		
Amphidinium sp.				320	80	
Sceletonema costatum					520	

Tab. 7. Verteilung der Algengruppen in der Unterwarnow, Stat. 1—6

	1	2	3	4	5	6
14. 3. 79						
Kieselalgen	0,098	0,061	0,141	0,764	0,217	n. e.
Grünalgen		0,012	0,012	0,012	0,012	
Flagellata s. l.	0,020	0,066	0,256	0,184	0,184	
Gesamt	0,118	0,139	0,409	0,960	0,233	
10. 4.						
Kieselalgen	15,760	7,343	7,230	5,172	3,283	0,413
Flagellata s. l.	0,885	0,308	0,460	0,147	0,066	0,005
Gesamt	16,645	7,651	7,690	5,319	3,349	0,418
23. 5. 79						
Kieselalgen	15,036	1,518	1,184	0,426	0,073	
Grünalgen	0,768	0,180	0,120	0,060	0,060	
Flagellata s. l.	8,580	0,176	0,601	2,389	1,353	1,025
Blaualgen				0,440		
Feueralgen				0,012		
Gesamt	24,384	1,874	1,905	3,315	1,486	1,025
27. 6. 79						
Kieselalgen	14,000	14,000	7,840	4,480		
Grünalgen	2,040	2,040	1,830	1,800	0,300	0,600
Flagellata s. l.				2,445	0,617	0,284
Feueralgen				0,240	0,132	0,320
Gesamt	16,040	16,040	9,670	8,965	1,049	1,204
26. 7. 79						
Kieselalgen	2,390	2,016	0,067			0,520
Grünalgen	0,925	0,333	1,320	7,200	0,564	0,005
Flagellata s. l.	0,157	1,635	5,747	7,920	3,048	6,740
Feueralgen				0,032	0,042	0,080
Gesamt	3,472	3,984	7,134	15,152	3,654	7,345
9. 11. 79						
Kieselalgen	n. e.	n. e.	0,146		0,072	
Grünalgen			0,024			
Blaualgen			0,053	0,009		0,028
Gesamt			0,223	0,009	0,072	0,028

Tab. 8. Verteilung des Phytoplanktongesamtvolumens in der Unterwarnow 1979, Stat. 1—6

	1	2	3	4	5	6
14. 3. 1979	0,1	0,1	0,4	1,0	0,2	n. e.
10. 4.	16,6	7,7	7,7	5,3	3,3	0,4
23. 5.	24,4	1,9	1,9	3,3	1,4	1,0
27. 6.	16,0	16,0	9,7	9,0	1,0	1,2
26. 7.	3,5	4,0	7,1	15,2	3,7	7,3
8.	n. e.					
9.	n. e.					
10.	n. e.					
9. 11.	n. e.	n. e.	0,2	<0,1	0,1	<0,1

formen, die Grünalgen Sommerformen. Die Verteilung der Populationen im Gewässerverlauf zeigt vor allem Beziehungen zum Salzgehalt, aber auch zum Trophiegrad und zur Gewässerdynamik. So liegen z. B. die Biomassemaxima der wichtigsten Oligohalobien an der Station 1 im Oberlauf, d. h. in einem Bereich mit niedrigem Salzgehalt, guter Nährstoffversorgung und wahrscheinlich allochthonem Algennachschub. Ihre Populationsstärken nehmen mit steigendem Salzgehalt, zunehmender Entfernung vom Entwicklungszentrum und zunehmendem Verbrauch der durch die Warnow zugeführten Nährstoffe ab. Zusammenwirken und zeitweilige Prädominanz der verschiedenen Einflußfaktoren lassen sich an der Populationsdynamik von *Stephanodiscus hantzschii* gut verfolgen. Im April erscheint der Salzgehalt als bestimmender Faktor. Der Halobität der Alge entsprechend, zeigen Biomasse und Salzgehalt bei ausreichender Versorgung mit N und P (Si leider nicht bestimmt) einen ausgesprochen reziproken Verlauf. Für das Fehlen (Stat. 4 und 5) bzw. die niedrigen Biomassekonzentrationen (Stat. 2 und 3) der Art im Mai können neben einem massiven Salzwassereinstrom auch die sehr niedrigen Si-Konzentrationen verantwortlich gemacht werden, weiterhin eine Massenentwicklung von *Eurytemora affinis* und Rotatorien, durch die offensichtlich auch an höhere Salzgehalte angepaßte Phytoplankter „kurz gehalten" wurden. Für den hohen Stellenwert der Gewässerdynamik spricht die Wiederausbreitung der Populationen im Juni nach erneutem Zustrom von Warnowwasser. Die mit steigendem Salzgehalt abnehmenden Biomassekonzentrationen weisen wie die gleichzeitige Zunahme offensichtlich besser an höhere Salzgehalte angepaßter nanoplanktischer Flagellaten und Chlorophyceen auf die vorrangige Bedeutung der Salzverhältnisse für die Verteilung von *Stephanodiscus hantzschii* in diesem Monat hin. Am Fehlen der Art im Bereich der Station 5 im Juni — wie an der niedrigen Biomassekonzentration überhaupt — könnte jedoch zusätzlich ein Nährstoffmangel ursächlich beteiligt gewesen sein.

Die von Mai bis Juli beobachteten Flagellaten- und Chlorophyceenpopulationen zeigen hinsichtlich ihrer Verteilung in der Unterwarnow engere Beziehungen zu den Nährstoffquellen als zum Salzgehalt. Die größten Biomassekonzentrationen dieser Algen wurden an der Station 1 (Mai) und an der Station 4 (Mai—Juni) registriert, d. h. im Einflußbereich des Warnowzuflusses und unterhalb des Kläranlagenablaufs.

Räumliche und zeitliche Variabilität des Phytoplanktongesamtvolumens im Verlauf der Unterwarnow ergeben sich aus der geschilderten umweltabhängigen Populationsdynamik der dominierenden Arten. Die ausgesprochene Reziprozität von Phytoplanktonbiomasse und Salzgehalt im April und Juni (s. Abb. 2) ist eine Folge der Prädominanz von *Stephanodiscus hantzschii* in diesen Monaten. Die entgegengesetzte Tendenz der beiden Parameter im Verlauf der Unterwarnow wird auch durch die an der Verschiebung des Artenspektrums erkennbare Nährstoffwirkung an der Station 4 im Juni nicht wesentlich gestört. Die von diesem Verhalten abweichende Verteilung der Phytoplanktonbiomasse im Mai und Juli beruht auf der überwiegend nährstoffabhängigen Populationsdynamik der in diesen Monaten dominierenden Flagellaten und Chlorophyceen. Wie ein Vergleich mit den Nährstoffkonzentrationen zeigt, ist die vor allem im Juli beobachtete auffällige Förderung

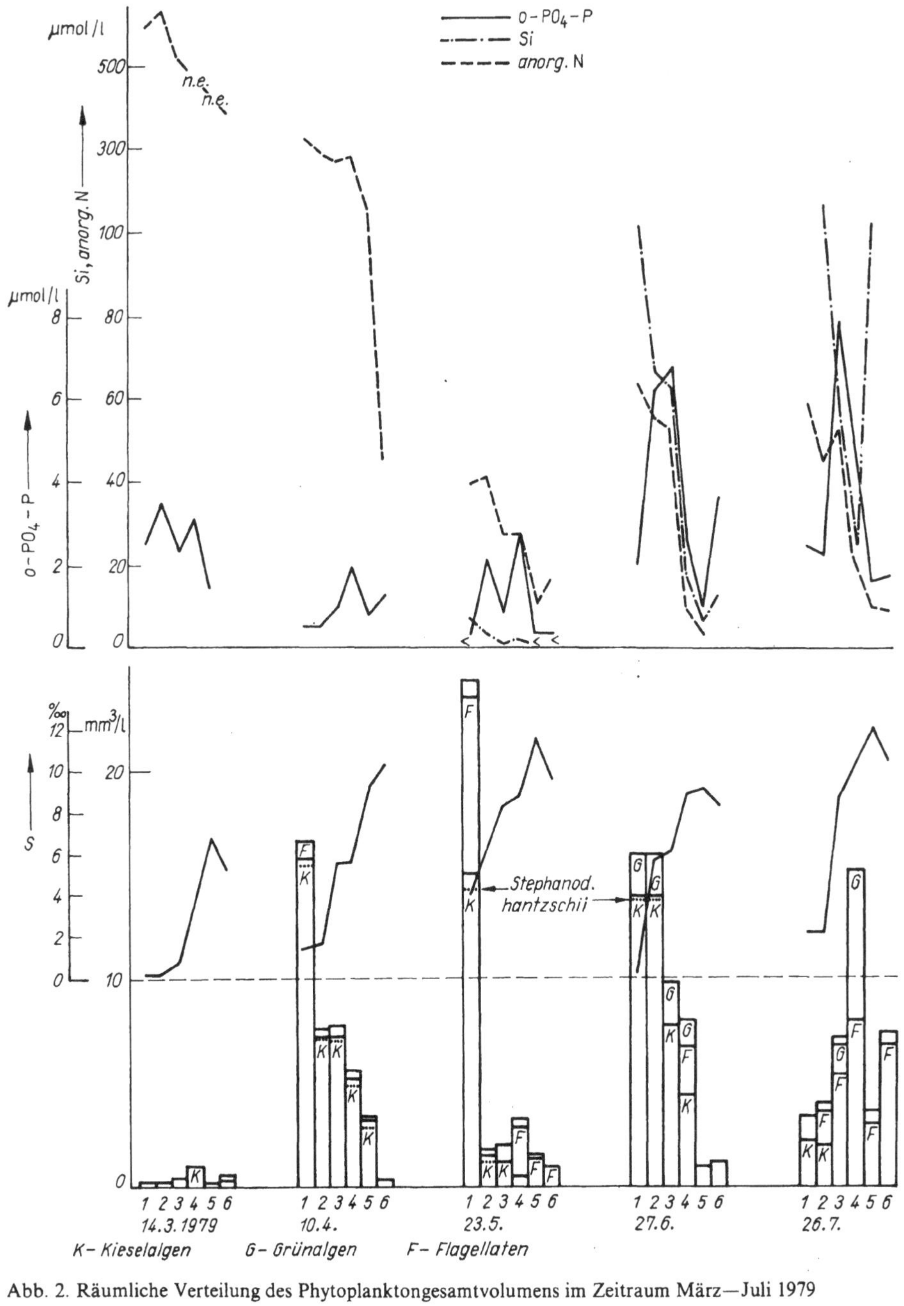

Abb. 2. Räumliche Verteilung des Phytoplanktongesamtvolumens im Zeitraum März—Juli 1979

der Phytoplanktonentwicklung im Bereich der Stationen 3 und 4 auf einen erhöhten P-Eintrag bei gleichzeitiger Verfügbarkeit von N und Si im Juni und Juli zurückzuführen. Der kleine Kieselalgengipfel an der Station 4 im März, repräsentiert durch die polyhalobe Art *Detonula confervacea*, könnte auf einem 3. Wege, nämlich durch lokal begrenztes Aufsteigen von salzreicherem Tiefenwasser, zustande gekommen sein, da die Alge im Oberflächenwasser der peripher gelegenen Stationen 5 und 6 trotz des dort höheren Salzgehalts nicht in nennenswerten Populationen beobachtet wurde.

Wie aus Abb. 3 ersichtlich, lagen die Biomassegipfel des Gesamtphytoplanktons im Untersuchungszeitraum ausschließlich an den Stationen 1 oder 4, d. h. im Einflußbereich der Belastungsquellen. Zumindest für die Zeit von April—Juli ist dabei ein ursächlicher Zusammenhang zwischen beiden Parametern anzunehmen. Von der Station 4 nahmen die Biomassekonzentrationen sowohl in Richtung Station 6 (Breitling) als auch Station 5 (Mündung) ab. Die Nährstoffkonzentrationen zeigten zwar mit Ausnahme des o-PO_4-P im Juni die gleiche Tendenz, Anzeichen für eine möglicherweise bestehende Nährstofflimitation sind aber nur im Mai erkennbar (Absinken der Si- und o-PO_4-Konzentrationen unter die Nachweisgrenze).

Die zeitlichen Veränderungen der Phytoplanktonbiomasse an den einzelnen Stationen (s. Abb. 3) lassen keinen einheitlichen Jahresgang erkennen. Das Maximum des Untersuchungszeitraumes lag an der Station 1 im Mai, an den Stationen 2 und 3 im Juni und an den sich anschließenden Stationen 4—6 im Juli. An der Station 1 war der Verlauf 1-gipflig. Zu- und Abnahme erfolgten kontinuierlich. Die Biomassekonzentrationen der Stationen 2—5 zeigen auf Grund der Depression im Mai einen 2-gipfligen Verlauf mit in Stromrichtung zunehmender Angleichung der beiden Gipfel. An der Station 6 schließlich fehlte der Frühjahrsgipfel, woraus eine durchgehend zunehmende Tendenz für den gesamten Untersuchungs-

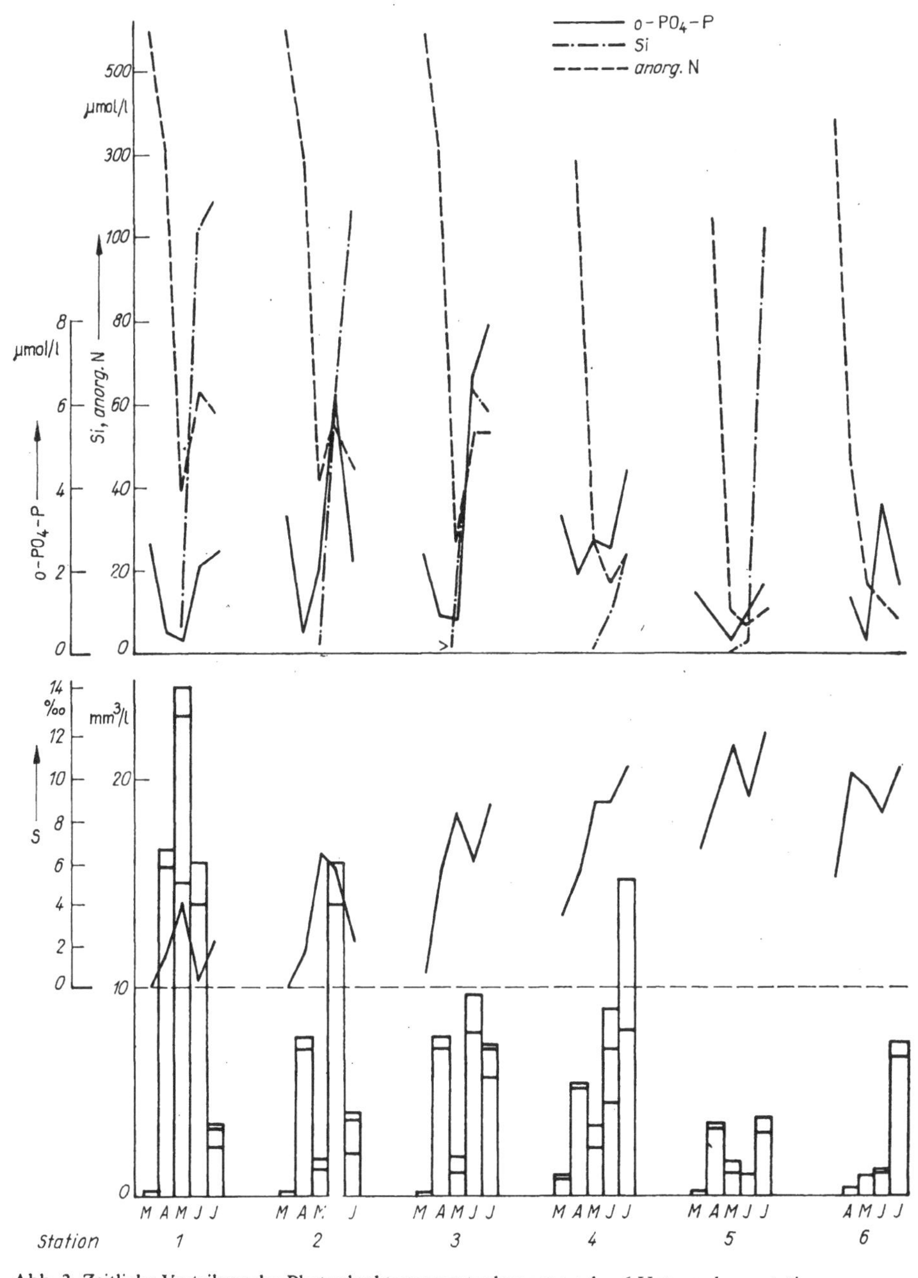

Abb. 3. Zeitliche Verteilung des Phytoplanktongesamtvolumens an den 6 Untersuchungsstationen

zeitraum resultiert. Ein Vergleich mit den entsprechenden Jahresgängen der Einflußfaktoren bestätigt die Interpretation der Populationsdynamik der dominierenden Phytoplankter. Weiterhin ergeben sich folgende Zusammenhänge: Der Biomassegipfel an der Station 1 im Mai fällt zeitlich mit einem N, P und Si gleichermaßen betreffenden Nährstoffminimum zusammen, das die weitgehende Umsetzung der mit dem Warnowwasser eingetragenen anorganischen Nährstoffe anzeigt. Daß diese intensive Bioproduktion trotz des von März bis Mai von ca. 0,1 auf $4^0/_{00}$ angestiegenen Salzgehalts möglich war, hängt einerseits mit der Reaktionsfähigkeit der Biozönose zusammen (vermehrtes Auftreten von Flagellaten), andererseits aber wahrscheinlich auch mit dem im Vergleich zu den Vormonaten erheblich geringeren Durchfluß (BACHOR 1989) und der damit verbundenen längeren Verweilzeit der Wassermassen. Die Abnahme der Biomassekonzentrationen im Juni/Juli, trotz der für Oligohalobien wieder günstigeren Salzverhältnisse, ist durch das Verhalten der Nährstoffkonzentrationen nicht zu erklären, denn anorg. N, o-PO_4-P und Si waren in ausreichender Menge und in einem günstigen Verhältnis verfügbar. Für *Stephanodiscus hantzschii* kann ein der Jahreszeit entsprechendes Ausklingen der Populationsentwicklung angenommen werden. Weshalb die frei werdende Nische nicht durch andere Arten besetzt wurde, ist mit Hilfe des zur Verfügung stehenden Datenmaterials jedoch nicht zu beantworten.

An den sich anschließenden Stationen im Stromverlauf sind die jahreszeitlichen Veränderungen der Nährstoffkonzentrationen kaum noch mit der entsprechenden Veränderlichkeit der Biomassekonzentrationen zu korrelieren. Das ist insofern plausibel, als die Nährstoffkonzentrationen in diesem Bereich nicht durch gewässertypische biologische Rhythmen, sondern durch die anthropogene Belastung verändert werden und weiterhin die von März bis Juli an einer Station festgestellten Biomassekonzentrationen z. T. ganz verschiedenen Wasserkörpern entstammen (s. Salzgehalt!). Die zeitliche Veränderlichkeit der Biomassekonzentrationen an den genannten Stationen entspricht daher nicht einem stationstypischen Jahresgang. Bei anderen Ein- und Ausstromverhältnissen ist auch mit anderen jahreszeitlichen Veränderungen der Bioproduktion zu rechnen.

4. Zusammenfassende Charakteristik

Das Phytoplankton der Unterwarnow ist artenreich. Ursache der Mannigfaltigkeit ist der Zustrom von Wassermassen aus Oberwarnow und Ostsee, die den unterschiedlichen Salzverhältnissen entsprechend von verschiedenen Arten besiedelt sind. Im Grenzbereich zwischen marinen und limnischen Einflüssen besteht ein Artenminimum (Stationen 4 und 6). Auf Grund intensiver Austauschprozesse im Stromverlauf ist die Artenreduktion an der Station 4 weniger ausgeprägt als an der abseits liegenden Station 6 im Breitling.

Mehr als 50% der festgestellten Arten sind Kieselalgen. Es folgen mit erheblich geringeren Anteilen Chlorophyta, Cyanophyta und Pyrrhophyta. 155 der determinierten Taxa gelten als Halobien, 53% davon als Oligohalobien, 16% als Mesohalobien, 31% als Polyhalobien. Hinsichtlich der Salztoleranz überwiegen meio-(48 Taxa) und meso-(43 Taxa)euryhaline Arten. Pleio- und holeuryhaline Arten (21 Taxa) kommen jedoch ebenfalls vor. 58 der determinierten Phytoplankter sind Saprobien, meist Indikatoren β-mesosaprober Verhältnisse mit geringem Zeigerwert. Einige wenige Arten schließlich sind als stenotherme Kaltwasserplankter eingestuft, doch zeigt auch die Mehrzahl der übrigen Taxa eine mehr oder weniger temperaturabhängige Bindung an eine bestimmte Jahreszeit. Im Gesamtartenspektrum überwiegen als Folge von Wasseraustauschprozessen und anthropogener Belastung Arten mit breiter ökologischer Amplitude.

Die Populationsdynamik des Phytoplanktons in der Unterwarnow entspricht der ökologischen Charakteristik der beteiligten Arten. Als maßgebliche Einflußfaktoren bestimmen Gewässerdynamik, Salz- und Nährstoffverhältnisse, Jahreszeit (Temperatur- und Lichtverhältnisse) und Zooplanktonentwicklung räumliche und zeitliche Verteilung der Planktonarten. Jeder der genannten Faktoren kann zeitweilig Prädominanz erlangen. In der Regel wirken jedoch mehrere Faktoren gleichzeitig, was vor allem die Aufdeckung kausaler Zusammenhänge sehr schwierig macht.

Die räumliche Verteilung der Arten unterschiedlicher Halobität und Salztoleranz in der Unterwarnow zeigt erwartungsgemäß enge Beziehungen zu den Salzverhältnissen. Die Gesamtzahl der an den einzelnen Stationen nachgewiesenen Oligohalobien nimmt mit dem in Richtung Ostsee steigenden Salzgehalt ab, die der Meso- und Polyhalobien zu. Die Oligohalobien dominieren zahlenmäßig an den Stat. 1—3 im Oberlauf, die Salzwasserarten an den Stationen 4—6 im Unterlauf. Die damit markierte ökologische Grenze zwischen Warnow- und Ostseeeinflüssen liegt also im gleichen Bereich wie das entsprechend zu interpretierende Artenminimum. Das Verhältnis von Süß- zu Salzwasserarten ist allerdings mit 3:1 bereits an der Station 1 relativ niedrig, und schon an der Station 2 beträgt der Quotient nur noch 1,7:1, d. h. Salzwassereinbrüche aus Richtung Ostsee erreichen die Station 1 und beeinflussen durch Verdrängung meioeuryhaliner Oligohalobien und Eintrag meso- und polyhalober Arten maßgeblich das Artenspektrum der unterhalb gelegenen Stationen. Auf Grund der mit Salzwassereinbrüchen meist verbundenen vertikalen Schichtung können in Netzproben, die unter diesen Bedingungen gezogen wurden (Vertikalhol), Arten unterschiedlicher Halinitätsbereiche nebeneinander vorkommen. Ihre Populationsstärke vermittelt einen ungefähren Eindruck von der vertikalen Ausdehnung der übereinander liegenden, verschieden salzigen Wasserkörper.

Die räumliche Verteilung des Saprobien zeigt keine entsprechenden Beziehungen zu den Hauptbelastungsquellen als den Steuergrößen der Saprobität. Ihre Abnahme im Stromverlauf ist salzgehaltsbedingt und kann nicht als Zeichen des fortschreitenden Abbaus der organischen Belastung gewertet werden.

Jahreszeitliche Sukzession und Populationsentwicklung der Phytoplankter werden durch Netz- und Schöpfproben unterschiedlich widergespiegelt. In den Netzproben dominieren größere Arten, die den gesamten Halinitätsbereich der abgefischten Wassersäule repräsentieren. In den bisher untersuchten Schöpfproben aus dem Oberflächenwasser bestimmen Nanoplankter das Bild, die vor allem produktionsbiologische Bedeutung haben. Sowohl im Netz- als auch im Schöpfplankton bilden wenige Arten die Hauptmenge des pflanzlichen Planktons.

Die jahreszeitliche Sukzession der Netzplankter verläuft in den verschiedenen Halinitätszonen unterschiedlich. Im Frühjahr dominieren bei höherem Salzgehalt ($\geqq$ ca. $10^0/_{00}$) polyhalobe Kaltwasseralgen wie *Detonula confervacea* und *Achnanthes taeniata*, bei niedrigem Salzgehalt ($\leqq 10^0/_{00}$) Brackwasserarten wie *Sceletonema subsalsum* und Süßwasserarten wie *Asterionella formosa*, div. *Melosira*-Arten etc. Gegen Ende des Frühjahrs, wenn die Kaltwasserformen — möglicherweise infolge ansteigender Wassertemperaturen — ihre Vegetationsperiode beenden, bildet die halophile Kieselalge *Diatoma elongatum* in allen Bereichen der Unterwarnow größere Populationen aus. Bei starker Vermehrung von Zooplanktern schließt sich an die Frühjahrsentwicklung der pflanzlichen Netzplankter im gesamten Gewässer eine bis zu 2 Monaten andauernde Periode an, in der zumindest die Netzproben nahezu frei von Phytoplanktern sind. Den sommerlichen Aspekt bestimmen bei höherem Salzgehalt polyhalobe Kieselalgen (*Rhizosolenia*-Arten, *Ditylum brightwelli* etc.) und Ceratien, bei niedrigem Salzgehalt oligohalobe Kieselalgen (z. B. *Melosira*-Arten) und Grünalgen. In den Herbstmonaten kann es vor allem bei höherem Salzgehalt zum erneuten Aspektwechsel kommen, wobei auch einzelne Frühjahrsformen (z. B. *Sceletonema*) wieder erscheinen können. An welchen Stationen die der Jahreszeit entsprechenden Repräsentanten der verschiedenen Salzgehaltsbereiche auftreten, hängt von der Gewässerdynamik ab. Die Dominanten des pflanzlichen Netzplanktons sind auf Grund des dargestellten spezifischen Verhaltens geeignete Indikatoren für die aktuellen Salzverhältnisse an den Stationen der Unterwarnow. Räumliche und zeitliche, qualitative und quantitative Veränderungen sind Parameter von Wasserbewegungen und Austauschprozessen innerhalb des Ästuars oder zwischen Unterwarnow und benachbarten Gewässern, sofern sie mit Salzgehaltsänderungen einhergehen.

Wegen der relativ kurzen Untersuchungsperiode sind die durch die Untersuchung von Schöpfproben gewonnenen Ergebnisse zunächst nur mit Vorsicht zu verallgemeinern. Trotzdem lassen sich auf Grund eindeutiger Beziehungen zwischen den Phytoplanktonbefunden und wichtigen Einflußfaktoren einige generelle Aussagen treffen. Träger der planktischen Primärproduktion in der Unterwarnow sind jeweils nur wenige Taxa, im Untersuchungszeitraum März—Juli 1979 maximal ca. 8, im Einzelfall weniger. Die Zahl der beteiligten Arten nimmt in Stromrichtung ab. Im Frühjahr ist *Stephanodiscus hantzschii* bei Salzgehalten bis ca. $9^0/_{00}$ und unter Ausstrombedingungen die dominierende Art. Bei Salzgehalten $> 5^0/_{00}$ können daneben Flagellaten eine Rolle spielen. Das Salzgehaltsgefälle wird im Frühjahr weniger durch die Dominanten als durch den Wechsel der Begleitarten widergespiegelt. Unter sommerlichen Bedingungen und bei einem niedrigen Salzgehalt um $2^0/_{00}$ besteht die Phytoplanktonbiomasse aus oligo- und mesohaloben Kieselalgen, Grünalgen und Flagellaten. Bei Salzgehalten $> 8^0/_{00}$ dominieren Flagellaten und Chlorophyceen.

Räumliche und zeitliche Variabilität des Phytoplanktongesamtvolumens werden durch die Populationsdynamik der dominierenden Arten bestimmt, die ihrerseits von den Umweltansprüchen der Dominanten und dem Zusammenspiel der maßgeblichen Einflußfaktoren abhängig ist. Beim Vorherrschen oligohalober Algen geringer Salztoleranz zeigen Biomasse und Salzgehalt im Verlauf der Unterwarnow eine ausgesprochene Reziprozität. Dominieren Arten mit überwiegend nährstoffabhängiger Populationsdynamik, entspricht die räumliche Verteilung der Biomassekonzentrationen dem von den Hauptbelastungsquellen ausgehenden trophischen Gefälle. Die räumlichen Biomassemaxima liegen im Frühjahr bei starkem Zustrom nährstoffreicher Wassermassen aus der Oberwarnow an der Station 1, im Juli an der Station 4, d. h. unterhalb des Kläranlagenablaufs. Die Biomassemaxima der Untersuchungsperiode (März bis Juli 1979) traten an der Station 1 im Mai, an den Stationen 2 und 3 im Juni, an den Stationen 4—6 im Juli auf. Ihre Höhe nahm von Station 1—3 zunächst ab, um dann bis zur Station 4 erneut deutlich anzusteigen. Biomasseentwicklung und -verteilung werden nach diesen Beobachtungen im Frühjahr überwiegend durch die Nährstoffeinträge aus der Oberwarnow bestimmt, wobei deren Einfluß mit zunehmender Entfernung abnimmt. Mit zunehmender Jahreszeit gewinnt dann offensichtlich der Kläranlagenablauf als Nährstoffquelle an Bedeutung. Niedrige Biomassekonzentrationen trotz vorhandener Nährstoffe im Mündungsbereich der Unterwarnow sprechen für eine unvollständige Umsetzung des Eintrags im Gewässer. Die Ursachen sind komplexer Natur. In Frage kommen Salzgehaltswirkung, relativ niedrige Verweilzeit und ein im Jahresverlauf zunehmendes ungünstiges N:P Verhältnis.

Die biologischen Verhältnisse im Breitling weichen in mehrfacher Hinsicht vom eigentlichen Strom ab. Niedrige Artenzahlen im Netz- und Schöpfplankton, fehlende Frühjahrspopulationen, und mit Ausnahme des Juni sehr niedrige Biomassekonzentrationen kennzeichnen im Untersuchungszeitraum den peripher gelegenen, seeähnlich erweiterten Gewässerabschnitt.

Eine klare Biomasse/Nährstoffbeziehung, wie sie für Standgewässer des Binnenlandes und auch für die Ostsee bekannt ist, besteht in der Unterwarnow offensichtlich nicht. Das Zustandekommen wird durch den kontinuierlichen Nährstoffeintrag, durch die Lage der Belastungsquellen und Wasserbewegungen wechselnder Richtung verhindert. Nährstofflimitationen spielten im Untersuchungszeitraum keine Rolle. Niedrige Konzentrationen eines oder mehrerer Grundnährstoffe im Oberlauf zeigen bei gleichzeitig hohen Biomassekonzentrationen eher eine gute Umsetzung des Angebots als einen Mangel an. Lediglich an den Außenstationen wird durch ein ungünstiges N:P-Verhältnis ein beginnender N-Mangel im Sommer angedeutet.

5. Zusammenfassung

An 6 Stationen im als „Unterwarnow“ bezeichneten Ästuar der Warnow wurden 1979 und 1980 — parallel zu chemischen Untersuchungen — an Netzproben aus der gesamten Wassersäule (Vertikalhol) und Schöpfproben aus dem Oberflächenwasser Phytoplanktonuntersuchungen durchgeführt. Registriert wurden insgesamt 190 Taxa, von denen 100 zu den Kieselalgen gehören. Die Verteilung der Halobien zeigt enge Beziehungen zu den Salzverhältnissen im Gewässer. Im Grenzbereich zwischen Flußwasser- und Ostseeeinfluß besteht ein Artenminimum. Sowohl in Netz- als auch in Schöpfproben bestimmen wenige Arten den jeweiligen Aspekt. Räumliche und zeitliche Verteilung der Dominanten des Netzplanktons werden hinsichtlich ihrer Beziehungen zu Salzgehalt und Jahreszeit analysiert. Die Netzplankter erwiesen sich dabei als geeignete Indikatoren der Salzverhältnisse und deren eng mit der Gewässerdynamik verbundenen Variabilität.

Bei den vor allem produktionsbiologisch wichtigen Dominanten der Schöpfproben handelt es sich überwiegend um Nanoplankter, die auf Grund ihrer geringen Größe im Netzplankton keine Rolle spielen. Im Untersuchungszeitraum März—Juli 1979 war *Stephanodiscus hantzschii* die beherrschende Art im oberen Abschnitt der Unterwarnow. Im Unterlauf erlangten zeitweilig auch diverse Flagellaten produktionsbiologische Bedeutung, Vertreter des Ostseeplanktons jedoch kaum. Räumliche und zeitliche Verteilung des Gesamtphytoplanktons werden durch die Populationsdynamik der Dominanten und deren Abhängigkeit von verschiedenen Einflußfaktoren (Jahreszeit, Salzgehalt, Wasseraustausch, Zooplanktonentwicklung) bestimmt. Die Biomassemaxima lagen im Einflußbereich der Belastungsquellen. Die Maxima während des Untersuchungszeitraumes wurden im Unterlauf später erreicht als im Oberlauf. Die Biomassekonzentrationen indizieren eine stromabwärts zunehmende Umsetzung der eingetragenen Nährstoffe.

Literatur

BACHOR, A. (1988): Untersuchung zum Salz- und Nährstoffhaushalt der Unterwarnow — einem Ästuar der westlichen Ostsee. Beitr. Meereskd., H. **59** (im Druck).

— (1989): On the variations of some chemical criteria in the Unterwarnow — an estuary of the Western Baltic Sea in 1977—1986. Limnologica (Berlin) **20** (1), 29—32

BREITIG, G., & VON TÜMPLING, W. (1982): Ausgewählte Methoden der Wasseruntersuchung. Berlin.

CASPERS, H. (1959): Vorschläge einer Brackwassernomenklatur ("The Venice-System"). Beschluß des Symposiums in Venedig, April 1958. Int. Rev. ges. Hydrobiol. **44** (2), 313—315.

CLEVE-EULER, A. (1968): Die Diatomeen von Schweden und Finnland. Lehre.

PANKOW, H., KELL, V., & MARTENS, B. (1976): Algenflora der Ostsee. II. Plankton. Jena.

PASCHER, A. (1915ff.): Die Süßwasserflora Deutschlands, Österreichs und der Schweiz. Jena.

PAULSEN, O. (1908): Peridiniales. In: BRANDT, K., & APSTEIN, C. (ed.), Nordisches Plankton. XVIII. Kiel und Leipzig.

Anschrift des Verfassers:

Dr. INGEBORG SCHMIDT
van Gosenstr. 1
Stralsund
DDR-2300

Limnologica (Berlin) **20** (1), 183—186, April 1990

Wilhelm-Pieck-Universität Rostock, Sektion Biologie, Rostock, DDR

Recent Changes in the Phytal Zone of Greifswald Bay

By ULRICH MESSNER & JÖRG-ANDREAS VON OERTZEN

With 3 Figures

Key words: Greifswald Bay; Phytal; Distribution pattern; Changes; Fauna.

Abstract

The phytal zone of Greifswald Bay, a coastal water in the northeast of the GDR was studied from spring to autumn in 1985. The submersal vegetation was mapped, and it was found that the coverage of the bottom had decreased from 90% (SEIFERT 1938) to about 10%. The authors consider increasing plankton production and the deterioration of light conditions at the bottom owing to eutrophication to are the main causes of this decline. Quantitative studies of the macrofauna of the phytal zone show that the zone covered by red algae [mainly *Furcellaria fastigiata* (L.) LAMOUR in the investigation area] is populated by a large number of species in various growth stages. The fauna inhabiting the less ramified *Potamogeton pectinatus* L. and *Zostera marina* L. stands was less diversified. It can be assumed that the decline of the red algae stocks to insignificant remains has led to major changes in the food web and energy flow. Although no quantitative data can yet be offered, it must be feared that loss of this habitat threatens the continued existence of a herring population there. The dense bluegreen blooms in mid-summer may restrict the use of these waters by holidaymakers for bathing. The changes that have taken place thus seem to jeopardize many of the uses to wich Greifswald Bay has been put.

1. Introduction

Human society is interested in lakes, rivers and the sea in many ways, using them for fisheries, transport, recreation, industry and waste water disposal. Their one-sided use in the past decades has, through eutrophication and pollution, destabilized the majority of aquatic systems. This has affected particularly the relatively closed and intensly utilized coastal waters, where the phytal zone plays an important role as both a component and functional element of the whole brackish water ecosystem. Macrophytobenthic communities, for instance are of major economic importance for fisheries as a spawning substrate, „kindergarten“ and habitat for several fish species. Furthermore, algal belts also act as sewage filters, wave breakers, etc.

2. Area of investigation

The area of investigation, Greifswald Bay, is located in the western part of the Baltic between the mainland and the island of Rügen. To the east it is bordered by an underwater slope and to the west by the mouth of the Strelasund (Fig. 1). Compared with its surface of 514 km^2, the Bay is relatively shallow with an average water depth of 5.6 m. The maximum summer temperature of the water is 20 °C, but in winter the bay can be covered with ice for several weeks. In terms of salinity the water is β-mixo-mesohaline, the mean annual salinity being about 7‰. According to SCHMIDT (1983) the bay is well oxygenated down to the bottom. As regards its nutritive status the bay is eutrophic to polytrophic (SCHMIDT 1983).

3. Methods

The main investigations were carried out from a sailing boat from April to November 1985. Underwater observation and sampling were performed by skin diving. Samples were collected with a net wich was pulled over the plants and then closed at the bottom. The plants were then freed from their substrate and sorted manually. Animals were fixed in formalin and later identified, counted and weighed. Dominances were determined on the basis of fresh weight. Altogether 73 samples were taken at 18 stations. GEISEL (1986) has described the plant sociological aspects of the vegetation in this area.

4. Results and Discussion

4.1. Growth conditions

The benthic vegetation in Greifswald Bay is clearly zonated (Fig. 1). The species and groups giving the zones their names are either dominant or form pure belts. *Potamogeton pectinatus* L. and *Zostera marina* L. are the most abundant species. Red algae have colonized the areas at the bottom of rocky cliffs. In some places relatively small belts of *Fucus vesiculosus* L. are present. Most regions with a high substrate mobility owing to wave action, currents or the nature of the bottom (sand) are free of vegetation. We have estimated that 10 to 15% of the submerged area of Greifswalder Bay is vegetated. Earlier descriptions by REIBISCH (1904), HENKING (1904) and SEIFERT (1938) give quite a different picture. According to SEIFERT (1938) about 90% of the bottom of Greifswald Bay was covered with plants. Interestingly, he states that red algae were found at depths between 3 and 8 m; they must therefore have covered most of the bottom. REIBISCH (1904) and HENKING (1904) gave an impressive description of the problems this caused for fishermen working with trawls, and REIBISCH (1904) described the "... extraordinary amounts of attached plants, especially red algae, sometimes forming a dense canopy which even the large, sharp fishing trawls could not penetrate. This was also the case where the most common alga, *Fastigaria furcellata*, was present as the non-attached *aegagropila* form. Since this form floats just above the bottom, the edges of the nets are soon completely fouled. As a result the net opening rises increasingly above the bottom."

In our opinion the poor light conditions at the bottom caused by increased plankton production is the main

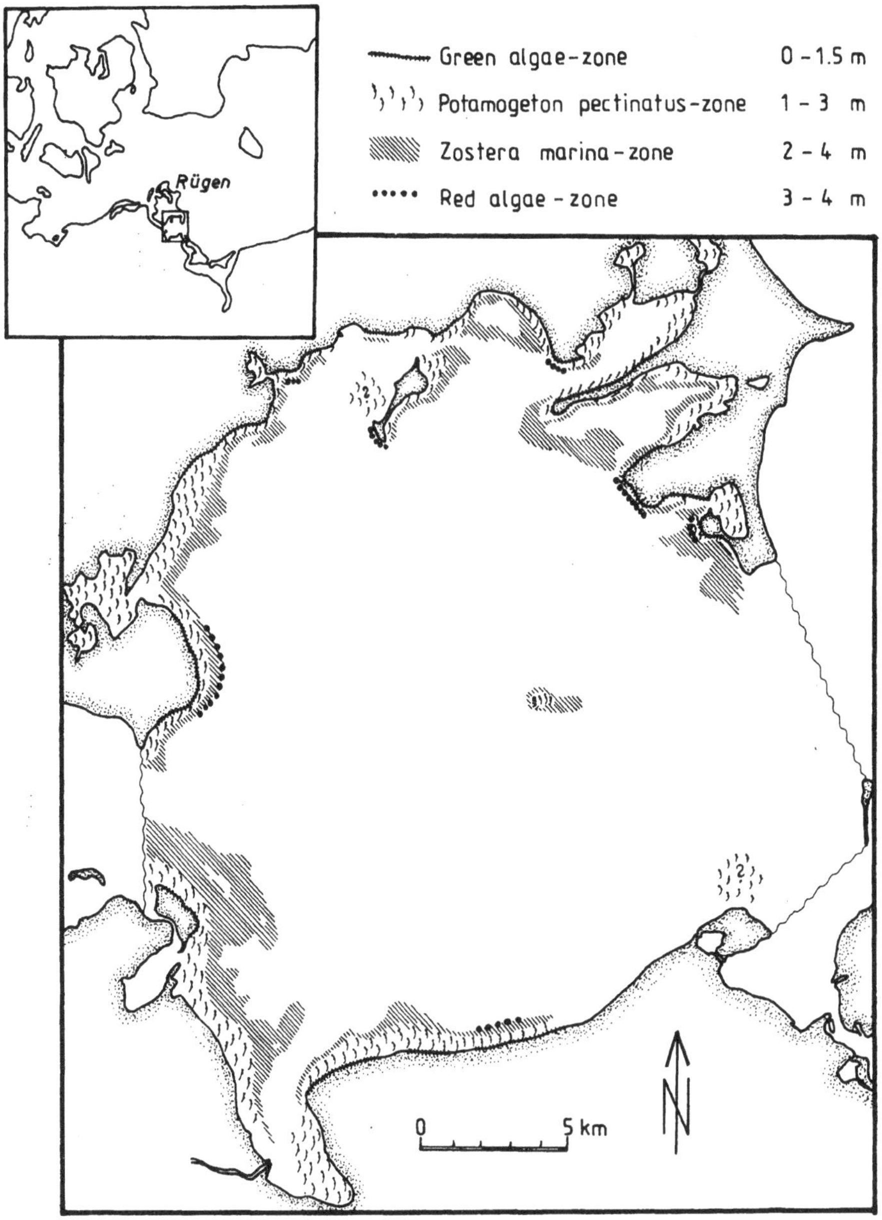

Fig. 1. Greifswald Bay, submersal vegetation (according to MESSNER & GEISEL, in press)

reason for the present drastic decline of the benthic vegetation. The few, but meaningful, comparative data available on this subject include transparency data from SEIFERT (1938) and SUBKLEW (1955) as well as information given by SCHMIDT (1983). SEIFERT (1938) wrote that the water transparency was never less than 2.5 m even during maximum plankton production. SUBKLEW (1955) reported a transparency of 1 to 1.5 m during the plankton bloom, and SCHMIDT (1983) gave a figure of only 20 cm. According to SEIFERT (1938) the phytal zone extended to a depth of 8 m. SUBKLEW (1955) stated that it reached down to about 6 m, and at present its mean depth is about 4 m. In calm and severely eutrophic inlets the submersal vegetation extends down to only 2 to 3 m.

4.2. The phytal fauna

The fauna inhabiting the phytal zone was studied quantitatively and qualitatively. The first step was to ascertain the species composition, and this led to the discovery of *Gammarus inaequicauda* STOCK for the first time in the GDR. The dominant species are *Idotea chelipes* (PALLAS), *Gammarus oceanicus* SEGERSTRÅLE, *Theodoxus fluviatilis* (L.) and *Mytilus edulis* L. Further typical species include *Jaera albifrons* LEACH, *Corophium lacustre* VANHOEFFEN, *Heterotanais oerstedti* (KRÖYER), *Leptocheirus pilosus* ZADDACH and *Jaera albifrons* LEACH, and *Membranipora crustulenta* (PALLAS). During the summer of 1985 we observed the mass development of *Laomedea flexuosa* ALDER, but this was not repeated in subsequent years. Quantitative samples were taken from *Cladophora*, *Potamogeton*, *Zostera*, *Fucus* and *Furcellaria* belts. One of our aims was to describe the distribution of the fauna inhabiting the phytal zone in different parts of Greifswald Bay. The results of these studies verified the link suspected by SEGERSTRÅLE (1943) and formulated by WIESER (1959) between the structure of plants and species composition of the fauna. In view

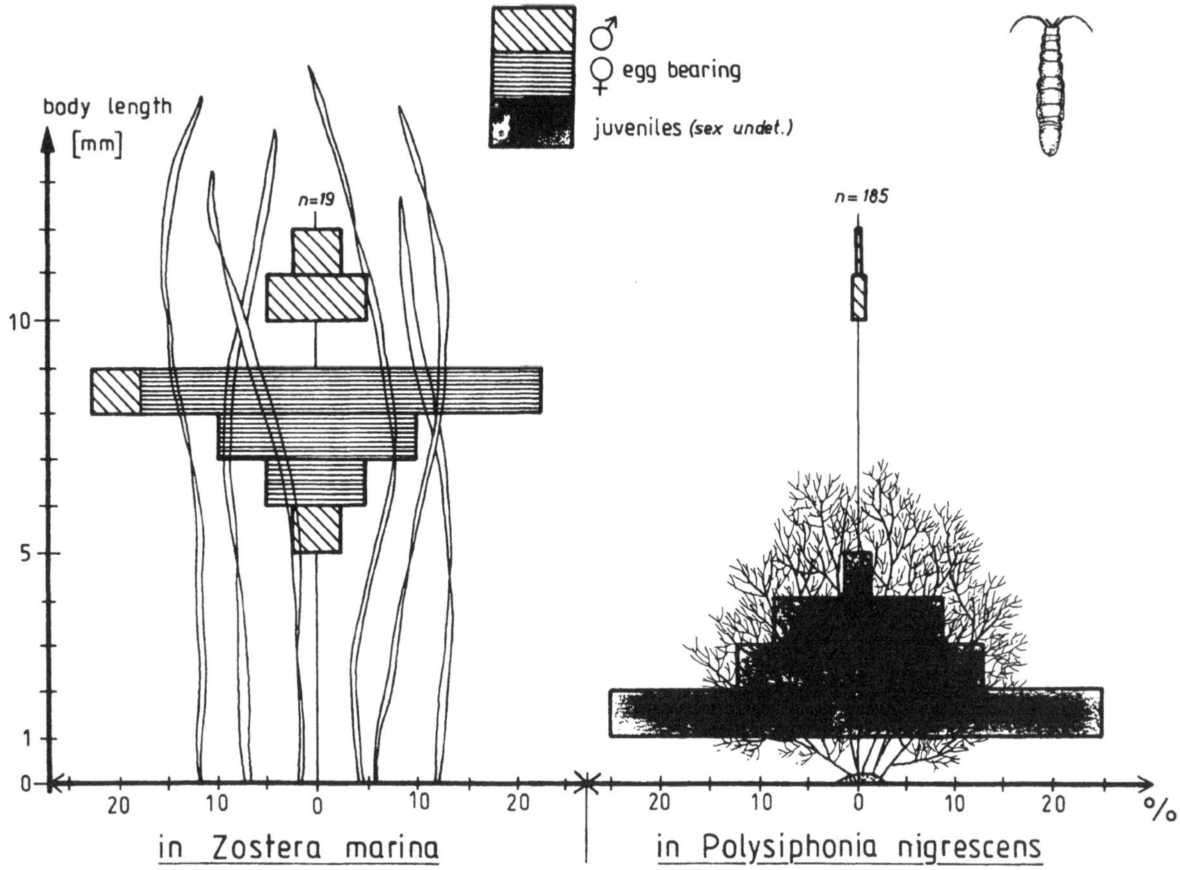

Fig. 2. Substrate preferences of different age classes of *Idotea chelipes* (PALLAS), 12. 8. 1985, St. 18, 3 m depth, 3 samples

of the great reduction of the phytal zone during the past 50 years, the biomass distribution and numbers of individuals within the phytal zone were of special interest. They varied in the course of the year, depending on the reproduction phases of the different animal species, and reached their peak in summer. It was evident that different age groups of a given species prefer different substrates. Fig. 2 demonstrates this in the case of *Idotea chelipes*, showing three samples taken from the same depth at one station. The sexually mature adults can be seen on *Zostera* leaves, whereas the juveniles find food, a suitable substrat and protection from predators only in the abundant and densely ramified *Polysiphonia* stands.

Even though population density changes with reproduction phase, it is possible to generalise the situation shown in Fig. 3 for August: The more ramified plant species growing at the greatest depth are most densely populated. It should also be considered that the *Cladophora* and *Potamogeton* belts exist for only a relatively short time and can therefore serve as a habitat for the fauna of the phytal zone only in summer.

5. Conclusions

The deterioration of light conditions in the bottom layers owing to eutrophication has reduced the maximum depth of the phytal zone to 4 m. This has led to the complete destruction of the former stocks of red algae in the deep water. However, as our survey shows, the red algae occupy that part of the phytal zone with the densest animal population. The present extensive *Potamogeton pectinatus* population exists only in summer and is only sparsely ramified. It provides a habitat for only a few individuals per unit area or biomass. The *Zostera marina* belt is perennial, but is also scarcely ramified. In other words: The decline of the dense red algae stands has greatly simplified the energy flux and circulation of matter in Greifswald Bay. The extensive phytal zone was similarly important for fish such as *Clupea harengus L.*, serving as a spawning ground, "kindergarten" and oxygen producer. In addition, the extremely abundant fauna of the phytal zone was also, and is still to a certain extent, food for many fish species.

In the last 50 years Greifswald Bay as an ecosystem has experienced an enormous change. The increased nutrient input to the Baltic as a whole (NEHRING 1984) manifests itself first in dense blooms of diatoms in the early spring and bluegreens in summer. These greatly reduce the transparency of the water and consequently restrict the growth of benthic vegetation. The effects of this on the food web and energy flux of the ecosystem and their quantification will be the subject of further investigations.

6. Zusammenfassung

Der Greifswalder Bodden, ein Küstengewässer im Nordosten der DDR, war Gegenstand von Untersuchungen am Phytal im Frühjahr bis Herbst 1985. Im Ergebnis einer Grobkartierung der submersen Vegetation wurde ein Rückgang der mit Pflanzen bedeckten Bodenfläche von ca. 90% (SEIFERT 1938) auf ca. 10% festgestellt. Die Hauptursache für diesen Rückgang wird in erhöhter Planktonproduktion und damit einer Verschlechterung des Lichtklimas am Boden im Zuge der Eutrophierung gesehen. Quantitative Unter-

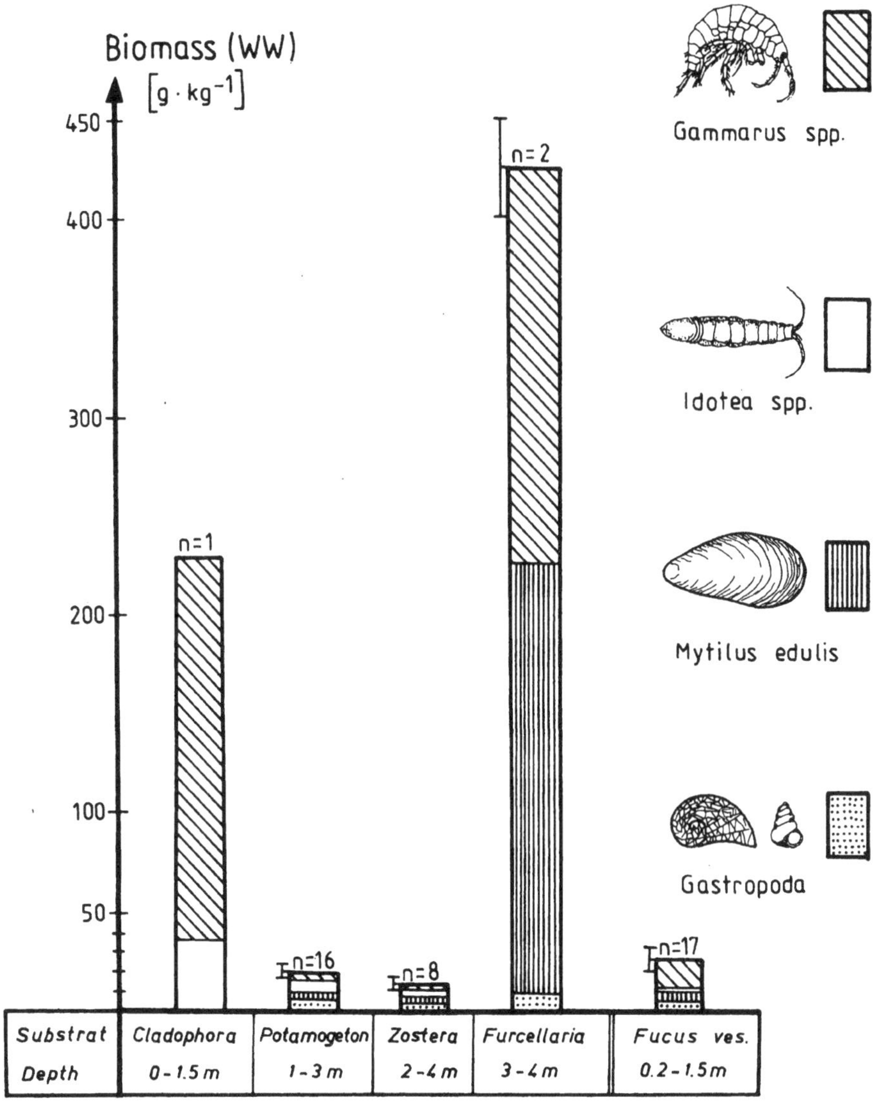

Fig. 3. Colonization of different substrates by the dominant animal groups in August 1985

suchungen an der Phytalfauna zeigten, daß die Rotalgenzone [speziell vertreten durch *Furcellaria fastigiata* (L.) LAMOUR] eine enorm hohe Besiedelung durch Tiere im Gegensatz zu den schwach strukturierten Arten *Potamogeton pectinatus* L. und *Zostera marina* L. aufweist. Mit dem Verschwinden der Rotalgenzone bis auf unbedeutende Restbestände ist anzunehmen, daß sich damit der Stoffhaushalt und Energiefluß im Gewässer stark verändert hat. Auch wenn diese Tatsache noch nicht quantifizierbar wurde, muß jedoch befürchtet werden, daß der Ausfall dieses Lebensraumes Konsequenzen in der Bestandsentwicklung einer Heringspopulation haben kann. Starke Blaualgenblüten im Hochsommer machen eine vollwertige Erholungsnutzung fragwürdig. Durch diese Veränderungen in den Ökosystemstrukturen des Greifswalder Boddens ist eine ausgewogene Mehrfachnutzung des Gewässers gefährdet.

References

GEISEL, T. (1986): Pflanzensoziologische Untersuchungen am Macrophytobenthos des Greifswalder Boddens. Dipl.-Arb., Wilhelm-Pieck-Univ. Rostock.

HENKING, H. (1904): Orientierungsfahrten im Greifswalder Bodden. Mitt. dtsch. Seefisch-Ver. **11**.

MESSNER, U., & GEISEL, T. (in press): Die Lebewelt des Bodens im Greifswalder Bodden. Meer und Museum. Stralsund.

NEHRING, D. (1984): Variations in the nutrient situation of the Baltic proper. *Limnologica* (Berlin) **15** (2), 277—287.

REIBISCH, J. (1904): Über das Vorkommen der als Fischnahrung wichtigsten Tiere im Greifswalder Bodden. Mitt. dtsch. Seefisch.-Ver. **11**.

SCHMIDT, I. (1983): Gutachten über Wasserbeschaffenheit und Unterwasserflora im Bereich des Standortes 10/Zudar. WWD Stralsund, unpubl.

SEGERSTRÅLE, S. G. (1943): Weitere Studien über die Tierwelt der *Fucus*-Vegetation an der Südküste Finnlands. Soc. Sci. Fenn. Comm. Biol. **IX** (4), 1—28.

SEIFERT, R. (1938): Die Bodenfauna des Greifswalder Boddens. Z. Morphol. Ökol. Tiere **34**, 221—271.

SUBKLEW, H.-J. (1955): Der Greifswalder Bodden, fischereibiologisch und fischereiwirtschaftlich betrachtet. Z. Fisch. Hilfswiss. N. F. **4**, 545—588.

WIESER, W. (1959): Zur Ökologie der Fauna mariner Algen mit besonderer Berücksichtigung des Mittelmeeres. Int. Rev. ges. Hydrobiol. **44** (2), 137—180.

Addresses of the authors:

ULRICH MESSNER
Lange Str. 7
Waren/Müritz
DDR-2060

Dr. JÖRG-ANDREAS VON OERTZEN
Wilhelm-Pieck-Universität Rostock
Sektion Biologie
Freiligrathstr. 7/8
Rostock
DDR-2500

Limnologica (Berlin) **20** (1), 187—190, April 1990

Institute of Oceanography, Gdańsk University, Gdynia, Poland

The Role of Common Mussel *Mytilus edulis* L. in the Biocoenosis of the Gulf of Gdańsk

By KRYSTYNA WIKTOR

With 4 Figures and 2 Tables

Abstract

As regard biomass, *Mytilus edulis* L. is the dominant element of the zoobenthos of Puck Bay — the western part of the Gulf of Gdańsk. It plays a very important role in the biocoenosis because it is the main filtrator in this area. For this reason the Institute of Oceanography of the University at Gdańsk has undertaken long term investigations of this species.

The following parameters were analysed:

— the composition of the *M. edulis* population (length and age)
— growth rate
— mortality
— distribution and total biomass of the species.

The investigations were carried out twice: in 1978/79 and 1986/87. No major differences were found in the measured parameters. The population of *M. edulis* in the Gulf of Gdańsk is stable. It was thus possible to calculate annual production and the energy accumulation for the whole population. The results of several authors were used as a basis to calculate the annual volume of filtrated water and sedimented suspended matter. The total biomass of the *M. edulis* population was 4,460 t dw, the shell constituted 3,7202 t dw and the body 758 t dw. The energy resource of the *M. edulis* population is $15.61 \cdot 10^9$ KJ, and annual energy accumulation is 83% of this value.

The annual volume filtrated water is 38.3 km^3, and sedimented matter is 88.560 t.

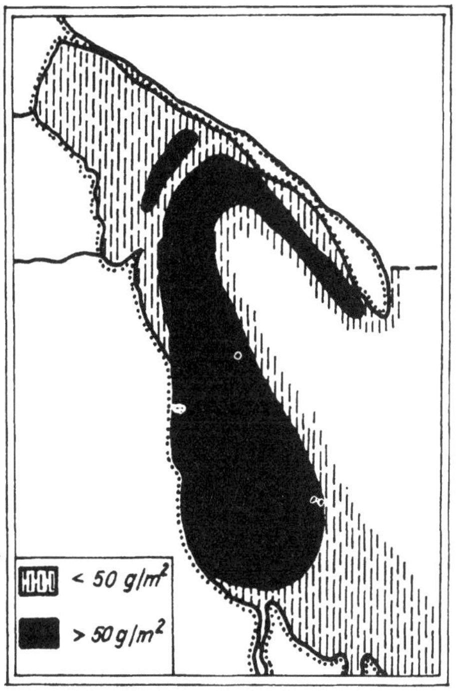

Fig. 1. Distribution of *Mytilus edulis* L. in the Gulf of Gdańsk

Introduction

The common mussel *Mytilus edulis* is a widely distributed species in the temperate seas of the North Atlantic. In the southern Baltic it is a dominant element of the bottom fauna (DEMEL & MULICKI 1954). In the Gulf of Gdańsk, especially in its western part, the Puck Bay, it is also the main component of the benthic fauna; which is why long term investigations into the population of this species and its role in the biocoenosis have been undertaken at the Institute of Oceanography of Gdańsk University. The present paper attempts to present a synthetic summary of the results of those investigations.

Results and Discussion

In the Gulf of Gdańsk *M. edulis* forms dense concentrations in its western part, the Puck Bay (Fig. 1). It occurs in the whole coastal water belt down to a depth of 35 m. The greatest concentrations, however, are encountered at depths of 8 to 20 m (Table 1). The remaining part of the Gulf of Gdańsk is not suitable as a habitat for *M. edulis* because of the character of the seabed (loose sand, mud).

Table 1. Differences in biomass of *Mytilus edulis* L. in the Puck Bay as a function of depth, after WENNE & K. WIKTOR (1982)

Depth (m)	3—5	8—10	20	30—35
Biomass (g · m^{-2})	13.44	45.74	76.89	2.97

In the Gulf of Gdańsk, *M. edulis* reaches a maximum length of 55 mm and an age of 6 years, although the majority of specimens are small (less than 15 mm in length). Over 40% of the population in shallow waters is made up of specimens with lengths ranging from 5 to 10 mm from age group II (BARRON & WOŁOWICZ 1981). It is only at greater depths (25—35 m), that *M. edulis* forms scattered concentrations, where older specimens with lengths of 24—40 mm and belongig to age-groups III and IV predominate (Figs. 2, 3). The growth rate of *M. edulis* in the Gulf of Gdańsk ranges from 14 mm in the youngest year-classes to 1.7 mm in the oldest (BARRON & WOŁOWICZ 1981), and is higher than off the Swedish coast (KAUTSKY 1981).

Studies of population structure and distribution of *M. edulis* in the Gulf of Gdańsk were carried out twice in the past ten years — in 1978/79 and in 1986/87. No great differences in the parameters studied were observed, so the population may be considered stable.

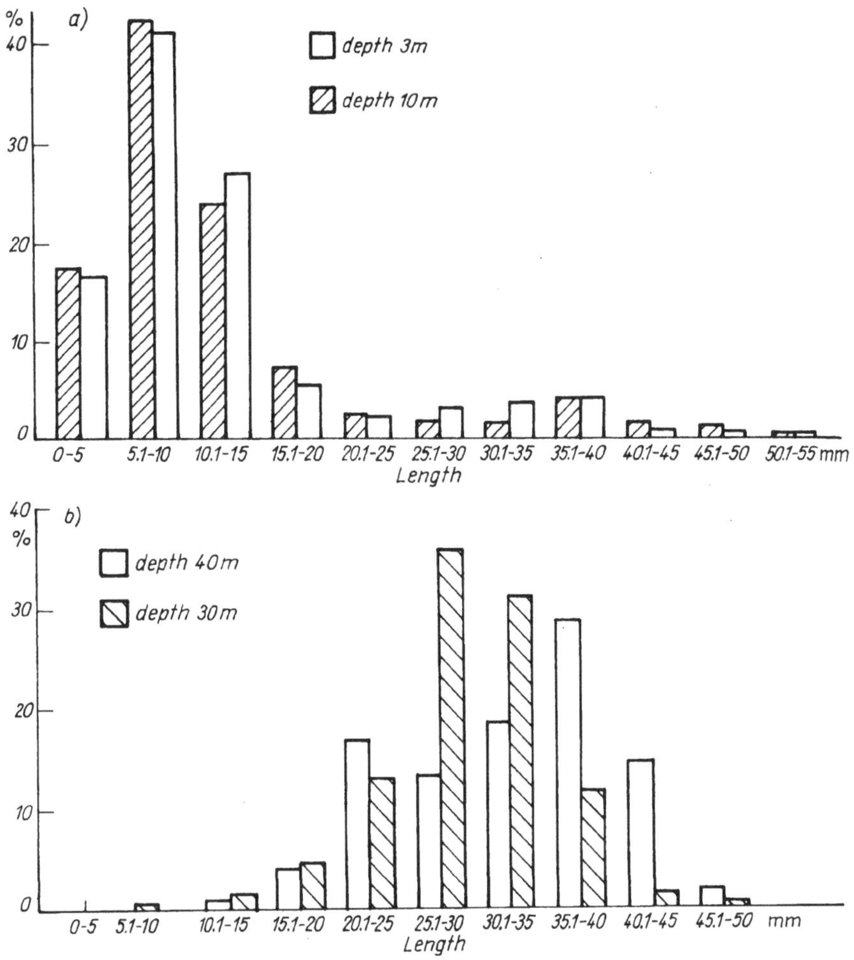

Fig. 2. Length composition of *Mytilus edulis* population
a: shallow-water zone (3—10 m)
b: deep-water zone (30—35 m)

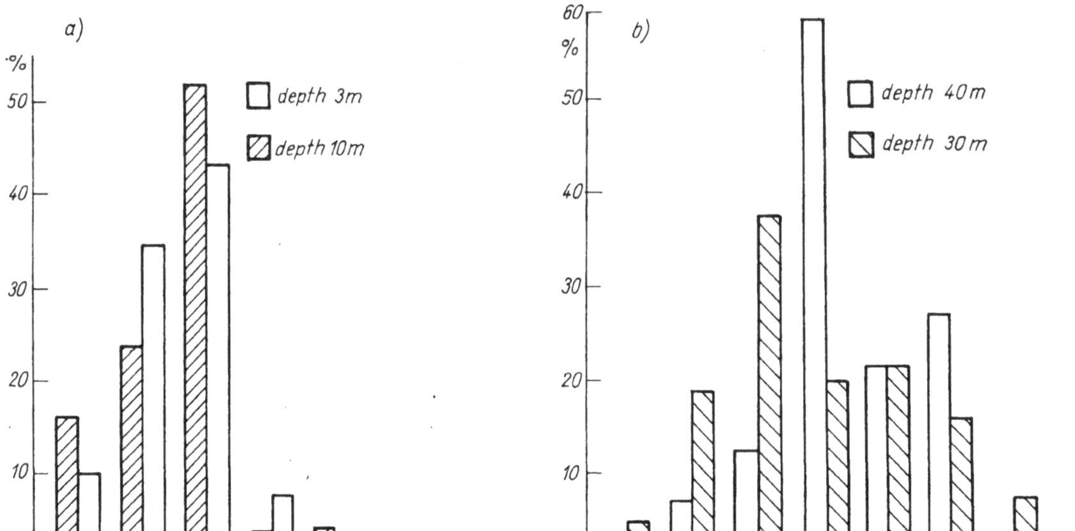

Fig. 3. Age composition of *Mytilus edulis* population
a: shallow-water zone (3—10 m)
b: deep-water zone (30—35 m)

The total biomass of *M. edulis* in Puck Bay was estimated at 13260.547 t (Table 2), which corresponds to 4460.004 tons dry weight. In the dry weight of *M. edulis*, the weight of the shell constitutes as much as 83%; thus, the dry weight of the shell equals 3.702 tons while the dry weight of the body is 758.2 tons.

Table 2. Total biomass (wet weight) of *Mytilus edulis* in the western part of the Gulf of Gdańsk at depth intervals (after DRGAS 1985)

Depth interval	Surface (km^2)	Biomass (t)
5—10 m	66.795	1937.055
10—20 m	138.978	8239.274
20—30 m	51.705	2722.950
30—35 m	24.915	361.268
Total	282.393	13260.547

The energetic value of *M. edulis* is very high and stable from season to season, equalling on average 20.5842 J · mg^{-1} dry weight (SZANIAWSKA & PAZIKOWSKA, in press). Having such a large biomass and high energetic value, the role of *M. edulis* is largely confined to accumulation of energy.

Total energy resources in the population amount to 15.61×10^9 KJ, and thus constitutes 35—68% of the energy resources of bottom fauna. Bearing in mind that annual mortality among this stable population is 80%, annual energy accumulation in the body alone amounts to $12.5 \cdot 10^9$ KJ. In addition, large amounts of energy are accumulated in the shells of *M. edulis*, in which concentrations of C_{org} in the dry weight range between 4.6 and 5.8% (WIKTOR, WIŚNIEWSKA & KORZENIEWSKI, in press).

Apart from energy accumulation, an important role of *M. edulis* consists in the transport of suspended matter from water layers into the bottom and accelerating the process of suspensions sedimentation. This takes place through filtration, which is connected with both feeding and respiration.

Filtration rates of *M. edulis* in the Gulf of Gdańsk have been studied by several authors (DRZYCIMSKI 1964; DUBRAWSKI 1979) either in a laboratory or in situ on single specimens. It is difficult to ascertain to what degree the results obtained reflect the relations existing in the environment, where *M. edulis* lives in dense concentrations. The filtration rate of *M. edulis* in the Gulf of Gdańsk is much lower than in seas with full salinity. For specimens with a lengths 5—10 mm it is 22.739 dm^3/year/g body wet weight and for older specimens belonging to age-groups IV and V — only 1.204 dm^3/year/g body wet weight (DUBRAWSKI 1979).

On the basis of our knowledge of the population structure and total biomass of *M. edulis*, it may be assumed that the whole population is able to filtrate annually 38.3 km^3 of water; the weight of matter filtrated and then sedimentated in the form of pseudofaces is 88.560 tons (DRGAS 1985). Thus, *M. edulis* in Puck Bay plays the role of a biofilter, being at the same time the main filtrator. In the process of active filtration, *M. edulis* contributes not only to the transport of suspended matter from water layers to the bottom, thus indirectly influencing water transparency, but also accumulates suspensions at the bottom, enriching the seabed with organic substances. It also contributes to the accumu-

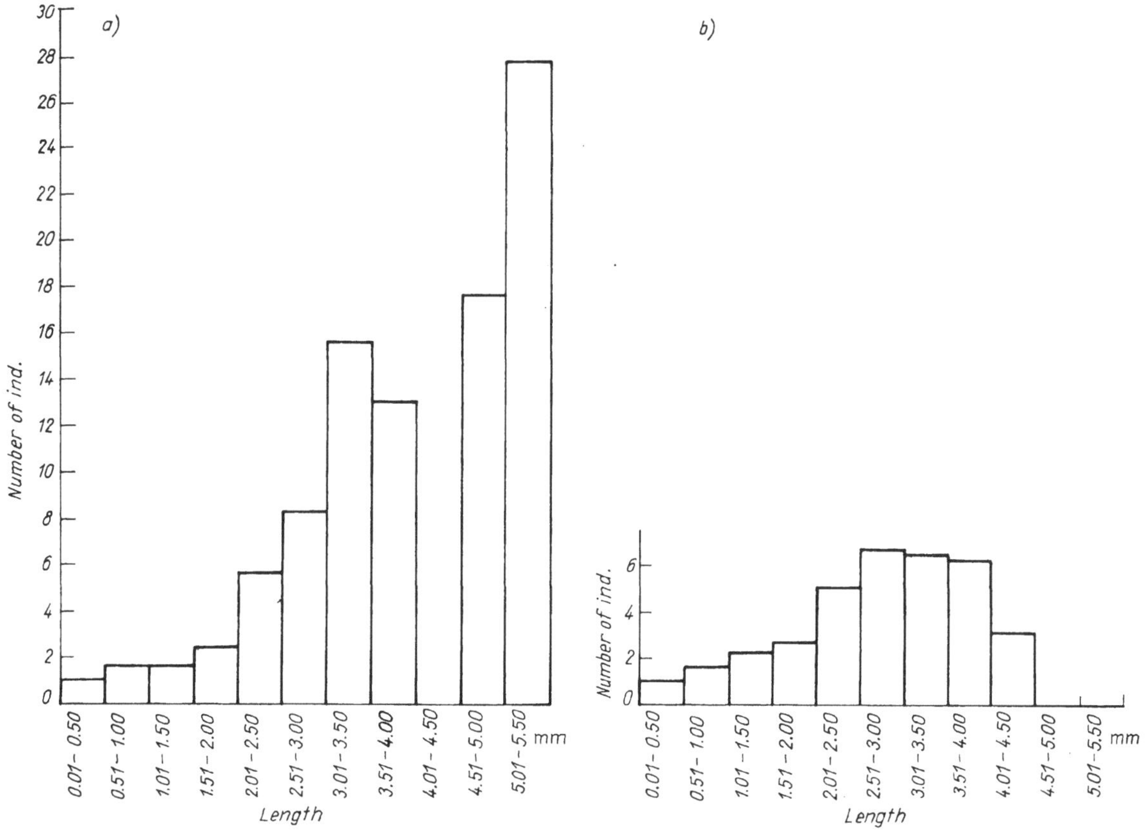

Fig. 4. Number of *Balanus* sp. individuals per one shell of *Mytilus edulis* of different length
a: shallow-water zone (3—5 m)
b: deeper water zone (10—20 m)

lation of toxic or harmful substances, such as trace metals, heavy metals and pesticides on the bottom.

Irrespective of their great role as a biofilter, dense concentrations of *M. edulis* create favourable living conditions for other bottom and near-bottom organisms. Multi-layer, porous shoals of this species provide shelter for near-bottom fauna and increase the area of the seabed for sedentary animals, giving them places to attach themself. Besides, enriching the bottom in detritus and organic substances (pseudofaces), they create favourable conditions for the feeding of detritusophagous animals. This role of dense concentrations of clams was emphasized by J. WIKTOR (1964) who analysed the population of *Dreissena polymorpha* in the Szczecin Lagoon.

Among sedentary animals, for which the shells of *M. edulis* become the substrate, it is necessary to mention *Bryozoa* and barnacles (*Balanus* ssp.), i.e. water-filtrators like the common mussel. The degree of overgrowth of *M. edulis* with barnacles varies and depends on its size and age, usually ranging from 2 to 8 barnacles per specimen (WIŚNIEWSKA 1978) (Fig. 4). Thus, the function of a *M. edulis* shoal as a filtrator is strengthened by the cooperation between the common mussel and barnacles. Although barnacles, especially in the shallow-water zone, are a permanent component of *M. edulis* shoals, the degree of attachment of other organisms varies, changing also with water depth. Permanent components (100% frequency) include, besides *Bryozoa* and *Balanus* mentioned above, *Nereis diversicolor* and species from the genera *Hydrobia* and *Gammarus*. The latter live exclusively within shoals of *M. edulis*, especially beyond the zone of the shallowest waters covered with bottom vegetation (WENNE & K. WIKTOR 1982). The remaining bottom organisms, which are not so closely connected with the shoals of *M. edulis*, are characterized by higher density in shoal areas than in those not occupied by dense concentrations of *M. edulis* (HERRA & K. WIKTOR 1985).

In conclusion it may be said, that although *M. edulis* constitutes, to quote KAUTSKY (1981) a "blind link" in the trophic chain, it nevertheless plays a significant role in the energy balance of the ecosystem.

References

BARRON, L., & WOŁOWICZ, M. (1981): A preliminary outline of the *Mytilus edulis* population from Gdańsk Bay. Zesz. Nauk. UG Oceanografia **8**, 127—140.

DEMEL, K., & MULICKI, Z. (1954): Studia ilościowe nad produktywnością biologiczną dna południowego Bałtyku (Quantitative investigations on the biological bottom productivity of the South Baltic). Prace Morsk. Instyt. Ryback. **7**, Gdynia, 75—126 (in Polish, with English summary).

DRZYCIMSKI, I. (1964): Zużycie tlenu i intensywność filtracji u omułka (*Mytilus edulis* L.) w południowym Bałtyku [Sauerstoffverbrauch und Filtrationsintensität der Miesmuschel (*Mytilus edulis* L.) in der südlichen Ostsee]. Zesz. Nauk. WSR w Olsztynie **17**, Nr. 340, 423—446 (in Polish, with English summary).

DRGAS, A. (1985): Próba oszacowania ogólnej biomasy omułka w Zatoce Gdańskiej i jego rola jako biofiltra (An attempt to determination the biomass of *Mytilus edulis* from Gulf of Gdańsk and its role as biofilter). Manuscript IOUG, Gdynia (in Polish).

DUBRAWSKI, R. (1979): Proces usuwania DDT ze środowiska morskiego przez małża *Mytilus edulis*. Prace Inst. Morsk. Gdynia, Nr. 649 (in Polish).

HERRA, T., & WIKTOR, K. (1985): Rozmieszczenie i biomasa fauny dennej w Zatoce Gdańskiej (Composition and distribution of bottom fauna in coastal zone of the Gulf of Gdańsk proper). SIMO KBM PAN **47** (7), 115—142 (in Polish, with English summary).

KAUTSKY, N. (1981): On the role of the Blue Mussel *Mytilus edulis* L. in the Baltic ecosystem. Dept. Zool. Asko Labor. University of Stockholm, 1—17.

SZANIAWSKA, A., & PAZIKOWSKA, G. (in press): The calorific value and lipid content of *Mytilus edulis* L. from Gdańsk Bay in relation to specimens age. Oceanologia **25**.

WENNE, R., & WIKTOR, K. (1982): Fauna denna przybrzeżnych wód Zatoki Gdańskiej (Benthic fauna of the inshore waters of Gdańsk Bay). SIMO KBM PAN **39** (6), 137—171 (in Polish, with English summary).

WIKTOR, J. (1969): Biologia *Dreissena polymorpha* (PALL.) i jej ekologiczne znaczenie w Zalewie Szczecińskim [The biology of *Dreissena polymorpha* (PALL.) and its ecological importance in the Firth of Szczecin. Studia i Mat. Morsk. Instyt. Ryback. **A-5**, Gdynia (in Polish, with English summary).

WIKTOR K. (1985): An attempt to determine trophic structure of the bottom fauna in the coastal water of Gulf of Gdańsk. Oceanologia **21**, 109—121.

— WIŚNIEWSKA, K., & KORZENIEWSKI, K. (in press): Zawartość C_{org} w muszlach omułka (*Mytilus edulis* L.) z Zatoki Gdańskiej [C_{org} contents in the shells of *Mytilus edulis* L. from the Gulf of Gdańsk]. Zesz. Nauk. UG Oceanografia **12** (in Polish, with English summary).

WIŚNIEWSKA, M. (1977): Charakterystyka ławic omułka Zatoki Gdańskiej jako biocenozy (The dense concentrations of *Mytilus edulis* L. from the Gulf of Gdańsk as a biocenosis). Manuscript IOUG Gdynia (in Polish).

Address of the author:

KRYSTYNA WIKTOR
Institute of Oceanography
Gdańsk University
Czołgistów 46
Gdynia
Poland

Limnologica (Berlin) **20** (1), 191—194, April 1990

Institute for Environmental Protection, Gdańsk, Poland; Institute of Oceanology PAS, Sopot, Poland

Degradation and Restoration of the Puck Bay (A Projekt)

By PAULIN CISZEWSKI & EWA STYCZYŃSKA-JUREWICZ

With 2 Figures and 2 Tables

Introduction

This contribution is a summary of a report by CISZEWSKI and co-authors (CISZEWSKI et al., in press) prepared in the framework of a project of restoration of Puck Bay. An outline of further studies on this water body and of main activities aimed in its restoration is also given. The complete list of references for the subject can be found in the above mentioned publication.

1. The present state of the Puck Bay

1.1. Hydrographic pattern

The Puck Bay, called more precisely the Puck Bay Proper Inner, is a lagoon at 91 000 ha, situated in the western part of the Gulf of Gdańsk and separated from the Baltic proper by Hel Peninsula and from the deeper waters of the Gulf by an underwater sandy barrier called "Rewa Mew". Hydrographic pattern of the lagoon is characterized by its shallowness (average depth 3 m), low salinity (oscillating from about 5 to about $7^0/_{00}$) and good oxygen conditions in the whole water column due to easy mixing by winds. Above the sandy barrier and through a collin in it, called "Depka", the water masses of the Gulf proper mix permanently with the water of Puck Bay; the mixing is stronger when the winds flow from the east. Main source of pollution of the Puck Bay are water currents transporting contaminants from the vicinity of the outfall of collecting pipe situated in Mechelinki (see Fig. 1) which discharges untreated municipal sewage from Gdańsk agglomeration directly into the Gulf. Around the Puck Bay there is a number of diverse local sources of pollution; two small rivers (Reda and Płutnica) transport a load of organic contaminants from adjacent agricultural areas, significant portion of municipal and industrial pollution originates from the town Puck, now with a population of about 30 000.

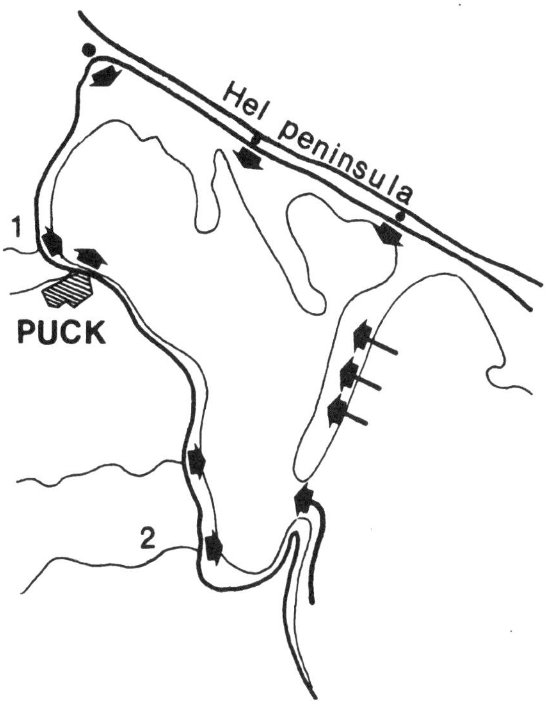

Fig. 1. Main and secondary coastal sources of pollution of the Puck Bay. Short arrows: discharge from small rivers and potential local contamination from coastal villages (black points); medium arrows: inflow of contaminated water masses above the sandy barrier from the Puck Bay Outer (= western part of the Gulf of Gdańsk proper); long arrow: main current of pollution from Mechelinki collecting pipe flowing through "Depka" collin. 2 m isobate marked with fine line. 1 — River Płutnica; 2 — River Reda.

1.2. Flora and fauna

Till the seventies of this century the Puck Bay was a very beautiful natural coastal area with rich brackish fauna and flora forming vast seaweeds, which covered the clean sandy bottom. The seaweeds consisted mostly of Characeae, the brown alga *Fucus vesiculosus*, red algae (*Furcellaria fastigiata* and *Phyllophora brodiaei*) as well as several vascular plants among which *Zostera marina* and *Potamogeton* sp. predominated. Rich fauna of invertebrates and fishes inhabited the seaweeds; some fish species (e.g. herring) have had there their spawning grounds. Rapid industrialization and human population increase which started in mid-seventies in the adjacent land area as well as increased use of artificial fertilizers and pesticides in the agriculture caused an enormous rise of pollution load discharged to Puck lagoon and a catastrophic ecological degradation of that water body. The previous natural ecosystem was transformed to a new one, the changes were remarkable first in the plant communities.

1.2.1. Macroflora

Before 1979, there occurred 100 species in the macroflora in the Puck Bay: 43 species of green algae, 23 brown algae, 19 red algae and 15 species of vascular plants. The most common were representatives of the genus *Chara*, *Potamogeton* and *Enteromor-*

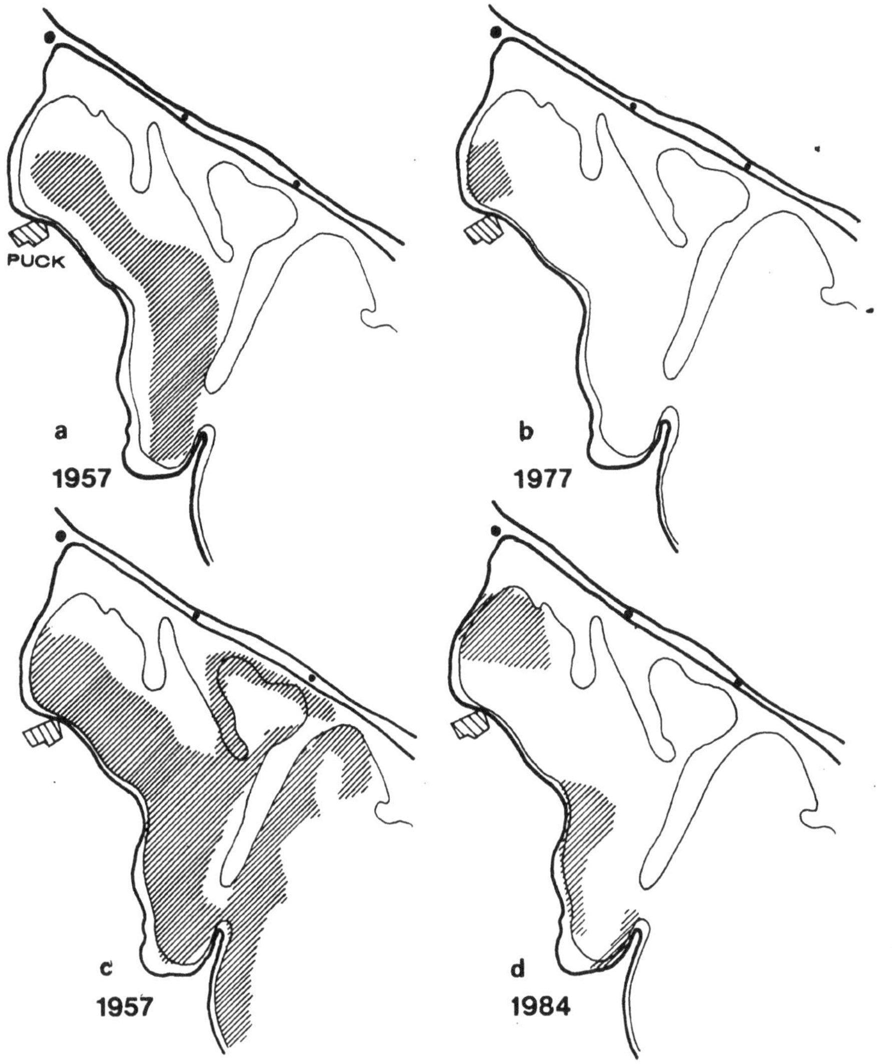

Fig. 2. Changes in the area covered with macrophytes in the Puck Bay. a, b — *Fucus vesiculosus*; a — in 1957, after CISZEWSKI (1962); b — in 1977, after PLIŃSKI (1982). Complete disappearance which occurred in 1979 was confirmed in 1987 monitoring, performed by the staff of the Institute for Environmental Protection.
c, d — *Zostera marina*; c — in 1957, after CISZEWSKI (1962); d — in 1984, after PLIŃSKI (1986); in the 1987 monitoring a significant reduction was recorded by the staff of the Institute for Environmental Protection

pha as well as *Fucus vesiculosus, Furcellaria fastigiata, Phyllophora brodiaei, Zostera marina* and *Zannichellia palustris*. In 1979 a complete disappearance of *Fucus vesiculosus, Furcellaria fastigiata, Phyllophora brodiaei, Polysiphonia nigrescens, P. sanguinea* was recorded. Some of these changes are demonstrated in Fig. 2. At the end of the seventies and in the eighties (it means also at present time) the filamentous brown algae *Pilayella litoralis* and *Ectocarpus siliculosus* predominate, the area covered by other species is significantly reduced. Disappearance of *Fucus vesiculosus* and *Furcellaria fastigiata* is particularly pronounced, since in the decade of the years 1963—1973 these two macroalgae were intensively harvested for industrial purposes without any remarkable damage to the standing crop. In further years a decrease in biomass and inhibition of growth in individual plants was documented by "in situ" experiments (KENTZER et al. 1976); as late as in 1982 TROKOWICZ (1987) found a few buoyant individuals of *F. vesiculosus* and *F. fastigiata* at a single sampling site, but no one specimen was found during the monitoring performed in 1987. Changes in total biomass of macrophytes are shown in Table 1.

1.2.2. Macrofauna

Starting from the mid-seventies a quantitative reduction was recorded in macrofauna connected with seaweeds. Less abundant became for example: *Palaemon adspersus, Cyathura carinata, Heterotanais oerstedti, Asellus aquaticus* and *Limnaea peregra*. But in general the biomass of macrobenthos increased and is still increasing in the central part of the Bay and in two deeper areas: "Jama Chałupska" and "Jama Kuźnicka". In the latter area the increase of biomass was particularly significant, by one order, from 10—30 g $\cdot$ m^{-2} in 1962 to 100—300 g $\cdot$ m^{-2} in 1977, mostly due to increased share of bivalves (Table 1). At the end the of seventies Crustacea and Polychaeta were somewhat reduced.

1.2.3. Plankton

Till the mid-seventies there were observed two maxima in the abundance of phytoplankton: in spring and in autumn with pronounced predomination of diatoms. In

Table 1. Main changes in the bicoenosis of the Puck Bay according to various authors

1.1. Occurrence of macroalgae

Author	Year	Number of species		
		Red algae	Brown algae	Green algae
LAKOWITZ (1907)	1845—1901	10	9	12
BURSA (1935)	1934—1935	8	8	20
KORNAŚ et al. (1960)	1958	6	8	11
BŁĘDZKI and KRUK-DOWGIAŁŁO (1983)	1979	4	3	10
PLIŃSKI (1986)	1984	1	2	5

1.2. Total biomass of macrophytes (calculated from data of cited authors)

Year	1957	1969—71	1977	1984
Author	CISZEWSKI et al. (1962)	KLEKOT (1980)	PLIŃSKI (1982)	PLIŃSKI (1986)
Biomass per area (in tons)	70×10^3	84×10^3	17×10^3	10×10^3

1.3. Biomass of benthic fauna (in $g \cdot m^{-2}$)

Year	1962	1977—78
Author	ŻMUDZIŃSKI (1967)	WENNE and WIKTOR (1982)
Range of variation	$30—100\ g \cdot m^{-2}$	$100—600\ g \cdot m^{-2}$

1.4. Abundance of zooplankton (in thousands indiv. $\cdot m^{-3}$)

Author	Year	Puck Bay	Gulf of Gdańsk	Southern Baltic Proper
Calculated after				
WIKTOR et al. (1982)	1977—78	100	43	no data
CISZEWSKI et al. (in press)	1978—79	255	41	21

1.5. Biomass of phytoplankton *) (calculated from the data by PLIŃSKI 1980)

Puck Bay Inner		Puck Bay Outer		
16.6	71.4	1875	191.7	375.2

*) in $\mu g \cdot m^{-3}$

summer predominated blue-green algae: *Aphanizomenon flosaquae* and *Nodularia spumigena*. However, total number and biomass of phytoplankton estimated during the seventies and the eighties always was significantly lower in the Puck Bay than in the adjacent part of the Gulf proper (see Table 1). The same pattern was observed concerning the content of chlorophyll **a** in the water.

Before the year 1958 information on zooplankton was scarce, as late as in the seventies a few authors described the structure and seasonal changes of zooplankton community. Its main components are euryhaline rotifers as well as copepods: *Acartia bifilosa* and *A. tonsa*. Freshwater Copepoda and marine Cladocera are less abundant, *Podon polyphemoides* is the most common representative

Table 2. Average contents (ppm) of trace metals in surface bottom sediments and in macroalgae (TROKOWICZ 1987, unpubl.)

Type of sediment	Year	Cu	Zn	Cd	Pb
Small grain sand	Decade	13.2	258.2	—	62.4
Medium grain sand	1968—1978	13.8	58.7	—	60.2
Mud		49.9	93.7	—	68.9
Organic mud		3201.3	4273.2	452.1	2891.2
Macroalga:					
Furcellaria fastigiata	1962—64	6.4	97.8	0.85	9.8
	1965—68	9.1	112.35	1.55	15.1
	1969—72	16.5	142.3	2.48	20.1
	1982	34.2	210.57	3.71	33.9
Fucus vesiculosus	1962—64	5.1	188.92	1.05	6.9
	1965—68	8.8	279.5	1.42	13.5
	1969—72	15.2	349.87	1.98	14.9
	1982	22.5	507.21	3.39	19.6

of the latter taxon. Abundance of zooplankton in the Puck Bay is higher than in the Gulf of Gdańsk and in the Baltic Proper, independently on phytoplankton scarcity (Table 1). Over the last three decades the share of *Eurytemora* sp., significant in the sixties, decreased, simultaneously the share and number of *Acartia bifilosa* increased 6 times as much.

1.2.4. Fishes

Starting from seventies the changes in ichthyofauna were pronounced by lower share of eel and pike in the catch and an increase in abundance of roach. Also out of fishery harvest, an increase of population of *Gasterosteus aculeatus* was remarkable in recent years.

1.3. Main contaminants

The fact of biological degradation of the Puck Bay is obvious but the list of contaminating agents is not yet completed. The most ample information concerns the accumulation of trace elements in bottom sediments and in macroalgae. While in bottom sediments 10% increase of concentration of heavy metals was recorded over the decade 1968—1978, bioaccumulation in the macroalgae was very high, as is shown in Table 2. Therefore there are serious reasons to suspect that contamination with heavy metals could play any role in the elimination of the two macroalgae *Fucus vesiculosus* and *Furcellaria fastigiata* from the polluted ecosystem.

Data on the concentration of other potential contaminants like nutrients, pesticides, detergents and hydrocarbons are still lacking.

2. Restoration and mangement

The project of restoration of the Puck Bay covers 3 stages of activity:

1st stage, to be realized probably within 2 years, will be focused on biological and chemical monitoring of the present state of the ecosystem. "In situ" ecological experiments will be also performed at this time. In 1987 the following variables were monitored: in the water — transparency, nutrient concentration, heavy metals concentration, oxygen, DOC, chlorophyll **a** and humic substances content; in the sediment — humic substances, carbon content, heavy metals, presence of free H_2S, thickness and type of sediment; in the biocoenosis — abundance and composition of macrobenthos (zoo- and phytobenthos), zooplankton and phytoplankton, trophic conditions, content of heavy metals in the macrofauna. Input of pollution (organic contaminants, trace metals, phenols) from the atmosphere was also monitored at 5 coastal sites.

At the 2-nd stage a hydrographic analysis of the bay and adjacent water masses of the Gulf of Gdańsk will be performed which has to supply necessary information on water balance and water exchange between these two water bodies, their natural morphometric configuration, wind direction and power, currents, and pollution load from different sources.

At the 3-rd stage the restoring management will be planned, aimed in improving the process of exchange of water in the bay. Activities like draining in key localities and/or sucking of the sediment are under consideration depending on the data collected during the 2-nd stage. All these activities should support the work of two puryfying plants, now under construction.

At present the project is at its initial phase, in 1987 chemical and biological monitoring was conducted, next year main hydrographic studies should give an answer concerning the reality of the 3-rd stage.

Acknowledgements

The authors are indebted to Dr. DANUTA TROKOWICZ from the State Institute of Geology in Sopot for kind permission for presenting her unpublished data on heavy metals content in sediments and in algae, as well as to Prof. M. PLIŃSKI from the University of Gdańsk for making his unpublished data available.

References

BŁEDZKI, L., & KRUK-DOWGIAŁŁO, L. (1983): Multiannual changes in the structure of the benthos of Puck Bay. Czlowiek i Srodowisko **7**, 1—2 (in Polish).

BURSA, A. (1935): Liste des algues recueilles dans les eaux de la Baltique Polonaise. Bull. Acad. Math. Nat. B **I**, 65—67.

CISZEWSKI, P., DEMEL, K., RINGER, Z., & SZATYBEŁKO, M., (1962): Resources of *Furcellaria fastigiata* in Puck Bay estimated by diving method. Prace MIR **11 A**, 9—36 (in Polish).

— ŻMUDZINSKI, L., KRUK-DOWGIAŁŁO, L., OSOWIECKI, A., RYBICKA, D., PYŚ-WOLSKA, M., WIKTOR, J., CISZEWSKA, L., & TROKOWICZ, D. (in press): Multiannual trends in changes of the ecosystem of Puck Bay. SIMO (in Polish).

KENTZER, T., BOROWCZYK, E., & SZCZEPKOWSKA, E. (1976): The growth of algae affected by pollution of the Baltic. SIMO **15**, 169—187 (in Polish).

KLEKOT, L. (1980): Qualitative investigations of seaweeds in Puck Bay. Oceanologia **12**, 125—139 (in Polish).

KORNAŚ, J., PANCER, E., & BRZYSKI, B. (1960): Studies on the benthic flora of the Gulf of Gdańsk in the vicinity of Rewa. Fragm. Florist. Geobot. **6**, 1—91 (in Polish).

LAKOWITZ, K. (1907): Die Algenflora der Danziger Bucht. Ein Beitrag zur Kenntnis der Ostseeflora. Danzig.

PLIŃSKI, M. (1980): Quantity and biomass of phytoplankton of the Gulf of Gdańsk in quartely samples collected in 1980. (Unpublished report of the monitoring programme, University of Gdańsk, in Polish).

— (1982): Distribution and biomass of phytobenthos in Puck Bay Inner. SIMO **6**, 197—217.

— (1986): Why should phytobenthos be an element of monitoring? Proceedings No. 19, Baltic Sea Environment Symposium, Tallin, USSR, March 10—15, p. 286—296.

TROKOWICZ, D. (1987): Distribution of bottom sediments in Puck Bay and estimation of the level of their contamination with trace metals. (Unpublished report prepared for the Institute for Environmental Protection, Gdańsk, in Polish).

WIKTOR, K., CYLKOWSKA, U., & OSTROWSKA, K. (1982): Zooplankton in coastal waters of the Gulf of Gdańsk. SIMO **39**, 137—172 (in Polish).

Address of the authors:

Dr. PAULIN CISZEWSKI
Institute for Environmental Protection
ul. Słupska 25
Gdańsk
Poland

Dr. EWA STYCZIŃSKA-JUREWICZ
Institute of Oceanology
PAS
ul. Powstańców W-wy 55
Sopot
Poland

Limnologica (Berlin) **20** (1), 195—198, April 1990

Wilhelm-Pieck-Universität Rostock, Sektion Biologie, Rostock, DDR

Fishery in Shallow Estuaries of the GDR, Possibilities and Limitations

By HELMUT MANFRED WINKLER

With 3 Figures and one Table

The eastern GDR coastline is characterized by an extensive system of "boddens" and bays with a long tradition of fishing which today, as in former times, is of great importance. As a result of the morphology, hydrography and in particular the high nutrient load from the intensively farmed hinterland the fish production is higher comparable to that obtained from marine regions (Fig. 1). In the last decades the load of nutrients and other substances has increased so that one could expect a higher catch, however there are also the negative effects to take into consideration.

How can be evaluate the current fishing situation in these waters, which special qualities have the fish communities and what are the problems in increasing the exploitation?

Due to the special location of these waters a rich fish fauna is to be found arising from the different degree of mixing of marine, fresh water and wandering fish species. The degree to which the waters are populated by a particular fish species depends on the salinity. Sea water species such as herring and garefish spawn in mesohaline regions and are intensively fished during this time. After the spawning season they return to sea. These waters are important nursery regions for larvae and young fish of the species concerned and are therefore also important to the catches in the Baltic itself. Reproduction of the marine species is limited through a decreasing salinity, on the other hand an increased salinity restricts the potential spawning regions for freshwater fish. The remarkable tolerance of young and adult fish of this species to salinity has opened up a large feeding area in which they are to be found spreading from the bay waters to the Baltic. This also means that the long term fishery of such species of fish along the whole coastline is only possible if the limited available spawning regions are preserved with a certain quality. The decline in the number of spawning places for fishes such as salmon, sea trout, shad and whitefish became noticeable already

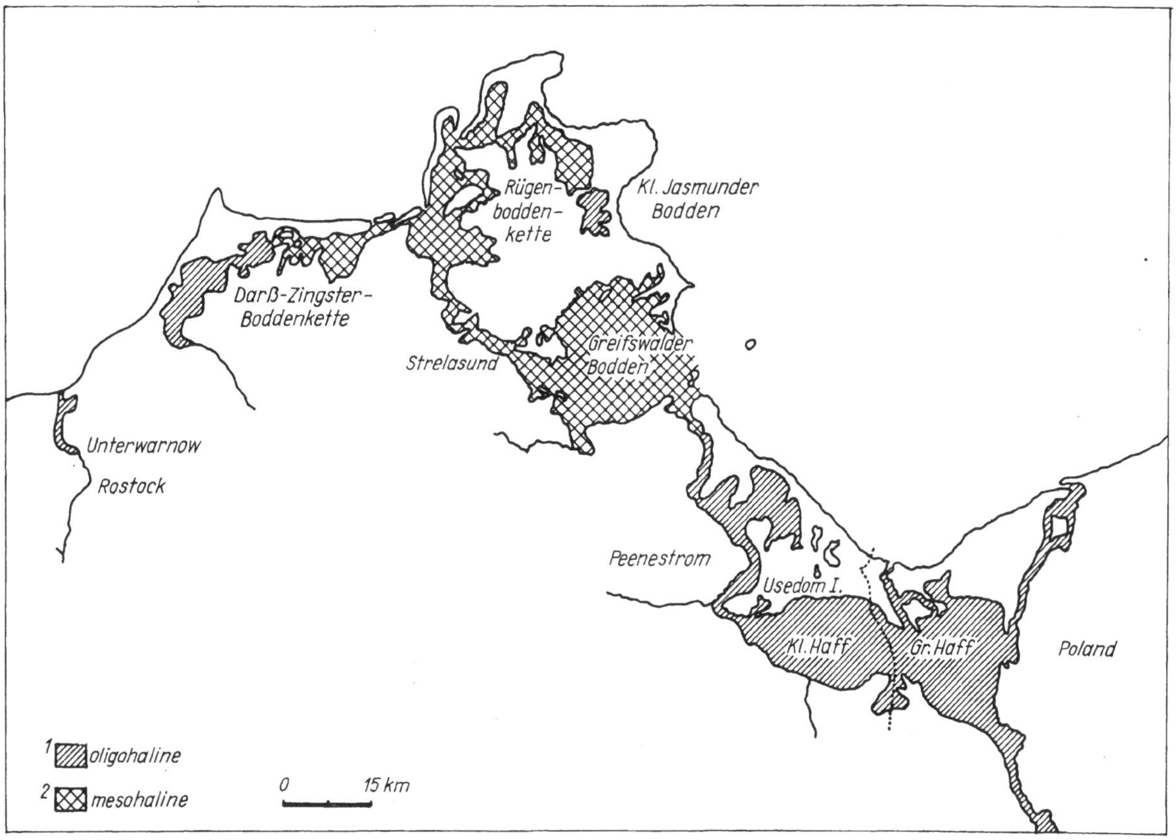

Fig. 1. Eastern part of the coast of the GDR

some time ago through the fall in the catch. This fact has, however, not been considered so important because these fish species constituted only a small percentage of the total catch. In general, the number of fish species in the bays is large, if we ignore the rarer types and likewise guests then this number is approximately 30—40. Only around 5—10 different species, however, compile the fishery yield.

The annual catch and its composition from the various coastal waters shows quantitative differences in yield, however there is uniformity in structure within the high eutrophic oligohaline and similarly within the eutrophic mesohaline waters. The catch from oligohaline waters is compiled mainly of freshwater fish (roach, bream, eel, whitefish and ruffe as benthivorous species, pikeperch, pike and perch as piscivorous species). Planktivorous species are not present in the catch, smelt is the only species to be found in abundance and can also be an important link in the food chain. Ruffe, the most abundant benthivorous fish found in these waters, is only fished in the „Kleines Oderhaff" and to a lesser extent in the "Peene", in other bays it is not caught. Pikeperch is the most dominant piscivorous fish and it is not unusual for it to give yields of 3—10 kg/ha/a and more in these waters (Table 1). In waters with a high pikeperch population perch yields are minimum. The high dominance of the latter is worth pointing out in the region of the „Oderhaff". Pike catches in the high eutrophic regions are less than 1 kg/ha/a. Piscivorous fish account for a relatively large proportion of the total catch lying around 30—50%. Where ruffe is more intensively fished this proportion is altered.

In mesohaline coastal waters in addition to freshwater species the catch consists of predominantly spring herring and less of gare- and flatfish. In these waters the same species of freshwater fish are caught as in the oligohaline regions however from the benthivorous fish roach is more frequently caught in comparison to bream and ruffe is not found in the same abundance. From the piscivorous species perch and pike dominate over pikeperch. If oligohaline waters are in close proximity and contain pikeperch spawning areas then it is possible to obtain a high yield of pikeperch in mesohaline regions (for example, near Greifswald, Usedom). Generally speaking, these waters represent our best fishing regions for pike and eel. It should be mentioned that pike is caught to a large extend by hobby anglers and these catches are not taken into consideration in the statistics. Smelt is also not found in abundance in mesohaline waters.

To explain such questions but even more for the management in fishery and manipulations of fish communities, besides an understanding of aut- and demecology knowledge of the trophical structure is necessary. Finally one must remember that a complete system has to be controlled in which inter- and intraspecific influences of species have to be taken into consideration.

Our investigations into feeding biology and to feeding ecology of piscivorous fishes have indicated several relationships which already had to be taken into consideration in classical management (Fig. 2). In oligohaline waters pikeperch and perch are the decisure predators which influence one another both directly and indirectly. In cases of food shortage cannibalism has a regulatory effect. Together with roach, perch and smelt, ruffe plays an important role in the food supply for piscivorous species. Should it arise that smelt is in short supply then it prevents any harm from occurring to the predatores. In such situations it has been proven that wrong decisions in fishery strategy have resulted in a loss in yields. Coastal waters in comparison to inland waters have

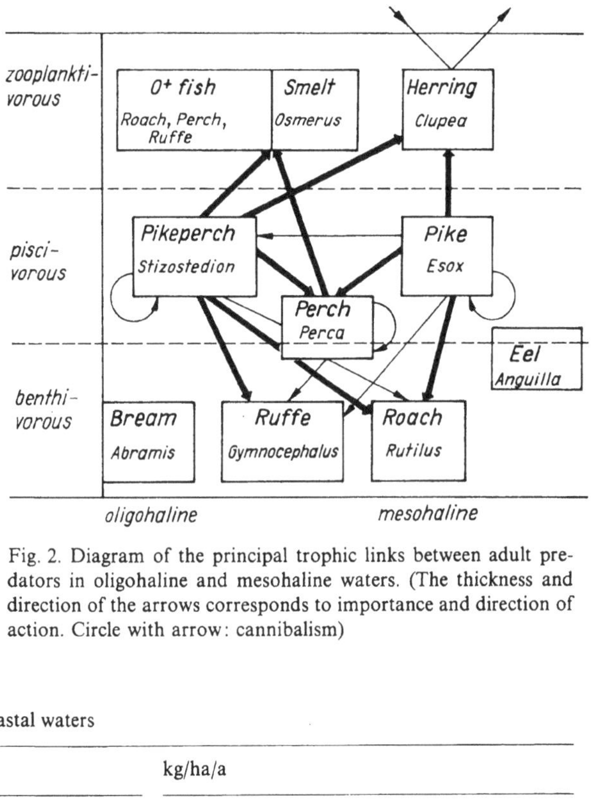

Fig. 2. Diagram of the principal trophic links between adult predators in oligohaline and mesohaline waters. (The thickness and direction of the arrows corresponds to importance and direction of action. Circle with arrow: cannibalism)

Table 1. The average annual catch and its composition from various coastal waters

Waters	n.sp.	ȳ, total kg/ha/a	% of ȳ			kg/ha/a			
			planktv.	benthv.	pisciv.	pikep.	pike	perch	time
Kl. Oderhaff	8	19.4	—	51.4	48.6	2.61	0.10	6.59	1975—86
	9	40.4	—	77.0	23.0	2.61	0.10	6.59	1975—86[1])
Gr. Oderhaff		60.6	—	72.9	27.1	4.47	0.18	9.62	1975—86[2])
Peenestrom	8	30.4	—	72.2	27.8	3.50	0.30	4.65	1977—86
Kl. Jasm. Bodden	5	20.4	—	49.5	50.5	9.30	0.50	0.50	1965—76[3])
DZBK, west	5	16.5	—	50.3	49.7	6.80	1.20	1.20	1977—86
DZBK, east	6	33.4	—	86.1	13.9	1.80	0.80	2.00	1977—86
	(3)	21.2	98.5	1.5	—				1977—86
Greifswalder Bodden	7	15.1	—	65.3	34.7	0.67	1.45	3.10	1976—85
	(4)	262.6	99.6	0.4	—				1976—85

n.sp. = number of fished species, in bracket = marine species; [1]) ruffe catches included; [2]) according unpublished Polish datas; [3]) according NOACK (1978) shown themselves more capable of overcoming the critical phase of various ecological situations. The larger number of different species and the possibility to move in a larger area make survival easier.

In mesohaline waters the pike replaces the pikeperch. Herring represents an additional food component and is intensively devoured by pike and pikeperch. In the management of these waters such relationships must be considered. The only fish unimportant in the supply of food to piscivorous fishes are bream and white bream and can thus be independently fished. It can be assumed that too intensive fishing of roach, perch and ruffe could have a possible feedback effect on piscivorous fish populations.

Referring once again to Table 1, even with knowledge of trophical relationships, it is difficult to explain the differences in yields and also in the species structure of the catch. This has various methodical reasons. One reason is, as already mentioned, the large catch of herring and freshwater species in the spawning regions of bays which arise through the large outside feeding areas and the sizes of which are not known. Furthermore the fishing intensity and the species caught is not always the same. Only in "Kleines Oderhaff" ruffe is intensively fished. Perch, bream and roach are caught to a greater or lesser extend depending on the situation in the fishing industry. In the case of eel, pikeperch and pike we can assume that variations in the catch correlate with population variations.

How can be rated the observed ha-yield relationship (Table 1)? In recent literature an existing relationship between primary production and fish yields has been frequently referred to. According to BULON (1983) the yield/primary production quotient (relative catch, % primary production) ranges from 0.02% (World ocean) and 0.10% (Baltic) to 0.4% for some southern inner seas. The highest value is found in shallow fish ponds where it reaches 1%. Significant deviations from these values result from the above mentioned methodical problems.

Considering the primary production value and using the BULON & VINBERG (1981) equation we would expect a catch of 23 kg/ha/a for the western part of "Darss-Zingster Bodden" and 10 kg/ha/a for the eastern region of this bay. In the first case the catch had a value of only 16.5 kg indicating that more intensive fishing of the waters was necessary to reach the value given in the literature of 20—30 kg/ha/a for Baltic pikeperch-lakes (BARTHELMES 1981). In the eastern part of "Darss-Zingster Bodden" however the catch is 33.4 kg/ha/a, approximately 2/3 over the expected value. The reason for this difference is the external supply of energy. This does however not mean that a further increase in catch is not possible. For example, following an analysis of bream populations in this region EL-GAMMAL (1982) concludes that the intensity of fishing can be increased.

Similarly, NOACK (1978) obtained a surprisingly small relative fish yield of 0.03% for the enclosed "Kleiner Jasmunder Bodden", furthermore, no relation between the catch of benthivorous fish and changes in benthos production could be ascertained.

BULON (1983) and other authors have indicated that deviations from the expected value most often occur when the catch is compiled mainly of benthivorous species — a poor correlation exists between benthos and primary production. Our little survey shows, however, that besides piscivorous only benthivorous species compile the catch from our shallow coastal waters.

These few examples demonstrate that the tested concept used to estimate production can only be employed with certain reservations due to the specific nature of the trophical structure and the open character of the system. Specific analyses can not be excluded.

A further possibility is the comparison of different estuaries. Old and new data on catches in Baltic bays show that, as soon as the species caught is of no interest, yields of 20—60 kg/ha/a based on autochthonic fish communities can be achieved. In the period between 1929 and 1938 the average catch in the "Kurisches Haff" was 60.6 kg/ha and that in the total "Oderhaff" was 35.6. Today the value for the Oderhaff-region in the GDR and Poland lies at more than 60 kg/ha/a. In the geographically more favourably situated Asow sea yields of 59 kg/ha/a were obtained in the 1930s and in the 60s 34.2 kg/ha/a. Approximately 2/3 of the catch is made up of small planktivorous fish species (MARTI & MARTINSEN 1969).

Particularly in the 70s a lot of discussion was given to the so-called cultural — or man made eutrophication, which most certainly, allowing for methodical limitations, showed that, from comparison of old with current catches, an increase in the total catch is to be seen (Fig. 3). Since the turn of the century the catch in the "Darss-Zingster Bodden" has been almost doubled. Of particular interest are the developments in the catches of pikeperch and pike. Whereas in the case of pikeperch the catch increased rapidly to expected level, the pike yield showed a decrease from the end of the 60s. This pattern corresponds to observations made in inland waters and is described in literature as a result of the eutrophication. Many questions still remain unanswered and some assumptions still require proof. In the "Greifswalder Bodden" the freshwater fish catch has increased by a factor of 3 and in the same time period the pike yield did not decrease but did in fact increase in the long term. This is also an example of the special character of these coastal waters.

Already at the end of the last century management methods for inland water fishery were extended to the coastal waters. They also included the introduction of allochthonous fish species to the coastal waters but the hopes of success were, however, seldom achieved. On several occasions, common carp was introduced to the oligohaline and trout to the mesohaline regions. In both

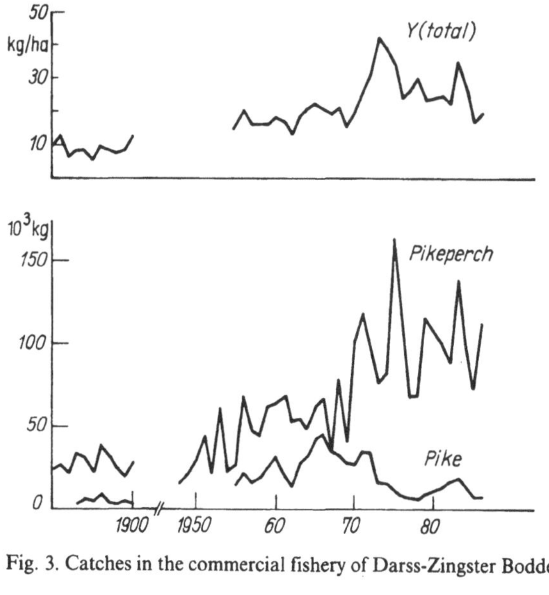

Fig. 3. Catches in the commercial fishery of Darss-Zingster Bodden

cases good and very good growth performances were seen. The economy of such steps was however questioned due to poor catches. This question is, however, currently being reconsidered and better yields are hoped with the support of an increased application of extensive aquaculture. Despite a certain optimism, the problems arising with the intensifying of fish production in these coastal waters should not be ignored.

The required water quality has already become a limiting factor in some regions. The results are the death of fish, decrease in the limited available spawning areas, reduced quality of captured fish etc.

We must also remember that our knowledge of the complete ecosystem and in particular the fish communities is so incomplete that any changes are performed on a trial and error basis. Especially inadequate are investigations of the synecological aspects. Due to the importance of coastal waters as areas for reproduction and growth utmost attention should be given to questions concerning the formation and development of young fish communities including the trophical relations. The yield from shallow waters comprising mainly benthivorous species demands investigations of the production potential of benthic community and their utilization by fish.

A step forward in the traditional management of freshwater fish in these waters would be the application of the multispecies concept.

References

BARTHELMES, D. (1981): Hydrobiologische Grundlagen der Binnenfischerei. Jena.

BULON, V. V. (1983): Zavisimost ryboproduktivnosti vodoemov ot pervičnoj produkcii. Sbornik naučnych trudov NIIORCH, Nr. **196**, 3—11.

— & VINBERG, G. G. (1981): Sootnošenie meždu pervičnoj produkciej i ryboproduktiwnostju vodoemov. Osnovy izučenija presnovodnych ekosistem, pp. 5—10. Moskva.

EL-GAMMAL, F. (1982): Die Bestandscharakteristik und Populationsdynamik des Bleis (*Abramis brama* L.) sowie seine Einordnung in das Trophiegefüge der Darss-Zingster Boddenkette. Diss. A, Wilhelm-Pieck-Univ. Rostock.

MARTI, JU. JU., & MARTINSEN, G. V. (1969): Problemy formirovanija i izpolzovanija biologičeskoj produkcii atlantičeskogo okeana. Moskva.

NOACK, B. (1978): Probleme der Ausnutzung des natürlichen Nahrungsangebotes durch Fische in den Gewässern der Insel Rügen und ihre Bedeutung für die fischereiliche Bewirtschaftung. Diss. A, Wilhelm-Pieck-Univ. Rostock.

Address of the author:

Dr. rer. nat. HELMUT MANFRED WINKLER
Wilhelm-Pieck-Universität Rostock
Sektion Biologie
Universitätsplatz 5
Rostock
DDR-2500

Limnologica (Berlin) **20** (1), 199—200, April 1990

Wilhelm-Pieck-Universität Rostock, Sektion Biologie, Rostock, DDR

Round Table Discussion on "Perspectives in Estuary Research and Management"

The round table discussion was dominated by the results presented at the symposium, which showed that

— most of the internal coastal waters in particular are eutrophic, but in some cases they are also contaminated and saprobic. Many of these ecosystems are already useless for some purposes. Their present state has also affected conventional fisheries and open water aquaculture, both of which need healthy shallow water regions for optimal catches;
— drastic changes have taken and are taking place in Baltic stocks of macroalgae and the colonization of the sea bed by animals;
— algal blooms become increasingly common in the inland and seaward coastal waters; and
— the polluted rivers flowing into the Baltic Sea influence large areas of the sea (satellite pictures).

The round table discussion reflected the participants' desire to ensure the estuarine and near-shore ecosystems, and therefore the Baltic Sea itself, remain intact and functionable. The course of the discussion was therefore marked by very concrete ideas concerning measures to be taken to preserve the "free benefits" afforded by these ecosystems and the costs involved as well as by the aims of basic research in this respect. In detail the discussion centred on defining general problems and problems of monitoring, basic research and management, identifying the causes of the problems and finding avenues for their solution.

1. General problems

Owing to their production potential and the purposes for which they are used, the estuaries and coastal waters along the southern and south-eastern shores of the Baltic are of major economic importance. The ecosystems of these waters are fairly similar, and the results of research into them are likely to be generalizable. In other words, results obtained at one place may also apply elsewhere. This facilitates the cooperation in basic research, the classification of waters, monitoring and management.

Meaningful steps to improve the situation are needed in addition to more intensive basic research orientated towards practical applications. The measures now being taken to the rehabilitation of the boddens south of the Darss-Zingst peninsula and the inner Puck Bay are first promising steps in this direction. The analysis and generalization of the results are of great relevance to further projects for the rehabilitation of the whole southern and south-eastern Baltic coast.

Close international cooperation and a constant exchange of information are further important prerequisites for solving the problem as a whole and for efficient research in similar parts of the Baltic Sea.

2. Monitoring and assessment

The different interests of the Baltic countries in the utilization of their inland and seaward coastal waters complicate the elaboration of a standard monitoring and assessment programme that would be accepted by for all countries. In addition, the availability of resources for such a programme varies.

In view of this, monitoring programmes should be designed to cater for the main purposes to which the coastal waters will be put in each particular case. The scope of monitoring activities will therefore be restricted, and this makes it even more important that suitable methods and sufficiently sensitive, representative indicator organisms will be used. Permanent exchange of experience and consultations between the Baltic countries are essential to maximize the efficiency of monitoring programmes.

The Department of Biology of the Wilhelm Pieck University Rostock has proposed a system for classifying the trophic status of inland coastal waters in the GDR. It seems probably that the criteria used in this system can be modified for use along the whole of the southern and south-eastern Baltic coast and in other coastal regions. After it has been checked and validated by the appropriate authorities in the GDR, details of this system should be forwarded to all relevant scientific institutions in the Baltic countries, where its adoption could be considered.

The monitoring of contamination and saprobation still involves considerable problems. In the past, too little attention has been paid to accumulation of contaminants and adaptation capabilities in ecosystems and organisms. Moreover, the methods used to measure saprobation are unreliable. This clearly diminishes the validity of results. Also, little is known regarding the flux of contaminants in ecosystems, i.e. their transport and distribution between organisms, their feed-back effects or their interactions. Our knowledge in this field can increase gradually only as a result of coordinated joint efforts.

3. Basic research

The main problem from the scientific standpoint is the lack of a generalized estuary concept for the ecosystems being considered. Such a concept would greatly facilitate communication and cooperation between biologists, chemists, physicists, hydrologists, economists and layers.

Constant variability and elasticity are the chief characteristics of the ecosystems under discussion. Contrary to earlier opinions, these estuaries are astonishingly robust to short term pollution loads. Apart from their great production potential, they have a considerable buffering capacity. However, there is some uncertainty in regard to both the causes of the buffering capacity and the magnitude of the loads it can adsorb.

Almost all waters in these regions are shallow. The interactions taking place between the pelagic and benthic zones, and particularly at the sediment-water interface, are of decisive importance. Therefore, the question of what is accumulated and recycled how, where and when must form one major topic of study.

There is a reason to hope that the causes of seasonal processes and the mechanisms governing them will be elucidated in the near future. The prediction and modelling of such processes would be an important step towards the management of waters based on a scientific background.

Assessments of system stability and the degree to which it can be influenced depend on the availability of quantitative information concerning the main processes involved in the circulation of matter and energy flux. Work performed at the Department of Biology of the Wilhelm Pieck University Rostock shows that the "microbial loop" predominates in the pelagic zone. Further studies must show whether this applies to other waters. Less is known concerning these processes in the benthic zone, and knowledge of the communication mechanisms in these ecosystems is very scant.

In estuarine systems the exchange of water and matter with the open Baltic and the input by river systems are very important. In most cases the exchange and dwell times have been calculated. However, efforts should be made to construct hydrological models of these processes and later to link these models with models of biological processes.

Owing to many open problems, the limited research resources must be concentrated on basic research projects that will yield practical benefits for society. Cooperation between research institutions should therefore be closer and better coordinated to ensure that uniform standards of work are achieved more rapidly and the results of joint ecosystem studies are available more quickly. The similarities between these ecosystems make it relatively easy to check whether results obtained for one ecosystem are equally valid for another. Staff and material resources can then be released to deal with other problems while simultaneously increasing the rate of progress.

Increased attention should be paid to the development of ecophysiological and experimental ecological capacity and approaches. Wherever possible, field studies and mesocosm experiments should be linked to laboratory experiments (ecophysiological and microcosm) to obtain more quickly reliable, generalizable data that will be also valid for the ecosystem. Experience has shown that functional redundancy is higher at species level. Therefore, studies should be based rather on functional groups. These will lead more rapidly to meaningful results.

4. Anthropogenic influences and management

The Helsinki Convention states expressly that each country is responsible for the ecological control of its coastal waters, regardless of differences in legislation and interests. The scientists whose job it is to study the ecosystems are to a large extent responsible to the society for their preservation and optimal utilization. These scientists must therefore seek ways and means to ensure that their advice is heard and that they are included in corresponding administration decisions making process.

Priorities must be set for the utilization of estuaries and coastal regions according to their location and importance. The actual purpose to which such a region is put must not exceed the tolerance limits of the ecosystem. Wherever possible, only schemes that take fishing, industrial, community, agricultural, recreational, nature conservation and shipping interests into account should be considered. Such schemes allow the potential of these ecosystems to be utilized most fully and safely.

Owing to their position between land and sea, estuarine systems are obviously amenable to rehabilitation. Nevertheless, restoration of the normal ecosystem function after a false decision will always cost many times the profit achieved under normal conditions of utilization. Closer links are needed between economists and ecologist to better quantify the "free" economic benefits afforded by the ecosystems. In the past little attention has been given to the field where ecology and economics overlap. But in this area progress is not only of scientific interest, but is also doubtless of major importance for a scientifically based sociopolitical ecosystem concept.

U. Schiewer and E. A. Arndt

Limnologica (Berlin) **20** (1), 201—212, April 1990

Rezensionen

N. N. Smirnov (ed.): **Lake Glubokoe** (Developm. in Hydrobiol. **36**). Dordrecht, Boston, Lancaster: Dr. W. Junk Publishers 1987. 164 S., 49 Abb:, 35 Tab., 3 Photogr. Geb. Dfl 150.00/US $ 64.50

Als gegen Ende des vergangenen Jahrhunderts Forbes (1887) sein Essay „The lake as a microcosm" und Forel (1982) den ersten Teil seines Werkes „Le Leman: monographie limnologique" veröffentlichten, begann die Limnologie ihre Entwicklung zu einer sachlich und methodisch selbständigen biologischen Disziplin.

Etwa zur gleichen Zeit (1891) wurde durch den russischen Biologen N. Yu. Zograf am Glubokoe-See in unmittelbarer Nähe Moskaus eine der ersten limnologischen Stationen der Welt gegründet, die heute auf eine fast einhundertjährige Forschungstätigkeit zurückblicken kann. Aus diesem Anlaß bringt der Band 36 der „Developments in Hydrobiology" eine Reihe zusammenfassender Arbeiten über die Hydrochemie, das Zooplankton, das Zoobenthos, die Litoralfauna, die Nahrungsbiologie der Fische und über die Paläolimnologie des Sees. Aber auch bestimmte Tiergruppen wie die Daphnien, die Rotatorien, die Fische und die Parasitenfauna werden behandelt.

Stellvertretend für insgesamt 18 Beiträge möchte ich den Inhalt von 5 Arbeiten etwas näher vorstellen.

Yanin, Kashina & Sayet charakterisieren den See als ursprünglich dystroph. Durch Melioration des Einzugsgebietes wurde der Zustrom braunen Wassers unterbrochen, infolgedessen erweiterte sich die euphotische Zone deutlich. Im übrigen ist das Gewässer von anthropogenen Einflüssen praktisch frei. Matveev berichtet über Langzeitveränderungen in der Gemeinschaft der Planktoncrustaceen. Nach seiner Auffassung spielen dabei Konkurrenzmechanismen und der Einfluß der Fische die entscheidende Rolle. Wie in vielen anderen Seen der Paläarktis, so ist auch im Glubokoe-See die Taxonomie der Gattung *Daphnia* schwierig. Glagolev versucht dieses Problem in seiner umfangreichen und mit vielen Abbildungen versehenen Arbeit zu lösen. Boikova untersucht die Nahrungsbiologie der Fische des Gewässers. Durch starken interspezifischen Konkurrenzkampf zwischen der Plötze und dem Barsch um die Hauptnahrungsquelle Zooplankton ist der Barschbestand in den letzten Jahren stark zurückgegangen. In einem Abschlußkapitel faßt der Herausgeber des Bandes die grundlegenden Merkmale der Biozönose des Glubokoe-Sees zusammen, und bringt in einem Anhang alle bisher gefundenen Arten.

Insgesamt erlaubt die Darstellung einen Einblick in die bisherigen Forschungsarbeiten der Station. Ein Mangel sei allerdings auch genannt. Keine der Abbildungen besitzt eine Legende, so daß es bisweilen unmöglich ist, die Aussage auch nur zu erraten. Mir ist unverständlich, weshalb der Verlag nicht darauf bestanden hat, diesen zweifellos leicht abzustellenden Fehler zu beseitigen.

P. Kasprzak, Neuglobsow

P. De Deckker & W. D. Williams (eds.): **Limnology in Australia** (Monogr. Biol. **61**). Dordrecht, Boston, Lancaster: Dr. W. Junk Publishers 1986. XIII, 671 S., 158 Abb., 63 Tab. Ln. Dfl. 250.00/ US $ 115.00

Wasser gehört in Australien zu den kostbarsten Naturgütern. Der Hydrologie und Ökologie der australischen Binnengewässer wird deshalb besonders große Aufmerksamkeit zuteil. In 30 Jahren limnologischer Forschung in Australien und 25-jähriger Tätigkeit der „Australian Society of Limnology" ist viel geleistet worden. Die australische Limnologie hat nicht nur den Anschluß an die traditionsreiche europäische oder nordamerikanische Limnologie gefunden, sondern sie dominiert diese Wissenschaftsdisziplin bereits in einigen Bereichen.

Was die australische Limnologie zu leisten vermag, welche Struktur und Funktion die australischen Binnengewässer aufweisen, wie sie sich im Laufe ihrer langen Geschichte entwickelt haben und welche Probleme sich im Zusammenhang mit ihrer Bewirtschaftung ergeben, erfährt der Leser des 61. Bandes der von H.-J. Dumont heruasgegebenen Serie „Monographiae Biologicae".

In 40 Arbeiten berichten 50 australische Wissenschaftler über limnologische Arbeiten auf diesem Kontinent. Der erste Teil des Buches ist physiko-chemischen Aspekten gewidmet (92 S.), u. a. mit Abhandlungen von B. T. Hart & I. D. McKelvie über den Chemismus und von J. T. O. Kirk über optische Eigenschaften der Gewässer. Der zweite Teil beinhaltet biologisch-ökologische Studien (156 S.). Die Palette der Themen reicht von mikrobiellen Benthosgesellschaften in Salzseen (J. Bauld), der Biogeographie und Ökologie australischer Rotatorien (R. J. Shiel & W. Koste), der aquatischen Insektenfauna (R. Marchant; D. H. D. Edward), der Biologie von Neunaugen (I. C. Potter, R. W. Hillard & F. J. Neira) bis hin zu Adaptionsmechanismen der Flora in Salzseen (M. A. Brock), meromiktischen Seen (R. L. Croome), der Osmo- und Ionenregulation der Fauna in Salzwasserseen (P. Greenway) und genetischen Aspekten des Zooplanktons (J. A. H. Benzie). Im dritten Teil des Buches stehen mehr die ökologischen Prozesse und die Ökosysteme selbst im Vordergrund (236 S.). Es werden dabei gleichermaßen Fließgewässer, Teiche, Seen, Salzseen und Ästuare einbezogen. Besonders interessant ist die Darstellung der „Mound Springs" (W. F. Ponder) und „Billabongs" (T. J. Hillman). W. D. Williams kommentiert abschließend in diesem Abschnitt „Limnologie" und „Limnologie der Salzseen".

Der vierte und letzte Teil des Buches ist der Bewirtschaftung der Gewässer vorbehalten (161 S.). Er enthält u. a. Arbeiten zur Entwicklungsgeschichte (P. De Deckker), zu Stauseesedimenten (R. L. Clark & R. J. Wasson), schließt die Eutrophierungsproblematik (P. Cullen), ökotoxikologische Probleme (D. W. Connell), sowie die Nutzung von Wasserpflanzen (D. S. Mitchell) mit ein und endet mit Beiträgen zur Erhaltung der Gewässerfauna (F. B. Michaelis; D. A. Pollard & J. J. Burchmore) und der Diskussion des „Murray"-Projektes (K. F. Walker).

Ein Sachindex beschließt das sehr gut gedruckte und ausgestattete Buch. Der Band ist ein weiterer Glanzpunkt der „Monographiae Biologicae".

R. Koschel, Neuglobsow

Gerold Thompson, Jennifer Coldrey & George Bernard: **Der Teich.** Stuttgart: Franckh/Kosmos Verlagsgruppe 1986. 256 Seiten, 414 Farbfotos u. 155 Farb- und Schwarzweißzeichn. im Text. Geb. DM 98,—

Der Kosmos-Verlag ist seit langem bekannt für seine reichbebilderten Bücher, in denen populär und zugleich wissenschaftlich korrekt einzelne Lebensräume mit ihrem Tier- und Pflanzenleben dem interessierten Leserkreis vorgestellt werden. Ein Buch dieser Art ist „Der Teich", von den englischen Autoren THOMPSON, COLDREY & BERNARD, übersetzt und bearbeitet von BRUNO P. KREMER.

In diesem Buch werden eine Vielzahl von Pflanzen und Tieren der Kleingewässer in Wort und Bild vorgestellt, Einzelheiten aus ihrer Biologie mitgeteilt, aber auch die Ökologie des Teiches wird abgehandelt. Es entsteht ein Bild vor dem Leser, das dem Lebensraum Teich in vielfältiger Weise gerecht wird. Das Buch gliedert sich in folgende Kapitel: Vorwort (4 Seiten); Kurzbestimmungsschlüssel der Tierwelt des Teiches (4 Seiten); die Welt des Teiches (10 Seiten); das Pflanzenleben im Teich (8 Seiten); Teichpflanzen und Pflanzengemeinschaften (46 Seiten); Tierleben im Teich (6 Seiten); Teichtiere (152 Seiten); vom Umgang mit Teichbewohnern (2 Seiten); Erklärung der Fachausdrücke (5 Seiten); tabellarische Klassifizierung der abgebildeten Pflanzen und Tiere (8 Seiten) und das Register (4 Seiten).

Dieses Buch dem deutschsprachigen Leserkreis erschlossen zu haben, ist ein Verdienst. Seine Konzeption, sein informativer Text und vor allem seine Bildausstattung machen es zu einer Besonderheit auf dem einschlägigen Büchermarkt. Viele der 414 Farbfotos suchen ihresgleichen. Überragend und z. T. vielleicht sogar bisher einmalig sind die Makrofotos der Teichtiere. Hier liegen offensichtlich auch die Hauptinteressen der Autoren. Wenn man allerdings den Anspruch im Vorwort bedenkt, von den wichtigsten Pflanzen und Tieren des Teiches zu berichten, erscheint die mikroskopische Welt des Teichlebens etwas untergeordnet. Sie ist auf 24 Seiten von 255 abgehandelt. So ist dieser prachtvolle Bildband vor allem für jene interessant, die die Vielfalt des Teichlebens mit dem unbewaffneten Auge, oder mit der Lupe erleben wollen. Hier vermittelt es verblüffende Entdeckungen und regt zu eigener Beobachtung an. Sicher trägt es auch bei vielen Naturfreunden zu der Erkenntnis bei, daß unsere in der Kulturlandschaft so gefährdeten Kleingewässer schützens- und erhaltenswert sind.

W. SCHEFFLER, Neuglobsow

R. P. MCINTOSH: **The Background of Ecology.** Concept and Theory (Cambridge Stud. in Ecol.). Cambridge: University Press 1985. 383 S. Ln. US $ 39.50

Der Begriff „Ökologie" und viele von ihm abgeleitete Wortbildungen sind gegenwärtig feste Bestandteile der öffentlichen Diskussion, obwohl damit nicht gesagt ist, daß der sachliche Gehalt aller dieser Schlagworte völlig klar ist. Im Gegenteil, seit die Auswirkungen der „Umweltkrise" etwa mit Beginn der sechziger Jahre ins Bewußtsein vieler Menschen und Politiker gedrungen sind, droht der Ökologie eine weitere Verschlechterung ihrer ohnehin oft unscharfen terminologisch-konzeptionellen Situation, denn längst nicht alle Umweltprobleme der Menschen sind ökologischer Natur.

ROBERT P. MC INTOSH versucht mit seinem Buch „The background of ecology. Concept and theory" Klarheit in das babylonische Durcheinander zu bringen, indem er den Entwicklungsweg der Ökologie zurückverfolgt bis in die Zeit ihrer Herausbildung als selbständige Wissenschaft etwa an der Wende vom 19. zum 20. Jahrhundert. Dabei beabsichtigt er nicht die Schaffung eines Geschichtsbuches der Ökologie, sondern möchte ihren Hintergrund in seiner Bedeutung für die gegenwärtigen Probleme der Ökologie als Wissenschaft darstellen.

In insgesamt 8 Hauptkapiteln werden folgende Themen behandelt: Antecendents of ecology; The crystallization of ecology; Dynamic ecology; Quantitative community ecology; Population ecology; Ecosystem ecology, systems ecology and big biology; Theoretical approaches to ecology; Ecology and environment. Ein umfangreiches Literaturverzeichnis mit etwa 950 Zitaten und ein Sachwortindex befinden sich am Ende des Buches. In den Abschnitten kann der Leser die Entstehung und Entwicklung wichtiger Konzepte und Theorien aus Tier- und Pflanzenökologie, Limnologie und Ozeanographie verfolgen. Als Beispiele seien hier Ökosystem, Gleichgewicht, Nische, Biozönose, Gemeinschaft, Konkurrenzausschluß und Koexistenz genannt. Ebenso wird die z. T. kontroverse Diskussion um Begriffe und Konzepte spürbar, für die es wohl zwei Hauptgründe gibt: die Suche nach einem theoretischen Fundament der Ökologie (vergleichbar etwa dem der Physik) und die Spaltung der Ökologen in das Reduktionisten- und Holistenlager.

Bei der Schaffung einer ökologischen Theorie ging es einerseits darum, die Ökologie gegen den Vorwurf der Unwissenschaftlichkeit und den der simplen Okkupation von Wissensgebieten der Physiologie zu schützen. Andererseits bemerkten viele Ökologen der sechziger und siebziger Jahre, daß die Ökologie als wissenschaftliche Disziplin ohne eine theoretische Basis keine entscheidenden Fortschritte erzielen könne. Allerdings waren die Hoffnungen auf ein der Physik vergleichbares Gebäude von Gesetzen, Prinzipien, Konstanten und mathematischen Modellen zu hoch gesteckt. Als Ausweg versuchten einige Ökologen die Gesetze der Thermodynamik (als Ersatz) in die Diskussion einzuführen. MC INTOSH bemerkt dazu: „Apply they must, but they are hardly ecological laws." Die Grundidee einer mathematischen Durchdringung der Ökologie wird inzwischen weitgehend akzeptiert. Ob es aber dabei in erster Linie auf die Entdeckung streng deterministischer Zusammenhänge oder auf die „Beherrschung" der Variabilität von Ökosystemen mit den Mitteln der Mathematik ankommt, ist wohl keine Frage von „entweder—oder". MCINTOSH zitiert in diesem Zusammenhang SLOBODKIN, der den nach meiner Auffassung einzig möglichen Ausweg aus dieser Situation zeigt: „The normal criteria of scientific quality which we use as biologists are not the same as those of the physicist and mathematician... Empirical sciences must develop their own standard of quality and cannot take refuge from the necessity of thought in the shadow of Newton or Euclid."

Von Anfang an war der Versuch einer theoretischen Begründung der Ökologie mit dem Streit um Holismus und Reduktionismus als Mittel der Erkenntnis verbunden. Auch hier bietet MCINTOSHs Buch eine ganze Reihe interessanter Fakten und Zusammenhänge. Bereits FOREL und FORBES waren in diesem Punkt (unbewußt) Kontrahenten, indem FOREL die Eigenschaften des Seenökosystems auf dem Niveau der Art zu erklären versuchte, während FORBES das organismische Konzept des Ökosystems zu entwickeln begann, das sich dann bei THIENEMANN zum „Überorganismus" wandelte. MCINTOSH zitiert CODY, der den Holismus gegen den Reduktionismus in Schutz nimmt: „This it seems to me is what ecology is all about, only thus can one keep one's nose above water in the seas of experimental reductionism and secular pedantry that currently threaton to swamp the field." Aber auch die Meinungen entschlossener Gegner des Holismus werden nicht verschwiegen. MCINTOSH interpretiert GAUSE: „He allowed that there was a difference between the whole and the sum of its parts, which was often variously stated but never effectively quantified."

Dieser Streit zwischen Holisten und Reduktionisten dauert bis heute an, und man kann dem Autor zustimmen, wenn er sagt, daß die Position der einzelnen Ökologen offensichtlich sehr von ihren (bewußt oder unbewußt eingenommenen) philosophischen Haltungen beeinflußt wird. Wie dem auch sei, der Streit war insofern fruchtbar, als er zum gründlichen Nachdenken über die Eigenschaften des Ökosystems geführt und damit zum Erkenntnisgewinn beigetragen hat.

Insgesamt stellt MCINTOSHs Buch ein interessantes Kapitel über die Entwicklung der Ökologie auf ihrem Weg zu einer sachlich und methodologisch selbständigen Wissenschaftsdisziplin dar, dessen Lektüre jedem Ökologen bestens empfohlen werden kann. Einerseits wird er an die Wurzeln seines Wissenschaftszweiges zurückgeführt, und mit den z. T. zu unrecht vergessenen Leistungen der frühen Ökologen bekannt gemacht. Andererseits bekommt er eine Hilfe, die es ihm ermöglicht, seinen eigenen Beitrag in den bestehenden ökologischen Wissensfundus verallgemeinernd einordnen zu können. Ob er sich dabei als Holist oder notgedrungen als „large-scale-reductionist" fühlt, ist ein Problem zweiten Ranges.

P. KASPRZAK, Neuglobsow

B. LENHART: **Limnologische Studien am Ammersee 1984—1986** (Informationsberichte Bayer. Landesamt für Wasserwirtschaft **2/87**). München: Bayerisches Landesamt für Wasserwirtschaft 1987. 112 S., 55 Abb., 5 Tab. Brosch. DM 10,—

Der vorliegende Bericht setzt die Tradition der Veröffentlichung aktueller limnologischer Forschungsergebnisse in der Reihe „Informationsberichte des Bayerischen Landesamtes für Wasserwirtschaft" fort. Es werden Untersuchungen zu den Nährstoffverhältnissen, zur aktuellen Nährstoffbelastung, zur Charakterisierung des Phytoplanktons sowie zur Entwicklung der Trophie an Hand stratigraphischer Analysen der Kieselalgen in einem kurzen Sedimentkern im Ammersee dargestellt.

Während der 1978 in der gleichen Reihe erschienene Ammersee-Bericht sich auf die Probleme im Zusammenhang mit der Eutrophierung des Sees konzentrierte, zeigt der neue Ammersee-Bericht eine erfreuliche Entwicklung auf: Dank umfangreicher abwassertechnischer Maßnahmen im Einzugsgebiet des Sees und aufgrund der inzwischen zurückgegangenen Nährstoffbelastung hat sich der trophische Zustand des Ammersees inzwischen deutlich gebessert. Diese Entwicklung wird anhand der Ergebnisse der hydrochemischen und -biologischen Untersuchungen belegt.

Instruktive Abbildungen und Tabellen ergänzen den übersichtlich gestalteten Text. Hinzuweisen ist vor allem auf die Phytoplankton(Kieselalgen)darstellung und die damit im Zusammenhang stehenden sehr anschaulichen licht- und rasterelektronenmikroskopischen Aufnahmen der häufigsten Planktonalgen des Ammersees.

R. KOSCHEL, Neuglobsow

ROBERT HENRY PETERS: **The Ecological Implication of Body Size** (Cambridge Studies in Ecology). Cambridge: Cambridge University Press 1986. 329 S., 59 Abb., 44 Tab., 1 Tabellenanhang. Paperback. $ 16.95

Die ökologische Forschung muß sich in jüngster Zeit mehr und mehr mit einer zwiespältigen Situation auseinandersetzen: Einerseits wird der Bedarf an allgemeingültigen, empirischen Theorien immer größer, weil mit ihrer Hilfe die wachsende Zahl von Umweltproblemen auf quantitativer Grundlage behandelt werden kann, andererseits findet man sehr viele Ausnahmen, die solche Theorien scheinbar nicht zulassen.

Der bekannte Limnologe R. H. PETERS unternimmt in seinem Buch „The ecological implications of body size" trotzdem den Versuch, einen Teil der vielfältigen Eigenschaften von Säugetieren, Vögeln und anderen Vertebraten, sowie von Wirbellosen, Protozoen, Algen und Bakterien auf der Grundlage eines sehr einfachen Merkmals zu beschreiben und zu erklären, nämlich der Körpergröße.

Viele biologische Daten, das Populationswachstum, die Bewegungsgeschwindigkeit der Tiere, die Tierdichte, das Individualterritorium oder die Beutegröße, so hat sich gezeigt, lassen sich oft als einfache Funktion der Körpergröße beschreiben. Der benutzte mathematische Apparat ist sehr klein und einfach und deswegen auch von Lesern ohne spezielle Kenntnisse erfaßbar. Außerdem wird er in einem kurzen vorangestellten Kapitel erläutert. Die folgenden Abschnitte befassen sich mit dem Einfluß der Körpergröße auf den Stoffwechsel, auf die Bewegung, auf die Futteraufnahme, auf die Produktion, auf den Stoff-Flux und die Tierabundanz, um nur einige Beispiele zu nennen. In einem Kapitel über ein spezielles allometrisches Simulationsmodell werden die einzelnen Erkenntnisse zusammengefaßt und in einem weiteren Abschnitt erklärt.

Je weiter man beim Lesen in den Text des Buches vordringt, um so offensichtlicher wird PETERS' vorbildliche Konzeption. Zwar umreißt er nirgends explizit seine Zielgruppe, aber die jeweiligen Kapiteleinleitungen und -zusammenfassungen, die Hervorhebungen und die zahlreichen Graphiken geben dem Band den Charakter eines Lehrbuches. Andererseits ist der große Reichtum an enorm komprimiertem Datenmaterial auch für den Berufsbiologen sehr attraktiv.

Welcher Umfang an Arbeit hier geleistet wurde, wird dem Leser erst vollkommen klar, wenn er den sehr nützlichen, 60 Seiten langen Tabellenanhang näher betrachtet, in dem PETERS den numerischen „Extrakt" von etwa 560 Literaturzitaten darstellt. Am Ende seines Buches schreibt er: „Finally I would stress my belief that the allometric relations listed in the appendix represent by far the greatest body of quantitative general theories in biology. I sincerely hope that this closing show of arrogance will stimulate others to prove me wrong. This can only be shown by amassing an even greater body of ecological theory".

Ich glaube, diese Rekordleistung wird wohl eine Weile Bestand haben.

P. KASPRZAK, Neuglobsow

MARTIN WORBES: **Lebensbedingungen und Holzwachstum in zentralamazonischen Überschwemmungswäldern** (Sripta Geobotanica **17**). Göttingen: Verlag Erich Goltze 1986. 112 S. mit 72 Abb. u. 14 Tab. Brosch. DM 36,—

Im Rahmen eines ökologischen Forschungsprojektes der Abt. Tropenökologie des Max-Planck-Institutes für Limnologie in Plön mit dem brasilianischen Instituto Nacional des Pesquias da Amazônia in Manaus ging der Autor der Frage nach, wie die Bäume der bisher kaum untersuchten Überschwemmungswälder des Amazonasbeckens die oft mehrere Monate dauernden Überflutungen ohne Schaden überstehen können. Die an 2 Untersuchungsflächen bei Manaus durchgeführten vegetationskundlichen, ökologischen und dendrologischen Untersuchungen ergaben, daß in den Überschwemmungswäldern die Zahl der Baumarten wesentlich geringer ist als in den benachbarten Wäldern ohne Überflutungen. Während der Überschwemmungen vermögen die Bäume den Stoffwechsel in ihren Wurzeln auf anaerobe Atmungswege umzustellen. Bei zunehmender Überschwemmungsdauer reagieren sie artspezifisch mit einer mehr oder weniger ausgeprägten Ruhepause des Kambiums.

Die angezeigte Arbeit enthält wichtige Befunde und Ansätze zu weiterführenden Forschungen über die Ökologie der in ihrem Fortbestand vielfach bedrohten Urwälder des Amazonas-Gebietes.

H.-D. KRAUSCH, Potsdam

Decheniana 140. Verhandlungen des Naturhistorischen Vereins der Rheinlande und Westfalens. Bonn: Selbstverlag des Naturhistorischen Vereins 1987. 221 S., 63 Abb., 39 Tab. Brosch. DM 80,—

Auch dieser Band der Decheniana weist unter den zahlreichen floristisch-faunistischen Abhandlungen und Kurzmitteilungen mehrere Arbeiten auf, die für den Limnologen von besonderem Interesse sein dürften.

Die von GEORG GEUERT vorgestellten Ergebnisse der limnologischen Untersuchung der Sieg enthalten nicht nur eine umfangreiche Artenliste der Invertebraten dieses Flußabschnitts, sondern neben physikalischen und chemischen Parametern wird vor allem auf die mit Hilfe biologischer Untersuchungsmethoden ermittelte Gewässergüte eingegangen (S. 148—163).

Auch GABRIELE MICKOLEIT nimmt in ihrer Bestandsaufnahme des Makrozoobenthos des Alpenbaches und seiner Nebenbäche (Oberbergischer Kreis, BRD) eine Beurteilung der Gewässergüte nach biologischen Parametern vor (S. 164—173).

Von Interesse ist auch die von THOMAS ZUMBROICH vorgelegte Arbeit über das Manganbakterium *Leptothrix echinata* in der Wiehltalsperre (Oberbergisches Land, BRD) sowie über den Manganeintrag und seine Wirkung auf die Biozönose des Sees (S. 184—189).

Einen Beitrag zur Limnologie eines Tagebausees gibt MARTIN DITTRICH mit seiner Arbeit über den Knapsacker See (südlicher Teil der Ville, BRD; S. 174—183).

A. LANGE, Jena

WERNER NACHTIGALL: **Lebensräume.** Mitteleuropäische Landschaften und Ökosysteme (Spektrum der Natur — BLV Intensivführer). München, Wien, Zürich: BLV Verlagsgesellschaft 1986. 223 S., 185 Farbfotos, 34 Zeichn. Flex. Kunstst. DM 34,—

Dieses Taschenbuch aus der Reihe der bewährten BLV-Führer enthält eine Beschreibung der charakteristischen Landschaften und Ökosysteme Mitteleuropas: Berg und Fels; Wald und Busch; Wiesen und Weiden; Moore und Heiden; Trockenfluren und Ödland; See und Teich; Bach und Fluß; Meeresküste und Watt. Neben natürlichen Lebensräumen werden dem interessierten Naturfreund auch die anthropogen geschaffenen vorgestellt, bis hin zu Kiesgruben, alten Gemäuern und Straßenpflaster.

Dem Autor gelingt es, die Biologie dieser Lebensräume erlebbar zu machen. Das wird dadurch erreicht, daß trotz aller Vielfalt der Wechselbeziehungen in den einzelnen Ökosystemen das Typische und Wesentliche in allgemeinverständlicher Form erläutert wird. Viele hervorragende Farbfotos und Zeichnungen, didaktisch und optisch wirkungsvoll eingeordnet, unterstützen die Erläuterungen und regen den Wissensdurst der Leser an.

Und ist die Neugier erst einmal geweckt — mit der im Anhang aufgeführten populärwissenschaftlichen Literatur (übersichtlich nach Lebensräumen geordnet) kann der Naturfreund sein Wissen vertiefen.

A. LANGE, Jena

WOLF-EBERHARD BARTH: **Praktischer Umwelt- und Naturschutz.** Anregungen für Jäger und Forstleute, Landwirte, Städte- und Wasserbauer sowie alle anderen, die helfen wollen. Hamburg, Berlin: Verlag Paul Parey 1987. 310 S., 33 Farbbilder auf Tafeln, 116 Abb., 27 Tab. Karton. DM 48,00

Behandelt werden Luftverschmutzungen, Fließ- und Standgewässerbelastungen, Moorzerstörungen, die ökologischen Probleme der Stadtplanungen, des Straßenbaus, der Land- und Forstwirtschaft, der Jagd, des Angelsportes und des Tourismus. Ferner werden einbezogen: Lärmbelastung, Biozid- und Salzausbringungen, Mülldeponien, Toxikologie, Flurbereinigung und Hecken, Biotop- und Artenschutz und vieles mehr.

Das Buch ist populär und agitatorisch geschrieben, ohne jedoch die sachliche Basis zu verlassen. Es ist relativ umfassend, seine Schwerpunkte sind im Untertitel genannt.

In leicht faßlicher Darstellung werden die biologisch-ökologischen Grundlagen der Umweltbelastungen und -zerstörungen vorgestellt, die notwendigen ökologischen Kenntnisse und Grundbegriffe vermittelt und die unökologischen Bewirtschaftungsweisen der Ökosysteme sowie Fehlplanungen im Städtebau deutlich gemacht. Es wird immer auf die durch die Eingriffe in die Ökosysteme ausgelösten Kettenreaktionen hingewiesen, durch die schließlich die aus ökonomischen Erwägungen gemachten Eingriffe und Belastungen auf die Gesundheit des Menschen, sein Wohlbefinden und auch auf die Ökonomie selbst zurückschlagen. Der Verfasser kritisiert nicht nur, sondern bringt Anregungen und vorliegende Kenntnisse über die Restaurierung beschädigter Ökosysteme und über ökologisch fundierte Städte-, Straßen- und Landschaftsplanungen. Er zeigt sehr umsichtig ökologisch vertretbare Nutzungsweisen der Ökosysteme auf, entscheidet sich aber, wenn diese nicht möglich sind, immer für die Beibehaltung des natürlichen Zustandes. Den Limnologen interessieren besonders die sehr guten Ausführungen über den ökologischen Flußbau und die Möglichkeiten des Rückbaus falsch ausgebauter Fließgewässer, aber auch über die Wiederbelebung und Neuanlage von Standgewässern und die Erhaltung von Mooren.

Der Verfasser steht auch auf seiten der ökologischen Jagdausübung, betont aber die Naturschutzfunktion der Jäger. Ähnliches gilt auch für die Angler, obwohl der heutige Angelsport ökologisch problematischer ist als die Jagd.

Das Buch demonstriert, daß der echte Naturfreund auf der richtigen Seite steht. Es vermittelt ihm Möglichkeiten, wie er aktiv Umweltschäden vermeiden, erkennen, bekämpfen und restaurieren kann. Da dies weder den verantwortlichen ökonomischen Planern noch den mit den Naturressourcen Arbeitenden imponiert, sei auf die lehrreichen Beispiele in diesem Buch hingewiesen, die zeigen, wieviel Geld uns unökologisches und die Natur ignorierendes Wirtschaften in Wirklichkeit kostet. Es ist ein engagiertes und kenntnisreiches Buch für den Umwelt- und Naturschutz, dem weite Verbreitung und Beachtung zu wünschen ist.

Nur noch die folgenden kritischen Hinweise: Nicht alle im Text zitierten Autoren erscheinen im Literaturverzeichnis. (Das Literaturverzeichnis bezieht sich fast ausschließlich auf Umweltschutzliteratur, nicht auf ökologische Schriften). Der Schlüsselstellung des Grundwassers in den Umweltbelastungen hätte noch etwas mehr Beachtung geschenkt werden können. Von ihr sagte schon M. VON PETTENKOFER: „Auf die Verschmutzung des Grundwassers hat die Natur die Todesstrafe gesetzt".

W. SCHÖNBORN, Jena

J.-G. KOHL & A. NICKLISCH: **Ökophysiologie der Algen.** Wachstum und Ressourcennutzung. Berlin: Akademie-Verlag 1988. 253 S., 129 Abb., 64 Tab. M 48,—

Die Ökophysiologie als junge, dynamische Grenzdisziplin zwischen Physiologie und Ökologie vermag unter Anwendung des Instrumentariums und der Erkenntnisse beider Wissenschaftszweige wesentliche Impulse für die Aufklärung und Steuerung von Prozessen in Ökosystemen zu geben und damit einen Beitrag zur Nutzung von Ressourcen zu leisten.

Die Autoren widmen sich im vorliegenden Buch hauptsächlich der Ökophysiologie von Algen aquatischer Ökosysteme, also der Organismengruppe, die etwa ein Drittel der Primärproduktion auf der Erde realisiert. Bodenalgen und Algen anderer Habitate, z. B. von Extremstandorten bleiben weitgehend unberücksichtigt. Im Mittelpunkt steht die Nutzung der Ressourcen Licht, Kohlendioxid, Phosphat und Stickstoff durch Algenpopulationen. Die behandelten Fragen sind von theoretischem und praktischem Interesse — so spannt sich der Bogen potentieller Leser vom Algologen, Physiologen und Ökologen bis zum Praktiker der Wasserwirtschaft, Hydrobiologie und Technischen Mikrobiologie.

Das Buch stellt eine gelungene, reich mit neuen Ergebnissen aus der Originalliteratur und eigenen Forschungsresultaten der Autoren ausgestattete Kompilation dar. Es reiht sich würdig in die Linie solcher Schriften wie Physiological Limnology (GOLTERMAN 1975) und Physiological Bases of Phytoplankton (PLATT 1981) ein.

Im 1. Kapitel (28 S.) haben die Autoren in straffer Form die im Laufe von Jahrzehnten von namhaften Phykologen gesammelten Erfahrungen über die Algenkultur zusammengestellt. Es beinhaltet wertende Beschreibungen von Nährmedien und Kulturmethoden sowie Möglichkeiten der Parameterermittlung des Algenwachstums.

Dem Lichtdargebot und seinem Einfluß auf Photosynthese und Wachstum ist das 2. Kapitel (71 S.) gewidmet. Breiten Raum nehmen darin Darstellungen über das Pigmentsystem der Algen und seine Adaptationsmöglichkeiten ein.

Die folgenden drei Kapitel beinhalten ökophysiologische Betrachtungen über Dargebot und Nutzung von anorganischen Kohlenstoffverbindungen (34 S.), von Phosphat (33 S.) und von Stickstoff (55 S.).

Das 6. Kapitel (3 S.) gibt einen kurzen Ausblick auf grundsätzliche Richtungen zukünftiger ökophysiologischer Forschungen im Hinblick auf die Klärung praktischer Fragen der Steuerung und der optimalen Nutzung von Algenpopulationen als Träger der Primärproduktion bzw. als Verursacher von Nutzungsbeschränkungen aquatischer Ökosysteme.

Nützlich für den Leser dieses anspruchsvoll geschriebenen Buches ist das Verzeichnis der verwendeten Abkürzungen und Symbole und ein gut ausgewähltes Literaturverzeichnis (21 S.). Das Sachregister konnte auf Grund des detaillierten Inhaltsverzeichnisses sehr knapp bemessen werden.

Einige kleine Kritiken:

Der Verlag hat das Buch etwas zu spartanisch ausgestattet. Einige Schreibfehler (z. B. wurde die Alge *Nitzschia* auf den S. 93, 160, 161 falsch geschrieben) sollten in zukünftigen Auflagen korrigiert werden.

Es ist zu akzeptieren, daß die Autoren keine nomenklatorische Revision der Algennamen aus der Originalliteratur vornehmen konnten. Dennoch wäre ein kleiner Hinweis auf die korrekte Zugehörigkeit mancher wichtiger Laborstämme, für die sich ungültige Namen eingebürgert haben, bestimmt vom Leser dankbar aufgenommen worden. Besonders erscheint dies bei „*Anacystis nidulans*" KRATZ-ALLEN, Bloomington 625 nötig, gerade angesichts solcher Beispiele wie auf S. 121, wo in den nebeneinanderstehenden Abbildungen 3—11 von „*Anacystis nidulans*" und 3—12 von *Synechococcus leopoliensis* gesprochen wird.

L. KRIENITZ, Neuglobsow

ROLF KLEE & CHRISTIAN STEINBERG: **Kieselalgen bayerischer Gewässer.** Informationsberichte Bayer. Landesamt für Wasserwirtschaft 4/87 (Loseblattsammlung). München 1987. 195 S. und 4 unpag. S., 255 Abb., Kunstst.

Die Idee dieses „Informationsberichtes" ist ausgezeichnet: Die Erkenntnisse der neueren Literatur über die Kieselalgen, speziell über deren Taxonomie und Ökologie, sollen den in der Wasserwirtschaft tätigen Wissenschaftlern und Praktikern zugänglich gemacht werden. Angesichts der Bedeutung der Kieselalgen für die Wassergütewirtschaft ist die rasante Entwicklung auf dem Gebiet der Taxonomie der Kieselalgen geeignet, mehr Verwirrung als Nutzen zu stiften. Das liegt daran, daß die klassischen Bestimmungswerke nach wie vor als Bestimmungshilfen benutzt werden müssen, obwohl deren mit dem Lichtmikroskop gewonnenen Erkenntnisse entweder gar nicht oder nur in begrenztem Maße mit denen in Übereinstimmung gebracht werden können, die mit Hilfe des Elektronenmikroskops erbracht worden sind.

Der „Informationsbericht" versucht nun, die im Elektronenmikroskop erkennbaren feinen Strukturen, die die heutige Taxonomie zur Artbestimmung heranzieht, für den Lichtmikroskopiker identifizierbar zu machen, in der leider oft vergeblichen Hoffnung, daß „man solche Strukturen, die man einmal im Rasterelektronenbild eindeutig identifiziert hat, auch im Lichtmikroskop sehen oder zumindest doch erahnen kann".

Die großzügig gedruckte und mit zahlreichen ausgezeichneten Bildern hervorragend ausgestattete Loseblattsammlung stellt klassische Zeichnungen aus den traditionellen Bestimmungswerken sowie lichtmikroskopische Fotografien einander gegenüber. Die vorliegenden Blätter behandeln zunächst die Ordnung Centrales mit den Gattungen *Stephanodiscus*, *Cyclotella*, *Cyclostephanos*, *Tallassiosira*, *Skeletonema*, *Melosira* und *Actinocyclus*. Dabei wird die den Autoren zugängliche wichtigste moderne Literatur berücksichtigt (leider nicht alle Arbeiten der sowjetischen und polnischen Diatomologen). Insgesamt wird überall das Bemühen deutlich, den verschiedenen Ansichten einigermaßen gerecht zu werden. Das ist gar nicht leicht, weil die Auffassung vieler Autoren weit voneinander abweicht. So kann es nicht ausbleiben, daß an einigen Stellen die Schlußfolgerungen, die die Autoren ziehen, unbefriedigend bleiben. Das gilt z. B. für *Stephanodiscus neoastraea* und *alpinus*, für *S. minutulus* und *parvus*; in begrenztem Maße auch für *S. hantzschii*, für *Cyclotella meneghiniana* (*atomus*!) und für die Behandlung der Gattung *Cyclostephanos*. Es wäre aber völlig ungerecht, die Unausgewogenheit der erwähnten Abschnitte den Autoren allein anzulasten. Hier ist viel durcheinander gebracht worden in denjenigen Arbeiten, die die Autoren als ihre wichtigsten Zeugen für ihre Auffassungen zitieren (*parvus*-Komplex). Ob es gut ist, im allgemeinen Teil so ausführlich auf die biologische Art und ihre Folgen für die Taxonomie im Rahmen dieser Loseblattsammlung einzugehen, ist zumindest fraglich. Sonst zeugt der Text von Fleiß und Einfühlungsvermögen und von großer Beherrschung der Technik. Der Begriff Alveole sollte stets deutsch mit Kammerung und nicht mit Verstrebung (S. 3.1/1) „übersetzt" werden. Die Angaben über die Ökologie bedürfen sicherlich noch starker Ergänzung. Es wäre wünschenswert, wenn bei Angaben über Wasserblüten die Zellzahlen pro Volumen angegeben würden. Was über die sogenannte *S. hantzschii* fo. *tenuis* ökologisch ausgesagt wird, gilt auch für die Nominatform.

Insgesamt ein bemerkenswertes und wichtiges Unternehmen, das, wie im Vorwort ausgeführt, sicherlich erfolgreich sein wird. Die Form der Loseblattsammlung ist für Ergänzungen günstig. Rezensent fürchtet allerdings, daß bei intensiver Nutzung das Material sehr leiden wird.

S. J. CASPER, Jena

M. MUNAWAR (ed.): **Proceedings of an International Symposium on the Phycology of Large Lakes of the World** (Arch. Hydrobiol., Beih. Ergebn. Limnol. 25). Stuttgart: E. Schweizerbart'sche Verlagsbuchhandlung (Nägele u. Obermiller) 1987. 256 S., 98 Abb., 44 Taf., 7 Abb. im Anhang. DM 98,—/US $ 56.—

Die großen Seen (Binnengewässer mit einer Fläche von über 500 km^2) verfügen über 88% des gesamten Süßwasservolumens der Welt. Diese gewaltigen Ressourcen gilt es verstärkt ökologisch zu erforschen, um einer in den letzten beiden Jahrzehnten sichtbar gewordenen Verschlechterung des Gewässerzustandes gezielter entgegenwirken zu können. Die wissenschaftliche Arbeit an diesen „Binnenmeeren" oder „Miniatur-Ozeanen" stellt ein Bindeglied zwischen Limnologie und Ozeanographie dar. Mit dem vorliegenden Band wird ein bemerkenswerter Beitrag auf diesem Feld geleistet. Er enthält die Berichte eines Symposiums zur Algologie der großen Seen im Rahmen des 1. Internationalen Phykologischen Kongresses in St. John's, Neufundland, Kanada 1981. Damit wird erstmals in dieser Form eine geschlossene Sammlung aktueller Forschungsergebnisse über die Phytoplanktongemeinschaften und ihre Rolle im ökologischen Bedingungsgefüge von etwa 20 der wichtigsten großen Seen herausgegeben.

In einer Einführung werden morphometrische und geographische Angaben zu den Untersuchungsgewässern zusammengefaßt (MUNAWAR). Die folgenden Beiträge sind der Struktur und Funktion von Phytoplanktongemeinschaften in den subarktischen Seen Kanadas — Großer Bärensee und Großer Sklavensee — (DUTHIE & HART), der nordeuropäischen Seen Ladoga und Onega (PETROVA) und des Baikalsees (KOZHOVA) gewidmet. Mit der Rolle von Blaualgenmassenentwicklungen in den Dnepr-Stauseen beschäftigt sich SVIRENKO. In 7 weiteren Beiträgen stehen die großen Seen Nordamerikas im Mittelpunkt: Phytoplanktonökologie des Ontariosees und des Oberen Sees (MUNAWAR et al.); Phytoplankton und Physikochemie des Michigansees (CLAFLIN); Phytoplankton und Metalltoxizität (MUNAWAR et al.); Picoplankton als Indikator für Gewässerverunreinigung (MUNAWAR et al.); Zooplanktonfreßraten (ROSS & MUNAWAR); Einwanderung von marinen Algen in die großen Seen (SHEAT); Chlorophyll a und Phytoplanktonbiomasse im Huronsee (EL-SHAARAWI & MUNAWAR). Den Band beschließen zwei Arbeiten über das Phytoplankton der 5 großen Seen Zentralafrikas (HECKY & KLING) sowie des Viktoriasees (TALLING).
Ein wertvoller und informativer Sammelband.

L. KRIENITZ, Neuglobsow

W. R. BARCLAY & R. P. MC INTOSH (eds.): **Algal Biomass Technologies.** An Interdisciplinary Perspective. Proceedings of a Workshop on the Present Status and Future Directions for Biotechnologies Based on Algae Biomass Production April 5—7, 1984, University of Colorado, Boulder (Beih. Nova Hedwigia **83**). Berlin, Stuttgart 1986. 273 S., 140 Abb., 93 Taf. Brosch. DM 120,—

Kultursysteme von Mikro- und Makroalgen können die Grundlage für alternative Technologien zur Produktion von Proteinbeigaben zur menschlichen und tierischen Nahrung, zur Herstellung diverser extrahierbarer Chemikalien sowie im Rahmen der Biokonversion zur Energieerzeugung darstellen. Nachdem die Biotechnologie bisher vor allem die Stoffwechselleistungen von Bakterien und Pilzen ausnutzte, gelangt mit den Algen eine auf der Sonnenenergie basierende Ressource in den Blickpunkt des Interesses.

Auf dem Workshop über Algenbiomassetechnologien berieten 118 Fachleute — überwiegend US-Amerikaner — aus den Bereichen der Wissenschaft und der Industrie über die Möglichkeiten der autotrophen Biomasseproduktion. Der Sammelband enthält 35 Beiträge über den gegenwärtigen Stand und die zukünftigen Richtungen der Algenmassenkulturen und ihrer Nutzung. Das Material wurde in 5 Kapitel gegliedert. Jedes Kapitel reflektiert eine Hauptsitzung des Workshops; als besondere Bereicherung ist jedem Abschnitt der Wortlaut der Diskussion, die im Auditorium geführt wurde, beigefügt. Die offensichtlich lebhaften Streitgespräche erhellen viele interessante Aspekte dieses jungen Zweiges der Biotechnologie, lassen allerdings auch trotz der optimistischen Grundtendenz und mancher handfester Erfolge erkennen, daß wir keine „Wunderdinge aus der Algenretorte" erwarten können.

Im Einzelnen wurde über folgende Schwerpunkte berichtet:

1. Algengenetik und Stammselektion. In diesem Kapitel geht es um die Anwendbarkeit der Genmanipulation an Mikroalgenkulturen (*Anabaena*, *Dunaliella*, *Chlamydomonas*) sowie an Meeresalgen, um Zellfusion, Protoplastenproduktion und Gewebekulturen.
2. Physiologische und biochemische Aspekte der Algenbiomasseproduktion. Charakterisierung bestimmter Stoffwechselabschnitte bei Algen unter standardisierten Hochproduktionsbedingungen sowie unter Streß; Optimierung der Kulturbedingungen. Auch in diesem Kapitel gehören Vertreter der Gattung *Dunaliella* zu den beliebtesten Untersuchungsorganismen.
3. Algen als Quelle von Chemikalien und Naturprodukten. Im Mittelpunkt dieses Abschnittes steht die Variabilität (sprich: Steigerung) der Ausbeute an verschiedenen organischen Komponenten, die von Kulturen mariner Algen (*Sargassum*, *Porphyra*, *Porphyridium*) und Mikroalgenkulturen produziert werden können.

4. Energie aus Algenbiomasse. Forschung und Praxis zur Produktion von Kohlenwasserstoffen, Lipiden und Wasserstoff durch Algenmassenkulturen sowie die Konversion von Algenbiomasse zu Alkohol und Methan.
5. Technologien der Algenmassenkultur. Design von Algenkultursystemen im Labor- und Freilandmaßstab. Fermentation. Erntemethoden. Kommerzielle Aspekte.

Ausführliche Literaturangaben zu den einzelnen Beiträgen und ein detaillierter Subjekt-Index mit Angaben über chemische Komponenten, Namen der Algenarten und technische Schlagworte gestalten das Buch benutzerfreundlich. Eine informative und anregende Sammlung aktueller Forschungsberichte.

L. KRIENITZ, Neuglobsow

R. W. EPPLEY (ed.): **Plankton Dynamics of the Southern California Bight** (Lecture Notes on Coastal and Estuarine Studies Vol. **15**). Berlin, Heidelberg, New York, Tokyo: Springer-Verlag 1986. XIII, 373 S. Soft cover. DM 69,—

Auf den ersten Blick scheint es sich bei diesem Band nur um eine der vielen neu erscheinenden Ökosystemstudien zu handeln, die sich darüber hinaus mit einem weit entfernten Objekt befaßt, der US-amerikanischen Pazifikküste. Das Buch ist dann aber in Inhalt, Form und Originalität so überzeugend, daß es über die begrenzte Region hinaus als richtungsweisend für die Dynamik pelagischer Prozesse in marinen Küstengewässern angesehen werden kann. Eine auf Initiative des berühmten JOHN STRICKLAND am ebenso bekannten Scripps Institution of Oceanography/La Jolla bereits 1963 begründete „Food Chain Research Group" von ca. 30 Wissenschaftlern — hier repräsentiert durch 8 Autoren — zeichnet verantwortlich für diese vorbildliche Monografie.

Die ersten drei Kapitel beschäftigen sich ausführlich mit den geografischen, klimatologischen sowie physikalischen und chemisch-ozeanografischen Besonderheiten der „Südlichen California-Bucht (SCB)". Reizworte wie DOM und POM leiten dann „organisch" über zur biologischen Einführung, in der ausgehend vom „Food-web-Konzept" die Struktur und Funktion des Lebens im Pelagial der SCB (also über den Titel hinausgehend auch Elemente des Nektons und des Benthos berührend) sehr illustrativ behandelt werden.

Moderne Termini wie „marine snow", „microbial loop" oder „bacterio- and virioplankton" werden dabei genauso sorgfältig interpretiert wie die mehr konventionellen Größen, z. B. Generationsdauer, Nahrungsmechanismen oder Aktivitätsmuster.

Entsprechend ihrer Bedeutung im Pelagial stehen im Mittelpunkt der nächsten drei Kapitel die Primärproduktion, die Raum-Zeit-Struktur sowie das mikrobielle Nahrungsnetz und die damit verbundenen Nährstoffkreisläufe.

Wegen seiner großen produktionsbiologischen Bedeutung und als potentielle Eingangspforte für Xenobiotica in die marine Nahrungskette rückt dieser Komplex immer mehr in das Zentrum des Interesses. Im letzten, absichtlich wohl etwas provokativ formulierten Kapitel „People and the plankton", werden vor allem Probleme der organischen Verschmutzung, der DDT-Dynamik und des Kühlwassers angerissen.

Kurioserweise fehlt auch diesem Band, der mit einem umfangreichen und übersichtlichen Literaturnachweis und einem Sachregister schließt, der Versuch einer interdisziplinären Zusammenschau — aber offenbar ist ein knappes Jahrzehnt kontinuierlicher objektbezogener Forschung auch unter optimalen Bedingungen absolut zu kurz für ein solches Anliegen.

J.-A. VON OERTZEN, Rostock

K. N. FEDOROV: **The Physical Nature and Structure of Oceanic Fronts** (Lecture Notes on Coastal and Estuarine Studies Vol. **19**). Aus dem Russischen übersetzt von N. DEMIDENKO. Berlin, Heidelberg, New York, London, Paris, Tokyo: Springer-Verlag 1986. VIII, 333 S., 106 Abb. Soft cover DM 98,—

Bedingt durch die in den letzten 10 Jahren in stärkerem Maße zur Verfügung stehenden, von Schiffen schleppbaren, kontinuierlich abtastenden Meßsysteme sowie hochaufgelöste Satellitenaufnahmen im sichtbaren und infraroten Spektralbereich wurde das wissenschaftliche Interesse an ozeanischen Fronten in starkem Maße stimuliert. Diesem Umstand sowie der Sprachbarriere Rechnung tragend, ist die vorliegende Monographie als englische Übersetzung des vor vier Jahren veröffentlichten russischen Originals des Autors (P. P.-Shirshov-Institut für Ozeanologie der Akademie der Wissenschaften der UdSSR, Moskau) mit der Zielstellung veröffentlicht worden, bekannte Beobachtungen ozeanischer Fronten mit neuen zu kombinieren und eine moderne physikalische Interpretation dieser Erscheinungen zu geben.

Der größte Teil der Monographie ist der Beschreibung von Fronten in den Randmeeren und den Randgebieten der Ozeane auf der Grundlage von Messungen des Autors und aus der Literatur zusammengestellter Beobachtungen gewidmet. In drei Sektionen werden mathematische Modellierungen von Fronten auf der Basis der hydrodynamischen Grundgleichungen diskutiert.

Im ersten Kapitel beschreibt der Autor die historische Entwicklung der Forschungen zur Problematik ozeanologischer Fronten und gibt eine Einordnung dieses Forschungsgegenstandes in die allgemeine Ozeanographie. Sodann definiert er die für die Beschreibung eines Frontsystems wesentlichen Begriffe und gibt eine Klassifikation ozeanischer Fronten entsprechend der Größe ihrer räumlichen Ausdehnung und ihrer Lebensdauer.

Es folgt eine Diskussion der modernen Meßinstrumente zur Beobachtung ozeanischer Fronten, an die auf Grund der räumlichen Strukturen der Fronten und ihrer schnellen Verlagerungen hohe Forderungen bezüglich der Abtastrate in der vertikalen Ebene und bezüglich der Meßgeschwindigkeit gestellt werden.

Im zweiten Kapitel gibt der Autor einen Überblick über die Häufigkeit von Fronten in verschiedenen Gebieten der Ozeane sowie in ausgewählten Randmeeren. Weiterhin macht er den Leser an Hand von Spektren und Darstellungen im physikalischen Raum mit den typischen räumlichen Skalen der Variabilität von Temperatur und Salzgehalt in der Deckschicht der Meere vertraut.

Daran schließt sich eine Diskussion der physikalischen Prozesse an, die für die Entstehung, die Erhaltung und den Zerfall von Fronten bestimmend sind. Die Ergebnisse der numerischen Modellierungen von Fronten im offenen Ozean zeigen, daß die Dichteadvektion eine Schlüsselrolle im Prozeß der Frontbildung spielt. Am Schluß der Kapitel diskutiert der Autor die Rolle von Ekman-Schichten und barotropen sowie baroklinen Instabilitäten in der Frontdynamik, wobei letztere, die oft in der Nähe von Fronten beobachtet werden, den nichtlinearen Charakter dieser Prozesse unterstreichen.

Ausführliche Beschreibungen der Erzeugung von Fronten durch mesomaßstäbliche Wirbel, der Wirbelgenerierung an Fronten und der mit Golfstromringen verbundenen Fronten werden im dritten Kapitel auf der Basis von Beobachtungen gegeben. Dieses Kapitel enthält darüber hinaus eine Diskussion der wesentlichen Eigenschaften von Fronten in verschiedenen Auftriebsgebieten des Weltmeeres, von durch Flußwasser erzeugten Salzgehaltsfronten in Küstennähe, von estuarinen Fronten und von durch Gezeitenreibung erzeugten Fronten. Das Kapitel schließt mit der Beschreibung der Eigenschaften von Fronten an der Meeresoberfläche.

Im vierten Kapitel macht der Autor den Leser mit der Feinstruktur von Fronten in den verschiedensten Skalen vertraut, bis hinab zu Zentimetern. Es werden dabei die physikalischen Prozesse diskutiert, die diese Feinstruktur erzeugen und die einen Stofftransport durch die Fronten hindurch ermöglichen.

Im fünften Kapitel wird die Bedeutung der Eigenschaften von Fronten für andere Disziplinen der Ozeanographie erörtert, wie Chemie und Biologie, und zukünftige Fragestellungen zur Erforschung dieses Problems werden formuliert. Ein ausführliches Literaturverzeichnis (273 Zitate) und ein Stichwortverzeichnis beenden die Monographie.

Der Text ist flüssig zu lesen, was auf eine gute Übersetzung des russischsprachigen Originals hinweist, und die Abbildungen sind gut erkennbar. Die Dimensionsbezeichnungen der Diagramme entsprechen jedoch nicht dem neusten Stand und an einigen Karten fehlt die Zuordnung des entsprechenden Erdquadranten (N, S, O, W).

Die Monographie wird für alle Studenten und Fachwissenschaftler interessant sein, die sich mit dem Problemkreis der ozeanischen Fronten, insbesondere im Randbereich der Ozeane und der Randmeere, auseinandersetzen müssen.

Da der Autor in der Einleitung um Anregung bittet, sei nicht verschwiegen, daß das Buch gewinnen könnte, wenn die Diskussion der Eigenschaften von Fronten mehr auf der Grundlage zumindest zweidimensionaler Darstellungen der entsprechenden Felder erfolgen und vor allem das Geopotential im Bereich von Fronten analysiert würde. Ein weiterer nicht behandelter Punkt ist die Frage, wie Fronten unter Abstrahlung von Rossby-Wellen modifiziert werden.

H. U. LASS, Rostock

MEI-E REN (ed.): **Modern Sedimentation in the Coastal and Nearshore Zones of China.** Beijing: China Ocean Press; Berlin, Heidelberg, New York, Tokyo: Springer-Verlag 1986. 466 S., 228 Figs. Hard cover DM 238,—

Mit dem vorliegenden Band werden neue Ergebnisse von Forschungseinrichtungen der VR China zur rezenten bzw. subrezenten küstennahen Sedimentation vorgestellt.

Es ist das Verdienst der Herausgeber und der Verlage, dem breiten Publikum mit diesem Werk eine Vielzahl von Quellen zugänglich gemacht zu haben, die bisher nur in chinesischer Sprache vorlagen. Dabei läßt die Einteilung des Buches Hauptforschungsrichtungen und thematische Schwerpunkte in den beteiligten Institutionen erkennen.

Nach einem kurzen einführenden Abschnitt gibt das 2. Kapitel ausgehend vom morphologischen Küstentyp an Hand von Fallstudien einen Überblick über die sedimentdynamischen Verhältnisse an Sandküsten verschiedener Provinzen der VR China. Etwa der gleiche Raum (ca. 70 Seiten) wird der Betrachtung von Küstenabschnitten mit Wattcharakter im Kapitel 3 gewidmet. Es wird zwischen Sedimentationsprozessen an offenen Wattküsten und an Wattküsten in Buchten bzw. in Flußmündungsgebieten unterschieden und eine Reihe von Informationen zu Zusammensetzung und Genese der anstehenden Sedimente werden vermittelt.

Von größererm Umfang sind die Kapitel 4 und 5, welche eine Darstellung von Sedimentationsprozessen in ausgewählten Ästuar- und Deltaregionen der VR China beinhalten. Eine besonders eingehende Behandlung erfährt in diesen Abschnitten das Mündungsgebiet des Changjiang, während zur Sedimentdynamik entsprechender Bereiche des Huanghe mit seinen außerordentlich hohen Sedimenttransportraten vergleichsweise wenige Daten vorliegen.

Jeweils etwa 40 Seiten umfassen abschließend die Kapitel 6 und 7, in denen Forschungsergebnisse zur Sedimentbildung und -dynamik im Bereich von submarinen Sandrücken bzw. von Korallenriffen zusammengetragen sind.

Die einzelnen Beiträge des Buches wurden von 14 Autoren aus verschiedenen Zentren der geowissenschaftlichen Forschung der VR China verfaßt. Daraus ergeben sich unterschiedliche Betrachtungs- und Darstellungsweisen der jeweiligen Thematik. So stehen Übersichtsbeiträge neben sehr ausführlichen Einzeldarstellungen bestimmter regionaler Einheiten. Dementsprechend differenziert ist die angewandte Untersuchungsmethodik. Sie umfaßt neben der sedimentologischen, ozeanologischen geomorphologischen Bearbeitung auch Beiträge zur Mineralogie, Geochemie und Biologie bzw. Paläontologie der betreffenden Sedimente. Kritisch anzumerken ist die Vielzahl der im Text auftretenden Druckfehler. Ungünstig auf das Verständnis der Abbildungen wirkt sich aus, daß die zugehörigen Legenden z. T. erst am Ende des Kapitels erscheinen. Schließlich fehlen bei den Fototafeln 3-1 bis 3-25 die Bildunterschriften vollständig. Trotz dieser Mängel ist der vorliegende Band mit seiner Fülle an bisher kaum bekannten Informationen eine wertvolle Bereicherung der vorhandenen Literatur zur rezenten Sedimentdynamik und bietet sich mit seiner Vielzahl an Primärdaten für vergleichende Untersuchungen in älteren Ablagerungen an.

W. LEMKE/D. LANGE, Rostock

Beiträge zur Meereskunde, Heft 57. Hrsgeg. von KLAUS VOIGT und Mitarbeitern. Berlin: Akademie-Verlag 1987. 112 S., 65 Abb., 18 Tab. Brosch. M 28,—

Im Heft 57 der renomierten Schriftenreihe werden 3 Untersuchungen aus dem Ostseeraum und 7 sich ergänzende Arbeiten aus dem Ostatlantik vorgestellt.

Barotrope (Wasserstand) und barokline (Salzgehalt) Druckgradienten sowie lokale Winde bewirken in der westlichen Ostsee Austauschvorgänge mit speziellen Reaktionsmustern. Es entsteht eine Asymmetrie, und die theoretischen Untersuchungen unterstreichen den erheblichen barotropen Bewegungsanteil, wobei auch auf Unterschiede zwischen theoretischen Modellen und experimentellen Resultaten eingegangen wird. Eine wichtige historische Darstellung der 30-jährigen Geschichte der Konferenz der Ostseeozeanographen (CBO) faßt die 15 bisherigen Konferenzen zusammen.

Die Arbeit PRANDKE & STIPS „Echostreuschichten in der Ostsee" dürfte bei Limnologen besonderes Interesse finden, wobei physikalische Dichtesprungschichten von wenigen Dezimetern, deren Entstehung widersprüchlich gedeutet wird, zu Planktonstau führen sollen. Spezialuntersuchungen mit mikrorechnergestützten Meßsystemen (freisinkende Meßsonde und Echolot) erbrachten dagegen einen offensichtlichen Zusammenhang zu suspendiertem Material.

Die Ostatlantikarbeiten basieren auf komplexen Untersuchungen des Akademie-Forschungsschiffes „Alexander von Humboldt" und des Fang- und Verarbeitungsschiffes „Arnold Zweig" im küstennahen Auftriebsgebiet des Nouakchott-Cañons vor der mauretanischen Küste im Jahre 1984 durch ein Wissenschaftlerteam des Instituts für Meereskunde, der Hochseefischerei und der Universität Rostock.

L. KREY, Jena

Isotope Techniques in Water Resources Development. Proceedings of an International Symposium, Vienna, 30 March—3 April 1987. Wien: International Atomic Energy Agency 1987. 816 S., 358 Abb., 93 Tab. Brosch. ÖS 1660,—

Isotopentechniken sind in der Ozeanologie, Limnologie oder Hydrologie unerläßliche Hilfsmittel zur Analyse verschiedener Transport-, Verteilungs-, Umsatz- und Wandlungsprozesse in den Gewässern geworden. So lassen sich z. B. solche Vorgänge wie Durchmischung, Diffusion, Dispersion, Aufenthaltszeit, Sedimentation, Grundwasserzu- und -abfluß, Produktion und Mineralisation weitgehend durch den Einsatz und die Erfassung verschiedener radioaktiver und stabiler Isotope quantifizieren.

Die IAEA (International Atomic Energy Agency) gibt der theoretischen und praktischen Seite der Isotopenanwendung auf diesen Gebieten seit 3 Jahrzehnten die richtungsweisenden Impulse. Unter der Schirmherrschaft der IAEA werden u. a. die „Technical Reports Series" und die „Proceedings Series" auf dem Gebiet der Isotopenanwendung herausgegeben (vgl. Rezensionen *Limnologica* **11**(2), 479—516; **12**(1), 153—187; **14**(2), 395—431; **18**(1), 227—251).

Der jüngste Band der IAEA Proceedings Series beinhaltet die Verhandlungen eines vom 30. 3. bis zum 3. 4. 1987 in Wien durchgeführten Symposiums „Isotope techniques in water resources development", an dem über 150 Wissenschaftler aus 41 Staaten teilnahmen. In 41 veröffentlichten Vorträgen und 40 Posterbeiträgen werden neueste Erkenntnisse der Isotopenhydrologie vorgestellt. Die Beiträge verteilen sich auf folgende Schwerpunkte: Thermalwasser, Grundwasserdatierung, Hydrologie arider und semiarider Zonen, Felduntersuchungen mit natürlichen Isotopen, Niederschlags-Oberflächen-Grundwasserbeziehungen, Verunreinigungen, künstliche Traceranwendung und Transportprozesse von Sedimenten. Ein Vorwort, eine Teilnehmerliste und zwei Indizes (Autoren, Beiträge) komplettieren das ansprechend gedruckte Buch.

R. KOSCHEL, Neuglobsow

A. B. VINER (ed.): **Inland Waters of New Zealand** (DSIR Bulletin **241**). Wellington: DSIR Science Information Publishing Centre 1987. IX, 494 S., 122 Abb., 14 Schwarz-weiß-Fotos, 14 Farbfotos, 71 Tab. Ln.

Neuseeland besitzt eine Vielzahl von Binnengewässern. Allein die Seen nehmen insgesamt eine Fläche von 3400 km^2 ein, das sind etwa 1,3% der Landflächen. Unter ihnen sind bekannte und große Seen: Tauposee (623 km^2, maximale Tiefe 163 m), Te Anausee (352 km^2, maximale Tiefe 417 m) und Wakatipusee (293 km^2, maximale Tiefe 380 m). Besonders die Typenvielfalt der neuseeländischen Gewässer ist beeindruckend. Man findet Gewässer tektonischen, vulkanischen und glazialen Ursprungs, Thermalgewässer, Gewässer der warmtemperierten bis gemäßigten und antarktischen Klimaregionen, Gewässer mit Hochgebirge und im Flachland. Ebenso beeindruckend ist, mit welcher Kontinuität und Akribie diese für die limnologische Forschung ungemein günstige Konstellation in verschiedenen Institutionen in Neuseeland seit Jahren umgesetzt wird. Einen nachhaltigen Eindruck dieser vielfältigen Arbeiten vermittelt das vorliegende Buch. 23 Autoren berichten über Struktur und Funktion der neuseeländischen Binnengewässer. In einem ersten Abschnitt werden die Grundlagen für die Mannigfaltigkeit der Seen behandelt (112 S.). D. Lowe & J. D. Green stellen Ursprung und Entwicklung der Seen vor. Der Einfluß des Klimas auf die Thermik der Gewässer wird von J. D. Green, A. B. Viner & D. J. Lowe dargestellt. M. H. Timperley widmet sich in seinem Beitrag der regionalen Beeinflussung der Wasserchemie der Seen.

Der zweite Abschnitt hat die Fließgewässer zum Inhalt (78 S.). M. J. Duncan analysiert die Hydrologie und den Sedimenttransport. Ergebnisse zum Stickstoff-, Phosphor- und Sauerstoffhaushalt sowie zu den Invertebratengesellschaften der Fließgewässer werden von J. C. Rutherford, R. B. Williamson & A. B. Cooper bzw. von M. S. Winterbourn vorgestellt. Ein dritter Abschnitt ist dem Seenplankton vorbehalten. Darin werden von A. B. Viner & E. White wichtige Aspekte des Phytoplanktonwachstums und von M. A. Chapman & J. D. Green Aspekte der Zooplanktonökologie diskutiert.

Zoogeographische Besonderheiten der Invertebratenfauna und der einheimischen Fische werden in einem vierten Abschnitt von D. J. Forsyth & M. H. Lewis sowie R. M. McDowall behandelt. Der fünfte Abschnitt trägt die Überschrift „Ökologische Invasionen". C. Howard-Williams, J. S. Clayton, B. T. Coffey & I. M. Johnstone berichten dabei über Makrophyteninvasionen und R. M. McDowall über den Einfluß exotischer Fische auf die einheimische Fauna. Geothermisch beeinflußte Gewässer und die Limnologie der antarktischen Gewässer sind Gegenstand des 6. Abschnittes „Extreme aquatische Lebensräume". W. F. Vincent & D. J. Forsyth sind die Autoren dieser interessanten Arbeiten.

Der letzte Abschnitt ist der Beeinflussung der Gewässer durch den Menschen gewidmet. Zum einen geht es dabei um die Sportfischerei (D. Scott) und zum anderen um die Gesamtnutzung der Binnengewässer (R. H. S. McColl & J. C. Ward).

Zwei Apendizes (Übersichtskarten der Seengebiete, Morphometrische Kenndaten der Seen) und drei Indizes (Gewässer-, Organismen- und Sachregister) beschließen das ansprechend gedruckte und ausgestattete Buch. Ein Buch, das sicher von vielen Ökologen, Limnologen, Hydrologen und Wasserwirtschaftlern mit Interesse und Gewinn zur Hand genommen wird und das auch in Text und Bild etwas von der ökologischen Faszination zu vermitteln weiß, die von der Vielfalt der neuseeländischen Binnengewässer ausgeht.

R. Koschel, Neuglobsow

J. I. Prosser (ed.): **Nitrification** (Special Publ. of the Society for General Microbiology, Vol. **20**). Oxford, Washington D. C.: IRL Press Ltd. 217 S., 58 Abb. 37 Tab. Brosch. 22.00/US $ 40.00

Das Buch ist aus Vorträgen auf einem Symposium der englischen Society for General Microbiology hervorgegangen. Es enthält in 10 Beiträgen Informationen über die mikrobiologischen Grundlagen der Nitrifikation, die methodischen Ansätze zu ihrer Untersuchung und ihre Bedeutung an unterschiedlichen Biotopen. Alle Artikel enthalten eine umfangreiche Bibliographie, so daß ein recht umfassender Einblick in den gegenwärtigen Kenntnisstand möglich ist. R. McL. Macdonald (Nitrification in soil: an introductory history) zeichnet den mühsamen Weg von dem Nachweis der bakteriellen Nitrifikation als Schlüsselprozeß im Stickstoffkreislauf bis zur Gewinnung von Reinkulturen der nitrifizierenden Bakterien nach, der auch heute, 100 Jahre nach den bahnbrechenden Arbeiten von Winogradsky, voller Unsicherheiten ist. E. Bock, H.-P. Koops & H. Harms (Cell biology of nitrifying bacteria) zeigen insbesondere mit cytologischen Befunden und Verwendung der DNA bzw. 16r RNA-Sequenzhomologie sowie physiologischen Untersuchungen die taxonomischen Beziehungen dieser Bakteriengruppe. P. M. Wood (Nitrification as a bacterial energy source) stellt die biochemische Charakteristik des Prozesses in den Mittelpunkt seines Beitrages. J. I. Prosser (Experimental and theoretical models of nitrification) behandelt vor allem die Einsatzmöglichkeiten und Aussagefähigkeit der Perkolatortechnik und der kontinuierlichen Kultur mit Bodensäulen. Die Hemmung der Nitrifikation durch eine Anzahl von Agenzien, insbesondere Nitrapyrin, besitzt für die landwirtschaftliche Praxis der Steuerung des N-Haushaltes in Böden große Bedeutung. St. J. Powell (Laboratory studies of inhibition of nitrification) berichtet über die physiologischen Grundlagen der Hemmwirkung von verschiedenen Chemikalien. Mit der gleichen Problematik befaßt sich auch D. R. Keeney (Inhibition of nitrification in soils), wobei hier mehr die Ergebnisse von Feldversuchen im Vordergrund stehen. Weiterhin unklar bleibt die tatsächliche Bedeutung der heterotrophen Nitrifikation, mit der sich K. Killham (Heterotrophic nitrification) auseinandersetzt. Eine sehr geschlossene Übersicht über die Nitrifikation im N-Haushalt von Seen gibt G. E. Hall (Nitrification in lakes).

Für marine Habitate analysiert B. B. Ward (Nitrification in marine environments) die gleiche Fragestellung und geht dabei besonders auf die experimentellen Möglichkeiten der Bestimmung ein. Im letzten Beitrag setzt sich H. A. Painter (Nitrification in the treatment of sewage and waste-waters) mit der beträchtlichen Rolle der Nitrifikanten in der Abwasserreinigung auseinander, wo deren Steuerung und sinnvolle Kopplung mit Denitrifizierern für die Effizienz der ganzen Anlage von großer Wichtigkeit ist.

Ungeachtet einiger Überschneidungen sind die Artikel zu einer gut gegliederten vielseitigen Darstellung der Nitrifikation geordnet. Es zeigt sich freilich auch, daß nach wie vor durch die geringen Wachstumsleistungen der nitrifizierenden Bakterien beträchtliche Schwierigkeiten bei ihrer mikrobiologischen Untersuchung zu überwinden sind.

H.-D. Babenzien, Neuglobsow

Helga Müller: **Hydrocarbons in the Freshwater Environment — A Literature Review** (Arch. Hydrobiol., Beih. Ergebn. Limnol. **24**). Stuttgart: E. Schweizerbart'sche Verlagsbuchhandlung (Nägele u. Obermiller) 1987. 69 S., 9 Abb., 16 Tab. DM 42,—/US $ 24.—

Mit dem Heft 24 der Reihe „Ergebnisse der Limnologie" liegt ein englischsprachiger Literaturbericht zur Thematik „Kohlenwasserstoffe in limnischen Gewässern" vor. Die Autorin hat diese Ausgabe gegenüber der 1985 erschienenen deutschsprachigen Fassung (Vorkommen, Verhalten und Wirkung von Kohlenwasserstoffen in limnischen Gewässern. Literaturbericht. 146 S., Landesanstalt für Umweltschutz, Baden-Württemberg) nur geringfügig verändert. somit erfaßt die Arbeit im wesentlichen die Literatur der Jahre 1970—1984.

Primär galt das Interesse der Limnologen den Kohlenwasserstoffen unter dem Aspekt der Umweltbelastung. Tankerunfälle, marine Ölförderanlagen, Motorbootverkehr führten immer wieder zu teilweise spektakulären Verschmutzungen aquatischer Ökosysteme. Interessanterweise können auch Rückstände des Straßenverkehrs zur Belastung von Gewässern beitragen, wie die angeführten Arbeiten von Wakeham (1977) verdeutlichen, der 46% des Eintrages aliphatischer Kohlenwasserstoffe in Lake Washington auf Abspülungen von Straßen und Brücken zurückführt.

Neben dieser Problematik der Gewässerbelastung erlangten Kohlenwasserstoffe breite Anwendung bei der Differenzierung von Mikroorganismen bzw. mikrobiellen Leistungen. Die Identifizierung der gebildeten Kohlenwasserstoffe läßt den Schluß auf die Leistungsträger der Prozesse zu.

Der vorliegende Literaturbericht stellt in gelungener Weise wichtige allgemeingültige Zusammenhänge dar, verweist auf gegensätzliche Gesichtspunkte in der Diskussion einiger Ergebnisse (z. B. bei der Biosynthese von polyzyklischen aromatischen Kohlenwasserstoffen durch Bakterien) und gewährt dank einer straffen Gliederung einen sehr guten Überblick über dieses weitgefächerte Thema. An treffenden Stellen wird auf spezielle zusammenfassende Arbeiten hingewiesen.

Die Arbeit ist neben Definition und Einleitung in 4 Hauptkapitel eingeteilt (Eintrag: 23 Seiten; Verteilung im Wasser: 12 Seiten; Verteilung in Wasserorganismen: 7 Seiten; Effekte auf Organismen und Ökosysteme: 8 Seiten). Eine Zusammenfassung und das 333 Zitate umfassende Literaturverzeichnis beschließen das Heft. Besonders hervorzuheben sind die Auswahl sehr guter tabellarischer Übersichten, das vorangestellte Abkürzungsverzeichnis, die klare Textunterteilung sowie die Darstellung der wichtigsten Arbeitstechniken. Etwas irritierend wirkt in Tabelle 1, daß die Formelnummern in Form der Literaturzitate dargestellt wurden.

Ein gelungener Band, dessen Lektüre jedem Limnologen sehr zu empfehlen ist.

P. CASPER, Neuglobsow

HELMUT TSCHÄPE: **Plasmide.** Biologische Grundlagen und praktische Bedeutung (Wissenschaftl. Taschenbücher, Reihe Biologie, Band **233**). Berlin: Akademie-Verlag 1987. 203 S., 50 Abb., 29 Tab. M 16,—

An das Auftreten von Plasmiden und ihre gentechnische Handhabung knüpfen sich gegenwärtig viele Erwartungen, aber auch Befürchtungen. Tatsächlich sind seit LEDERBERGS frühen Untersuchungen über den F-Faktor bei Bakterien und der Definierung extrachromosomaler DNA-Elemente als Plasmide (1952) Erkenntnisse von großer theoretischer und ökonomischer Tragweite erarbeitet worden. Der Autor bietet zu diesem inzwischen umfangreichen und in seinen Hypothesen verwirrenden Gebiet eine übersichtliche und informative Einführung. Das zunächst sehr spezielle Teilgebiet der Bakteriengenetik und Molekularbiologie erweist dabei schnell seine hohe Brisanz und aktuelle medizinische, biotechnologische und ökologische Bedeutung. Das einleitend entwickelte Bild von Plasmiden als den „Trägern für Leistungen für alle Fälle" zeigt, welche ungeheure Erweiterung der Adaptationsfähigkeit der Bakterien damit genetisch determiniert ist. Anschließend werden Grundlagen der Klassifikation von Plasmiden, Terminologie und Eigenschaften dargestellt. Als „Lebensweise der Plasmide" werden dann die Mechanismen der autonomen Replikation, Mobilität und Integration der Plasmid-DNA in den betroffenen Bakterienpopulationen behandelt. Hier findet man auch Informationen über die ökologisch interessanten Prozesse der Übertragung bei grampositiven, gramnegativen oder Archaebakterien. Unter geeigneten Bedingungen kann über die konjugative Replikation eine anfangs plasmidfreie Population „durchinfiziert" werden, wobei die Übertragung nicht vor Art-, Gattungs- oder Familiengrenzen haltmacht. Im folgenden Kapitel werden Plasmide als Träger bakterieller Genreserven analysiert und 3 Aspekte in den Vordergrund gestellt: plasmidvermittelte Antibiotikaresistenz, Virulenz und metabolische Leistung. Hieraus leiten sich die in den nächsten Kapiteln behandelten Probleme ab — ökologische Aspekte der Plasmidverbreitung sowie medizinische wie praktische Konsequenzen. Es folgen Literaturverzeichnis, Abkürzungsverzeichnis, Glossar und Sachregister. In seinen Ausführungen beschränkt sich der Autor nicht nur auf eine allgemeine theoretische Darlegung, sondern leitet aus konkreten Beispielen, etwa der Resistenzentwicklung bei antibakteriellen Therapien — Gentamycinapplikation in der DDR — die enorm wichtigen und grundsätzlichen Probleme der Chemotherapeutika-Strategien ab. Auf der anderen Seite steht die biotechnologische Nutzung dieser Form der Genmanipulation erst am Anfang. Hier, wie auch bei der Behandlung der ökologischen Relevanz für „Umweltbakterien", sind die Ausführungen leider ein wenig zu kursorisch geraten und deuten nur die schon jetzt vorhandenen Entwicklungen an. Dennoch kann man dieses Taschenbuch als eine gelungene, vielseitige Zusammenfassung unserer Kenntnisse über Plasmide ansehen, die gleichzeitig damit einem breiten Leserkreis grundsätzliche Wesensmerkmale der Bakterien vorstellt.

H.-D. BABENZIEN, Neuglobsow

H. KINDL.: **Biochemie der Pflanzen.** Ein Lehrbuch. 2., völlig neubearb. Aufl. Berlin, Heidelberg, New York: Springer-Verlag 1987. XII, 379 S., 323 Abb. Geb. DM 98,—

Die Ergebnisse der biochemischen Forschung haben im letzten Jahrzehnt unsere Kenntnisse zur Biochemie der Pflanzenzelle wesentlich erweitert und auf einigen wichtigen Gebieten erst richtig erschlossen. Diese Entwicklung macht verständlich, daß von der 1975 erschienenen 1. Auflage des Buches „Biochemie der Pflanzen" (Autoren: H. KINDL & G. WOBER) nur noch relativ wenig in eine den gegenwärtigen Stand repräsentierende Neuauflage fließen konnte. Geblieben ist aber die Intention des Buches, die pflanzliche Zelle mit den Schwerpunkten Strukturierung, Regulation und Informationsfluß aus der Sicht des Biochemikers darzustellen.

In 8 Kapiteln werden dazu die pflanzliche Zelle und ihre Kompartimente (22 S.), die Katalysatoren der Zelle: Enzyme (21 S.), der Informationsfluß und seine Regulation (44 S.), die Energie-Konversion an Membranen (41 S.), der Stoffwechsel der Chloroplasten (61 S.), der anabole Stoffwechsel (81 S.), der Katabolismus: Mobilisierung von Reservestoffen (38 S.) und das extrazelluläre Kompartiment (23 S.) behandelt. 2 Appendizes (Prochirale Substrate — Aspekte der Stereochemie; Strukturformeln von Hemmstoffen und Hilfsverbindungen) sowie ein Sachverzeichnis beschließen ein hervorragendes Lehrbuch. Das Werk besticht besonders durch die klare und übersichtliche Gestaltung des Textes, die sehr instruktive graphische Ausstattung, einen hervorragenden Druck und vor allem durch einen in diesem Rahmen kaum zu überbietenden Informationsgehalt zur Biochemie der Pflanzenzelle.

R. KOSCHEL, Neuglobsow

Biologia Acuatica. Schriftenreihe des Instituts für Limnologie „Dr. Raul A. Ringulet" der Universität La Plata. Hrsgeg. von HUGO L. LOPEZ. **Heft 8** (1986): 60 S., Brosch. US $ 10.00; **Heft 10** (1987): 74 S., Brosch. US $ 3.00

In Heft 8 der Schriftenreihe „Biologia Acuatica" werden die Gebisse von 34 Arten argentinischer Characidae beschrieben. Von besonderem Interesse dürfte der Vergleich der Gebisse der Subfamilien Characinae, Bryconinae, Glandulocaudinae, Tetragonopterinae, Thoracocharacinae, Cheirodontinae und der Gattung *Hoplias* sein.

Heft 10 enthält die Ergebnisse limnologischer Untersuchungen von 8 Seen und 11 Teichen Feuerlands, das durch die Magellan-Straße vom südamerikanischen Festland getrennt ist. Die Studie enthält neben wichtigen morphometrischen Daten Angaben über physikalisch-chemische Faktoren, die Primärproduktion und die in diesen Gewässern vorkommenden Arten des Phytoplanktons und Zooplanktons.

A. LANGE, Jena

TOM FENCHEL: **Ecology of Protozoa.** The Biology of Free-living Phagotrophic Protists (Springer Series in Contemporary Bioscience). Madison: Science Techn. Inc. Berlin, Heidelberg, New York, London, Paris, Tokyo: Springer-Verlag 1987. X, 197 S., 47 Abb., Hard voer. DM 94,—

Die Protozoen spielen in den Stoffumsätzen aquatischer und edaphischer Ökosysteme eine führende Rolle. Sie scheinen nach den Bakterien die zweite Stelle im turnover dieser Systeme einzunehmen. Daneben sind sie durch viele Interaktionen mit der übrigen Zönose verbunden. Diese Erkenntnisse setzen sich in der neueren Literatur immer mehr durch. Im Gegensatz dazu steht nicht nur die geringe Würdigung, die diese Zusammenhänge in der Allgemeinen Ökologie bisher erfahren haben, sondern auch der Mangel an einschlägigen Zusammenfassungen.

T. FENCHEL gilt als sehr guter Kenner ökologischer Probleme bei Protozoen. Eine zusammenfassende Darstellung über die Ökologie der Protozoen aus seiner Sicht wird daher allgemein begrüßt werden. Aus dem bereits vielschichtigen Gebiet der Protozoenökologie wählt der Autor deutliche Schwerpunkte für sein Buch aus, die auch weitgehend seine eignen Arbeitsgebiete tangieren: Bewegung und Orientierung, Physiologie und Ökologie der Nahrungsaufnahme und des Energiehaushaltes, Protozoengemeinschaften mariner Habitate, Bedeutung der Sexualität bei Protozoen, ökologische Nischen und das in letzter Zeit immer faszinierender werdende Phänomen des Polymorphismus von Lebenszyklen und -strategien, das zu einem neuen Verständnis der Formenmannigfaltigkeit bei Protozoen führt. Ferner wird das Spezieskonzept für Protozoen kritisch dargestellt und die Beziehungen zwischen Ökologie und Mechanismen der Speziation werden aufgezeigt, doch auch der Mangel unserer Kenntnisse über diesen wichtigen Gegenstand.

Ein interessantes Buch, das aus der praktischen Erfahrung seines Autors schöpft.

W. SCHÖNBORN, Jena

Avifauna der Deutschen Demokratischen Republik. 5 Bände. **Band 1: Die Vogelwelt Mecklenburgs.** Hrsg. von G. KLAFS und J. STÜBS, unter Mitwirkung von W. STARKE, H. PRILL und H. ZIMMERMANN. 3., neubearb. Aufl. Jena: VEB Gustav Fischer Verlag 1987. 426 S., 6 Diagr., 177 Verbreitungskarten, 32 Fototafeln, 1 mehrfarbige Übersichtskarte. Ln. M 43,50

Bereits 10 Jahre nach der Erstauflage [vgl. *Limnologica* **13** (1), S. 225, 1981] legen die mecklenburgischen Ornithologen eine wesentlich erweiterte Neuausgabe ihrer stark beachteten Avifauna vor. Der Text wurde in fast allen Teilen gründlich überarbeitet und besonders im speziellen Teil durch das kritisch ausgewertete Faktenmaterial der letzten 10 Jahre ergänzt. Der Gesamtbestand fast aller Brutvogelarten wurde auf Grund von Hochrechnungen von Bestandszählungen und Siedlungsdichtewerten in Kombination mit weiteren quantitativen Methoden in geschätzten Zahlenwerten ausgedrückt, wodurch eine höhere Aussagefähigkeit gegenüber den früheren Angaben in Häufigkeitsklassen gegeben ist.

Die entscheidende Neuerung der vorliegenden Auflage ist jedoch die Darstellung der Brutverbreitung fast aller Arten auf Rasterkarten. Dazu wurde in den Jahren 1978—82 eine flächendeckende Kartierung des Gesamtgebietes auf der Basis von Meßtischblatt-Quadranten durch einen großen Mitarbeiterstab von Freizeitornithologen durchgeführt. Das Ergebnis sind vertiefte und teilweise auch überraschende Einblicke in das feinere Verbreitungsmuster vieler Brutvogelarten. Freilich darf dieser Erfolg der vergleichsweise groben Rasterkartierung nicht darüber hinwegtäuschen, daß für die kommunen Sperlingsvögel, wie Kohlmeise, Amsel, Buchfink usw., nicht mehr als das zu erwartende Ergebnis flächendeckender Verbreitung herausgekommen ist. Für die Beurteilung der vorgelegten Rasterkarten wesentlich ist die ausführliche Beschreibung der angewendeten Kartierungsmethode. Die Nachweise auf den Kartierungsflächen werden in die Kategorien „Mögliches" (B), „Wahrscheinliches" (C) und „Sicheres Brüten" (D) eingeteilt. Auf den Karten sind die Gesamtwerte der absoluten Zahlen sowie der Rasterfrequenzen und deren Aufschlüsselung auf die einzelnen Kategorien angegeben. Leider sind die B-Nachweise, die zweifellos in nicht unbeträchtlicher Zahl lediglich unsicheres Brüten bedeuten, graphisch nicht deutlich genug (durch leere Kreise oder auch Fragezeichen) herausgehoben worden. Dem Benutzer des Buches bleibt deshalb eine sehr sorgfältige Analyse des Informationsgehaltes vieler Verbreitungskarten nicht erspart, um sich ein klares Bild über die Verbreitungsverhältnisse der entsprechenden Arten machen zu können. Stellvertretend hierfür sei der Flußuferläufer genannt, dessen gegenwärtige Verbreitung in Mecklenburg zweifellos zu optimistisch gesehen wird.

Trotz dieser kritischen Bemerkungen gebührt den mecklenburgischen Ornithologen Dank und Anerkennung für ihre aufopferungsvolle und vorbildliche Arbeit, mit der sie erstmals in unserem Lande eine neue Qualität avifaunistischer Dokumentation erreicht haben.

D. FLÖSSNER, Jena

Avifauna der Deutschen Demokratischen Republik. 5 Bände. **Band 3: Die Vogelwelt Thüringens.** Hrsg. von D. VON KNORRE, G. GRÜN, R. GÜNTHER und K. SCHMIDT. Jena: VEB Gustav Fischer Verlag 1986. 339 S., 24 Diagr., 8 Tab., 24 Verbreitungskarten, 4 Karten, 53 Fotos. Ln. M 35,—

Obwohl Thüringen, zusammen mit dem angrenzenden Anhalt, durch die bahnbrechenden Forschungen CH. L. BREHMS die Wiege der deutschen Vogelkunde ist, hat es bislang merkwürdigerweise an einer zusammenfassenden Darstellung seiner Vogelwelt gefehlt. Die vorliegende erste Landesavifauna Thüringens (Bezirke Erfurt, Gera und Suhl) wurde deshalb allseits mit Spannung erwartet. Die Gliederung des umfangreichen Stoffes folgt den bereits vorliegenden Bänden der Avifauna der DDR [vgl. *Limnologica* **13** (1), 225, 1981 u. **16** (2), 352, 1985]. Neu sind kurze, aber anregende Kapitel zur zoogeographischen Stellung der Thüringer Vogelwelt und über Thüringen als Durchzugs- und Überwinterungsgebiet. Hervorzuheben ist ebenfalls der ausgezeichnete Überblick der naturräumlichen Gliederung der geomorphologisch und geobotanisch reich ausgestatteten Thüringer Landschaft. Die konzentriert geschriebenen Artkapitel, an denen 35 Bearbeiter beteiligt waren, stellen eine Fülle von faunistischen Daten in klarer und übersichtlicher Form zusammen, wobei besonders die umfassende und kritische Berücksichtigung auch der älteren Angaben bis zurück in die klassische Periode der Thüringer Vogelkunde zu den Zeiten BECHSTEINS und BREHMS hervorgehoben sei. Zugdiagramme und Tabellen mit brutbiologischen Angaben ergänzen wirkungsvoll die textliche Aussage. 53 Fotos stellen berühmte Thüringer Ornithologen sowie typische Landschaftsausschnitte und eine Auswahl charakteristischer Vogelarten in durchweg guter Qualität vor. Die Darstellung der Brutverbreitung ausgewählter Arten auf 24 Punktkarten ist zahlenmäßig etwas zu kurz ausgefallen und nicht ganz frei von Mängeln (Nachtigall, Vorkommen im Werratal). Leider enthält auch die kommentierte Artenliste mehrfach widersprüchliche Aussagen gegenüber den Angaben in den einzelnen Artkapiteln (z. B. Rohrdommel, Krickente, Kolbenente, Kleinralle, Beutelmeise). Das umfassende Literaturverzeichnis enthält etwa 900 Titel. In einem Nachwort werden noch neueste bemerkenswerte Nachweise aufgeführt.

Insgesamt wird bei der Behandlung des Stoffes eine stärkere Berücksichtigung der Brutvögel als dem Kernstück einer Avifauna und eine vertiefte ökologische Durchdringung des Datenmaterials (z. B. Höhenverbreitung) vermißt. Besonders Thüringen fordert mit seiner starken landschaftsökologischen Gliederung auf engem Raum zu einer solchen Betrachtungsweise heraus. Trotz dieser kritischen Bemerkungen bildet die vorliegende Vogelfauna von Thüringen eine solide, weil kritisch aufgearbeitete Grundlage für künftige avifaunistische Untersuchungen, deren Hauptziel eindeutig die ökologische Erforschung der Brutvögel des Gebietes sein sollte.

D. FLÖSSNER, Jena

M. BOURNAUD & H. TACHET (eds.): **Proceedings of the 59 International Symposium on Trichoptera, Lyon (France) 21.—26. July 1986** (Series Entomologica 39). Dordrecht: Dr. W. Junk Publ. 1987. XXIII, 396 S. Geb. Dfl. 250.00/£ 99.50

Seit 1974 treffen sich die Trichopterologen der Welt im Abstand von drei Jahren zu internationalen Symposien. Das bisher letzte fand im Juli 1986 in Lyon (Frankreich) statt und hatte 89 Teilnehmer aus 18 Ländern. Es wurden 17 Poster gezeigt und 70 Vorträge gehalten.

Zwei Geländeexkursionen gaben den Teilnehmern Gelegenheit, Gebirgslandschaft und Fließgewässer der Region um Lyon kennenzulernen. Während der ersten Exkursion hatten einige Teilnehmer Gelegenheit zum Fang einiger Individuen von *Thremma gallicum* McL., einer morphologisch ungewöhnlichen Larve, die zum Symbol des Symposiums wurde und deren Abbildung auch die Titelseite der hier annotierten Tagungsberichte schmückt. Das 6. Symposium wird im Jahre 1989 in der Volksrepublik Polen stattfinden, seine Vorbereitung liegt in den Händen von Herrn Dr. C. TOMASEWSKI. Die Publikation der Arbeitsergebnisse von Lyon erfolgte in drucktechnisch anspruchsvoller Form, mit guten Abbildungen, grundsätzlich in englischer Sprache.

Nach einem Gruppenfoto und kurzem Vorwort folgt die Liste der Teilnehmer sowie die der Poster und mündlichen Vorträge. Die Beiträge sind nach 8 Hauptthemen geordnet. Wir geben diese nachfolgend wieder und fügen in Klammern die jeweilige Anzahl hinzu: 1. Morphologie und Anatomie (3); 2. Taxonomie (6); 3. Fossile Formen (3); 4. Biogeographie (10); 5. Phylogenie (4); 6. Biologie und Physiologie, 6.1. Reproduktion (5), 6.2. Stoffwechsel (2), 6.3. Interspezifische Verhältnisse (2); 7. Lichtfallen und Flugzeiten, 7.1. Sammlungen (3), 7.2. Flugzeiten (4); 8. Ökologie, 8.1. Entwicklungszyklen, Habitate und Nahrungsformen (11), 8.2. Zonierung und Verbreitung (6), 8.3. Gemeinschaften (2), 8.4. Umweltstörungen (5).

Für den Limnologen und Wasserwirtschaftler dürften in erster Linie die ökologischen Arbeiten von Interesse sein, ihr Anteil ist mit etwa einem Drittel aller Beiträge auffallend groß.

Am Schluß des Buches finden sich Autorenverzeichnis, geographische Übersicht und Sachverzeichnis, sowie ein taxonomischer Index. Das Werk stellt ein nützliches und aktuelles Nachschlagwerk dar, nicht nur für Trichopterologen, sondern für alle Fließwasser-Ökologen.

W. ZIMMERMANN, Gotha

MICHAEL J. ROBERTS: **The Spiders of Great Britain and Ireland.** Vol. 2. Linyphiidae und Check List. Leiden: E. J. Brill Publishers 1987. 204 S., 94 Fig., 4 Farbtafeln. Ln. Gld. 190.—

Der vorliegende 2. Band beschließt das monumentale, einzigartige Bestimmungswerk der Spinnen Großbritanniens und Irlands [vgl. *Limnologica* **19** (2), 1988]. Er behandelt allein die umfangreiche Familie der Linyphiidae, die fast die Hälfte aller britischen Spinnenarten stellt. Die Bestimmung beruht, wie allgemein üblich bei den Webspinnen, auf dem Bau der Genitalien. Sie erfolgt nicht mit Hilfe eines dichotomen Schlüssels, sondern durch Vergleich der in Tabellenform zusammengestellten diagnostisch wichtigen Merkmale. Der Autor gibt genaue Anweisungen über das Vorgehen bei der Bestimmungsarbeit, so daß mit Hilfe der sehr gut durchgearbeiteten Beschreibungen und der klaren Strichzeichnungen der männlichen Palpen und der Epigynen die sichere Bestimmung der zahlreichen Arten auf keine unüberwindlichen Schwierigkeiten stoßen dürfte. Dem Speziellen Teil ist ein kurzes Literaturverzeichnis beigefügt. Ein Anhang behandelt 6 in den letzten Jahren für Großbritannien neu nachgewiesene Arten der in Band 1 behandelten Spinnenfamilien. Eine Check List der britischen Spinnenarten, Fachworterklärungen und Index beschließen das Buch.

Die vorliegende dreibändige Spinnenfauna stellt ohne Zweifel für längere Zeit ein Standardwerk der Arachnologie dar, um das man die britischen Zoologen eigentlich nur beneiden kann. Ob es gelingt, ein ähnlich hervorragendes Werk über die Spinnenfauna anderer Teile Europas zu schaffen, bleibt abzuwarten. Zu wünschen wäre es.

D. FLÖSSNER, Jena

GUSTAV HEGI: **Illustrierte Flora von Mitteleuropa: Band I:** Gymnospermae, Angiospermae, Monocotyledones 1. Teil 3: Gramineae. **Lieferung 4, S. 241—320.** 3. völlig neu bearb. Aufl. Hrsgeg. u. bearb. von HANS-JOACHIM CONERT. Berlin, Hamburg: Verlag Paul Parey 1987. 37 Abb., 2 farb. Tafeln. Brosch. DM 35,—

Die Fortsetzung der Neubearbeitung der Süßgräser umfaßt aus der Unterfamilie der Pooideae die Gattungen *Helictotrichon* (Schluß des speziellen Teils), *Gaudinia*, *Ventenata*, *Koeleria*, *Rostraria*, *Trisetum*, *Aira*, *Deschampsia* sowie den ersten Teil der Gattung *Corynephorus*. Darstellung und Ausstattung sind ausgezeichnet. Warum eigentlich *Deschampsia littoralis* (S. 310), wenn die ursprüngliche Schreibweise mit „tt" offensichtlich ein orthographischer Fehler ist? Wichtig ist, daß CONERT *D. littoralis* und *D. rhenana* taxonomisch nicht voneinander trennt.

S. J. CASPER, Jena

Berichte des Geobotanischen Institutes der ETH, Stiftung Rübel. Zürich (Heft 53). Zürich 1987. 106 S. mit 23 Abb. u. 7 Tab. Brosch. SFr 20.—/US $ 15.00

Das vorliegende Heft enthält den Institutsbericht über die Jahre 1985 und 1986, englische Zusammenfassungen der in diesen Jahren abgeschlossenen Dissertationen und Diplomarbeiten und eine Übersicht über die im Freien Geobotanischen Kolloquium 1985 und 1986 gehaltenen Vorträge (darunter auch 3 über Wasser- und Röhrichtpflanzen). Die drei wissenschaftlichen Beiträge behandeln die Chromosomenzahlen einiger Pflanzen aus Albanien (S. 47—63; keine Wasserpflanzen), das Verhalten einiger alpiner Gräser in Einzeltriebklonierungs- und Begrünungsversuchen oberhalb der Waldgrenze (S. 64—90) und einen systematisch-ökologischen Vergleich von Gentianaceen im westlichen Kaukasus und in den Schweizer Alpen (S. 91—106).

H.-D. KRAUSCH, Potsdam

Veröffentlichungen des Geobotanischen Institutes der ETH, Stiftung Rübel, Zürich (Heft 99): Genetisch-ökologische Untersuchungen an alpinen Pflanzenarten auf verschiedenen Gesteinsunterlagen: Keimungs- und Aussaatversuche. MARTIN SCHÜTZ. Zürich 1988. 154 S. mit 63 Abb. u. 12 Tab. Brosch. SFr 48.—/US $ 34.—

Diese Arbeit entstand im Rahmen der Forschungen des Geobotanischen Institutes zur Ökologie alpiner Pflanzen und Pflanzengesellschaften. Sie untersucht Keimverhalten und Jungpflanzenentwicklung bei einer größeren Zahl alpiner Pflanzenarten. Das Verhalten der untersuchten Arten ist sehr unterschiedlich und zeichnet sich durch große Diversität und Variabilität aus. Daraus ergeben sich wichtige Anhaltspunkte zur biologischen Erosionsbekämpfung in der alpinen Stufe, wie sie vor allem durch den Wintersportbetrieb notwendig geworden ist. Die Ergebnisse dieser Arbeit sind über die Schweiz hinaus auch für andere Länder mit höheren Gebirgen von Bedeutung.

H.-D. KRAUSCH, Potsdam

G. GOVI: **Introduzione alla Micologia.** Bologna: Edagricole 1986. XI, 292 S., 414 Abb. Geb. L. 35000

Mit dem Werk von GOVI, Ordinarius für Mykologie an der Universität Bologna, gelangt ein mykologisches Lehrbuch in italienischer Sprache in unsere Hände, das als Einführung in die Mykologie sowohl für interessierte Amateure als auch für Studenten gedacht und geeignet ist.

Erst einmal besticht die gediegene Aufmachung des Buches. Das abwaschbare Hardcover wurde attraktiv mit einer farbigen Reproduktion aus einem alten Pilzbuch versehen. Der Druck auf schwerem, holzfreien Papier ist sehr sauber, und die erfreulich zahlreichen Abbildungen in Form von Schemata, Strichzeichnungen, Schwarzweiß- und Farbfotos sind meist von hervorragender Qualität. Bedauerlicherweise werden die Quellen der Abbildungen nur teilweise wiedergegeben, so daß der Eindruck erweckt wird, daß es sich dabei größtenteils um Originale handelt. Das ist jedoch nicht der Fall. Fig. 8.7 (S. 76) beispielsweise wurde aus AINSWORTH & BISBY, Dictionary of the Fungi, ed. 7, S. 174 entnommen, was aber nirgends vermerkt wird.

Der inhaltliche Schwerpunkt des Buches liegt interessanterweise nicht auf der Pilzsystematik, sondern auf dem, was man mit „Allgemeine Mykologie" umschreiben könnte. Die Palette der Kapitel reicht dabei von einem kurzen Abriß der Geschichte der Mykologie in Italien über cytologische und ökologische bis hin zu physiologischen Fragestellungen. Besonders umfangreiche und instruktive Kapitel sind beispielsweise solchen Themen wie „Thallus und cellulare Organisation" oder „Mykorrhizische Symbiose" gewidmet. Der didaktische Ansatz des Buches bietet trotz der Sprachbarriere manch Anregendes für den deutschsprachigen Leser.

Der systematische Teil umfaßt etwa die Hälfte der Gesamtseitenzahl des Werkes, doch werden auch hier noch einige Themen allgemeineren Charakters abgehandelt, z. B. aus der Pilzgenetik. Der Verfasser übernimmt weitgehend das System von WEBSTER, erwähnt diesen Autor aber eigenartigerweise nicht. Bei der Abhandlung und Beschreibung einzelner taxonomischer Gruppen findet sich viel Kritikwürdiges. Man kann dem Verfasser wahrlich nicht bescheinigen, daß er im Detail auch nur einigermaßen auf dem neuesten Stand wäre. Die Verwendung solcher antiquierter Gattungsnamen wie *Trachypus* (statt *Leccinum*) oder *Ixocomus* (statt *Suillus*) charakterisiert das hinreichend. Die Lektüre des Buches wird zudem noch dadurch beeinträchtigt, daß eine Vielzahl von lateinischen Nomina, Autornamen und zitierten Literaturtiteln mit orthographischen Fehlern behaftet sind.

Trotz der erwähnten Mängel, vor allem im systematischen Teil, liegt mit dem Buch von GOVI ein reich illustriertes, didaktisch gelungenes Mykologie-Lehrbuch vor, das sicherlich in Italien auf einen breiten Interessentenkreis trifft.

G. HIRSCH, Jena

D. J. MABBERLEY: **The Plant-Book.** A Portable Dictionary of the Higher Plants. Cambridge, New York, New Rochelle, Melbourne, Sydney. Cambridge University Press 1987. XII, 706 S. Kunstl. DM 75,00

Das vorliegende „tragbare" Verzeichnis aller gegenwärtig taxonomisch anerkannten Gattungs- und Familiennamen sowie der allgemein gebrauchten englischen Namen der Blütenpflanzen und Farne ist, wie im Untertitel ausgeführt, „arranged largely on the principles of editions 1—6 (1896/97—1931) of WILLIS's „A dictionary of the flowering plants and ferns“, ja es will eben den „WILLIS“ er- bzw. fortsetzen. Und da sei gleich festgestellt, daß auf den 700 engbedruckten Seiten eine Fülle von Daten ausgebreitet wird, die sich durch Genauigkeit (Literaturzitate! Artenzahlen!) und Knappheit auszeichnen. Auffällig die klaren Familien„diagnosen“ und die z. T. aufschlußreichen Anmerkungen zu wichtigen Gattungen, sich dabei eng CRONQUIST's „An integrated system of classification of flowering plants“ (1982) anschließend (CRONQUISTS System ist auf den Seiten 627—636 abgedruckt). Der Benutzer wird für das Verzeichnis der Floren, Handbücher und Zeitschriften dankbar sein und die Listen der Abkürzungen (Termini, Zeitschriftentitel, Autoren usw.) begrüßen. Wer aber soll das Buch benutzen? Botaniker von Beruf oder aus Neigung, Gärtner und Ökologen? Ja und nein. Denn trotz aller immensen und bewundernswerten Bemühungen ist doch kein neuer „WILLIS“ entstanden. So wird z. B. der Herbarbotaniker viele „seltene“ Gattungsnamen vermissen und dann eben wieder in den alten „WILLIS“ schauen. Der an den „Volksnamen“ interessierte Laie wird eine gewisse Unausgewogenheit bemerken, so wenn er z. B. unter „fern“ (S. 225) keinen „water fern“ findet, wohl aber unter „water“ (S. 612) die „w. fern(s)“ *Azolla* spp. und *Salvinia* spp. entdeckt. Solche kleinen Schönheitsfehler ändern nichts daran, daß dem Autor tatsächlich eine bemerkenswerte „Annäherung“ an den „WILLIS“ gelungen ist. Dem Verlag gebührt Anerkennung für den vorzüglichen Druck.

S. J. CASPER, Jena

ALASTAIR FITTER: **Blumen: Wildblühende Pflanzen;** Biologie + Bestimmen + Ökologie (Pareys Naturführer). Übersetzt und bearbeitet von LARS NEUGEBOHRN (Hrsg. DAVID ATTENBOROUGH). Hamburg, Berlin: Verlag Paul Parey 1987. 320 S., 1515 farb. Abb., 7 Tab. Kart. DM 34,—

Pareys Naturführer Plus erheben den Anspruch, eine völlig neue Generation naturkundlicher Bücher zu sein. Sie sollen mehr bieten als die herkömmlichen Bestimmungsbücher, und zwar vor allem deswegen, weil sie das herkömmliche Bestimmungsbuch mit dem ökologisch ausgerichteten Sachbuch kombinieren. Der Bearbeiter und Herausgeber betont, daß es ihm darum geht, dazu beizutragen, ökologisches Denken zu verbreiten und tatkräftiges Handeln zum Schutz der Natur anzuregen. Das war auch das Anliegen des Autors der Originalausgabe, ALASTAIR FITTER, die im Verlag William Collins Sons & Co. Ltd., London, ebenfalls 1987, erschienen ist.

Im Bestimmungsteil werden über 1100 der wichtigsten Wildpflanzen Mittel- und Norwesteuropas abgebildet und beschrieben. Im Anhang finden sich die Beschreibungen von weiteren über 300 seltenen oder sehr unscheinbaren Arten. In ihrer Anordnung folgen die Autoren der Flora von GARCKE (1972). Es ist schon erstaunlich, was alles auf den wenigen Seiten übersichtlich, ausgezeichnet gedruckt und sehr gut illustriert dargeboten wird. Eine kurze Einführung behandelt wichtige Aspekte der Evolution der Blütenpflanzen. Der zweite Teil bringt die Bestimmungsschlüssel und das Verzeichnis der Arten samt Abbildungen, und der dritte Teil befaßt sich mit der Naturgeschichte der wilden Blütenpflanzen. Bleibt der zweite Teil noch ganz im Traditionellen verhaftet, so bringt der 3. Teil tatsächlich Neues: Morphologie, Anatomie und Verbreitung von Samen und Früchten sowie die Keimung der Samen, botanisch wichtige Tatsachen über Sprosse, Blätter, Parasitismus, Symbiose, Mycorrhiza, Insektivorie sowie Abwehrmechanismen. Schließlich wird noch ausführlich auf Vermehrung, Wachstum, vegetative und sexuelle Fortpflanzung eingegangen. Die Darstellung wird vervollständigt durch eine Erklärung der botanischen Ausdrücke, durch eine Liste der wichtigsten Pflanzengesellschaften nach ELLENBERG (1982), durch Register der deutschen Gattungs- und lateinischen Arten-Namen sowie durch Angaben über weiterführende Literatur.

Ein durchaus gelungenes und anregendes Buch, das man auch auf Exkursionen mit sich tragen kann. Im deutschsprachigen Raum existiert ein vergleichbarer Bestimmungsführer in die Welt der Blütenpflanzen nicht.

S. J. CASPER, Jena

Der Personalcomputer PC-1715. Hrsgeg. von einem Autorenkollektiv des Kombinates ROBOTRON unter Ltng. von R. ZETH. 1. Aufl. Berlin: Verlag Die Wirtschaft 1987. 224 S., 11 Fotos, 40 Abb., 43 Tab. Brosch. M 14,—

Die Zielgruppe der Hersteller von Personalcomputern ist nicht die relativ kleine Menge von EDV-Experten, sondern eine große Anzahl von Anwendern, die von Programmiermethodik und Softwaretechnologie nichts oder nur sehr wenig verstehen, deren Aufgabenfeld jedoch eine nutzbringende Verwendung dezentraler Rechentechnik erlaubt. Für diesen Zweck hat das Kombinat ROBOTRON seit 1985 eine große Anzahl des Personalcomputers PC-1715 hergestellt.

Inzwischen gibt es mit der Broschüre „Der Personalcomputer PC-1715“ auch ein Literaturangebot, das den Leser mit dem Gerät und seiner Software in zusammenhängender Form bekannt macht. Die einzelnen Kapitel beschreiben u. a. die Hardware, die Betriebssysteme, die vorhandene Standardsoftware, die bisherigen Einsatzbereiche und die Serviceleistungen für den PC-1715. Literaturhinweise und ein Sachwortverzeichnis ergänzen das Heft.

Bezüglich des angestrebten Leserkreises schreiben die Autoren im Vorwort: „Neben dem Leser, der Erstinformationen zum Personalcomputer PC-1715 sucht, wird auch der fortgeschrittene Nutzer eines Personalcomputers wichtige Informationen finden.“ Für den fortgeschrittenen Nutzer kann ich nicht sprechen. Aber für den Sucher von Erstinformationen, der ich bin, teilt sich das Buch in zwei Abschnitte. Bis zum vierten Kapitel wird man durch eine überwältigende Menge vom Fachtermini und Abkürzungen stark demotiviert. Hier liegt nach meiner Auffassung eine entscheidende didaktische Schwäche.

Von den folgenden Abschnitten über Programmiersprachen, Standardsoftware und Einsatzbereiche des PC-1715 kann auch der Laie profitieren. Für mich bleibt deswegen die Frage offen, ob das Buch insgesamt gesehen nicht doch für den Experten zu simpel und für den Laien zu kompliziert ist.

P. KASPRZAK, Neuglobsow

ISBN 3-05-500642-9
ISSN 0075-9511

Erschienen im Akademie-Verlag Berlin, Leipziger Straße 3—4, Berlin, DDR-1086

Lizenznummer: 202 · 100/502/89
Printed in the German Democratic Republic
Gesamtherstellung: VEB Druckerei „Thomas Müntzer“, 5820 Bad Langensalza
Umschlaggestaltung: Wolfgang Janisch
LSV 1305
Bestellnummer: 763 948 5 (2111/20/1)
08200

Proceedings of the Estuary Symposium,
Kühlungsborn, German Democratic Republic,
November 16—21, 1987

Inhalt/Contents

Band 20 der LIMNOLOGICA umfaßt 2 Hefte (Volume 20 of LIMNOLOGICA contains two numbers).

ISBN 3-05-500642-9
ISSN 0075-9511